THE HAMPDEN FILE

By Harry Moyle

An Air-Britain Publication

Published in the United Kingdom
by Air – Britain (Historians) Ltd
1 East Street, Tonbridge, Kent

ISBN 0 85130 128 2

Typeset by Arioma Editorial Services
Gloucester House, High Street,
Borth, Dyfed, SY24 5HZ

Printed by Hollen Street Press Limited
141/3 Farnham Road, Slough, Berks

Cover drawing by George Burn

CONTENTS

The first production Hampden, L4032, first flew in May 1938 and was official named "Hampden" on 24 June 1938 by Viscountess Hampden. The outer wing dihedral is obvious as is the greatly improved design of the nose with its flat panel through which the bombsight was aimed. (Photo: Handley Page via Dennis Thompson)

FOREWORD

The relatively small number of Hampdens (1,430) and Herefords(150) that were built, allied to the fact that the torpedo – bomber version was the only variant, presented a unique opportunity to include details of the crews and to give accounts of the manner in which many Hampdens came to grief.

Personal accounts from aircrew who flew on Hampdens provide the main background to the narrative which should enable the reader to reach a better understanding of this early period of the war, during which many lessons were learned that contributed in no small way to the eventual success of the bomber offensive.

The book provides ample evidence of the valiant service ofthe men from the Commonwealth, many of whom enlisted in the RAF before the outbreak of war; the contribution of the RAF Volunteer Reserve also receives the acknowledgement it so richly deserves.

Every effort has been made to ensure the accuracy of the information contained in this book but it is possible that errors have been inadvertently made; if this is so, the author wishes to apologise in advance and to welcome any corrections so that any later edition of the book may be amended.

The author served for the first twelve months of the war as an air observer on Hampdens at Waddington. His experiences during that time and his enthusiasm for the task will, it is hoped, compensate for his lack of experience as a writer which must have compounded the problems of the editor, Jim Halley,to whom the final words of this foreword must be a most sincere "Thank you".

DEDICATION

This book is dedicated not only to the aircrew who flew the Hampden but also those others, in the Wellingtons and Whitleys in particular, with whom the Hampdens shared the night sky over Germany during the darkest days of the war when Britain fought alone after the fall of France.

No dedication to the Hampden aircrew would be complete without including the Training Command units and the four Hampden torpedo – bomber squadrons of Coastal Command, which reinforced their Beaufort colleagues at a critical period of the war in a task that required a particular brand of courage in the performance of which their life expectancy in terms of operational flying hours was reckoned to be about half that of the Bomber Command Hampden crews.

The aircrew lads of those early days, wearing brevets and insignia of rank that often belied their immaturity, laid the foundations upon which the massive air offensive of later war years was built. Their achievements would not have been possible without the efforts of the factory workers who built the aircraft and of the "erks" who serviced them.

Full tribute must be paid to the latter who often worked under atrocious conditions; a number of them lost their lives and others were injured in accidents on the Hampden bases.

Posterity should also be reminded of the contribution of the lady "erks" of the Women's Auxiliary Air Force in a wide variety of tasks. The aircrew will particularly remember them as the girls who served their meals, issued them with parachutes, drove the lorries that took them to their aircraft at dispersal and, if their luck was in, collected them when they returned.

For the unlucky crew who did not come back, the smiling face of their WAAF driver, then the banter from the ground crew and finally the farewell waves from the cluster of figures by the flare – path Chance Light as they began their take – off, would have been their last contact with this world.

Prototype H.P.52 K4240 was first flown at Radlett on 22 June 1936 by Major J.B.L.H.Cordes, Handley Page's Chief Test Pilot. Note the pitot head mounted on aerial mast above fuselage, the "birdcage" nose, sliding canopy over rear gun position and the "dustbin" shape of the lower gun position (A&AEE photo MH6430)

The prototype H.P.53 L7271 which was originally the second Hampden prototype. The pitot is now mounted below the nose and the nose and lower rear position have been modified. Powered with Napier Dagger VIII engines, it first flew on 6 October 1938 with Major Cordes as pilot. (A&AEE photo MH4860)

Another view of L4032 showing the leading edge slots, flaps on the trailing edge, retractable tail wheel, stub exhaust from engine exhaust collector ring and pilot's fixed Browning machine gun. (Photo: Rolls – Royce 8520 via Dennis Thompson)

PART ONE: THE AIRCRAFT

Introduction

The Air Ministry requirement for twin – engined medium bombers as replacements for the RAF's antiquated bombers resulted in Specification B.9/32 being put out to tender in October 1932. Handley Page's design team at Cricklewood, headed by Dr Gustav V Lachmann, set to work to produce the design for such a twin – engined bomber and in it they embodied some of the features of the H.P.47 which, like the Hampden, was all – metal construction. In May 1933 the Air Ministry gave approval for Handley Page to build a prototype and at the same time instructed Vickers to proceed with their Type 271 from which the Wellington emerged. The power units for the H.P.52 were originally intended to be Rolls – Royce "Goshawks" but the design proved to be a failure and it was replaced by the nine cylinder Bristol Pegasus P.E.55a radial engine that developed 815 hp; it had a single stage blower and used three – bladed De Havilland controllable – pitch propellers.

The Swedish Hampden

In 1934, the Swedish Government ordered a version of the H.P.52 with an undercarriage that could be changed from normal land use to floats or skis.

In the event, only the fifth of the production aircraft, fitted with Bristol Pegasus XXIV engines instead of the normal Pegasus XVIIIs, was delivered to Sweden on 22 September 1938. It flew with the Swedish Air Force (Code I – 90) from Nykoping until late 1945 after which it became a flying test – bed for electronic equipment for SAAB with the civilian registration SE – APB. It was scrapped on 17 November 1947, by which time it had become the last example of a Hampden in airworthy condition.

The Hereford

The Hereford differed from the Hampden only in that it had the H – section 24 – cylinder Napier Dagger VIII engines in place of the Hampden's Pegasus XVIIIs. Prototype Hereford L7271, which had originally been the second Hampden prototype, made its first flight on 6 October 1938 and subsequently went to the Aeroplane and Armament Experimental Establishment at Martlesham Heath for performance trials, a report on which on 11 December 1939 ended thus: "Although L7271 was at this establishment for three months, it was only serviceable for about eight days, the remaining period being used for engine adjustments. Owing to the frequent trouble with the engines, trials were not completed by the time the first production Herefords had arrived and commenced tests and it was therefore decided to discontinue further tests on L7271." An inauspicious start indeed.

L6002, the first production Hereford, flew on 17 May 1939. 150 Herefords were built by Short and Harland at their Belfast works, all having been completed by September 1940. The problems with the Dagger engines were never overcome and those Herefords that did see any flying service in the RAF only did so in a training capacity; many of those built were struck off charge without ever reaching a service unit.

Herefords first appeared in service with 185 Squadron, which operated as a training squadron, in the autumn of 1939 and was disbanded to form the basis of 14 OTU in April 1940. No.16 OTU and 5 BGS also operated some Herefords but no Hereford ever flew on operations and by mid – 1941 they were being withdrawn from service altogether. Air Transport Auxiliary pilots who flew the Hereford were well aware of the unreliability of its engines and Peter Garrod recalled another reason for remembering the Hereford. "The Dagger engines had take – off revs of 4200 against the

The second production Hereford, L6003, a view which clearly shows the Handley Page slots in the leading edge of the wing. This feature had been dispensed with by mid – 1940 when balloon cable cutters were installed in the leading edge. (Photo: Handley Page via D.M.Hannah)

2650 of the Hampden's Peggies and the noise from the Hereford's engines was incredible." Thirty Herefords were involved in crashes, 17 of which were attributed to engine failure. Ten of the others were due to undercarriage collapses which were not necessarily due to any weakness in the aircraft.

Four of the Hereford crashes resulted in a total of twelve deaths, six of which were in N9074 which crashed into the sea after an engine caught fire (see 6D3). Four more perished when L6067 crashed near Bicester (see 6D2). The general opinion of the Hereford's Dagger engines is evident in the remark which appears on the crash card for Hereford N9091 (2) after a non – fatal accident due to engine failure at 5 MU Kemble which reads: "Test pilots of this unit have no confidence in the Napier Dagger engine".

Hereford Conversions to Hampdens

There were 23 "official" conversions of Herefords to Hampdens i.e. L6011, L6018, L6019, L6020, L6049, L6055, L6069, L6076, L6085, L6089, L6090, L6096, N9062, N9064, N9065, N9070, N9089, N9086, N9090, N9096, N9101, N9105 and N9106.

The words "Reduced to spares for repair of Hampdens" that appear on many Hereford "Movement Cards" confirm that they became a valuable source of wings and fuselages which enabled many mangled wrecks of Hampdens to be brought back to life.

The Production Hampden

L4032, the first production H.P.52, flown by Major Cordes first took to the air from Radlett in May 1938 and on 24 June the H.P.52 was officially christened "Hampden" by Viscountess Hampden. After the naming ceremony Major Cordes took the opportunity of displaying his own prowess as a pilot and of demonstrating the Hampden's manoeuvrability.

L4032 and L4033, the first production Hampdens, were used for performance trials at the Aeroplane & Armament Experimental Establishment at Martlesham Heath and at the Royal Aircraft Establishment at Farnborough. Both became Ground Instructional Airframes in May 1940 and were struck off charge in May 1943.

X3115 fitted with Wright Cyclone engines in November 1940 at the Aeroplane and Armament Experimental Establishment, Boscombe Down. L4032 had been used for similar trials in October 1939 but no Cyclone – engined Hampdens saw service. (Photo: A&AEE MH4859)

The fifth of the production line Hampdens was delivered to the Swedish Air Force on 22 September 1938. Coded I – 90, it was in use until November 1945 when it was sold to SAAB for use as a flying test bed for electronic equipment. It was powered by Pegasus XXIVs instead of the normal Pegasus XVIIIs. (RAF Museum PC71/66/759)

The Torpedo Bomber Hampdens

The Torpedo Bomber version of the Hampden was basically different from the original production aircraft only in that the bomb bay and the lower gunner's position were modified to allow a torpedo to be carried. The TB Hampdens, including those that were fitted with ASV (Air to Surface Vessel radar), are dealt with in Part 2J, as is the detachment of the Hampdens of 144 and 455 RAAF TB Squadrons to Russia in September 1942.

The Hampdens were handed over to the Russian Navy in October 1942 and they wasted no time in converting one to dual control for instructional purposes; they are also reported as having fitted an extra rear – firing fixed machine gun under the lower gunner's position (see 2J).

Drawing for the installation of dual control in the Hampden had been made in 1938 but the Air Ministry was apparently not convinced of the need for such a modification and the drawings were cancelled.

Modifications

During the five years of the Hampden's service life the appearance of the aircraft remained virtually unchanged except for minor details such as the substitution of twin Vickers "K" guns in the rear upper and lower positions in place of the original single gun. An improved type of D/F loop in an external faired housing replaced the original loop aerial which retracted into the fuselage. Bristol Pegasus XVIII engines provided the power for every service Hampden. The stub exhaust in use at the outbreak of war emitted an alarming amount of flame and sparks and was replaced by a longer exhaust fitted with flame dampers soon after night operations commenced.

Trials were carried out on L4032 in October 1939 and later in X3115 (7) to test the suitability of the American – built Wright Cyclone engine as an alternative to the Pegasus XVIIIs in the event of production of the latter being interrupted by bombing of the Bristol company's factory. It was also intended that eighty of the Hampdens being built in Canada should be powered by Cyclones, which did in fact prove to be an ideal alternative power unit.

The demand for Cyclones for other military aircraft proved to be so great that there were now none to spare for the Hampden for which, fortunately, the supply of Pegasus XVIIIs never dried up.

The Rolls – Royce Merlin X was also considered as a possible power unit; it would almost certainly have been a big improvement on the Napier Dagger but the demand for Merlin engines for Hurricanes and Spitfires in particular meant that there were no spare engines to experiment with on the H.P.52.

Method of Production

The technique evolved by Handley Page for the manufacture of the Hampden was quite revolutionary when compared with other contemporary methods of producing similar – sized aircraft. It consisted of splitting the aircraft into sections which, after being constructed in jigs as sub – assemblies, were brought together and assembled without the use of an intricate final assembly jig.

The fuselage comprised three main sections and was of stressed skin monocoque construction in which alclad aluminium sheet was riveted to a framework of hoops and longitudinal members; the sections were:

a. The nose assembly, extending from the rear of the perspex nose to just aft of the astro – hatch, was built as a single sub – assembly to which the perspex nose was added on completion.
b. The centre assembly, from just aft of the astro – hatch to aft of the wing trailing edge. This was built in three parts i.e. the port and starboard side sub – assemblies and a "roof" section which was not fitted until the sides, complete with electrical wiring and fuel and hydraulic pipework, had been joined together.
c. The rear tail boom section. Manufactured in vertically opposed halves which were joined together after the fitting of appropriate equipment including rudder control mechanism.

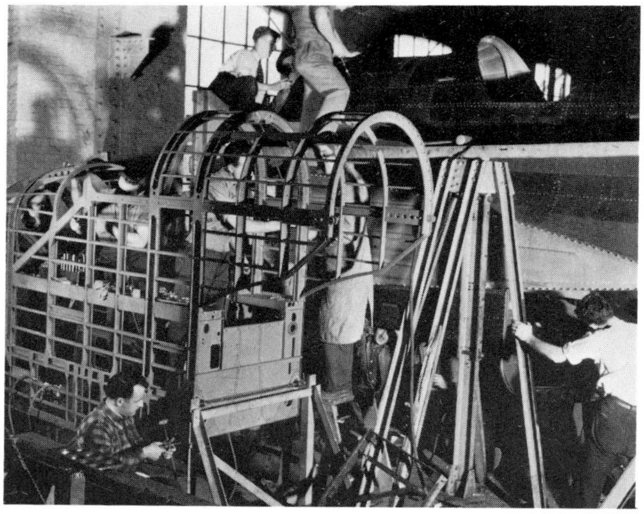

Hampden assemblies being worked on at Canadian Vickers Ltd, Montreal. In a country renowned for its wealth of space, it is surprising to see the workforce operating under such cramped conditions, as is the profusion of workers, all male, in some of these photographs.

Top left: Nose section jig in foreground, part – completed rear fuselage section behind it.

Top right: Rear fuselage jig. Equipment is installed in the port and starboard halves of the rear fuselage before they are joined together.

Centre left: Nose sections being completed.

Centre right: Final inspection: is the discussion about "Where does this bit go?"

Bottom left: Finishing touches to three assembled fuselages.

"The Hampden Road Show" at Canadian Vickers Ltd, Montreal, in 1941. The three sections of the fuselage have been assembled, with the forward section being temporarily fastened together and a pair of wheels fitted to facilitate movement from the fuselage assembly section to the main assembly area. The forward part of the fuselage was later separated from the rear and then re-assembled around the wing centre secion/main spar. The same method of construction, and jigs, were used in England and Canada. (Canadian-Vickers/Public Archives Canada C32441)

The mainplane also consisted of three main sections, all of similar construction and built around a main spar which took all the bending loads. Forward and aft of the main spar the stressed skin wing was formed around ribs constructed of alloy extrusions braced vertically and diagonally with duralumin tubes. The inner wing section, which in the completed aircraft contained the fuel tanks, engines, undercarriage, etc, had all the associated wiring and controls installed along the leading front wall of the main spar box section. The wing nose section which eventually covers them was easily detachable for access to the systems. The two other sections were the port and starboard outer wing sections.

The tailplane assembly was of fixed cantilever type and stressed skin construction; its nose section was detachable to allow easy access to the enclosed rudder controls and servo mechanisms. The interchangeability of wing and fuselage section assemblies facilitated the repair of damaged aircraft and when the Herefords, which except for their engines were identical to the Hampdens, were withdrawn from service in 1941, the considerable number of Hereford airframes that became available resulted in many "miracle" repair jobs being carried out on Hampdens that would otherwise have been written off.

* * * * * * * * * * * * * * * * *

General Dimensions & Performance Data 1939

The Hampden and Hereford were identical in construction except for the power plant which was as follows: –

Hampden

Two Bristol Pegasus XVIII engines

Maximum power:	900 hp at 17,500 ft/5334 m
Cruising:	545 hp at 15,000 ft/4572 m
Take-off:	980 hp

Hereford

Two Napier Dagger VIII engines

Maximum power:	1,000 hp at 4,000 rpm at 8,750 ft/2667 m
Cruising:	925 hp at 9,000 ft/2743 m
Take-off	955 hp at 4200 rpm

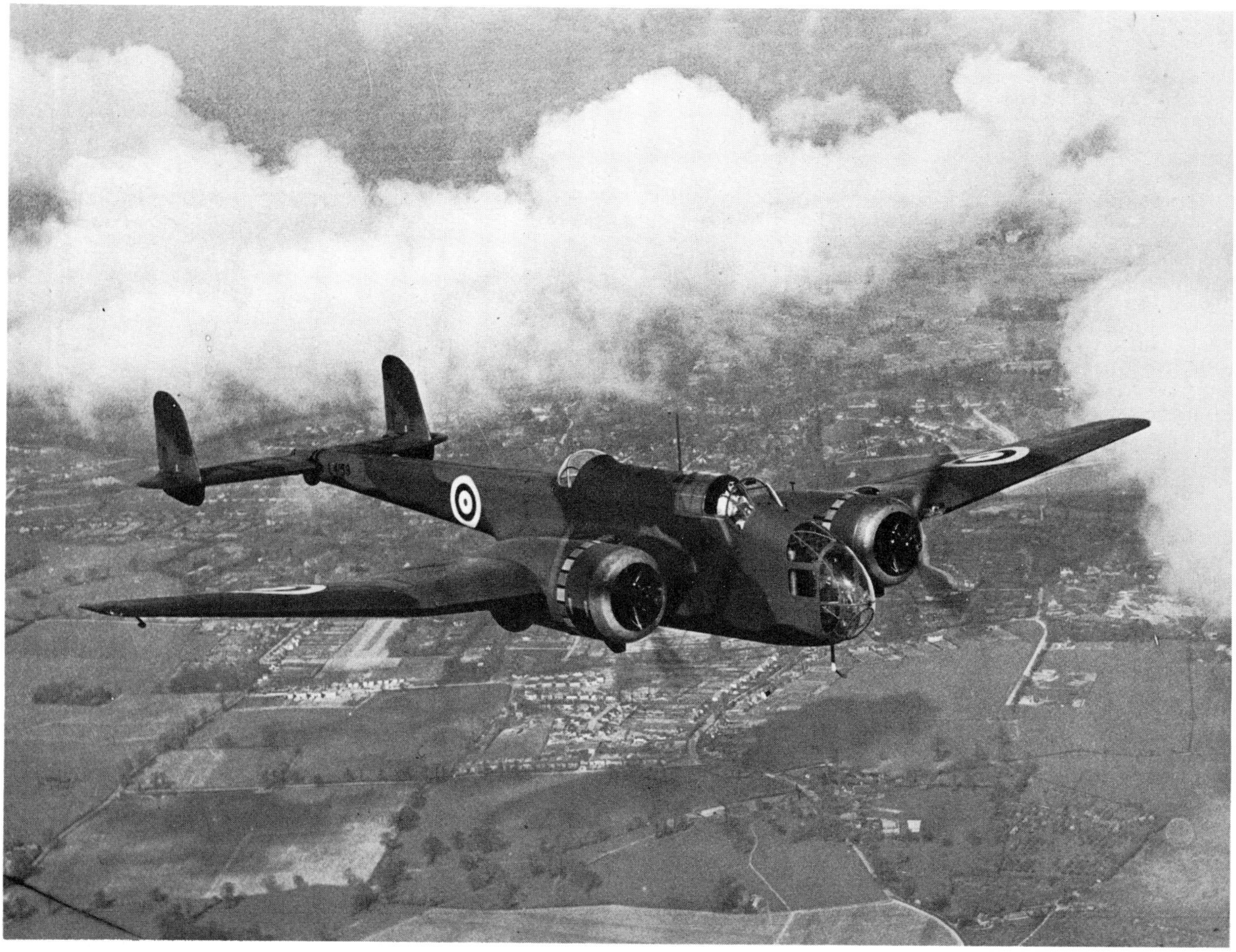

L4159 was delivered to No.7 Squadron on 28 April 1939 but hit trees on approach to Upper Heyford on 6 November 1939 (Photo: Charles E Brown)

Fuselage Dimensions
Maximum width: 3.00 ft/0.915 metres
Maximum height: 8.7 ft/2.65 metres
Maximum length: 53.3 ft/16.25 metres

Wing Dimensions:
Total area: 668 sq ft/62.1 sq metres
Span: 69.33 ft/21.1 metres
Mean aerodynamic chord: 11.33 ft/3.46 metres
Chord (root): 16.29 ft/4.96 metres
Chord (tip): 3.85 ft/1.17 metres
Aspect ratio: 6.58
Tail plane span: 21.17 ft/6.45 metres

Weight Data
Gross weight: 18,756 lb/8525 kg
Empty weight: 11,780 lb/5354 kg Useful load: 6,976 lb/3171 kg
Distribution of Useful Load
Fuel: 3,173 lb/1442 kg
Oil: 216 lb/98 kg
Military load: 3,587 lb/1631 kg
(including crew of 4 + 5 parachutes)

Fuel
Petrol: maximum 654 Imp gal/2975 litres
Oil: maximum 36 Imp gal/164 litres

Landing Gear Track: 17.25 ft/5.26 metres

Performance Data with Pegasus XVIII engines at normal flying weight of 18,750 lb/8540 kg

Maximum speed at
15,500 ft (4730 m): 265 mph/427 km/hr
Cruising speed at normal
rpm at 15,000 ft (4580 m): 217 mph/350 km/hr

Economical cruising speed
at 15,000 ft (4580 m): 167 mph/269 km/hr
Service Ceiling: 22,700 ft/6920 m
Time to 15,000 ft (4580 m): 18.9 min
Rate of climb at sea level: 980 ft/min/4.98 m/s
Take-off run, clearing
50 ft (15.2 m barrier): 550 yds/503 m
Landing speed: 73 mph/117 km/hr

Military Load and Range

Load		Range at Normal Cruising speed		Range at Economical Cruising speed	
lb	kg	miles	km	miles	km
2,587	1176	1,475	2,380	1,179	2,880
3,587	1631	1,095	1,760	1,335	2,159
5,587	2549	360	580	440	710

At maximum permissible flying weight of 21,000 lb/9550 kg
Cruising speed at 15,000 ft/4580 m: 212 mph/341 km/hr
Economical speed at 15,000 ft/4580 m: 172 mph/277 km/hr
Service Ceiling: 19,500 ft/5950 m
Time to 15,000 ft/4580 m: 27.2 min
Rate of climb at sea level: 760 ft/min/3.87 m/s
Take-off run, clearing 50 ft/15.2 m barrier 810 yds/741 m

Military Load and Range

Load		Range at Normal Cruising speed		Range at Economical Cruising speed	
lb	kg	miles	km	miles	km
3,587	1631	1,725	2780	1,990	3210
5,587	2540	1,165	1875	1,340	2160
6,587	2994	788	1270	870	1400

Armament

At the outbreak of war this comprised:

One fixed 0.303 – in Browning machine gun in the nose, operated by the pilot.

A Vickers "K" gun for use by the navigator; this gun was fitted into a mounting and fired through a small gun hatch in the floor at the front of the aircraft's nose. It was stowed inside the fuselage when not in use.

Single Vickers "K" guns in the upper and lower rear gunner positions.

Soon after the outbreak of war twin "K" guns were fitted in both rear positions and after the experience of 12 April 1940 when Bf 109s carried out beam attacks, against which the Hampden had no defence, rectangular apertures 12 inches high by 10 inches wide were cut in each side of the fuselage just aft of the wing spar and in the perspex side panels of the navigator's position. The navigator would operate his "K" gun from one of these positions if beam attacks developed.

By June 1940, 58 Hampdens had beam guns and 38 had been fitted with a modified mounting for the rear upper guns to allow a wider traverse while a modified mounting for the lower guns was approved in July 1940.

In January 1941 the upper gun mounting was fitted with an electric motor to power the traverse. Rose Bros were the contractors and the motorised mounting became known as the RB3 mounting.

The vulnerability of RAF bombers to fighter attacks, in which aircraft were set on fire through petrol tanks being hit, resulted in self – sealing tanks being fitted. Hampdens were receiving these tanks by March 1940 and extra armour plating was also fitted to protect the fuel tanks and crew positions.

The torpedo bomber role of the Hampden, described in Part 2J, gives further proof of the advantage of the bomb – bay design. The ability to carry six 250 – lb depth charges in the bomb – bay and another on each wing made the Hampden a useful aircraft for carrying out attacks against submarines and it also enabled it to operate as an Air Sea Rescue aircraft for dropping the Lindholme dinghy apparatus (see 5B).

The 2,000 lb bomb, introduced in mid – 1940, was used by Hampdens against warship targets at Kiel and other ports, and when the German warships were at Brest in 1941 – 2 it carried a load of two 2000 – lb bombs. On these occasions the rear outer fuel tanks were not used so the fuel load was reduced to 434 Imp gal, mainly to compensate for the fact that the heavier bomb load moved the aircraft's centre of gravity too far aft for safety.

Bomb Load

At the outbreak of war this comprised:

Internally: 2 cradles of 3 x 250 lb bombs or 2 cradles of 2 x 500 lb bombs or 2 cradles of small bomb containers

Externally: 2 bombs of up to 500 lb each on wing racks.

The design of the Hampden's capacious bomb – bay, in which the bombs were carried horizontally, proved to be an unforeseen advantage when the RAF began its minelaying campaign in April 1940, as the Hampden turned out to be the only Bomber Command aircraft capable of carrying the 1500 – lb mine and, as may be seen in the "Minelaying" section (see 2D), it played a major part in the minelaying campaign until the advent of the Manchester and Stirling.

Final assembly of Hampdens on the Handley Page production line at Radlett early in 1939. The five aircraft down the centre of the photograph are in final stages of assembly. The first and second are almost complete; fuel pipes and electrical wiring are visible along the front of the main spar. On the third aircraft, the port outer wing section is being prepared for fitment. The fourth aircraft has engines but no outer wing sections and the rear aircraft has no engines yet. A completed Hampden can be seen on the right and beyond it are three completed tail units awaiting fitment on the line. In the far right corner of the shed a completed wing centre section is standing on its leading edge and to the left is a fuselage mounted on its temporary wheels. On the left is another completed tail unit on a trailer and to its right another Hampden fuselage, also on its trolley. (Photo: RAF Museum 5836 – 13)

A Hampden of 24 MTAP of the Russian Naval Air Force. Captain A.F.Stoyanev is in the foreground. (via Carl – Frederik Geust)

Canadian – built torpedo – bomber AN127 which arrived at No.7 Aircraft Assembly Unit, Hooton Park, on 20 March 1942 and went to 489 RCAF Squadron on 20 July, being coded XA – Y. It was briefly tested at A & AEE Boscombe Down and later, after a crashlanding at Wick on 17 October 1942, it was repaired and re – coded XA – B. (Photo: A&AEE MH6433)

The Swedish Hampden as used by SAAB from November 1945 until being removed from the civil register on 17 November 1947. It bore the civil registration SE – APD. (SAAB – Scania photo 15/76)

Front view of SE – APD showing electronic equipment installed in the nose for test purposes. A somewhat primitive – looking directional aerial can be seen in the D/F aerial position. (SAAB – Scania photo 15/73)

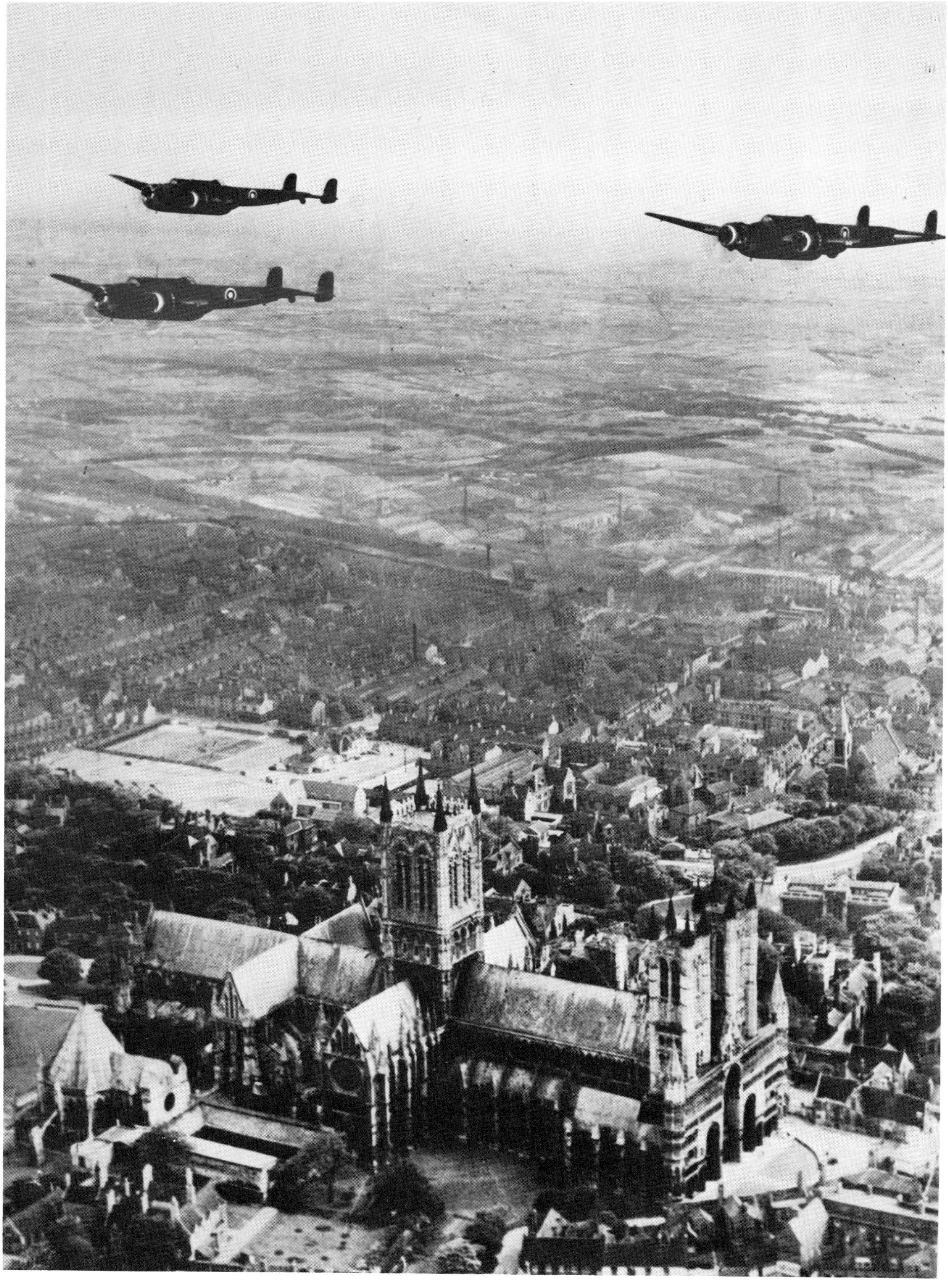

Three Hampdens flying over Lincoln Cathedral before the war. The Cathedral was a prominent landmark and a reminder to the crews of the pubs, cinemas, dance halls, etc. which they frequented on their off duty hours. *(Photo via Julian Badcock)*

The engines of AE288 of 408 Squadron RCAF being run up. The navigator's escape hatch can be seen, open, behind the pitot head. The astrohatch, also open, is partly covered by the sliding cover of the pilot's cockpit. (Photo: Public Archives of Canada)

PART TWO : OPERATIONS

2A OPERATIONAL HISTORY SUMMARY

Outbreak of War to End of 1939

At the outbreak of war on 3 September 1939, six Hampden squadrons were operational, Nos.44 and 50 Squadrons at Waddington, 49 and 83 at Scampton and 61 and 144 at Hemswell. As a sensible precaution against the Luftwaffe carrying out massive attacks on aerodromes, as had happened when Poland was invaded, the Hampden squadrons not on immediate stand-by were dispersed to aerodromes further inland, whilst the aircraft that were required for stand-by were spread out in fields around their own bases so as not to present too easy a target. Nos.44 and 50 Squadrons dispersed to Tollerton, Nottingham's airport, a much appreciated change of scenery with an abundance of young ladies and a fresh round of ale houses for the crews to visit. Hemswell went first to Speke but, possibly because the AA defences in the Mersey area had taken pot-shots at the Hampdens, they later used Doncaster. Scampton was allotted Newton near Nottingham and Ringway, Manchester.

During these first four months of the War the Hampden squadrons were only involved in operations on any scale on three occasions, 3 September, 29 September and 21 December, which between them resulted in eight Hampdens being lost. (See 2C)

JANUARY TO JUNE 1940

By the end of 1939, even the most optimistic of the staff officers at Bomber Command HQ had good reason to ponder upon the merits of sending unescorted bombers to attack targets in daylight that were defended by German fighters. The debacle of 18 December, when 12 Wellingtons, out of a total force of 22, were shot down by the Luftwaffe's fighters as they attempted to attack targets in the Wilhelmshaven-Heligoland area, must have been final proof of the futility of continuing with such attacks and from January 1940 Bomber Command's strategy turned towards the use of the bombers on night operations.

The Luftwaffe learned the same lesson about the prohibitive losses that bombers without escort would suffer during the Battle of Britain. Another factor that often goes unmentioned and of which the British High Command was apparently unaware was the existence of the German radar stations on Heligoland, Borkum, Sylt and Wangerooge, which were able to pick up incoming bombers from up to 90 km range. The "Freya" radar equipment that the British captured in the daring raid on Bruneval in 1942 was in fact the same type that was in use along the coastline of Germany in 1939.

The Hampden's introduction to night operations began with leaflet raids in January 1940, on which 18 packages of leaflets weighing a total of 1,080 lbs were carried. Ten packages were in bomb containers in the bomb bay, the other eight being stacked inside the fuselage for disposal by the navigator and upper and lower rear gunners. Each package contained twelve bundles of 1,500 leaflets so, if all the leaflets wafted safely down on the Fatherland, they provided over 300,000 sheets of bum-fodder for the citizens of the Glorious Third Reich.

These were soon followed by "Security Patrols" of German seaplane bases which were intended to hamper the activity of the He 115 minelaying seaplanes. On the night of 19/20 March, the RAF made its first attack on a land target, when Hampdens took part in the much-publicised, but not very successful, attack on military installations on the island of Sylt. Reconnaissance flights were also made of the areas in which it was intended that the RAF's own minelaying effort was to take place and, in mid-April, the Hampdens became the spearhead of the minelaying campaign (see 2D).

The German invasion of Norway presented a brief diversion from these nocturnal activities when, on 12 April, six out of a force of twelve Hampdens from Waddington were lost in an attack in daylight on German shipping at Kristiansand (see 2C).

Five Hampdens had been lost on night operations before the minelaying campaign began in mid-April; a fortnight after minelaying began a further twelve had been added to the total. The effort in May shifted towards the bombing of German industrial targets that were not situated close to residential areas. By the end of May "railway targets" in Germany, damage to which might hamper troop movements, had been added to the list and in June, road traffic and troop concentrations were attacked as the Germans invaded the Low Countries. After France had fallen, the bombing of industrial targets and minelaying operations had been resumed. The first Hampden to be shot down by a German fighter at night was almost certainly P1319, on 25/26 April 1940 at 1100 hrs near Sylt. It fell to Fw (Sergeant) Hermann Forster of IV/JG2 who was flying a Bf 109 with the cockpit hood removed to allow better night vision and he was assisted by ground radar to contact his prey. The same pilot claimed the only other Hampden to fall to a night fighter in the period up to the end of June, P1178 on 3/4 June. The loss of P1340 and P4345, brought down by the balloon barrage at Harwich in the space of ten days in June, illustrates another peril in the night skies at that time (see 5E5,5E6).

JULY 1940 TO THE END OF 1940

After France had surrendered, the British stood alone and the Germans had two main courses of action to consider if the British could not be persuaded to agree to a negotiated peace settlement. The most positive solution would be the invasion and occupation of Britain; the alternative was a war of attrition against the civilian population by bombing and attacking the shipping which carried the food and munitions to beleaguered Britain. Whichever strategy the enemy chose, it was obvious that the German Navy would play a considerable part in putting it into operation, and Bomber Command aircraft were sent to bomb German dockyards and naval bases where new warships such as the *Bismarck* and *Tirpitz* were being fitted out and new vessels, including the aircraft carrier *Graf Zeppelin*, were being built. The attack, using modified sea mines, on the *Tirpitz* at Wilhelmshaven is described in Part 2E.

The coastline of Hitler's territories now extended from Norway's Arctic north southwards to the border of neutral Spain and the Hampdens made a major contribution towards the minelaying effort (see 2D). Important inland waterways, such as the Dortmund – Ems canal, were attacked on a number of occasions, and on one of these attacks the bravery of Flt Lt "Babe" Learoyd earned the award of the Victoria Cross. An attempt was made to block the Kaiser Wilhelm (Kiel) Canal by dropping specially – modified mines against the lock gates. The mines had a salt water – activated fuse which unfortunately failed to operate in the less saline water of the canal.

By mid – August the "Battle of Britain" had been raging in the skies of southern England for a month, and the hard – pressed fighter pilots and ground defences were having a hard time of it. Fate, in the form of a German bomber crew which bombed London – apparently in error – and caused civilian casualties, took a hand in the proceedings. Winston Churchill ordered Hampdens, Wellingtons and Whitleys of Bomber Command to bomb Berlin as a reprisal and, although the RAF's attacks on the German capital appear not to have caused serious damage, they infuriated Hitler to such an extent that he ordered the Luftwaffe to bomb London. This change of tactics resulted in the RAF fighter organisation being able to recover and regroup and inflict such losses on the Luftwaffe in its attacks on London that the Fuhrer abandoned the idea of invading Britain because of the Luftwaffe's failure to establish mastery of the skies. There is a school of opinion which considers that the Bomber Command attacks on Berlin, by causing this alteration in the German plans, contributed in no small way to the outcome of what has now become known as the "Battle of Britain". In September 1940 the assembly ports for the barges in which the German troops were to be transported to Britain, were also attacked and Hampdens were lost in attacks against Calais, Ostend, Boulogne, Le Havre and Antwerp.

After the threat of invasion receded, the Hampdens resumed minelaying activity and distant targets such as Berlin, Leipzig, Stettin and Danzig were added to the bombing lists. These forays to comparatively faraway places resulted in a number of Hampdens failing to reach England on the return journey and going into the sea, out of fuel (see 5B). Luftwaffe "Intruder" aircraft also appeared on the scene, a special night intruder Gruppe I/NJG2, based at Gilze – Rijen in Holland and equipped with Ju 88s and Do 17s, having been formed in August 1940. Only one Hampden (X3027) appears to have been a definite victim of these nocturnal prowlers in 1940 but the confusion that their presence created over the areas of the RAF's bomber bases no doubt led to many accidents that might have been avoided if brighter flarepath lighting could have been used, but which the presence of enemy aircraft precluded.

As 1940 drew to a close, the Luftwaffe was carrying out night bombing of British towns and, in an attempt to augment the RAF's relatively meagre night fighting force, Hampdens were sent up to patrol selected areas in the hope of intercepting German bombers. It is not surprising to find that the Hampden was not a great success as a night fighter, as it was no faster than the enemy that it sought and the experiment was soon discontinued. X3065 was lost on operations over Bristol (see 5A3).

During 1941 the Hampdens made night bombing attacks on targets that ranged from the U – boat bases on the west coast of France, to ports along the coast of France, Belgium, Holland and Germany, with Bremen, Hamburg, Kiel and Brest each receiving a number of visits. Brest became a focal point of bombing and minelaying activity after the *Scharnhorst* and *Gneisenau* arrived there in March 1941, and on some of these operations Hampdens carried a load of two 2,000 – lb bombs. The shorter distance to the target meant that less petrol needed to be carried, thus making the extra bomb load possible, one of the bonuses for the RAF in having the battlecruisers at Brest. Bombing of inland targets in Germany continued; Dusseldorf was attacked a number of times as were most of the other targets in the Ruhr, whilst during the months when the nights were longer Berlin again received attention.

The problems of bad weather on return are highlighted by the loss of five Hampdens on the night of 1/2 March; all crashed due to bad visibility with the loss of life of 13 crew members and injuries to the other seven. The attacks on the battlecruisers at Brest included a number of fighter – escorted sorties and Hampdens joined in with the bombers from other Groups in these attacks. Other daylight activity included a brief attachment of B Flight of 61 Squadron to No.2 Group; these Hampdens operated from Bodney, a satellite of Watton, from 13 April to 29 April and lost two aircraft in an attack on targets near Cherbourg. Towards the end of 1941 a few Hampdens operated in cloud – cover nuisance raids over north – west Germany.

Minelaying continued throughout 1941. Brest attracted more than a fair share of mines, which the Hampdens laid by night and day, losing a number of aircraft in the process. On 7/8 November four Hampdens were lost in a minelaying attack on Oslo harbour, one of the particular hazards in this operation being the narrow, steep – sided fjords which allowed the German light flak guns to be positioned so that they could fire downwards at the low – flying Hampdens as they flew in to drop their mines. 1941 ended with 50 Squadron being involved in the first "Combined Operations" attack against Vaagso in Norway (see 2G).

J – VN of 50 Squadron taxying out through the snow for ops in January or February 1942. The bomb on the port wing shows up well.
(Photo: Cee Wadsworth)

The bombs loaded in pairs on the bomb trolleys in the foreground are 500 – lb GP (General Purpose). The two men on the right are handling 250 – lb GP bombs which have been transported in groups of three, secured to wooden carrying cradles. Three more 250 – pounders are on a cradle, which is itself on a bomb trolley, behind the tractor at the extreme right of the photograph. The aircraft is "EA – F", P1333 of 49 Squadron, which arrived on the squadron on 11 June 1940 and went missing on 16/17 August 1940. (Photo via Dennis Thompson)

The period from July 1941 until the end of the year was the climax of Hampden operational activity. Two new squadrons, 408 RCAF and 455 RAAF, began operations in July, and 97 and 207 Squadrons operated Hampdens in July and August whilst attempts were being made to remedy snags on Avro Manchesters. In October 1941, 61 Squadron began converting to Manchesters and at the end of the year 44 Squadron handed over its Hampdens to the newly – formed 420 RCAF Squadron and converted to Lancasters. Many of the crews on the Manchester squadrons had already survived a tour of operations on Hampdens; it is quite apparent, when one talks to survivors who had done so, that most of them would have preferred to have done a second tour on Hampdens.

1942

By the end of January 1942, 83 Squadron had joined 61, 97 and 207 Squadrons as a Manchester squadron. Two Coastal Command squadrons, 415 RCAF and 489 RNZAF, were being equipped with Hampdens that had been modified for use as torpedo – bombers (see 2J), but had only just commenced their torpedo – bomber training on Hampdens at the time of the breakout of the German warships *Scharnhorst, Gneisenau* and *Prinz Eugen* from Brest on 12 February 1942. The participation of 5 Group's Hampden squadrons in the attacks on the German warships is described in Section 2H.

Targets towards which Hampdens directed their bombs included Wilhelmshaven, Hamburg, Emden, Munster and other Ruhr targets such as Essen, Cologne, Dortmund and Dusseldorf. Minelaying also continued and once the German warships had left Brest the minelaying emphasis moved with them towards German waters in which the warships might move.

On 21 April 1942, 144 and 455 RAAF squadrons transferred to Coastal Command and moved to Leuchars to train for torpedo – bombing operations. They flew their aircraft to aerodromes in North Russia, near Murmansk, on 4/5 September (see 2J)

Nos.50 and 106 Squadrons also ceased operating Hampdens in March 1942 and a month later 49 Squadron, after suffering seven losses in a fortnight, handed over its Hampdens to 144 Squadron. The last squadrons on 5 Group to relinquish their Hampdens were the two RCAF Squadrons, 408 and 420. No.420 left Waddington on 1 August for Skipton – on – Swale where it re – equipped with Wellington IIIs.

L4109 and AD845 (1), from 14 OTU Cottesmore, were the last Hampdens to be lost on night bombing operations on the night of 13/14 September 1942. L4109 was shot down by a night fighter over the Netherlands whilst AD845, returning from the same operation, an attack on Bremen, crash – landed out of fuel near Cottesmore and could justly claim to be the last Hampden to be written off on a night bombing operation. Happily the crew of AD845 escaped without injury.

1943

At the beginning of 1943 the Hampden only remained in operation on the four torpedo – bomber squadrons. Nos.415 RCAF, 455 RAAF and 489 RNZAF squadrons operated them until the end of that year; 144 Squadron, which had been flying Hampdens since March 1939, said farewell to the Hampden when Flight Sergeant J W King brought AT195 in to land at Lossiemouth at 0105 on 19 January 1943, and began to re – equip with Beaufighters.

Hampdens were used by meteorological units from April 1943 until October 1943. Twelve were lost on meteorological reconnaissance flights, mainly from operations in the Bay of Biscay area. No.1407 Met Flight, based at Reykjavik in Iceland in June 1943, expressed an opinion of the Hampden in its Operations Record Book. "Hampden aircraft are found to be most unsuitable for Met sortie work, due to the old age of these aircraft which are ex – Bomber Command, their lack of de – icing equipment and their general unserviceability."

2B SUMMARY OF OPERATIONAL LOSSES

Out of the total of 1433 Hampdens almost exactly half, 714, were lost on operations. The term "lost on operations" is applied to any Hampden that was lost when detailed for an operation. A crew that crashed on take – off, ran out of fuel and abandoned their aircraft or crashed on return were just as much an operational loss as one that "Failed to return". The term "Failed to return" is applied where neither crew nor aircraft returned to this country. Aircraft that crashed into the sea off the English coast or, as in one case, crashed into high ground in Southern Ireland with the loss of the crew, are considered as being "Crashed on return" (see 2K). Only aircraft actually written – off as a result of a crash on take – off for, or on return from, operations are taken in to account in the statistics. Many more Hampdens crash – landed on return and were subsequently repaired.

Of the 714 Operational losses, 451 (63% of the overall total) were Hampdens that failed to return, 237 of which can be accounted for. German fighters shot down 128 Hampdens, 92 at night and 36 in daylight. Flak claimed 108 Hampdens but there is only one known victim of German balloon barrage defences (P4375) which is surprising when one realises that there were at least five Hampdens lost to our own "friendly" balloons (see 5E). This leaves 214 losses due to unknown causes and some possibilities as to the manner in which they came to grief are considered in Part 3F. The remaining 263 operational losses (37% of the overall total) were due to a variety of causes.

2C EARLY DAYLIGHT OPERATIONS

A pre – war line – up of No.44 Squadron Hampdens at Waddington in July 1939. The tarmac area in the foreground only extended along the front of the hangars, the aerodrome otherwise having a grass surface. The "JW" code letters were changed to "KM" at the outbreak of war. (Photo: Harry Moyle)

3 September 1939

Detailed "Operations Orders" for the Hampdens of No.5 Group had been drawn up long before the outbreak of war and within six hours of the actual declaration of war, two formations, each of nine aircraft, were on their way across the North Sea towards Wilhelmshaven to implement Operational Order B2, the intention of which was "to cause the maximum damage to enemy surface warships located in the Jade or Schillig Roads or in the harbours or vicinity of Wilhelmshaven". The attacks were to be carried out in daylight, from a low level, and were "to be made only on targets which are well clear of residential areas". The nine Hampdens from Scampton, six from 83 Squadron, three from 49 Squadron, were led by Squadron Leader L Snaith AFC, whilst 44 Squadron's nine aircraft from Waddington were under the command of Wing Commander Boothman AFC. Both formation leaders had been involved in the Schneider Trophy contests a few years earlier, Wing Commander Boothman having won the Trophy outright for Britain. Two of the Scampton pilots were later to become holders of the Victoria Cross, Flying Officer Learoyd of 49 Squadron and Flying Officer Guy Gibson of 83 Squadron.

Indecision about the type of bomb to be used in the attack had caused delays in the time of take – off, and it was after 1800 hours before the two separate formations had become airborne. Neither had reached the target area before darkness descended so no aircraft was able to make an attack. Most of the eighteen pilots involved had never flown a Hampden at night before, none had actually taken off with a full war load, so that even though the operations achieved no military success, and the return flight across the North Sea was an undignified "follow my leader" straggle, with navigation lights switched on, it is a tribute to the pilots that all of the aircraft were safely back on the ground at their bases by around midnight.

29 September 1939

From the outbreak of war, routine reconnaissance flights over the Heligoland Bight had regularly sighted enemy warships, but by the time that their sighting reports had been sent back to base it would be too late for bombers to be despatched to attack them. The enemy warships appeared to operate to a set programme, being at sea on most days except Sunday and Monday when they remained at their bases. They had been seen on a number of occasions in the vicinity of Heligoland at around 1000 hours after apparently exercising at night and were usually heading for their home port, Wilhelmshaven, when sighted and it seemed reasonable to assume that if, instead of a lone reconnaissance aircraft, a squadron was sent, then any warship sighted could be attacked at once.

On 29 September 1939 twelve Hampdens of 144 Squadron took off from Hemswell in two formations of six aircraft. Wing Commander J C Cunningham's formation (Raid EH3) took off at 0640 but was soon reduced to five aircraft as one aircraft developed engine trouble and returned to base, landing at 0705, just after the second formation led by Squadron Leader W H J Lindsay (Raid EH4) had taken off at 0700. The two formations proceeded independently across the North Sea towards Heligoland, Lindsay's six Hampdens, armed with 500 – lb bombs for attacks on surface vessels, being the first to arrive in the Heligoland area, flying at a height of 250 feet with good weather conditions and when south of Heligoland, two destroyers were sighted about ten miles away heading east at a speed of 25 knots. The first of the two sections of three aircraft carried out an attack in line astern from 300 feet during which the leading aircraft was hit by light flak and its navigator slightly wounded; the destroyers turned and avoided the bombs. The second flight of three did not make an attack and the six Hampdens of EH4 headed north and after

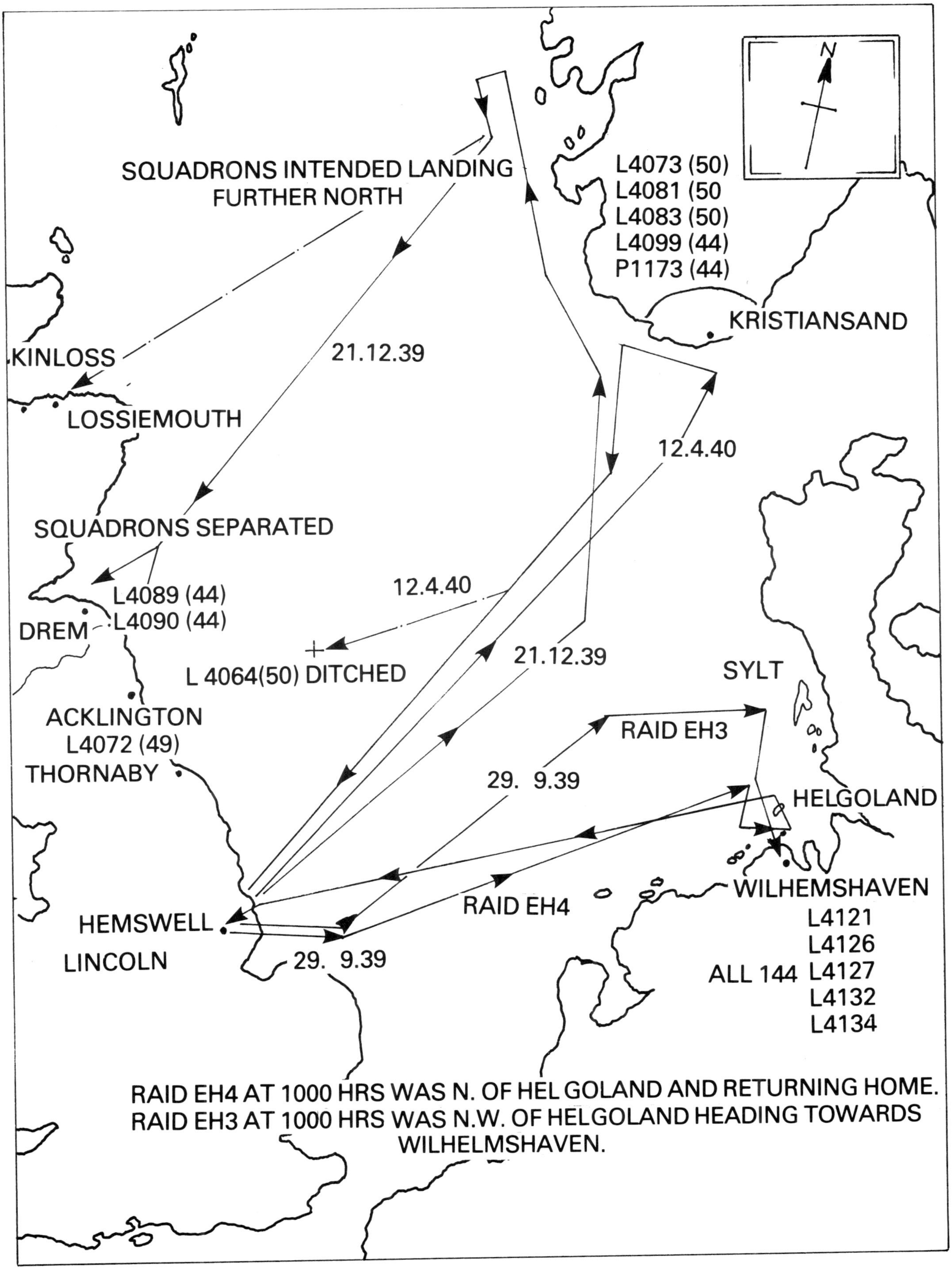
SQUADRONS INTENDED LANDING
FURTHER NORTH
L4073 (50)
L4081 (50
L4083 (50)
L4099 (44)
P1173 (44)
KRISTIANSAND
KINLOSS
21.12.39
LOSSIEMOUTH
12.4.40
SQUADRONS SEPARATED
L4089 (44)
L4090 (44)
12.4.40
DREM
L 4064(50) DITCHED
21.12.39
SYLT
ACKLINGTON
L4072 (49)
RAID EH3
THORNABY
29. 9.39
HELGOLAND
WILHEMSHAVEN
RAID EH4
HEMSWELL
L4121
L4126
LINCOLN
29. 9.39
ALL 144
L4127
L4132
L4134
RAID EH4 AT 1000 HRS WAS N. OF HEL GOLAND AND RETURNING HOME.
RAID EH3 AT 1000 HRS WAS N.W. OF HELGOLAND HEADING TOWARDS
WILHELMSHAVEN.
N

being subjected to heavy anti-aircraft fire as they flew past Heligoland, flew north for a few minutes more before turning for home. All landed safely after making landfall at Spurn Head at 1200.

The five Hampdens of Wing Commander Cunningham's formation (Raid EH3) each carried four 500-lb General Purpose bombs fitted with an instantaneous fuse for use against submarines, which meant that the bombs had to be dropped from a height of at least 1,000 feet. This formation, although the first to take off, had proceeded by a more northerly route and arrived in the Heligoland area at about the same time that the other formation was beginning to turn for home. The Wing Commander, heading east, led his formation past the south of Heligoland, flying at 500 feet as look-out aircraft, with the other four Hampdens 250 feet below him. As they passed Heligoland, the heavy coastal defence batteries opened fire and although no shells landed anywhere near them, the distant plumes of water gave notice that the enemy was aware of their presence. The crews would have been much more ill at ease had they realised that their comrades in the other formation had flown through this same area about twenty minutes before and that a hornet's nest, in the form of the Bf 109 fighters of I/ZG26, had been stirred up and that at least twenty enemy fighters were already swarming out to meet them from Jever.

The memory of the last moments of L4121 (1) remains vividly impressed in the mind of Jock Galloway, a sergeant pilot flying as the aircraft's navigator, and at the rear of the "vic" formation. He recalled: "Just before the attack commenced, one of the fighters came right up and closed in until he was tucked in between our mainplane and tailplane, just forward of the maximum arc of fire of our rear upper gun and shielded from the fire of the other Hampdens which dared not open fire on him for fear of hitting L4121. I clearly remember looking sideways and upwards and seeing every feature of the German pilot's face. It occurred to me that if I could have got to the astro hatch and opened it, I would have had a very good chance of shooting him with a revolver shot, but even as the thought passed through my mind he peeled off to rejoin his formation. From my position in the nose I had no idea how the attack from the rear developed; all I could see was the hail of bullets thrashing the sea ahead and below our aircraft but immediately after this first attack our Hampden dived towards the sea, only 200 feet beneath us. I instinctively lay down flat on the floor in the hope that I would better be able to withstand the crash that seemed so imminent. Just as I had done this a burst of tracer fire came up through the bomb bay and passed inches away from my body and although it could only have been seconds it seemed to remain there in a steady stream until we hit the sea and I lost consciousness".

Jock's next recollection is of momentarily regaining his senses to realise that he was floating on the sea, being kept afloat by a tiny pocket of air in the collar of his Mae West which he had not had time to inflate. The voices of German sailors calling from a rowing boat had summoned Jock back to the land of the living, but he did not remain conscious for long and next remembers finding himself in the wheelhouse of a small German patrol vessel. When the Germans picked him up there was no trace of the aircraft, nor any sign of the other three crew members; all had disappeared beneath the waves. One can only speculate upon the manner in which Jock himself had been spared a watery grave as, in his position in the fragile nose of the Hampden, he would have seemed to have been the least likely candidate for survival.

It is known that when P4358 crashed onto mudflats at Wilhelmshaven, the perspex nose collapsed and the navigator walked out. It is quite likely that when L4121 hit the sea the same thing happened, so that as the aircraft sank, taking the other three crew members who had possibly already been killed in the attack with it, the unconscious Jock floated out of the shattered nose and up to the surface.

The other four Hampdens suffered the same fate as L4121, each crashing in flames on to the sea. One member of each crew survived the crash, AC1 Liggett, the wireless operator of L4134 (2) and the pilots of L4126 (3), L4127 (4) and L4132 (5)

German patrol boats picked up the four survivors and the bodies of the crew members of L4132. FO Beck of L4127 died before the rescue boat reached Wilhelmshaven. An extract from the post-war report of PO R M Coste described how his aircraft (L4132) came to grief. "We had been detailed to make an attack on Wilhelmshaven harbour. We reached the area without incident and several ships were seen in the harbour approaches. The enemy seemed to be expecting an attack as the vessels had an escort of twenty Bf 109s. Bombs were dropped and a course set for home. The enemy fighters attacked in "vics" of five, diving from 1,000 ft and closing in astern to fifty yards. My aircraft, "R", was the fourth to be attacked; it was set on fire and the tailplane damaged. I crashed into the sea and an explosion occurred which threw me clear of the aircraft, still sitting in my seat. I was injured but managed to gather together the bodies of my crew, all of whom had probably been killed in the combat. A German vessel picked me up after I had been in the water for six hours and took the bodies of my crew on board".

An enquiry into the operations of that day concluded that the raids had been badly planned and made the point that is obvious from the map, i.e. that the first formation roused the German fighters which arrived in time to annihilate the second flight. When it became known that fighters had shot the Hampdens down, steps were taken to improve the armour plate protection for the crew, particularly at the back of the pilot's seat. The rear single Vickers "K" guns were gradually replaced by twin guns of the same type and the question of fitting self-sealing petrol tanks was considered, as were improvements to gun mountings, reflector sights and the supply of tracer ammunition.

The method of carrying out an attack against moving ships, particularly if they were of the destroyer type, i.e. fast and manoeuvrable, was also investigated. The tactic of going into line astern and bombing individually, on the same heading, was clearly not a good one, yet the lesson was not learned as will be seen in the account of the April 1940 attack against shipping at Kristiansand later in this section.

From the German point of view the attack on 29 September was a great success. Five Hampdens had been shot down in flames for the loss of two fighters which had to ditch because of battle damage and as the pilots of both machines were saved the losses were even more one-sided. In all fairness it has to be pointed out that the losses were not due to the Hampden being particularly vulnerable to fighter attacks, Blenheims and Wellingtons having suffered similar treatment during this "phoney war" period.

21 December 1939

The 49 Squadron Operations Record Book gives the background to the next operation worthy of note, in the entry for 21 December 1939. "At about 2100 hours on 20 December 1939, the unit received orders to locate and bomb the pocket battleship *Deutschland* which had been reported to be in the vicinity of the Norwegian coast". The plan was to carry out a search from Lister in Southern Norway to a point near Bergen and from there to return to land at Lossiemouth in Scotland. Twelve aircraft took off from Scampton, nine were from 49 Squadron, three from 83 Squadron; they met up with twelve more Hampdens of 44 Squadron from Waddington over Lincoln. Wg Cdr W C Sheen of 49 Squadron was in command of the 24 aircraft which were heading out across the North Sea by 0825 hours.

Landfall was made just to the east of Lister and the aircraft turned to fly northwards along the Norwegian coast. Spread out in line abreast, they hunted for the *Deutschland*. Fishing boats and coastal merchant vessels appeared; the Hampdens closest to the coast caught an occasional glimpse of the snowclad rocky headlands of neutral Norway, but the quest was in vain and at the extreme limit of their range the Hampdens turned to head for Kinloss and Lossiemouth aerodromes in the north of Scotland on the shores of the Moray Firth.

The weather throughout the flight so far had been poor, with cloud base down to 500 feet at times and rain and sleet showers reducing visibility. On the return flight conditions worsened and about two-thirds of the way back the two squadrons became separated. The Scampton aircraft made landfall at Blyth on the Northumberland coast and, after being intercepted by fighters of 43 Squadron which identified them as friendly, most of the formation landed at nearby Acklington at 1547. L4072 (7) crashed as it went in to land, two of the crew being killed and one seriously injured, the failure of an engine due to fuel shortage being the most likely cause of the crash. Another of the 49 Squadron Hampdens force-landed at Belford twenty miles away and one landed at Leuchars in Fifeshire. The Waddington Hampdens, under the command of Squadron Leader J J Watts had made a slight alteration of course northwards after becoming separated and when land was sighted at around 1500 hrs, the formation turned on a northerly course, believing that they were near Peterhead, when in fact they were south east of Dunbar so their course took them towards the Firth of Forth. The Forth was at this time an important Naval base.

The incoming Hampdens had been picked up by the shore radar stations and Hurricanes of 72 Squadron, from Drem, took off to investigate unidentified aircraft at 1508 and when two miles east of North Berwick at 1,000 ft were told that the aircraft were believed to be Hampdens. The Hurricanes almost immediately intercepted a large straggling formation of Hampdens and informed their controller of this.

No.602 (City of Glasgow) Squadron, equipped with Spitfires, was also based at Drem and the whole squadron was scrambled; two of the 602 Squadron pilots shot L4089 (8) down whilst a third sent L4090 (9) into the sea. A Court of Inquiry was held at Drem next day and decided that the bomber pilots should be impressed with the fact that "The onus of establishing identity rests with themselves", which would allow the fighter to shoot first and identify his victim afterwards. Air Chief Marshal Dowding, the Air Officer Commanding, Fighter Command, disagreed with the Court's findings, however, saying that "he found it an intolerable doctrine, to which he was surprised that responsible officers should have set their names" and ended by saying that "It is the duty of the Fighter Pilot to ensure that he does not open fire on a friendly aircraft."

The Court of Inquiry certainly highlighted aspects of the confusion that existed. The Observer Corps, for example, identified the aircraft as Hampdens, so did the Hurricanes of 72 Squadron, whilst the aircraft recognition of the 602 Squadron Spitfire pilots seems to be based on the premise that if it had two engines it was German, as the statements of two of the attacking pilots demonstrated.

No.602 Squadron's "Red Leader" called up on the R/T as he went in to attack L4089 "Tally Ho. Six Heinkels -- No, Dorniers, proceeding west,

1 mile north of Berwick", while the Flying Officer who shot L4090 down described how, in the middle of his attack "my windscreen was suddenly coated with oil. I tried to keep behind the bomber, firing all the while, but eventually I was afraid of colliding, and pulled away to one side and recognised the aircraft as a Hampden". It was also apparent that the Royal Navy at Donibristle was aware that Hampdens were due to return to Lossiemouth and Kinloss on the Moray Firth but the controller in the Operations Room at Turnhouse was not and that the RDF (radar) that plotted the incoming aircraft believed that the only RAF bombers that they could have been were "Wellingtons that were due back over Norfolk".

The official Court of Inquiry had not been attended by any of the crew members of the two Hampdens that had been shot down but when these returned to Waddington on 23 December the Station Commander, Group Captain L H Cockey, obtained detailed statements from them which were submitted to Headquarters No.5 Group.

The Waddington report on the operation arrived at a number of conclusions. It pointed out the problems of arranging the rendezvous of the two squadrons, in spite of the ideal weather conditions of 30 miles visibility and 3,000 feet cloud – base. The Waddington aircraft had been in position by 0750 and had then to circle for twenty minutes, using up valuable petrol, before the Scampton Hampdens arrived. The formation flying involved in the operation meant that pilots had to make use of their engine controls to maintain position and the report concluded that "Hampdens working in large formations under operational conditions are limited to about seven hours effective flying time".

Neither of the crews of the shot – down aircraft had made use of their dinghy, mainly because of the rapidity with which the aircraft filled with water and sank. L4089 had gone down within fifteen seconds and, although the dinghy pack came out of the aircraft via the astro – hatch, it could not be operated because a sufficient pull could not be exerted on the operating cord, as the dinghy pack and those who were trying to pull the cord were all wallowing about in the sea, the Hampden having sunk. The crew were in the water for half an hour before a fishing boat picked them up.

L4090 stayed afloat for about a minute. The pilot and navigator were still trying to extricate the wireless operator, who was caught up under the twin guns, from the mid – upper position although he had every appearance of having been killed in the action, when the aircraft sank with the dinghy still inside the fuselage. The fourth crew member had been wounded in the attack but he and the pilot and navigator were kept afloat by their Mae Wests until a fishing boat picked them up and took them to Port Seton.

The practice of the pilot firing the recognition cartridge from the Very pistol which fired downwards was not considered satisfactory and it was suggested that it would be better if the wireless operator fired the signal from the rear of the aircraft where it would be easier seen by an attacking fighter. The general system of visual identification, including comment on the criticism that the bombers ought to have been flying with their undercarriages down, also feature in the Waddington report and no doubt many lessons were learned from the incident, although, if the experience of Tony Reid is anything to go by, the fighter pilots continued to make quite appalling mistakes which often had tragic consequences.

L4089 was shot down by two 602 Squadron Spitfires. It crashed onto the sea and sank within fifteen seconds, jamming the observer/lower gunner in the astro/escape hatch as it went down and he was extremely fortunate to manage to free himself as the Hampden reached its last resting place on the sea – bed, some fifty feet below the waves. The dinghy had failed to inflate and the four men were kept afloat by their Mae Wests until a North Berwick fishing boat arrived and picked them up just as daylight began to fade. Without their aid, this book would not have been written!

After the battle, 12 April 1940. Damage caused by a 20 mm cannon shell to the tail unit of P4290 of 44 Squadron. Waddington hangars in the background. (Photo: Harry Moyle)

On arrival later at Station Sick Quarters at Drem, the crew were met by their Squadron Commander. The enthusiasm of his greeting to their pilot, "Damn good experience for you, George!" hardly reflected their own feelings, particularly as they had just learned that the wireless operator of L4090 had been killed and the observer/lower gunner injured when they too had fallen victim to 602 Squadron and ended up in the sea.

The other Hampden, L4090 had actually been flying on one engine when it was shot down and could hardly have presented a very aggressive appearance. Its navigator, RAF Volunteer Reserve pilot, Sergeant J A M Reid, who before the war had been an airline pilot with British Airways as co – pilot on Lockheed 14s, escaped unhurt. Jammy, like the other survivors, was given leave for Christmas and whilst in London called in to see how his pals at Heston were faring. He met up with one of his former first pilots, "Million – mile" Slocum, now also in the RAF as a Flight Lieutenant flying Hudsons, the military version of the Lockheed 14. Tony Reid, no doubt somewhat disenchanted with the Hampden scene, needed very little persuading when Flight Lieutenant Slocum asked if he would like to join him on the Hudson flight at the Photographic Development Unit, forerunner of the photographic reconnaissance squadrons, and in no time at all he was flying alongside his old airline friend.

On 3 March 1940 they were flying together on a routine flight, taking photographs of RAF airfields in the south east of England so that the effectiveness of the camouflage could be assessed. Their Hudson was over Gravesend, flying at 7,000 feet in perfect weather, when three Hurricanes of 32 Squadron arrived on the scene and attacked the Hudson, which went down in flames. Tony Reid was the only one to escape from the blazing aircraft, being badly burned before he managed to bale out and having to spend four months in hospital undergoing skin grafts before returning to duty. To be shot down by Spitfires into the sea and then to be shot down ten weeks later by Hurricanes and have to bale out would have qualified any Luftwaffe pilot for a Knights Cross, or something similar. Tony Reid had to be content with membership of the Caterpillar, Goldfish and Guinea Pig Clubs. In the same period he was awarded the Distinguished Flying Medal for a daylight photographic reconnaissance flight to Kiel.

Bomber Command had obviously decided not to abandon completely the use of bombers in daylight, however, and when a reconnaissance aircraft reported that German warships had become trapped in the ice off Heligoland every available Hampden was ordered to stand by to take off to attack them. That was on 17 February 1940, the weather during the previous six weeks having been particularly severe, with heavy falls of snow and sub – zero temperatures. The bitter weather probably saved the Hampdens from another mauling as the reconnaissance aircraft that flew out to confirm the previous report failed to locate the ships due to the blizzard conditions that it encountered.

12 April 1940

On April 8, Hitler's troops invaded Denmark and moved northwards to attack Norway. On 12 April 1940 a composite squadron of twelve Hampdens, under the command of Squadron Leader J J Watts and consisting of seven aircraft from 44 Squadron and five from 50 Squadron, took off from Waddington at 0820 to attack shipping off Kristiansand, the principal target being a battlecruiser and a cruiser which were expected to be in that area at 1100 hours.

The weather on the flight to the target area was 10/10 low cloud, with heavy rain, which forced the formation to fly at sea level until 1130 hours. As they approached the Norwegian coast the weather suddenly changed to conditions of no low cloud, a clear sky and visibility of fifty miles, which enabled shipping in Kristiansand Bay to be seen from twelve miles away. Squadron Leader Watts, using his R/T, ordered the formation to carry out a high level bombing attack from 9,000 feet. The Hampdens, each carrying four 500 – lb semi – armour – piercing bombs, bombed in sections of three in line astern. The sections followed one another towards the target on the same heading and at the same height, with the result that the fire from anti – aircraft guns on ships and ashore became increasingly more accurate as successive sections made their bombing runs. Immediately after the first section had made its attack, enemy fighters were sighted and the squadron leader put the nose of his aircraft down and dived to sea level, accompanied by the other two 44 Squadron Hampdens in his section, and headed for home.

By the time the second section of three, also 44 Squadron aircraft, made its bombing run, the anti – aircraft gunners, having established the range and height, were firing with deadly accuracy. L4099 (10) piloted by Flying Officer W G Taylor suffered damage and although it reached sea level it lagged behind the other Hampdens and soon fell victim to the fighters, crashing into the sea in flames. The wireless operator of another Hampden reported seeing his colleague in L4099, Corporal Harold Brown, still firing his guns as the blazing aircraft hit the water.

Pilot Officer Homer's aircraft, L4074 (11) also of the second section, was badly damaged by cannon and machine gun fire from the attacking fighters; one cannon shell entered the aircraft and passed within six inches of the pilot's head without exploding. One shell that did explode blew a

Fieseler Fi 156 Storch CK–KL and Bf 109s of II/JG 77 at Kjevik in April 1940. (Photo: Knut Maesel/Helmut Immich)

large hole in the starboard wing, another exploded in the oil tank of the port engine and machine gun bullets left a number of holes in the wings and fuselage, including damage to the perspex nose. Corporal Coldicott, the wireless operator/upper gunner, shot down one of the attacking 109s and was later awarded the Distinguished Flying Medal for his part in the action. It is not surprising that, after receiving this battering from the fighters, L4074's port engine began to overheat, causing the pilot to nurse the Hampden back across the North Sea towards Acklington, the nearest aerodrome, where it landed at 1535. The leader of the second section caught up with No.1 Section and returned safely to Waddington.

The third section came through the flak without apparent damage but suffered the attentions of the fighters as they dived towards the sea. Corporal Wallace, the wireless operator/upper gunner, in the aircraft of Squadron Leader Good, the section leader, shot down one of the 109s, but as they reached sea–level to join up with the leading section, Flying Officer Robson's aircraft P1173 (12), the No.2 aircraft, was hit and set on fire. The flames streamed out of the fuselage from the upper and lower gun positions as the luckless Hampden fell behind and crashed into the sea.

L4064 (13) of 50 Squadron, the No.3 aircraft of this third section piloted by Pilot Officer J B Bull, had also suffered damage from the fighters and was forced to land on the sea 100 miles east of Flamborough Head when he ran out of fuel due to his petrol tanks having been punctured in the action. Its crew were seen to have taken to their dinghy and Squadron Leader Good remained over them until an accurate position had been established by wireless bearings, then flew his own aircraft, which had suffered damage in the attacks, back to Waddington.

12 April 1940. Bf 109Es of II/JG 77 ready for action at Kjevik. The nearest aircraft belonged to Gruppe–adjutant Oblt Carganico, who was killed in action on 27 May 1944. (Photo: Knut Maesel/Helmut Immich)

None of the aircraft of the fourth section survived. One, probably L4073 (14), was seen to be hit by flak and go down in a vertical dive, whilst L4081 (15) fell victim to the fighters as it tried to catch up with the main formation. The last aircraft of this section, L4083 (16), was piloted by a Canadian, Flying Officer M Wilson Donaldson, known to his friends as "Weasel", on account of his small frame, although the Germans later, in his Colditz days in particular, may have called him other things. "Weasel" described how, despite opening up both engines to "Emergency full power" he had been unable to close the gap between his aircraft and the main formation, which to use his own words "was making off as if the hounds of Hell were after it". The inevitable result was that 109s had soon transformed the Hampden into a blazing wreck, with its lower gunner killed in the encounter and the wireless operator badly burned. The events that followed "Weasel's" crash landing on a small island off Tregve on the Norwegian coast, are described later in this section.

Fortunately for the Hampdens, shortage of fuel and ammunition forced the enemy fighters to abandon the chase as otherwise, in the words of one of the pilots who returned safely: "It was only a matter of time before they shot the lot of us down". Seven Hampdens had survived the onslaught. L4074, because of its overheating port engine, soon altered course to head for Acklington and L4064 went down in the sea. Only four Hampdens remained flying together towards Waddington.

Squadron Leader Watts in L4171 (17) was the first to touch down at 1600 hrs. P4290 (17) which had been hit on the starboard side of the tailplane by a cannon shell and L4091 and L4154 (18) landed a few minutes later. The crews were soon back in the crew room comparing their harrowing experiences, amongst them Sergeant Pilot J Clayton, navigator in L4154, who explained how he had taken his Vickers "K" gun from the nose of the aircraft to the astro–hatch, which he opened, then in an attempt to discourage the fighters from making beam attacks, had fired the gun from his shoulder as if it had been a rifle. He was later awarded the Distinguished Flying Medal for this inventive feat of air firing.

Squadron Leader Good's aircraft arrived after the other four, the only one of the five 50 Squadron Hampdens to return and his news that the crew of L4064 had taken to their dinghy raised hopes that they would soon be back at Waddington but sadly this was not to be the case. The crew were found next morning by a 220 Squadron Hudson which remained circling the dinghy until after midday, but when the aircraft left the scene that was the last that was seen of the crew. The dinghy was located again a few days later in the same position, i.e. 100 miles east of Flamborough Head, but this time it was empty. The only evidence of the crew lies in Kiel War Cemetery, where the observer/lower gunner is buried; the bodies of his three companions were never found.

The search for the missing crew had involved the Ansons of 44 and 50 Squadrons which were normally used for general non–operational flying. The 50 Squadron Anson N5193 took off at 1745 on 13 April for such a search and the pilot seems to have forgotten that an Anson would not stay in the air for as long as a Hampden did, with the result that after

The remains of L4083 being inspected by Generaloberst Erhard Milch (in centre of photo). Major Harry von Bulow – Bothkamp, Commander of II/JG 77, is also in the picture. The fuselage of L4083 lies on its side on the left of the photo; the pilot's windscreen is shattered. A Ju 52/3m stands in the background. (Photo: Knut Maesel/Helmut Immich)

being in the air for over five hours, he ran out of fuel and crashed near Grimsby; the Anson was a write – off and its four crew were all injured. An hour before the Anson crashed, 50 Squadron Hampden L4065 (19) took off from Waddington for a minelaying operation, its pilot, Flight Lieutenant R J Cosgrove, being the son of the Premier of Tasmania. The aircraft apparently completed its mission and had been picked up by our radar as it approached Mablethorpe but never reached the coast. No trace was ever found of crew or aircraft. (See 5B33)

The post – mortem on the operation, carried out by Group Captain C T Anderson, the Officer Commanding RAF Station Waddington, concluded that it had been a mistake to bomb in line astern, at the same height and on the same line of approach and that the leader should have collected his followers and formed a defensive formation as quickly as possible. There was criticism of the manner in which the aircraft that did gather together had flown in a wide "vic" formation, very close to the water. Because of this the lower rear gunners were unable to use their guns and the flank machines were subjected to individual attacks without the support of the remainder of the formation. The width of the formation and proximity to the water meant that it was impossible for the formation to make a turn. A height of 200 – 300 feet, with aircraft in "box" formations, would have made them more manoeuvrable and at this height the fighters would not so easily have been able to use the bullet splash on the sea as a guide to the accuracy of their aim.

Some of the fighters' attacks had been made by flying parallel to the bombers until slightly ahead of them and then turning in to make a beam attack against which none of the Hampdens' guns could be brought to bear. The crews who flew the Hampdens were well aware that this form of attack was possible, having experienced it on numerous occasions on fighter affiliation exercises, but their reports had been dismissed by the "experts" who still maintained such attacks were impracticable. Experience on this occasion did at least teach something and the installation of beam positions was proceeded with as a matter of urgency.

Group Captain Anderson's report mentioned that aerial photographs showed that several bombs obtained near misses on ships. They also revealed that one of the ships was the gunnery training ship *Brummer*, the obvious inference being that the *Brummer* was responsible for at least some of the heavy and accurate flak. Knut Maesel, of Kristiansand, supplied other details that illustrate how, had the attack been carried out 24 hours previously, the outcome may well have been much less of a disaster for the Hampdens. Knut makes the point that the cruiser *Karlsruhe*, one of the intended targets, already lay at the bottom of the Skaggerak, having been torpedoed by the submarine *Truant* on the evening of 9 April as it led the assault fleet of Kriegship – gruppe 4 towards the attack on Kristiansand. The German attack did not proceed without its own problems of co – ordination, bad weather and spirited resistance from the Norwegians, but by April 10 they had occupied Kristiansand and the adjacent airfield at Kjevik. On 11 April, 20 Bf 109Es of the elite unit II/JG77 arrived at Kjevik, escorting a formation of Ju 52/3m transport planes. The fighters had originally been intended to remain at Esbjerg in Denmark to cover the air operations over that country, but no real fighting took place so they moved on to Kjevik, arriving less than 24 hours before the Hampden raid.

The 88 mm guns of the heavy anti – aircraft units had been shipped on board the freighter *Kreba* which arrived at Kristiansand in the evening of 11 April. The guns were hurriedly bought ashore and by the morning of 12 April were manned and ready for action, a matter of only hours before the Hampdens arrived. According to eye – witnesses of the raid, no German ships were hit; a few windows on the shore were smashed and two Hampdens were shot down by flak before the rest of the Hampdens disappeared over the horizon, with the 109s in hot pursuit.

Some of the fighters' attacks had been made by flying parallel to the bombers until slightly ahead of them and then turning in to make a beam attack against which none of the Hampdens' guns could be brought to bear. The crews who flew the Hampdens were well aware that this form of attack was possible, having experienced it on numerous occasions on fighter affiliation exercises, but their reports had been dismissed by the "experts" who still maintained such attacks were impracticable. Experience on this occasion did at least teach something and the installation of beam gun positions was proceeded with as a matter of urgency.

Official Luftwaffe reports claimed that II/JG77 had brought down six Hampdens and a Hudson during the battle, five Hampdens being claimed by Fw Menge (two) and Ofw Herfeld, Uffz Opotski and Ofw Arnoldy one each. Four Bf 109Es were lost, including Ofw Herfeld and Uffz Opotski, two of the victorious pilots, and Oblt Ruthammer and Ofw Stierle. One Bf

109E force landed at Mandal out of fuel.

The only survivors from the six Hampdens were three of the crew of L4083, its pilot, "Weasel" Donaldson, relating the extraordinary series of events that followed the crash of his blazing aircraft on an island which lay just off the coast near the small fishing village of Tregve.

Two Norwegian fishermen rowed out from the mainland and picked up the three survivors who were then taken by car to the hospital at Mandal, where, as the Huns had not yet reached this part of Norway, the local Norwegians looked after them very well. An extract from "Weasel's" account shows what happened next!

"On what I believe was our second day in the hospital, the doctor, who seemed to hold some military rank, came to see me and asked if I could fly a Bf 109. Upon learning that I believed I could, he bundled me into his car and away we went out into the forest. Lo and behold, when I arrived there what do I see but a 109 hidden under the branches of some bushes. I checked it out very carefully and could find nothing wrong with it. As it turned out it was one of the aircraft that was in action with us that had run out of fuel. The Norwegians had arrested the pilot and stuck him in the local gaol. The doctor and I then went around the town and picked up all the petrol we could find, returned to the aircraft and refuelled it. I decided to try to take the other two of my crew with me, so we took out the pilot's seat so that only the seat braces remained. We placed Corporal Henry on the floor and Don Middleton sat on the seat frame whilst I sat on his knee, so that the three of us were in the single – seat fighter.

The field we were in was only about a hundred yards long and was very narrow. We roared down the field a couple of times but she just would not develop the necessary horsepower on car petrol. On our third attempt we ended up shearing the starboard wing off against a tree, so back to hospital the three of us went. Two nights later the doctor came to see me in the middle of the night and said that the Germans were coming into the town the next morning so that if I felt well enough to walk he would send someone with me and we should get the Hell out of it forthwith. Middleton and myself got dressed and were led off into the woods by our guide. After walking for what seemed like an eternity we came to the coast, where there were several summer cottages. Our guide settled us down in one of these, handed us some food and said that he would be in touch with us later. We never saw him again.

In the meantime the Germans had moved into Mandal and, finding Corporal Henry in the hospital, soon learned that we were still skulking around somewhere. They sent for the local sheriff and ordered him to turn us in, or face the consequences. The poor fellow had no option but to tell them where we were and the Germans arrived at our cottage and arrested us."

"Weasel" met the sheriff and the chap who had been their guide when he visited Norway after the war. The sheriff had been most upset at having to turn the two men over to the Germans and found it hard to face Weasel, who reassured him that it was all part of war. His meeting with his former guide revealed that after he had left them he had walked thirty miles to the east of Kristiansand, where he arranged with a Swedish shipping line to send a small boat to pick the two officers up. The boat had arrived two days after the Germans had taken them captive. In view of the merry dance that Weasel, and others, led the Germans in their prisoner of war days, there were probably quite a few of their gaolers who often wished that he had made a successful escape from Norway. One thing is for certain and that is that Weasel was the only pilot who could boast that he had pranged a Hampden and a Bf 109 in the space of such a short time. Sadly his valiant effort met with little recognition.

The wreckage of L4083 of 50 Squadron at Kjevik, Kristiansand. (Photo: Knut Maesel/Helmut Immich)

Hampden torpedo – bombers of 489 Squadron in echelon to starboard. The cover of the port beam midships gun position, situated just forward of the egg – shaped Direction Finding Loop, has been removed in each of the aircraft. The lower guns do not appear to be installed.
(Photo: RAF Museum P7181)

2D MINELAYING

On Monday, 8 April 1940, representatives of the Admiralty and RAF Bomber and Coastal Commands met at the Air Ministry to discuss Plan WA 15, which detailed the method of laying mines from the air and the areas in which they were to be dropped. Air Vice Marshal A T Harris, the Air Officer Commanding 5 Group, expressed confidence in the ability of his crews to drop mines with the required degree of accuracy from 500 feet in the North Sea and Baltic areas which had already been reconnoitred by 5 Group.

The distribution of mines had already taken place, 155 to the Hampdens of 5 Group and 45 to Beaufort squadrons and these 200 mines were intended to be laid during the April moonlight nights. It was felt that the first lay of mines might well escape notice because the areas had been frequently visited at night throughout the last four weeks. The dropping of 500 – lb bombs, with long delay fuses, was considered as a plan to confuse the enemy as to what was happening, and it was agreed that twelve bombs could be dropped in Kiel Fjord and another twelve in the Schillig Roads off Wilhelmshaven, on clear nights after the moonlight period.

The No.5 Group Hampdens were allocated the following main areas:

Hemswell (61 and 144 Squadrons)
Kiel approaches 48 mines; Warnemunde 12 mines.

Waddington (44 and 50 Squadrons)
Elbe approaches 42 mines; Swinemunde approaches 18 mines.

Scampton (49 and 83 Squadrons)
Neustadt/Travemunde 24 mines; Swinemunde approaches 18 mines.

This was based on three sorties, each by 20 aircraft per station, during moonlight nights within a period of seven days.

Bomber Command Operation Order No.93, dated 9 April 1940, gave details which included the method of dropping the "M" mine. It had to be dropped into a depth of between 5 and 12 fathoms (10 – 24 metres) of water and would then lie on the sea bed until fired by any violent change in the surrounding magnetic field e.g. a passing ship. The mine weighed 1500 lb (680 kg), of which 750 lb (340 kg) was the explosive charge. It had a parachute attached to prevent it from disintegrating when it went into the water and had to be dropped from a height of between 400 and 1000 feet (122 to 305 metres).

At the time of release the aircraft speed did not have to exceed 200 mph (322 kph) as above these conditions the parachute would tear away. The mines were to be laid with as much secrecy as possible, i.e. not closer than 1 mile (1.6 km) to the shore or nearest surface vessel or lighthouse.

The code name "Gardening" was given to the minelaying operations and the mines were referred to as "vegetables" which were "planted" in areas at first identified by the names of vegetables and flowers, to which the names of fish and shrubs were later added. Some of the first areas to be mined bore such names as "Yams" (North Heligoland), "Daffodils" (Baltic Sound), "Quince" (Kiel Bay), and "Eglantine" (Elbe Estuary). Over sixty such code names were eventually in use but in "The Hampden File" reference is made to the true name of the area.

Birger Hansen from Copenhagen, who has made an intensive study of RAF minelaying, particularly in the Baltic, provided information confirming early successes achieved by the Hampden minelaying campaign.

14 April 1940: *Nyborg* 2555 BRT: Damaged at Sprogo island (between Nyborg and Korsor) in the Great Belt, but was later repaired.

17 April 1940: *Odin* 1581 BRT: Damaged in Great Belt.

20 April 1940: *Christian IX* 1590 BRT: Damaged in Great Belt, and sank, but was later raised and repaired. These were the three train ferries.

Other losses in April, confirmed by Birger, included:

15 April: German *Sperrbrecker XI* 4845 BRT: Mined 6 km SW of Korsor harbour. Taken to Kiel and repaired.

24 April: *Cronhagen* 1787 BRT: Mined and sank 2 km south of Drogden lightship; not raised.

Birger also reported that the troop carrier *Pionier*, 3285 BRT, was mined north of Skagen (the Skaw) in an area where no British mines are reported as being dropped so, although the Germans said it had hit a British mine, it may in fact have hit a German mine that had drifted from the Skagerrak. Of the 823 Germans on board the *Pionier*, 346 were killed and 89 injured. On 12 November the German ship *St Jurgen* was mined and sank in Little Belt. It had on board "815 tons of slaughtered pigs –

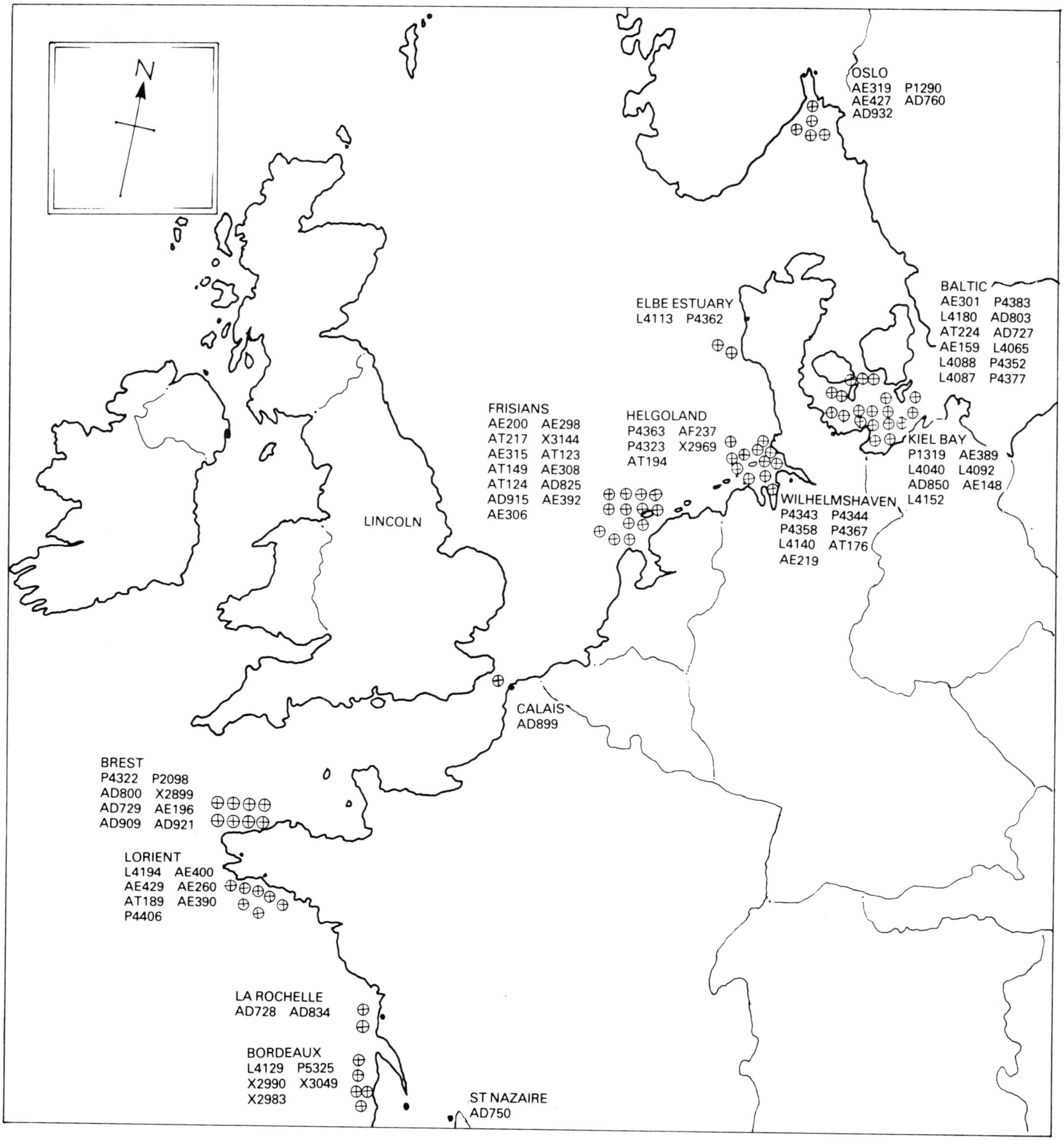

enough to feed the population of Berlin for 5 days. 36% of the cargo was later recovered and used as human food". Were there any complaints, one wonders, about the saltiness of the salvaged bacon rations?

In October 1940 the Germans reported that they had found well over 100 mines in Danish waters, which included mines from the British submarine *Seal* which had been captured in the Kattegat in May 1940. By the end of the first six months of the minelaying campaign in that area around forty ships had struck mines.

Another aspect of the minelaying campaign on which Birger Hansen commented concerned the attacks on ships by Hampdens using the two 250-lb bombs that were carried, one under each wing. On 7/8 August 1941, Pilot Officers Thompson and Anekstein, both of 44 Squadron, reported successful bombing attacks on ships whilst on minelaying operations in Danish waters. Birger's reports confirm the results. "The German ship *Viola* was bombed in the Great Belt and strafed with machine gun fire. A German seaman was killed. The ship went to Kiel for repair."

The Swedish ship *Vanersberg* (1200 BRT) was hit by two bombs, one on the starboard side near the bridge, the other amidships. A fire broke out and the crew took to the lifeboats. The ship's Captain, the Danish pilot and a boy seaman were killed. The German ship *Christian Russ* arrived, put the fire out, picked up the *Vanersberg*'s crew and brought the stricken ship back to Malmo. The *Vanersberg* must have been PO Anekstein's victim, as he reported two hits which set the ship on fire and that the fire could be seen from 50 miles away as they flew back towards home.

The Germans were quick to react to the minelaying threat. They established flak batteries along the shores of estuaries, such as the Elbe, on the steep sides of the Norwegian fjords and in the Baltic on the many islands between which the navigable channels flowed. In the more open sea areas large fishing trawlers were armed with flak up to 88 mm, but mainly with the lighter 37 mm and 20 mm cannon, while patrol craft kept watch for the incoming minelaying aircraft. Radar ashore, and later afloat, facilitated the tracking of the minelayers and guided patrolling fighters towards them.

In the course of enquiries into operations around Wilhelmshaven, contact was made, through the "Wilhelmshaven Zeitung", with Friedrich A Greve who most kindly supplied the information that follows and some photographs.

The Germans converted a number of old warships into "Flakkreuzer" also known by the fascinating title "schwimmende Flakbatterie". They operated in the mouths of rivers and inlets where the areas of water could not be covered by shore batteries. There were two of these flak cruisers in the Wilhelmshaven defences. Both had originally been built for the German Navy and were already antiquated at the time of the First World War. They were *Arkona*, built 1901 and defending the Aussenjade (from July 1940) and the *Medusa*, built 1900, protecting the Innenjade (from August 1940).

In 1940 the armament of both *Arkona* and *Medusa* consisted of five 105 mm guns and one 30 mm and four 20 mm cannon, the latter having a firing rate of 500 rounds per minute. A searchlight was also part of the equipment which by 1943 had been modernised three times and included Wurzburg Type 39 TD II gun – laying radar and Freya aircraft detection radar, with an increase in armament.

The other five "Flakkreuzer" were:

Niobe (ex – Dutch cruiser *Gelderland*)
Ariadne (ex – Dutch coast defence ship *Hertog Hendrik*)
Undine (ex – Dutch coast defence ship *Jacob van Heemskerck*),
Nymphe (ex – Norwegian coast defence battleship *Tordenskold*),
Thetis (former Norwegian coast defence battleship *Harald Haarfagre*).

The *Niobe, Ariadne* and *Undine* each had eight 105 mm, five 40 mm and sixteen 20 mm guns and the *Nymphe* and *Thetis* carried six 105 mm, two 40 mm and fourteen 20 mm as armament. They had their own means of propulsion but were only capable of a speed of 10 knots, sufficient to steam out to their allotted position.

Friedrich also provided information about German use of Ju 52/3m aircraft, equipped with a "Stromschleife" loop under its wings. With this equipment they would explode the magnetic mines. This device presumably operated on the same principle as the RAF's mine destroyer Wellington DWI with its large circular magnetic impulse generator. The Germans increased their sea – borne minesweeping forces to counter the threat to shipping and it is obvious that, whatever the direct results of the minelaying campaign so far as the sinking of ships was concerned, there was a tremendous indirect consequence in the manpower and material that had to be diverted to meet the threat. Later Bv 138 three – engined flying boats and four – engined Ha 139 seaplanes were fitted with cables looped round the aircraft for degaussing mines.

There is evidence that the minelaying activity became the subject of tribal warfare between 5 Group and Bomber Command, between Bomber Command and Coastal Command and between the RAF in general and the Admiralty in particular. A letter on 9 August 1940, from Air Commodore A Durston, Director of Operations, Naval Co – operation, Air Ministry to the Air Officers Commanding Bomber and Coastal Commands was obviously intended to regularise the somewhat confused situation about who was actually running this part of the war. It advised that the Admiralty would pass direct to Coastal Command the details of the areas to be mined while Coastal Command would be the mine – distributing authority and would be responsible for passing information of areas to be mined to Bomber Command.

As 1940 drew to its close, the early euphoria about the effectiveness of the minelaying operations had obviously diminished as by the end of December it had been decided that Hampden "learner" crews would lay the mines and on 27 December 1940, Manchesters and Stirlings were brought into the minelaying effort to lay mines during their working up period. A modified mine fitted with a fuse that operated on impact, the "Imp" mine, was used as a land mine on the first raids on Berlin in the autumn of 1940. The misgivings concerning its reliability are mentioned in Part 3F4 in connection with the losses of X2981 and X2996 (1).

Seventy – five Hampdens went missing on minelaying operations, the first on 13/14 April 1940 and the last on 12/13 July 1942. Of the 303 crew members involved (three aircraft had a crew of five), 260 were Missing or Killed, 42 became Prisoners of War and one evaded capture and came back to England via Spain – FO Whitecross from AD384 (see 5A26).

Most of the Hampden minelaying operations were carried out at night but six Hampdens were lost on daylight minelaying, two off Brest in December 1941 and four off the Frisians on 6 and 7 February 1942 (see 2H1).

Where the cause of Hampden losses on minelaying can be ascertained, the vast majority were shot down by flak from ship or shore batteries. A few instances of such losses, received in reports from Birger Hansen and personal accounts, illustrate what happened.

L4088 21/22 April 1940 (3) Pilot Officer Dutton "... after releasing his mine in the Kattegat, saw ahead of him a fleet of small vessels which he believed were fishing boats. He dived to attack them with his two 500 – lb wing bombs. Unfortunately they were flak ships which gave him the full benefit of their concentrated fire as he came within range. An engine was put out of action and, in spite of the crew throwing out every movable piece of equipment, the stricken Hampden could not be made to gain height. After crossing the coast near Kiel, the pilot had to make a belly – landing. None of the crew suffered serious injury and all became Prisoners of War."

L4087 Sgt Farrands, 19/20 July 1940 (4) "Shot down by light flak from Frederikshaven and crashed into the sea off Skagen. The observer/navigator and the wireless operator were killed but the pilot, who had been in L4089 when it was shot down by Spitfires, and the lower gunner, an air gunner officer, Pilot Officer "Pop" Green MC (the MC was from the previous war) were taken prisoner." Both are mentioned in the "Kriegie" section.

AE301 FO Harwood DFC 26/27 August 1941 (5) "...was brought down by Flakzug (AA train) *Hals* from a height of 100 metres, with 51 rounds of 20 mm gunfire. The aircraft crashed 850 metres SW of the train in shallow water. The wireless operator was the only survivor and he became a Prisoner of War. The other crew members are buried at Frederikshaven. Radio code books, wireless equipment, maps, a log book and other important equipment were found and when the aircraft was recovered one 325 – kg mine was found and brought to the Marine HQ at Kiel."

AD850 FO Hartley 5/6 November 19411 (6) "Hit by machine gun fire from the searchlight position at Skallingen Ende, and from "Boat 5" from HSFI Esbjerg. The wreckage was found on the SW coast of Jutland" ... The crew are all buried at Esbjerg cemetery, Denmark.

Wreckage of AD915 of 420 RCAF Squadron on Schiermonnikoog. It was shot down by flak on a minelaying operation on the night of 18/19 February 1942 (See 2D 10). (Photo: Wyb Jan Groendijk)

Things that go bang in the night. One of the German floating flak batteries ("Schwimmenden Flakbatterie") which accounted for many mine – laying aircraft. The former light cruiser Arcona was a light cruiser launched in 1902 and converted to a hulk in 1930. She was rebuilt as a flak ship armed with five 105 mm, one 37 mm and four 20 mm AA guns. The removal of the forward boiler room meant that its funnel could be cut down and topped with a searchlight. Moored in the approaches to Wilhelmshaven, her armament was later increased and radar installed. (Photo: F.A.Greve)

Five pilots from Waddington who took part in the first minelaying operation on 13/14 April 1940. They are, left to right, back row: Flying Officer Duncan Corr (50 Squadron), and Sgt Smythe (44 Squadron); front row: Flight Lieutenant Dutton (44 Squadron), Squadron Leader Weir (44 Squadron) and Flight Lieutenant Bennett DFC (50 Squadron). Jimmy Bennett was the first Bomber Command pilot to be awarded a DFC for a night bombing operation, the attack on Sylt on 19/20 March 1940. Flt Lt Dutton became a prisoner of war in L4088 within a few days of the photograph being taken. Sgt Smythe was killed in a low level attack against Bremerhaven docks in P1338 on 11/12 September 1940; in the five months after the photo was taken, he had been granted a commission, been awarded the DFC and had attained the rank of Flight Lieutenant. Flying Officer Carr survived his Hampden operations but failed to return later in the war from an attack on Le Creusot. (Photo: Jimmy Bennett)

AD803 and AT224 15/16 May 1942 (7) "Both of these Hampdens were from 408 RCAF Squadron and had taken off to lay mines in the Sejro Deep in the West Baltic. The same German naval vessel shot them both down into the sea between Odder (on the east coast of Jutland) and Samso Island. Wreckage from both aircraft was found 2 – 3 km west of Endelave Island. Sgt Copeman, the Canadian pilot of AD803, was the only member of the two crews to survive and he was picked up from the water, badly injured. Of the other seven crew men the bodies of three were never found. The body of the pilot of AT224, F/Sgt Raymond J Dillon, an American in the RCAF, was picked up next day; his lower gunner's body came ashore a month later on Samso Island. The sea reluctantly gave up the body of Sgt Copeman's lower gunner and he was buried at Aarhus Vestre Cemetery on 15 September but it was not until 25 February 1943 that the body of the wireless operator, Sgt N W Smith, was found at sea south of Vestborg lighthouse."

One of the "perks" of the minelaying operations was the pair of bombs that were carried on the wing racks, either 250 – pounders or 500 – pounders and generally of the GP (General Purpose) type. These bombs could be used on any target that looked likely to hinder the enemy war effort and on a number of occasions it was the target for these bombs and not the actual mine drop that brought trouble. Fred Bailey was wireless operator in X2983 (8) on 14/15 February 1941 and after they had dropped their mine successfully in the Gironde estuary the pilot looked around for some suitable target that could be livened up with his 250 – pounders. A building that appeared to be a water tower seemed to be worth a bomb and the pilot turned inland to make a bombing run. The German flak gunners opened up with a deadly stream of cannon fire which tore into the aircraft, causing the pilot to make an immediate crash – landing in which he, and his navigator who was in his position in the nose of the aircraft, were both killed. Fred Bailey had got away with a broken rib, the lower gunner Sgt Dixon had broken both ankles. Fred wasted no time in getting clear of the wrecked Hampden, dragging his crewmate with him. After checking that there was no hope of his other two crewmates being alive in the front part of the Hampden, Fred became aware that other Hampdens were flying above the now – blazing X2983. He rushed down to the beach and waved his comrades above away from the scene in case they, too, fell victims to the flak battery. After the war Fred met up with one of the lads who had been in one of the Hampdens that flew over him. He remembered seeing Fred, standing on the beach and waving. His interpretation of the wave was that Fred was waving "Goodbye", maybe a new form of an "Airman's Farewell", to complement the soldier or sailor's immortal salutation!

The author, historian and television and radio commentator, Robert Kee, was a wartime Hampden pilot. Robert's book "A Crowd In Not Company" (published by Jonathan Cape Ltd) gives an engrossing account of his experience as a Prisoner of War. It begins with a graphic description of how he came to grief on a minelaying operation, after over twenty ops with 420 RCAF Squadron.

Pilot Officer Kee's troubles began as AD915 made its landfall on the north coast of Holland on the night of 18/19 February 1942 (10). The navigator had confirmed that the island ahead of them was Schiermonikoog from which their timed run had to be made. Robert adjusted his speed and height and turned the Hampden on to the correct heading, his navigator prepared to start his stop – watch. The German flak gunners on the land beneath them had made their preparations too and, as AD915 flew low towards them, their 20 mm light ack – ack guns unleashed a devastating blast of cannon shells that ripped into the hapless Hampden. Before Robert had time to open the throttles and gain speed to escape from this murderous onslaught, the port engine was hit and put out of action. The immediate result of this was that Robert momentarily lost control of the aircraft; as he relates: "If we had been 300 feet higher we might have recovered and flown away. As it was we crashed in something like the landing position... There was a grinding, sliding sensation and more noise that I have ever heard in my life. Flashes broke across my eyes. I waited for pain and fire. There was only silence. I uncovered my head. It was difficult to believe that the wreckage in which I was sitting had once been an aircraft. The wings seemed to have disappeared completely. The useless broken control column lay ridiculously across my legs. All around me blotches of metal were heaped upon the beach. The cockpit itself had disintegrated and I was sitting on the ice... I was still strapped in my seat ... I pulled the release pin of my safety harness, stepped over some wreckage and began to waddle across the ice ... past a blotch which was not metal. It had a hand and a flying boot. It groaned. The face was swollen and painted with blood. I recognised my navigator ... he did not seem to be too badly hurt. I went to see what had happened to the other two. There was enough left of both of them to tell me that they were dead ... I could not believe that what I had found had very much to do with the two men I had known."

Postscript to AD915. One of the carrier pigeons from the aircraft was picked up at Spurn Head, on the east coast of England, at 1330 hrs on the 22 February. The message that the bird had dutifully brought back read "SOS AD915 – F. Time despatched 10 o'clock 21.2.42. Position Oosterbierum, Friesland, Netherlands", and written on the side of the message were the letters "OZO". The report about the message appeared in the 420 Squadron Operations Record Book on 23 February. The significance of the letters "OZO" was not apparent until it was discovered that they were the symbol of the defiance of the Dutch resistance against their Nazi oppressors. They were the initials of the Dutch words that stood for "Holland will rise again" and appeared in all sorts of places as a reminder to the Germans that they had not broken the spirit of the Dutch people. The message that the pigeon brought back is really most

Problems for mine – layers. German light anti – aircraft guns were located on high ground that looked down on Oslo Fjord. This particular gun was situated on top of Oslo Town Hall. The gunners proudly display the RAF roundel trophy from a Hampden victim which may have been from AE427, AD760 or P1290 which were all shot down by flak in early November 1941. (Photo: Emil Rabe via Knut Maesel)

mysterious. The aircraft number and serial letter are correct but the date 21.2.42 is wrong and the aircraft actually crashed on the island of Schiermonnikoog which is 50 km NE of Oosterbierum, at about 22.00. The first part of the message could have been written by the wireless operator just before he was killed in the crash on Schiermonnikoog, where he is buried with the lower gunner. It really seems most improbable that any of the crew would know what the letters "OZO" meant and, as none of them could have written it after the crash, the only answer seems to be that the pigeon, with the part – completed message, was freed from its basket in the crash, and fluttered about in the sand dunes, possibly flying as far as Oosterbierum until it was found by someone with connections with the Dutch resistance, who completed the message and added those three proud letters "OZO".

The design of the capacious bomb – bay of the Hampden enabled it to carry a 1,500 – lb mine. It was the only Bomber Command aircraft capable of doing so during 1940. (Photo: RAF Museum P21820)

The wreckage of X2996 of 44 Squadron in the outskirts of Berlin on 15 November 1941. The lack of damage to the eaves of the building on the right and to the windows and door of that on the left suggests that the aircraft may have blown up in the air. This possibility is dealt with in the "Minelaying" section because the aircraft was carrying an "Imp" (Impact) mine which was basically a sea – mine fitted with a nose impact detonator. (Photo via John Foreman)

The sky at night, Wilhelmshaven, 29/21 July 1940. The rectangular tower to the right of the centre of the photograph was the Town Hall, headquarters of Navy Flak Group 262, which was responsible for the defence of Wilhelmshaven. It had over 400 light flak guns available for use against low flying aircraft and many batteries of 105 mm and other heavy guns to deal with high level attacks. (Photo: F.A.Greve)

2E WILHELMSHAVEN 20/21 JULY 1940

The No.5 Group Operations Record Book records that on 20/21 July 1940 "fifteen aircraft from Hemswell carried out a special attack on the warships *Tirpitz* and *Admiral Scheer* lying in harbour in Wilhelmshaven, on verbal instructions from the Air Officer Commanding ("Butch" Harris)." Three from each squadron (61 and 144) were to make diversionary attacks, whilst three from each squadron attacked the *Tirpitz* with "M" bombs, and two from 61 Squadron and one from 144 Squadron attacked the *Admiral Scheer* with "M" bombs.

The general plan was for the "M" bomb Hampdens to attack from a very low height just after midnight and within the space of half an hour, while the diversionary high level attacks were taking place. They were to bomb individually and approach the docks from different directions. The "M" bombs were modified sea mines with their normal detonating mechanism replaced by a fuse that was activated by salt water after the mine had been in the sea for forty minutes. The intention was to drop the mine as close to the battleship as possible so that when it exploded on the sea-bed it would cause serious damage to the hull bottom which was less well protected with armour plating than were the sides of the vessel. Each mine weighed 1500 lb (680 kg), of which the explosive change was 750 lb (340 kg) of TNT. It had to be dropped from between 400 and 1000 ft and at a speed less than 200 mph, or the parachute, which was designed to retard the descent of the mine, would tear away and cause immediate disintegration on impact.

Of the 61 Squadron diversionary aircraft "S/Ldr England unhappily missed his way over northern Holland and returned with his bombs". PO Webster and Sgt Wooldridge were successful in drawing AA fire and attacked various targets. The three 144 Squadron diversion planes fulfilled their mission. PO Collins flew over the area for one hour and twenty minutes. PO Skehill "encountered such accurate AA fire at 13,000 feet that the two rear members of his crew were shaken from their seats by one burst" and Sgt Baker also drew heavy AA fire.

The attack on the warships in the inner docks at Wilhelmshaven, probably the most strongly defended area in the whole of Germany, was carried out in the face of intense flak such as had never been experienced in the history of either of the squadrons. In such an attack, heroism coupled with tragedy was inevitable. Of the three aircraft from 61 Squadron detailed to attack the *Tirpitz*, F/Sgt Saunders reached Emden at 10,000 feet and descended steadily until he reached Jever aerodrome, then set a north easterly course before turning south at Hooksiel to head for Wilhelmshaven at a height of fifty feet. The dazzling multitude of searchlights and a seemingly impenetrable barrage of AA fire thwarted his first run in; on his second attempt he saw one aircraft go down in flames just in front of him and decided that nothing could get through. Accordingly he turned and dropped his bomb close to ships lying off the Mole north of the harbour. On return his aircraft was found to have been hit in six places. PO Jones (P4344) and PO Davis (P4343) of 61 Squadron failed to return.

Three aircraft of 144 Squadron attacked the *Tirpitz*. "PO Robson dropped down almost to water level along the Jade river. Despite the searchlights and flak which was just indescribable, he ran straight in for the Bauhafen, where the *Tirpitz* was lying, at over 200 mph and at a height of barely fifty feet. His luck was atrocious; half way there he had to swerve to avoid a church steeple and before he could recover course he had flashed past the *Tirpitz* 300 yards away to his right, below funnel level. He dropped his bomb from twenty feet almost alongside the two warships at the southern end of the wharf of Hipper Haven. Despite the fact that from the New Mole to the Jade basin, he was never out of AA fire of every description and colour, his aircraft, as if by a miracle, was never hit. He was awarded the DFC".

"PO Tudhope's machine was hit over Emden by an AA shell which burst in the starboard mainplane. Despite this he ran in over the Jade river, skimming the roof tops in the face of the same flak as had met PO Robson. Unluckily he missed the basin by some 300 yards. As he circled to make a second run his plane was hit again several times. Spent shell splinters lodged in the observer's clothing but he was unhurt. One engine was not functioning properly, the bomb doors were open but would not close and the plane was almost uncontrollable. In these circumstances it was impossible to penetrate the AA barrage so he turned north and dropped his mine from 200 feet south of Schillig Point. His hydraulic system had been shot to pieces and, on reaching home, he had to blow the undercarriage down by using the emergency compressed air bottle. Over 150 holes were subsequently discovered in the aircraft and his own gunners had used over 2000 rounds of ammo. Tudhope was awarded the DFC but did not live long to enjoy it, as he was killed when P4368 (5) crashed into the Ysselmeer three weeks later.

F/Lt Edwards piloted the third of the 144 Squadron Hampdens, P4367 (1), to attack the *Tirpitz*. On his return journey he obtained a fix when some eight miles from the English coast and reported by W/T that his plane had been badly damaged by flak. At 0322 his call sign and the letters SOS were heard; after that there was silence. Two search parties, one of four aircraft, the other of three, combed a wide area of sea from 0630 hours to 1630 hours but no trace of the crew or machine was ever discovered nor did they receive any recognition of their gallant effort.

In the attack on the *Admiral Scheer*, Sgt Fuller of 144 Squadron encountered the same terrific AA fire over the Wilhelmshaven area. He made his run in at fifty feet but was so blinded by searchlights that he could not see the ship. The splash of his bomb as it dropped into the basin where the ship lay was clearly observed by his rear gunner. As an encore, Sgt Fuller made a machine gun attack on Borkum aerodrome on his way back to Hemswell, where he landed unscathed. For this he was awarded the DFM. Sgt Massey of No.61 Squadron was compelled to take such evasive action over Oldenburg that he lost his way. Being unable to reach the target area he dropped his bomb from about thirty feet midway between the Mole and a dredger at the mouth of the Ijssel.

PO Gould of 61 Squadron failed to return. Wireless Operator/Air Gunner Sgt Neil Prendergrast recalled how P4358 (2) came to grief. Neil's Hampden made its approach from about 50 miles to the south of Wilhelmshaven. Hedge-hopping across the moonlit countryside the crew could see every detail of the landscape; electricity pylons and even telegraph poles were easy to see. On nearing the target they climbed to 600 feet. Fire from the light flak was everywhere, searchlights caught them in their glare as the tracer shells hammered home. Neil shouted to the pilot that the port engine was on fire. The words had hardly left his lips when he saw that the starboard engine was also ablaze and he screamed over his intercom "The other bugger's on fire too ..."

A shell took away Neil's radio and part of the perspex canopy. All Hell broke loose. The lower gunner, PO David Carnegie, motioned that he wanted to come up as he had been hit. Neil bent down to open the perspex dividing screen and as he released the bolts he passed out from the intense heat from the flames. The lower gunner must at this moment have decided to get out of the blazing aircraft before he was burned alive but his parachute had not had time to open before he tumbled on to the mud flats. That their softness saving him from being killed outright became apparent when his body was recovered next day, as it lay at the end of a trail of marks in the mud which clearly showed that he had died in an attempt to drag himself towards the blazing wreckage of P4358 and his comrades.

The pilot jettisoned the mine after the aircraft caught fire, then tried to get clear of the target area but the blazing Hampden presented an unmissable target for the flak and it was hit continuously until both engines faded away. He made a wheels-up crash-landing on the mud flat, leaving a trail of wreckage as the tailplane, both propellers and the port engine broke away before the flying furnace that had been P4358 came to rest. Sgt Cowan, the navigator, had his escape made easier by the fact that the top half of the perspex nose of the aircraft had disappeared; thus he was able to walk straight out through this unorthodox exit. Pilot Officer Gould, the pilot, was also unhurt and before he made his own escape had seen movement in the rear of the aircraft, which he assumed meant that those behind him were also making their way out over the side of the upper gun position.

Neil Prendergrast had been overcome by the heat just before the Hampden crashed, and it was possibly the impact that brought him back to consciousness. He struggled to his feet, but the whole of the interior of the fuselage was on fire and there was no easy way out, as the guns and perspex canopy over the rear upper gun position were a tangled mess. There only remained one chance, which was to find a way out – through the front end and through a door in the fuselage that closed off the rear end of the aircraft would only open part way. Neil squeezed himself through the gap, only half aware of the exploding ammunition, and for a few seconds he again lost consciousness because of the searing heat. The pain from his burns urged his mind and body to a fresh effort of self-preservation and he recovered to face the nightmare of the fiery fuselage.

P4343 ends in a blaze of glory. This photograph, taken from the south side of the Grosser Haven, shows ships lying alongside the Nord-Gazelle-Brucke opposite, with the Kaiser Wilhelm Bridge clearly visible to the right of the picture. The source of light by which the photograph was taken was the blazing P4343, which was about to crash into the water a few metres to the west of the Kaiser Wilhelm Bridge. At this exact moment, the pilot, Dudley Davis, was being forced by the flames to abandon the aircraft at a height of fifty feet and the other three crewmen were about to perish. Divers recovered the wreckage and bodies next day. (Photo: F.A.Greve)

The wreckage of P4344 which crashed on an emergency landing ground at Irmenhof, near Jever, 10 km west of Wilhelmshaven. (Photo: F.A.Greve)

The way out, he was sure, lay in the direction of the pilot's position; if he could lower the armour plate extension he would be able to scramble out over the the back of the pilot's seat, or maybe he could lower the back of the whole seat. Such thoughts vanished with the realisation that the flames had already burned his hands so badly that he could not grip the bolts that would free either the armour plating or the seat back. In despair he tried to slide down under the pilot's seat in the hope that he might be able to escape through the nose of the aircraft, but found the way barred by a mass of mud and wreckage.

His last chance lay in getting out through the astro hatch in the roof of the fuselage behind the pilot's seat. To do this he crawled back on to the main spar and in a final despairing effort, accompanied by a bitter outburst of curses and cries of pain, Neil wrenched one of the astro hatch's double doors open. There was just enough room for him to haul himself out on to the starboard wing and then hasten away from what had seemed destined to be his funeral pyre. Once clear of the flames and heat Neil dropped to his knees and crawled away to a safe distance, determined that any final explosive gesture from the Hampden should not, even now, claim his life.

The sky was still alight with streams of tracer and he could hear other Hampdens roaring low across the docks and at least one had already suffered the same fate as his own aircraft and was on fire. Neil's recollections of the next few hours are not too clear, as he was near collapse and could not speak at all because of the burns to his face. He vaguely remembers a journey in a lorry, someone showing him his blackened face in a mirror, and seeing the charred remains of his flying helmet, then lying on a stretcher and having his clothes cut away as someone administered morphia from a syringe. It was several days before Neil was capable of taking any active interest in his surroundings and when he did, he realised that he was in the Naval Hospital at Wilhelmshaven, together with two of his squadron mates from P4344, Sergeants Wright and Cain. All three remained there for over six months.

P4343 (3) The pilot, Pilot Officer Dudley Davis, was the sole survivor from this aircraft and he too encountered a veritable wall of cannon fire which transformed his aircraft into a searing mass of flame as shells hit the engines and fuel tanks. Ahead of him in the docks Dudley could see the sleek, sinister shape of his target, the *Tirpitz* and, although he knew that his aircraft was doomed, he decided to press on with his attack. He throttled back to reduce speed before releasing his mine, then when it had gone pushed the throttles wide open in a bid to escape from the murderous flak. But it was too late as the flames were already eating into the inside of the fuselage and as the aircraft's controls became sloppy and Dudley slid the cockpit hood back and clambered out on to the wing to escape from the blazing aircraft. Holding on to the side of the fuselage with one hand, he struggled to control the aircraft with the other. The desperate notion that even now he might be able to crash the blazing Hampden into one of the ships in the dock had hardly had time to form when the aircraft began to roll to starboard. Dudley knew that this would be the last manoeuvre that P4343 would ever perform; he must also have realised that from the height at which the blazing aircraft was flying, a mere fifty feet, there would be little chance of his parachute opening. There would be no chance at all of survival if he remained with the aircraft so, choosing the lesser of two evils, he pulled the ripcord of his parachute, still holding on to the side of the fuselage as he did so. The parachute canopy streamed out in the slipstream and plucked him off the aircraft and as he began to swing downwards below the opened 'chute, his body thumped against something hard, which to his indescribable relief turned out to be the dockside road at Wilhelmshaven. Dazed, and hardly able to believe his good fortune, he staggered to his feet; the only damage he had suffered in this miraculous escape was a singed moustache. The splutters of flaming wreckage on the water of the dock a few yards away marked the end of P4343 and his three comrades.

P4344 (4) was the third of the 61 Squadron Hampdens to be lost. This also suffered a severe mauling by the German light anti-aircraft gunners

A heavy flak battery of four 105 mm guns in action at Sande, Wilhelmshaven, on a winter's night when snow lay on the ground. (Photo: F.A.Greve)

A 20 mm flak gun. The towers in the background surmount the Wilhelmshaven Town Hall. (Photo: F.A.Greve)

The smouldering remains of P4358 on the Rutersiel mudflats, just north of Wilhelmshaven. (Photo: F.A.Greve)

so that its pilot, Pilot Officer Jones, was forced to make a crash landing on the decoy aerodrome at Irmenhof, near Jever. The crew, two of whom were injured, managed to get clear of their Hampden which they set on fire before troops arrived to take them prisoner.

All the men who survived as prisoners of war, eight in all, were awarded the DFC or DFM. Neil Prendergrast became a leading light in the escape organisations in various camps. He was also instrumental in sending much coded information back in his letters to home, having been instructed in the use of codes a fortnight before he became a prisoner. For his work when a "Kriegie", Neil was awarded the MBE.

The Hemswell Operations Record Book summed up the result thus; "The half success of this spectacular operation was purchased at a heavy price – four valuable crews". From the point of view of causing damage to German warships, the raid achieved nothing. The heroism of the pilots who actually carried out the attack is beyond question, as the density of the light flak over Wilhelmshaven must have been greater than above any other target in Germany. Over 400 guns, of between 20 mm and 40 mm of Marine Flak Division 262 defended this naval dockyard; in addition there were many heavier calibre guns in use against aircraft flying at a higher altitude.

37 mm and 20 mm light flak defences on the harbour breakwater at Wilhelmshaven. (Photo: F.A.Greve)

A 37 mm flak position on the north–east side of Wilhelmshaven. On the shed, to the right of and below the gun barrel, are painted the words "Heil Hitler" and beyond the right hand top corner of the shed a hauled–down barrage balloon can be seen. (Photo: F.A.Greve)

P4493, EA – M, was flown by Flt Lt Learoyd when he won his Victoria Cross during the attack on the Dortmund – Ems Canal on 12/13 August 1940. The motif on the fuselage side is Pinnocchio, armed with bow and arrow. (Photo: Peter Green/Douglas Garton)

2F OPERATIONAL INCIDENTS, INCLUDING HAMPDEN VCs

Operational losses are dealt with under a number of headings, Minelaying, Torpedo – Bombing, descriptions of specific operations such as the "Channel Dash", Escapes by Parachute, Ditchings, Crashes on Return and particular incidents that led to aircrew becoming Prisoner of War. The following accounts include incidents in which Hampdens suffered damage due to enemy action yet were subsequently repaired and so do not appear in the serial list as being written off. There are also occasions when a crew member was mortally wounded by a piece of shrapnel or a stray bullet which caused no serious damage to the aircraft.

The most publicised of these incidents resulted in the awards of the Victoria Cross to Flt Lt R A B Learoyd of 49 Squadron and to Sergeant John Hannah of 83 Squadron, both based at Scampton. "Babe" Learoyd won his VC on the night of 12/13 August 1940 when eleven Hampdens from Scampton were detailed for operations against the Dortmund – Ems Canal. Two aircraft failed to locate the target, four carried out diversionary attacks whilst the remaining five attacked a viaduct on the canal near Munster where it crossed the River Ems.

Flt Lt Pitcairn – Hill DFC, in P4420, made the first attack from 100 feet, being badly mauled by the intense fire from the flak gunners as he did so. The second and third aircraft to attack, both piloted by Australians, Pilot Offr Ross in P4410 and Flg Offr Mulligan DFC in P4340, were both shot down in flames. The fourth, piloted by Pilot Offr Matthews, made a successful attack in the course of which his aircraft was damaged and had an engine put out of action so that he had to fly back to Scampton on one engine (1).

Learoyd, in P4403, made the final attack, suffering the full blast of the light flak and being almost blinded by searchlights as he roared along the canal at 150 ft to release his bombs. "Babe" banked P4403 away from the maelstrom of cannon shells and machine gun bullets and once clear of the target area took stock of the damage. Although P4403 had been hit repeatedly, the engines were still going strong. The main problem lay in the damage to the hydraulic system and Babe was able to fly the Hampden back to Scampton where he circled around until it became daylight before making a safe landing. P4402 and P4403 were both repaired and met their ends in crashes on return, P4402 being abandoned near Scampton and P4403 overshooting at St.Eval (1)

Just over a month after Learoyd's valiant effort, another incident earned the second of the Hampden Victoria Crosses. Again the recipient was from Scampton, this time an 83 Squadron wireless operator/air gunner, 18 – year – old John Hannah who became the youngest member of the RAF to receive this highest award. John was Pilot Offr C A Connor's wireless operator in P1355. They took off at 2235 hrs on 15/16 September 1940 from Scampton to bomb barges assembling at Antwerp for the invasion of Britain. The target was located without difficulty and they flew at 2,000 feet to drop their bombs. The first run in towards the target failed to satisfy the navigator/bomb aimer, Sgt D A E Hayhurst, and under increasingly heavy fire from the flak defences, they circled to make a second bombing run, P1355 being hit several times by cannon shells as they did so. The bomb aimer had just given the "Bombs Gone" signal to the pilot when a shell burst in the Hampden's bomb bay. Within seconds the rear half of the Hampden's fuselage was transformed into a searing Hell of flame and smoke. The intensity of the heat disintegrated the lower gunner's position, its occupant, Sgt George James, being lucky to escape by parachute before being roasted alive. The navigator had heard Hannah report to the pilot that the Hampden was on fire and, believing the aircraft was doomed, also baled out. Sergeants Hayhurst and James both became prisoners of war. The navigator was awarded the DFM for having completed a tour of 39 operations; this one was to have been his last before being rested.

Hannah, in the upper rear position, found himself in the midst of an inferno. Photographs of the damage to the fuselage illustrate far better than words the intensity of the heat which actually melted part of the floor and sides of the rear fuselage. Exploding ammunition added to the chaos as John grabbed the portable fire extinguisher and used it on the flames. The fact that he noticed his parachute was already smouldering and would probably be useless if he tried to bale out no doubt spurred him on in his efforts to subdue the fire. In a last desperate effort he even resorted to beating at the flames with his bare hands. The pilot struggled to control the blazing aircraft, now a perfect target for the German gunners until, after a few minutes that seemed to him like hours, he flew clear of the deadly barrage. His feeling of relief at being out of range of the guns increased almost at once as John Hannah reported on the

L4057, "Admiral Imaz Dryazel". Guy Gibson flew at least one of his ops with 83 Squadron on this aircraft, his regular aircraft being L4070 "Admiral Foo Bang". The reason for the nautical nicknames is obscure.
(Photo: Peter Green/Douglas Garton)

intercom that the fire was out. Without the assistance of a navigator and with the wireless set a charred wreck, he flew the Hampden back to Scampton and landed safely.

On 1 October 1940, it was announced that John Hannah had been awarded the Victoria Cross and that Pilot Officer Connor was to receive the Distinguished Flying Cross. John was discharged from the RAF on medical grounds in 1942 and died at his home in Leicester of tuberculosis early in 1947. Connor, one of many Canadians serving in the RAF at this time, perished with his crew when L4093 (2) crashed into the sea off Spurn Head on 4 November 1940 as they returned from Kiel. P1355, the Hampden in which Hannah won his VC, was repaired and was flying again in July 1941 at No.5 Bombing & Gunnery School at Jurby, after which it served on Met Flights until it finally went to the Hampden "knacker's yard" at Edzell at the end of 1943.

In his book "Gunner's Moon", John Bushby graphically describes his experiences and, whilst he was not on Hampdens himself, comments on the award of the Victoria Cross to John Hannah. "There were some ... who maintained that Hannah, with his parachute burned and unusable anyway, had no option but to stay with the aircraft and attempt to put out the fire and that the real hero of the whole incident was the pilot, who could have baled out but chose to stay at the controls because of Hannah." That statement is a fair reflection of the opinion amongst the aircrew on the Hampden squadrons at the time. Nobody begrudged the eighteen-year-old Hannah his award; it did after all, reflect credit on the Hampden crews in general. What was difficult to understand was the manner in which such awards were made, particularly as all the Hampden squadrons had had their share of low level attacks against heavily-defended targets in the summer and autumn of 1940. In many cases the valour of pilots who flew into an intense barrage of flak to press home a low-level attack was witnessed by other pilots.

These losses included some of the most experienced Hampden crews and the attacks were made against such targets as concentrations of barges intended for the invasion of England e.g. P1183 Le Havre, P4371 Calais, P2121 Antwerp, L4097 Ostend, X2922 Boulogne and P4360 Mittelland Canal (3).

Squadron Leader James A Pitcairn-Hill DSO DFC and his crew from 83 Squadron, flying in P1183, failed to return from a low level attack on Le Havre on 18 /19 September 1940. This gallant officer had been awarded the DSO for leading the attack on the Dortmund-Ems Canal on the night of 12/13 August, his aircraft being badly shot up in the attack as was that of his colleague Flt Lt Learoyd who was awarded the VC on this occasion. The deadly umbrella of cannon shells from the light flak over Le Havre five weeks later was so intense that even the exceptional skill and experience of such a courageous pilot proved of no avail and the Hampden was practically shot to pieces as it made its attack. That P1183 must have crashed in the sea in the Seine estuary, probably a mile or two south of Le Havre, is evident from the separate burial places of three of the crew, the fourth of whom, the wireless operator, has no known grave and his only memorial is the name Sergeant Charles Gordon McCarthy on the Runnymede Memorial Register.

The Squadron Leader was buried at Luc-sur-Mer churchyard on the Normandy coast, ten miles north of Caen. His navigator, PO Arthur P Lindsell, lies at rest further to the east along the same coast at Villerville on the opposite side of the Seine to Le Havre while the air gunner, PO Victor A Rendell, is buried at Honfleur, a small harbour ten miles up the Seine on its south bank.

The Hemswell attack on Wilhelmshaven in which P4343, P4344, P4358 and P4367 were lost is described separately (see 2F4). In an attack on the liners *Bremen* and *Europa* at Bremerhaven, X2913 and P1338 were shot down. The pilot of P1338, F/Lt T L S Smythe DFC, had been flying with 44 Squadron since before the war. He was regarded as an exceptional pilot, had earned promotion from Sergeant Pilot to Commissioned rank and been awarded the DFC. A third Hampden pilot decided that discretion was the better part of valour after he had witnessed his two squadron mates fall victim to the intense light flak and bombed from a safer height. There were no survivors from either aircraft and it has not been possible to find a contemporary German account of the attack but there is little doubt that there would have been a terrific concentration of flak to defend such important targets as the two ocean liners. The loss of such invaluable crews no doubt helped to convince Bomber Command that however valiant and skilful the pilots, it was tantamount to ordering them to commit suicide to send them on low level attacks against targets that were protected by a veritable wall of light flak.

The details in the Operational Record Books are, of necessity, brief but even from these sparse accounts a wealth of drama and heroism shines forth, as examples from the various squadrons show. In April 1941 106 Squadron lost their Wing Commander P Polglase in AD738 on a low-level night attack on the battlecruisers *Scharnhorst* and *Gneisenau* at Brest.(4)

In the same attack Sgt Topping's Hampden, AD756 (5), was hit and the navigator Sgt Purnell, although wounded in the legs by tracer bullets, continued to navigate the aircraft and earned a DFM. A few days later S/Ldr Tudor, bombing Kiel in AD735 (6), had heavy flak burst near the aircraft and the wireless op, Sgt Stables, was wounded in the leg by shrapnel. The third Hampden of 106 Squadron to return with a casualty within ten days was PO Paramore in AD758 (7) on 13/14 April 1941, whilst on a minelaying op off La Rochelle. They were flying at 300 feet when they were hit by light flak. The lower rear gunner, Sgt W G Sandom, was wounded but in spite of this he continued to fire at the ground defences but despite the efforts of the Wireless Operator the lower gunner died before reaching base.

The last six months of Sergeant Pilot George E Cowan's life were quite eventful. On 4 June 1940 he was at No.16 Operational Training Unit and carrying out solo night flying practice in P4295 when it stalled into the ground at Brackley, one of Upper Heyford's satellite airfields. He nevertheless completed his training and went to 61 Squadron at Hemswell. On 16/17 October he crash-landed X2989 at base with a wounded rear gunner on board (Sgt Woodward) after being shot up by a night fighter over Holland; X2989 was repaired. Three weeks later George ran into intense flak over Nuremburg in X3005, which may have been the reason why "when he lowered his undercarriage on return over Scampton, his port wheel fell off. He made a successful belly landing". Although the "port wing was completely smashed", X3005 lived to fly again.

Sergeant Cowan took off from Hemswell at 2147 on 16 December 1940 for Mannheim. It was to be the operation that would complete his 200-hour tour of ops, but X3128 failed to return. The names of the crew are all recorded on the Runnymede Memorial Register and as in so many other cases the reason for their loss was never discovered. Four days after their disappearance, it was announced that Sgt G E Cowan had been awarded the Distinguished Flying Medal.(8)

When George Cowan crash-landed X2989 he had, as was common practice, a less experienced pilot flying as his navigator, Sergeant R S C Ivatt. The navigator's experience with 61 Squadron had already included a crash on take-off in P4401 and he must have been relieved when he in turn became a first pilot. On 22 November 1940, at 2315, returning from an op, he undershot the flarepath, hit the bomb dump and the Hampden, X2987, burnt out. Sgt Ivatt and his navigator were killed, the other two crew members injured (9).

In one of 44 Squadron's last Hampden ops on 13/14 December 1941, daylight minelaying off Brest, the Squadron Commander, Wg Cmdr S T Misselbrook DSO in AE196, failed to return (10). Squadron Leader Burton-Gyles DFC in AE202 was attacked by Bf 109s and the following damage was done to his aircraft: "Starboard aileron damaged, starboard wingtip shot away. Mainplane behind the starboard engine damaged, starboard engine fairing torn off, nacelle, and airscrew damaged. The starboard rudder was shot away, the wheel of the starboard undercarriage was punctured and the tail unit holed. The Hampden still landed safely". Squadron Leader Burton-Gyles was awarded the DSO and AE202 was repaired. It is obvious that, had the attacking fighter trained his sights a little more to the left, AE202 and its crew would have been blasted out of existence. No.44 Squadron, incidentally, became No.44 (Rhodesia) Squadron with effect from 12 September 1941 (10).

On 10 June 1940, P4367 had one of its crew, Sergeant W F G Hollister, killed by a piece of shrapnel but the pilot, Pilot Officer Reynolds, had no difficulty in landing the aircraft at Bircham Newton (11). Pilot Officer W C Hartop and his crew escaped without injury when they crash-landed L4131 at West Raynham after suffering flak damage over Berlin, L4131 being repaired. A few months earlier, PO Hartop and his crew abandoned X3056 over Lincolnshire, again without coming to any harm (12).

On 1 April 1941, 144 Squadron, operating from St Eval, took off to carry out a daylight attack on the *Scharnhorst* and *Gneisenau* at Brest. "The squadron commander decided that there was insufficient cloud cover and decided not to continue towards the target, but Pilot Officer Hartop, in X3129, left the formation and carried on. He is believed to have made an attack, but failed to return. "The pilot was the only one to survive and became a Prisoner of War; his three crew including his regular Wireless Operator, Flight Sergeant R G Lambourne, are buried together in a churchyard 12 miles north of Brest.

The battlecruisers were still at Brest as 1941 drew to its close. No.144 Squadron had lost AE353 and AD921 (13) on daylight "cloud cover" operations against them on 11 and 13 December and on 24 December Sgt P A C McDermott tried his luck in AE309 (13). He received an unwelcome Christmas present when an exploding flak shell resulted in his navigator Sergeant J S Boucher being injured in the face and eyes, but in spite of these injuries he continued to navigate the badly-damaged Hampden which made a crash-landing at Exeter. AE309, repaired and converted to torpedo-bomber, went to No.5 (Coastal) OTU where it suffered damage in an undershoot when landing and on 14 October 1943 went to a watery grave when it ditched off Greenland after engine failure on a flight to Canada. AE309's last resting place is over 2,000 feet below the surface of that Arctic sea, which had to qualify for some sort of record for a Hampden and gave its lucky crew a good story to tell (see 5K14). An account of AE309's attack on Brest appears in Volume 1 of Denis Richard's "RAF, 1939-45" according to which "three Hampdens were involved of which one failed to return, one hit a balloon cable and lost half its tailplane and Sergeant Boucher was injured in the third aircraft". There is no record of a Hampden being missing from this operation, and the balloon cable incident is also a bit of a mystery.

Two of 420 RCAF Squadron's sergeant pilots who took off from Waddington just before 2200 on 4 May 1942 to attack Stuttgart were destined to be awarded the Distinguished Flying Medal for bringing their damaged Hampdens back to England. Sgt W J Maitland RCAF, in P1314 (14), was crossing the Belgian coast near Ostend when accurate flak put forty holes into the Hampden, injured the pilot in the chest and set the port engine on fire. Sgt Maitland did not need any encouragement to abandon his mission and make a hurried return to Martlesham Heath, where he landed at 0024. P1314 was repaired and saw service with 408 RCAF Squadron before being converted to torpedo-bomber.

Sergeant F S Hiley RCAF in P1187 (14) had paid his visit to Stuttgart and reached Luxembourg on his return flight when his aircraft was hit by flak. A short time later it was attacked by a Bf 110, whose onslaught killed the rear gunner, Sgt J F Halward RCAF, wounded the pilot, set the starboard engine on fire and severely damaged the luckless Hampden. The Canadian pilot must have heaved a sigh of relief when he made a crash-landing at St Mary's Farm, Great Bentley, Essex, without causing further injuries to himself or his crew, even though P1187 did have to be written off. Six weeks later, with the same wireless operator and navigator and a new lower gunner, F/Sgt F S Hiley DFM, in AD786, he crashed twenty minutes after take-off from Waddington, at Boothby Pagnall, and caught fire. Only F/Sgt Hiley, seriously injured, survived. AD786, in which F/Sgt Hiley crashed on 23 June, had literally been in the wars on 19/20 June, when a Ju 88 delivered an attack which seriously wounded the lower gunner, Sgt C Collard, and inflicted less serious injuries to the wireless operator Sgt F S Moritz.

F/Lt D S Jacobs in AE401 (15) had carried out an attack on Osnabruck and having just crossed the Dutch coast and set course across the North Sea, the crew must have been feeling that the worst of their troubles were over. If so they reckoned without taking into account the alert flak gunners on a ship below them; their height of 4,000 feet was insufficient to save them from the burst of shells that ripped into the Hampden. The lower gunner, Sgt R Davidson RCAF, was wounded so badly that he later died from his wounds. The undercarriage and flaps were put out of action by the same accurate flak and F/Lt Jacob's wheels-up and no-flaps landing at Waddington must have been a somewhat more speedy affair than he would have preferred. In spite of this AE401 was repaired and went back into service as a torpedo-bomber with another Canadian unit, No.415 RCAF Squadron.

Quite a number of the names of the aircrew who were involved in Hampden losses on the operational squadrons feature in the records of the Operational Training Units. Sergeant W K McDermid, for example, had two mishaps during his final training at 16 OTU. He failed to get airborne in P5313 and damaged P2138 when he developed a swing on landing. He arrived on 420 RCAF Squadron on 13 March 1942 and had trouble with icing-up over England on return from an op, his rear gunner baling out (see 5A45). Then a few weeks later in AE399, when landing from an op, he developed a swing when landing and crashed into a 44 Squadron Lancaster. This resulted in his being seriously injured and his navigator being injured. His wireless operator, Sergeant J N Waddington, had been with him in the icing-up incident but his change to another pilot after Sergeant McDermid became a casualty did not change his luck. He failed to return in AE248 and is buried at Koudum, in Holland (16).

The Operations Record Book of 408 RCAF Squadron is a fine example of how such a record ought to be kept and a marked contrast to some of the contemporary RAF Hampden squadron records. The record's first mention of a Hampden accident was on 10 September 1941, when Pilot Officer Caldwell had brake failure as he taxied AE244 and collided with the YMCA tea wagon; fortunately no-one suffered serious injury. A few weeks later, in AE437 (17), the same pilot went down in the sea 370 miles west of Brest (see 5B9).

John Hannah and his pilot, Pilot Officer C.A.Connor who together shared the blazing Hampden experience on 15 September 1940 which resulted in John being awarded a Victoria Cross and PO Connor a DFC. The latter lost his life soon afterwards when L4093 went down in the sea off the Humber on 2/4 November 1940 returning from Kiel. (Photo: Ron Lowe)

No.408's first operational crash occurred on 30 September 1941 when AE287 had its starboard engine cut on landing at Acklington which resulted in a collision with a Wellington. The crew escaped with minor injuries and AE287 was repaired. The crew on that occasion was Sergeants Farrow, May and Norton and an air gunner officer, Pilot Officer Maitland. Sergeant Norton was lost in AE438 a few weeks later. PO Maitland had been promoted to Flight Lieutenant, been awarded the DFC and was flying with the Squadron Commander in AE197 when they were shot down on the Squadron's last Hampden operation (18).

Sergeant J P Farrow and Sergeant T I May were together again, in AD754, on the night of 20/21 January 1942 in an attack on Bremen. They were flying at 15,000 feet when they became caught in at least 30 searchlights. It proved impossible to open the bomb doors to release the bombs because the "Power Bolt" which controlled the hydraulic system had frozen solid. The bombs were eventually dropped from 7,500 feet and course was set for base. Their troubles had only just begun, because vapour trails being made by their Hampden attracted the attention of two Bf 110s, which put five bursts of cannon fire into the aircraft. The pilot ordered the crew to bale out but could get no reply. He made a guess at a suitable course and flew on it until the navigator came to life and assisted in the process of directing the stricken Hampden towards the English coast. As they neared the coast, they began flashing SOS on the downward recognition light and when this was seen, the searchlight pointer organisation directed them towards West Malling. Sergeant Farrow's troubles were not yet over as the undercarriage failed to work and a crash-landing had to be made. All of the crew were wounded. Sergeant T I May and Sergeant V R Baker died within 48 hours, the wireless operator Sgt Millward had a cannon shell wound in the leg. Sergeant Farrow, slightly injured, was awarded an immediate Distinguished Flying Medal and became Pilot Officer J P Farrow DFM on 5 February 1942. The battered wreck of AD754 was restored to life as a torpedo-bomber and was flown to 32 OTU at Patricia Bay, via Iceland and Greenland. One wonders whether the crew who flew it to Canada had any idea what that Hampden had suffered (19).

Three photographs illustrate the damage caused by fire to P1355, the aircraft in which John Hannah won his Victoria Cross. The aircraft was subsequently repaired. The top photo shows the lower rear starboard side of the fuselage. The heat has melted the perspex panel at the side of the lower gunner's position, seen at top left. The protrusion beneath the panel is the tube through which the W/T set trailing aerial was streamed. The aerial drum lies on the grass at the lower left of the photo.

The centre photo shows the burnt-out floor of the wireless operator's position. The stowage for the W/Op's parachute, situated on the bulkhead door, is at the left. His seat support is at top left and the raised platform either side of his position is at top right and bottom left. A spare ammunition pan is stowed at upper right, two empty pan stowage pegs can be seen left of upper centre and at the lower right corner of the photograph.

The bottom photos shows the lower gun position from the port side. The door was jettisoned when the lower gunner baled out. The height-adjustable seat, operated by the handle which sticks up almost vertically at the far side of the seat, shows up clearly. It is quite obvious that the heat which blasted through this position made it impossible for the lower gunner to remain in the aircraft. (CH1347, CH1349 and CH1352)

The proportion of Hampdens lost at night by fighter attacks had gradually increased until by 1942 the majority of losses were attributable to this cause. Other 408 Squadron aircraft that survived such attacks included P5334, which Pilot Officer F Leach brought back from Hamburg on 26 July 1942 with his Canadian lower gunner, Sergeant L F Kelly, dead and many holes in the Hampden (20).

Another account in the 408 Squadron Record Book concerns Sgt R G Bell, Canadian pilot of AE366, his navigator of the same name and also a Canadian, Sgt J K Bell, and his two wireless operators in the rear, Sgt J S Murray RCAF and Sgt A Norman. They took off from Balderton at 0020 on 6 August 1942 to attack Duisburg and their ordeal began when they were attacked by a night fighter at 0230 as they approached their target. The fighter's cannon shells "blasted three large holes in the Hampden's elevators, one in the port aileron, one large hole in the port tail fin, one huge rent at the intersection of the tail boom and fuselage, another large hole in the bottom of the port engine nacelle. The top gunner's cupola was smashed and the magazines shot away from his guns. All port wing fuel tanks were riddled, the hydraulics shot away and the sides of the fuselage seamed and scored by cannon shells. One cannon shell struck the spar behind the pilot's left shoulder, inside the fuselage. The top wireless operator, Sergeant Murray, was wounded in this encounter and had shell splinters embedded in his scalp. His face was all scratched up by perspex and flying splinters and he was rendered unconscious for thirty seconds. The two wireless operators changed places; the pilot attempted another run over the target and successfully unloaded his bombs as detailed.

Approximately ten minutes after leaving the target, explosions were heard from the port engine, which stalled completely and stayed dead for the rest of the flight. The Hampden fell into a spin from which the pilot managed to pull out at 4,000 feet. He then climbed to 6,000 feet and told his crew to prepare to abandon the aircraft and possibly land in the sea. Whilst over the sea the pilot's right leg became so tired that he was unable to hold the rudder any longer and his navigator helped by taking hold of the rudder bar with his hands to take the strain off the pilot. They crash-landed at Lakenheath with no further injuries to the crew. The pilot and navigator were both recommended for an award of the Distinguished Flying Medal; the pilot was granted it immediately." The incident demonstrates the amount of punishment that a Hampden could take and its ability to maintain height, even to climb a little, with an engine out of action. The fact that, even after this catalogue of catastrophe, AE366 was repaired, in the course of which it became a torpedo-bomber, is nothing short of incredible (21).

Flight Sergeant R J Dillon, a native of the United States of America, had, like many of his fellow-countrymen, enlisted in the Royal Canadian Air Force. He trained as a pilot, and had managed a "wheels up" landing in P2142 at 16 OTU during the last period of his training prior to his posting at 408 RCAF squadron at Balderton. The precise reason why this "Yank in RCAF" came to be flying at 100 feet over the uneven terrain of North West Brittany is somewhat obscure but it very nearly resulted in the entry "Failed to Return" being made against AE150 in the Record Book. Instead there is a brief account of their encounter with the French electricity grid system, which might almost have been headed "A shocking experience".

"F/Sgt Dillon took off from Balderton in AE150 on 2/3 May 1942 ... at 0029 he collided with a High Tension Cable in position 48.20N:04.07W, whilst flying at 100 feet, the starboard wingtip and aileron were damaged, the pilot's windscreen was smashed and ten feet of cable left dangling from the starboard wing tip. The pilot noticed a huge shape in front of the aircraft which turned out to be a steel tower cable support and hit the cable as he banked to avoid the tower ..."

AE150 and its crew returned safely to Balderton. The Hampden was soon back in action and went missing in August, but F/Sgt Dillon and probably the same crew flying in AT224 were shot down by a flak ship only two weeks after the cable incident and all perished. F/Sgt Dillon and another of the crew are buried in Denmark; two were never found (22).

George A C Potts, who ended up as a Wing Commander DFC with over 1,000 hours flying on Hampdens which included two tours of ops with 50 Squadron, recalled how he had put P2124 down on the sea off Scarborough after returning on one engine from a flight that had lasted over ten hours. P2124 is listed as a write-off whereas incidents he described involving AD852 and P1166 are not (23).

AD852 was shot up over Brest when flying at low level on 26 May 1941 and Pilot Officer Seeley-King, a pilot flying as navigator, received a bullet wound in the leg which was serious enough to warrant an emergency landing at Boscombe Down. There was no brake pressure registering as George approached to make a landing and he recalled with amusement that the ambulance and fire engine that awaited his arrival had to chase the Hampden across the aerodrome and along the far boundary as George swung the aircraft to avoid hitting the boundary hedge. Just under a year later, Flying Officer Seely-King was flying as pilot of AT216 when it crashed soon after take-off with the loss of its crew.

George and his crew obviously had a close call on the night of 28 December 1941 when flying in P1166. In an attempt to escape from the attention of flak and searchlights, George dived just a little bit too close to the ground and hit the top of a tree with the result that "the whole of the front of the aircraft disappeared and with it all the maps. The W/T aerials were also removed. We still managed to get back to Skellingthorpe on time; I was so cold that I had to be lifted out of the aircraft, frozen stiff". The navigator must have had a pretty good tale to tell about this particular trip too! George's habit of bringing his aircraft back with damage that required patching up was a standing joke with his ground crew but it is doubtful whether their expertise extended to carrying out whole nose transplants. There are many more examples of Hampdens being brought back safely after suffering damage that would have caused the loss of a less robust aircraft. The skill of the pilot was, of course, a major factor in bringing about the safe return of these crippled Hampdens; the incidents often went unmentioned in the Squadron ORBs.

The final account in this section, related by Peter Tunstall, illustrates the manner in which a crew could escape death after what must have been a terrifying experience. The comparison with the loss of AD969 underlines how narrow was the dividing line between one crew returning with something to joke about and another crashing to their deaths in Germany (24).

On the night of 13/14 June 1940, the 61 Squadron Hampden in which Peter Tunstall, a pilot, was flying as a navigator, arrived over Reims to bomb units of the advancing German Army. According to intelligence reports, the enemy had not yet got their flak defences into position but as the Hampden roared in at 2,000 feet, it became evident that Intelligence had got it wrong. Searchlights sprang to life and light flak opened up. Watching the streams of tracer snaking slowly up from the ground until, after gradually increasing speed, they flashed past just out of range of the Hampden, Peter reflected on the anomaly that the fireworks display of these deadly streaks of light could look so harmless and pretty.

Alec Webster, the pilot, deciding that it was time to take evasive action, banked the aircraft and pointed the nose skywards as he did so. The dazzle from the searchlights, however, blinded Alec to such an extent that he lost control of the aircraft. There seemed no possible alternative but to order the crew to bale out so, after giving the order "Jump! Jump!", Alec slid his cockpit hood back and released his safety straps. He was halfway out of the aircraft when the cockpit hood slid forwards and pinned him by the stomach so that his feet remained planted on his seat while the top half of his body was battered by the slipstream outside the aircraft.

Having thus immobilised its pilot, the Hampden seemed to go wild with delight. After rolling over onto its back, it dived towards the ground then came back onto an even keel and began to climb. As the Hampden's nose pointed upwards, the cockpit hood slid back and the gravitational force obligingly thrust Alec back down on to his seat and placed him back at the controls. To his utter amazement, he was back in command of his headstrong Hampden. The manner in which Peter, in the nose of the aircraft, had experienced this chapter of events was even more of a nightmare. He recalled:

"As the Hampden rolled over onto its back I sailed off the floor amid a shower of ammunition pans, charts, night flying rations and other aeronautical impedimenta and hit the roof with a thump. All the things I was plugged into, such as oxygen and intercom, came adrift. I felt completely isolated, aware only of the noise of rushing air and the rattle of light flak hitting us. The searchlights shone down from above me and then tipped sidewards and slid underneath as I was slammed back violently onto the floor and I felt a stabbing pain in my left ankle. My observer-type parachute lay only a few feet from me but the dive into which we now descended was so steep that the invisible force of gravity held me so firmly in its grip that I could not push my hand out to reach it. I felt suddenly certain that death was not merely waiting for me but that I was being hurtled towards the scythe of the Grim Reaper. My mind was not filled with fear, only a black despair that I would never see my girl friend Ann again".

In the rear of the Hampden, the wireless op and the lower gunner had been equally certain that their names were about to be made available for the Squadron Roll of Honour. They were as surprised and relieved as was Peter to realise that Alec had regained control of the aircraft. The latter, when he found that his crew were still in the aircraft, decided that this must be a night when his star signs were particularly favourable and under Peter's guidance, made a low run over the target and dropped the bombs before heading for Hemswell.

As he restored some semblance of order to the navigator's domain, Peter noticed that the charts were stained by an evil-smelling liquid which he soon decided was not blood, lubricating oil or hydraulic fluid.

Rod Learoyd V.C. (Photo: Rod Learoyd)

The solution to the mystery began when, in the faint light of the chart table lamp, Peter discovered an unfamiliar object rolling about on the floor beneath the chart table. Gathering his already-tattered nerves together to deal with this sinister canister, he wondered whether it might turn out to be some new implement of Teutonic frightfulness that was about to blast them to Kingdom Come. After opening the escape hatch in the floor behind him, Peter reached out with trembling hands to grab the mystery missile. As he dropped it out of the hatch he read the words "Refill for Elsan Chemical Lavatory".

Peter had suffered a badly-sprained ankle in the aerobatic demonstration and hobbled about on crutches for several days, an easy target for much ribald comment, the main theme of which centred around brown stains, nasty smells and secret weapons. The terror of the experience had been implanted in Peter's mind; it surfaced as nightmares when the same situation would be re-created. There would be the same feeling of being trapped in the nose of the aircraft, floating helplessly to disaster and struggling against invisible bonds like a fly in a web whilst he flew down..down..down. He would awaken, kicking and yelling and wet with sweat and even when he flew as a pilot during the following months, he experienced sensory delusions that the aircraft's nose was going up into that near-vertical and uncontrolled climb.

Peter and his crew became prisoners of war when P4324 (24) force-landed on Vlieland on 26/27 August 1940. He was fortunate to survive the war as his escape activities incurred the wrath of his captors and he ended his POW days in Colditz.

2G DAYLIGHT OPERATIONS 1941-42

AD801 PL – C of 144 Squadron at Predannack in Cornwall on return from a daylight bombing raid on German warships at Brest on 24 July 1941 from which AE225 failed to return. AD801 was hit by a shell which passed through the port wing without exploding, fracturing the fuel pipes in the wing and causing petrol to drain out, fortunately without igniting. The open gun port of the port beam gun can be clearly seen.
(Photo: Jimmy Bennett, pilot of AD801)

Almost a year elapsed after the debacle of 12 April 1940 before the Hampdens again took part in daylight operations; when they did so it was once more against German warships, the *Scharnhorst* and *Gneisenau* at Brest.

The Hampdens were on occasions part of a larger force of assorted bombers with a substantial fighter escort, at other times they operated in small formations using cloud cover to avoid enemy fighters. Either way, they still had to face an intense flak barrage, the accuracy of which may be judged from the fact that in one attack every one of the six 144 Squadron Hampdens involved returned with damage caused by shell fragments. X3129 (see 2F12) and AD753 (see 5F8) were lost in the attacks on Brest (1).

On 13 April 1941, B Flight of 61 Squadron was detached to Watton to operate under No.2 Group on cloud cover raids, AD732 (see 3D16) and AD825 (see 5A13) being lost in attacks on airfields near Cherbourg before B Flight returned to Hemswell on 29 April (2).

Twenty – five sorties by single, or pairs of, Hampdens were flown, without loss, against targets in Germany between April 19 and May 1. The cloud cover, of which advantage was to have been taken, frequently failed to materialise, causing many of these missions to be aborted.

The next daylight attack was on 24 July when Hampdens joined Wellingtons and Halifaxes in attacks on the battlecruisers at Brest and La Pallice. AD962 (see 3D17) and AE225 failed to return (3) as did eight Wellingtons and five Halifaxes.

In August and September, Hampdens participated, without loss, in "Circus" operations, the main object of these attacks being to bring enemy fighters into the air so that the impressive escort of No.11 and No.12 Group fighters could shoot them down. At least that was the theory.

October and November were blank months as far as daylight operations were concerned but in December two of 106 Squadron's Hampdens, AE391 and AE151 (4), failed to return from "Sneaker" hit and run cloud cover attacks over North Germany. AE391 crashed near Oberhausen – Osterfeld, AE151 at Schoonebeek. As the Luftwaffe's fighters had become quite expert at shooting down bombers at night, these daylight intrusions should have presented no great problem.

Cloud cover bombing attacks against the battlecruisers at Brest and daylight minelaying operations off Brest during December resulted in the loss of AE353 (see 2F13), AD921 (also 2F13) and AE196 (2F12) (5).

The final daylight operation of 1941 took place on 27 December when 50 Squadron carried out bombing attacks in support of the first Combined Operations raid on Vaagso and lost AE369 and AE428 (see 5B32), both shot down by flak (6).

Sergeant Nairn C Robertson ("Robbie"), the wireless operator in AE422 (6), had vivid memories of the attack:

"We flew to Wick on December 14 to carry out formation and low low level flying training. On Christmas Day we were briefed and did our night flying test in anticipation of taking off before dawn next morning. The traditional Christmas Day visit of the officers to serve meals in the Airmens' Mess was carried out in an unusually Puritan manner and it was a very dry and sober day until around half past seven in the evening when the 50 Squadron Commanding Officer, Wg Cdr Potts, arrived at the Senior NCOs' billets with a fire bucket full of beer, which everybody refused to imbibe until he broke the news that the trip was postponed for 24 hours because the flagship, the cruiser *Kenya*, had screw trouble. Thereafter a good time was had by all.

It was around 0530 on 27 December when we took off for Vaagso where our target was to be the three big guns on the Point of Ragsundo, covering Vaagso and located to the east of it. The fjord had two entrances of which the northern one was to be used by HMS *Kenya* and the assault craft, so we approached from the southern entrance with the intention of flying over the gun positions then turning and making our bombing run from NE to SW.

The German gun crews had obviously become aware that something was afoot because we could see them running from a hut up a path to the gun emplacement and I let them have a good burst from my twin Vickers "K" guns. It was therefore no surprise to find the 20 mm flak gun banging away at us as we made our bombing run. John Day, our navigator/bomb aimer, got the full blast of one of the cannon shells as he was about to drop the bombs but dropped them nevertheless and gave the pilot a course to steer for the return flight before collapsing onto the floor in the nose of the aircraft.

Once the lower gunner and myself had finished firing our guns at the German positions I went up to the front of the aircraft to attend to the navigator, leaving the lower gunner/wireless operator to take over the wireless set until we landed at Sumburgh where John received the urgent medical attention that he required.

John Day, from his home in South Africa, provided a little more detail, adding that the raid itself was not so important in regard to the target but more so as a trial of tactics to be used in subsequent Combined Operations.

"Our target was a gun on an island in the fjord but nobody had told me that everything would be white with snow...the flak gun protecting the big gun opened up, our rear gunners replied and I was aiming the bombs when the shot hit me. I hope I pressed the "tit" in time. I felt no pain at all but had no control of my legs. I lay on the floor of the aircraft until Robbie was able to come up and give me morphia and fixed a tourniquet (see also 3C2).

AE428 was hit as it was about to drop a stick of phosphorus smoke bombs to cover the approaching landing craft, with the immediate result that one of the smoke bombs dropped into one of the landing craft. The Hampden crashed onto the water soon afterwards and sank very quickly, taking with it one of its crew. Only the pilot, Sgt R N Smith, survived the immersion in the Arctic waters (see 5B32).

In 1942, the Bomber Command Hampdens suffered their heaviest loss on any sort of operation when nine Hampdens went missing on 12 February 1942 in the "Channel Dash" operations which are described in Part 2H, which also covers the four Hampdens missing from daylight minelaying operations on 6 and 7 February.

Other daylight activity by Hampdens of Bomber Command included some fifty cloud cover operations against north – west Germany and minelaying on 4 April. The only loss was AT227 (7) (see 5C24).

The activities of the Hampden torpedo – bomber squadrons also resulted in daylight losses and these are covered in Part 2J.

2H THE CHANNEL DASH

The general impression one forms, when studying accounts of the breakout of the German battlecruisers from Brest on 12 February 1942, is of a catalogue of catastrophes as one failure follows another. The Coastal Command radar surveillance broke down at a critical time, the submarine that was on patrol off Brest was out of position when the ships emerged and reports of severe jamming of British coastal radar stations were ignored, as was the build up of German light naval forces such as E-boats, minesweepers and destroyers.

The first visual sighting of the fleet as it approached Le Touquet was made by patrolling Spitfires but the report of the sighting was delayed until the Spitfires had landed. It was not until the German force neared the Strait of Dover, already having reached the halfway point of their 30-knot race towards the Fatherland, that they came under attack. The coastal batteries at Dover opened fire, directed by radar, against the unseen enemy but the speeding ships steamed on unscathed. The MTBs from Dover attacked next but failed to penetrate the German E-boat escort screen; their torpedoes had to be fired from too great a range and no hits were scored.

The six Swordfish of 825 Squadron, Fleet Air Arm, led by Lt Cdr Esmonde, were the next to arrive on the scene, only one of the promised five squadrons of Spitfires, 72 Squadron, being in attendance. Twenty minutes after they had taken off from Manston the six "Stringbags" had all fallen victim to either the intense flak from the ships or the shells and bullets of the escorting German fighters. Their epic attack earned a posthumous Victoria Cross for Lt Cdr Esmonde but none of their torpedoes struck home. In the course of the action along the French and Belgian coasts, Fighter Command lost 16 aircraft and claimed about the same number of German fighters.

Captain Pizey, in *Campbell*, exercising off Orfordness with five other destroyers from the Harwich-based 21 and 16 Flotillas, received orders to intercept the enemy and at noon began a hazardous full speed race through British mine-fields towards the Dutch coast and the enemy. At a quarter to four, the destroyers attacked from about two miles range with torpedoes. The last to attack was *Worcester*, and as she closed to about 2,000 yards, three heavy shells from *Prinz Eugen* hit her but she nevertheless managed to fire her three torpedoes at *Gneisenau*. With her engines out of action, *Worcester* became motionless and presented *Prinz Eugen* and *Gneisenau* with an unmissable target. Eventually both ships ceased their shelling, each believing that *Worcester* would soon slip beneath the waves; she refused to oblige and soon afterwards the destroyers *Campbell* and *Vivacious* came on the scene and began the rescue of *Worcester*'s survivors. Whilst so engaged Campbell narrowly missed being hit by a torpedo from a 42 Squadron Beaufort. In the finest Naval tradition , the crew of *Worcester* gradually brought one engine back to life and she limped back home under her own steam. The utter confusion that reigned over the general battle area was evident in the attack by the Beaufort on *Campbell* but this was not an exceptional case. Two Wellingtons had earlier dropped bombs on *Walpole*, another of the destroyers, and a Hampden dropped some of its bombs so close to *Mackay* that the crews of the after guns were drenched with spray; it then made another run and just missed *Worcester*. Bombing attacks by Hudsons cost two aircraft and three Beauforts failed to return, so Coastal Command lost five aircraft. It is to the contribution of Bomber Command that the historians fail to give just and proper recognition, to the losses of the Hampden squadrons in particular. In their description of the action, in "Mission Accomplished", P A Rickson and A Holliday state: "The RAF were prepared to throw into the battle a total of 300 bombers, a few of which were minelaying Hampdens".

John Deane Potter, in his book "Fiasco", presents a graphic account of the German warships' progress along the Channel and the succession of valiant but vain attempts to cause at least some damage. As far as the Hampdens' participation is concerned, the only mention is in regard to one of the incidents in which one of our own destroyers was attacked. John does sum up the awards given to Coastal Command aircrew rather aptly... "There seems little doubt that to cover their own unforgivable inadequacies, Coastal Command rushed to recommend as many medals as they could to the courageous pilots". All of the Swordfish survivors were decorated, as also were the five destroyers captains. There does not seem to have been any comparable distribution to the Bomber Command aircrew.

A study of the aircraft involved in the attack, and the losses incurred, might have led to more emphasis being paid to Bomber Command's contribution. From German records which were, of course, not available to the British High Command in 1942, it is apparent that all the Hampdens lost on that day were lost through enemy action as they tried to attack the battleships.

AE310 of 144 Squadron in which Sqn Ldr Jimmy Bennett returned from a daylight minelaying op. south-west of Heligoland on 7 February 1942. The squadron descended through cloud in loose formation and came out over a convoy; enemy fighters shot down AB824 and AE392. AE310 was attacked by three Bf 109s which put the starboard engine out of action and damaged the wing. It took four hours for the aircraft to limp back to North Luffenham. (Photo: Jimmy Bennett)

Bomber Command's part in the action of 12 February 1942 was as follows:

Group	Aircraft Involved		Total Losses
1	37	Wellingtons	1 Wellington
2	37	Blenheims	2 Blenheims
2	10	Bostons	
3	44	Wellingtons	3 Wellingtons
3	11	Stirlings	+1 Wellington on return
4	13	Wellingtons	No losses
4	13	Halifaxes	
5	64	Hampdens	9 Hampdens
5	15	Manchesters	+1 Hampden on return
Total	244		Total 15 missing and 2 crashed on return

A total of about forty aircraft of No.16 Group, Coastal Command, took part inthe action; three Beauforts were lost in torpedo attacks and two Hudsons in bombing sorties. Coastal Command aircraft from No.18 Group's Scottish bases and 19 Group from Plymouth did not take part, nor did any aircraft from Flying Training Command.

Fighter Command lost six Hurricanes, six Spitfires and four Whirlwinds in the course of the air battles, mainly off the Belgian and Dutch coasts and, of course, 825 Squadron, Fleet Air Arm, had its six Swordfish annihilated.

The German warships reached their destination without being damaged by any of the attacks made by British Naval or Air Force action on 12 February but - and it is a very big "but" - *Scharnhorst* hit one mine at 1430 hrs when off Flushing and another at 2134 when north of Terschelling, where a little earlier *Gneisenau* at 1955 had also suffered damage from a mine explosion. So, the only damage to the German battlecruisers had been caused by mines yet none of the accounts describing the Channel Dash investigate how those mines came to be waiting on the sea bed. Messrs Rickson and Holliday in "Mission Accomplished" even contend that the mine that the *Gneisenau* struck had been "laid by the RAF only an hour before". The Operations Record Book of No.5 Group reveals that daylight minelaying operations took place on 6 and 7 February as follows:

6 February 1942: 34 Hampdens, 15 Manchesters on daylight minelaying off Friesians; 26 Hampdens successful; 11 Manchesters, 70 mines laid. One Hampden of 455 Squadron lost (AE308)

7 February 1942: 32 Hampdens laid mines off Friesians. 25 were successful. 3 Hampdens missing (AD834, AE392 and AE306)

Thus 95 mines were laid, 51 by Hampdens, 44 by Manchesters, for the loss of four Hampdens (1). These minelaying operations, making use of cloud cover, were nevertheless hazardous, being well within range of German coastal radar and the three Hampdens that were lost on 7 February all fell victim to Ofw Luth of II/JG1. Hauptmann Lent of II/NJG2 shot AE308 down on 6 February. All four went into the sea "West of Terschelling". F/Sgt R N Smith, the 50 Squadron Rhodesian pilot of AE306, had been the sole survivor from AE428 at Vaagso, six weeks earlier, and three of the crew of AE392 had survived a crash landing in AD923 (1).

Reports in the Record Books from crews who did return include positions in which the mines were laid. It is an indisputable fact that they were concentrated in an area to the north and west of Terschelling, where both *Scharnhorst* and *Gneisenau* suffered damage through striking mines on 12 February. No other Bomber or Coastal Command Groups report minelaying activity in the week prior to the breakout, so it does not seem unreasonable to claim the mine damage off Terschelling as attributable to the efforts of 5 Group. It was also marginally more likely that the Hampdens had laid the mines than their colleagues in the Manchesters, who would mostly be Hampden "Old Boys" anyway.

In an attempt to obtain a balanced view of the events of 12 February, several seemingly authentic books on the subject were read and whilst most give dramatic accounts of the heroic deeds of that memorable day, they also contain statements that were a complete travesty of the truth. How can one reconcile "The British had not anticipated that Hitler's famous intuition would send his powerful raiding squadrons dashing East" (Page 196, "Mission Accomplished") with reports from Official Records that plainly prove the opposite?

In April 1941, for instance, Bomber Command examined the possible routes that the German naval units might take if they left Brest. Bomber Command Operational Order 133, dated 1 May 1941, covered the possibility of the units attempting to pass through the Strait of Dover by daylight. Plans to cover such a breakout were drawn up, under code name Operation "Fuller" and this Plan was put into operation on 12 February 1942.

In the minutes of discussion at Chief of Staff Level, on 19 December 1941, the transfer of torpedo-bombers to the Middle East was opposed by Coastal Command, the prophetic words appearing that "In the event of a breakout of the German heavy ships from Brest, Coastal Command would need a greater force of torpedo-bombers than are at present available".

On 8 February 1942, HQ Coastal Command made an "Appreciation of the (break-out) situation". The document considered the situation and clearly shows that the Commander-in-Chief, Coastal Command, Sir Philip Joubert de la Ferté, was fully aware of the problems involved. Amongst the points made were the following:

"During the past three days, all three big ships have been exercising and should be ready for sea. There are four large destroyers and a number of small torpedo boats and minesweepers in Brest." Weather and tidal conditions are considered and the options that were open to the ships if they did leave Brest. They could break out and make for the Mediterranean or go into the Atlantic. Whilst regarding this as feasible, it concludes: "Owing to the conditions under which the three ships have been existing at Brest, they cannot be regarded, even now, as fully efficient fighting units capable of taking on British capital ships and the number of destroyers that have been brought to Brest seem to be in excess of those required to merely get the ships out to sea".

The second option was to break through up the Channel, "With the British Home Fleet in its present position, there would be no important naval force which could seriously threaten this passage, which could be carried out under the protection of land-based Air Forces, shore batteries and the strong flotillas that have been assembled in the Channel ports. In addition the weather and hour of sailing can be chosen to cause the maximum difficulty in intercepting the force. Finally, the large number of destroyers and small torpedo boats that have been concentrated at Brest would seem to indicate an attempt to force a way up the Channel" and Sir Philip concluded "The plan to break through the Channel will be put into execution at any time after Tuesday, February 10 1942". How right he was.

The 408 RCAF Squadron Record Book records that on 4 February 1942: "Nine crews were briefed at 0630 for an attack on the battlecruisers *Scharnhorst* and *Gneisenau*, which were believed to be going to try to escape from Brest up the English Channel ... the crews were warned of this op at 1700 on 3 Feb. but, owing to adverse weather and lack of information from reconnaissance aircraft, the crews were stood down". Three aircraft were again detailed at 1930 on 11 February to stand by from 0700 on 12 February. At 1143 on 12 February, information came through that the German naval force had left Brest and was proceeding up the Channel at high speed. At 1255, all available aircraft were ordered to take off at 1445. Eight Hampdens took off between 1442 and 1452. One landed at Horsham St Faith with engine trouble. Three were unable to locate the battleships due to bad visibility and returned with their bombs. Another two located the ships but could not bomb because of low cloud at 300 feet. Three bombed from 800 ft but could not see results. S/Ldr Constance's aircraft received considerable damage in attacking the warships. He was awarded the DFC. No.408 Squadron had not been operating on the previous night, whereas 106 Squadron had and was subsequently only able to despatch four Hampdens. One pilot claimed to have attacked the *Scharnhorst* but his bombs overshot. The other three failed to locate anything due to the bad weather conditions.

Nine Hampdens of 455 RAAF Squadron took off to attack the warships. W/Cdr Lindeman "bombed a ship from 800 feet and climbed into cloud before seeing his bombs burst". Sgt Wincott, who bombed at 1706, and S/Ldr French were hit by flak whilst making their attacks; neither saw the results. None of the other aircraft was successful and P1156 failed to return (2).

No.420 RCAF Squadron sent six Hampdens in to the attack. F/Lt Smith attacked cruisers at 1618 from 2,800 feet. Results were not seen nor were any protecting fighters. Three aircraft made no contact. AT134 and P4400 failed to return (3). No.49 Squadron had only been carrying out night flying practice on the previous night and "after being detailed for *Scharnhorst* and *Gneisenau* attacks at 1000 hrs, twenty aircraft took off at around 1400 hrs. All aircraft reported very poor visibility and four Hampdens failed to return". These were P5324, AE132, AE240 and AE396. There are names of sixteen aircrew in the casualty section of the Operations Record Book but they are not related to any aircraft, the serial numbers of which are not entered. It is apparent from the Movement Cards which four aircraft were lost and an attempt has been made to compile the names into crews. There is only one officer amongst the sixteen names, and he, Pilot Officer S W A Way, a 21 year-old wireless operator from Shoreham, had only been appointed to a commission on 4 February (4).

No.144 Squadron had not been on ops on 11/12 February and nine Hampdens entered the fray. Five brought their bombs back, two dropped their bombs but saw no result. AT175 failed to return and Sgt A Nightingale was killed when AE141, of which he was pilot, crashed near Norwich on return after being damaged by flak whilst attacking warships; the other three crew members survived the crash. Sgt Nightingale had survived another crash on return in AE311 when he crashed at North Luffenham on 9/10 November (5).

No.50 Squadron contributed eight Hampdens, AT177 failing to return. Its pilot, F O Carter, had previously crashed in fog on return on 2/3 September 1941 and the navigator, Sgt Symes, had lived to tell the tale after P1152 had crash landed onto a hillside at Guisborough in Yorkshire when returning from minelaying (6). The fact that 50 Squadron had been on ops the night before, and that some of the crews were dragged from their well-earned post-operational slumber, is described by Nairn Robertson who had been sleeping off the effects of a trip to Mannheim. "We were rudely awakened and told to report to the briefing room at 10 am in full flying kit". The news that they were to have a go at the battleships did not exactly help to brighten up that cold, cloudy February morning, nor did the fact that their allotted aircraft was plagued with communications failures which necessitated a transfer to another aircraft. Nairn eventually became airborne in AD852, piloted by Pilot Officer Norman Goldsmith. They succeeded in finding the German warships, went in to attack *Prinz Eugen* and were hit by flak which "knocked a hole 4 feet long by 18 inches wide in our starboard wing. The drag caused us to lose height and swing to starboard but Norman managed to get it straight and level at about 250 feet. He nursed it gradually up to 2,000 feet and I obtained fixes, etc. We were diverted to Horsham St Faith where we landed safely and came back to Lincoln by train". One wonders how many of their fellow passengers had any idea what the young, and no doubt untidy-looking, aircrew had suffered to merit their train journey. Norman Goldsmith received a DFC for his effort.

Other accounts from Hampden aircrew who took part in the action emphasise the atrocious weather, with low cloud, very poor visibility, problems with icing up if they flew into the cloud, uncertainty as to just where the German ships were and how many of them there were and the utter chaos of the whole business. The fact that the bombers had been sent out to attack the German warships with 500-lb bombs was in itself somewhat ridiculous. "We might as well have thrown bricks at them" seemed to be a pretty fair comment.

February 12th was the day that the Hampden squadrons suffered their heaviest loss. It is also a very sad fact that not one of the crew members of any of the Hampdens survived to become a Prisoner of War. German fighters claimed eight Hampdens and ships' flak two more, making a total of ten, which in view of the fact that several Hampdens were damaged, including AE141, make it almost certain that all the losses were due to enemy action.

Two more Hampdens were written off on operations on the night of 12/13 February. AD783 of 455 Squadron, returning from a Friesian Isles minelaying op, crashed while in the circuit at Wigsley with the loss of its crew. AE425 of 106 Squadron crashed after its port engine failed on take-off from Woodhall Spa for a Mannheim bombing op, its crew escaping injury. Neither AD783 or AE425 is included in the casualty summary (7). A breakdown of the casualties on 12 February reveals 31 missing and 6 killed on the nine failed-to-return Hampdens, one of which had a crew of five, plus Sgt Nightingale in AE141, a total of 38. On 6 and 7 February, two of the Hampdens had crews of five and of the total of 18 casualties, 15 were missing and 3 killed. So 56 men died in the Hampdens' contribution to the Channel Dash engagement. The total Naval casualties were 27 dead in *Worcester* and 13 dead or missing from 825 Squadron's Swordfish.

The awards for gallantry in the action are obviously in no way related to the sacrifice of life involved or the results obtained. The valour of the Bomber Command crews in general, and of the Hampden crews in particular, seems to have gone unrewarded. The fact that the only damage caused to the two battlecruisers was almost certainly the result of the daylight minelaying operations of 6 and 7 February seems to have been conveniently overlooked by all concerned.

The break out of the battlecruisers, and their arrogant cruise through OUR English Channel, brought forth howls of protest and, of course the traditional Court of Enquiry. However, it is not unlikely that although the Germans can never be criticised for the brilliant planning and execution of the operation, there will be an increasing volume of opinion that the transfer of the ships to German ports ultimately favoured the British.

"Torpedo gone"! A training drop by a Hampden of No.32 OTU into Patricia Bay, BC, on 19 February 1943. The stabilising monoplane air tail on the torpedo shows up well. (Photo: Public Archives of Canada PA – 144775)

2J TORPEDO-BOMBERS

On 2 January 1942 it was decided that 415 RCAF and 489 RNZAF Squadrons, already with Coastal Command, would re – equip with Hampdens. English Electric successfully completed the trial torpedo – bomber modifications to a Hampden and, by 24 March 1942, forty Hampdens had been modified after which they were to be supplied at the rate of 12 per month.

No.415 Squadron suffered the first Hampden TB operational loss on the night of 29/30 May 1942 when AT236 (1) failed to return from a bombing attack on shipping off the Friesian Isles. No.489 entered the fray and on 13 June 1942, P5335 (1) returned from an early morning anti – shipping patrol off the Norwegian coast with its navigator seriously wounded after an encounter with two Fw 190s. On 21 April 1942, two of No.5 Group's Hampden squadrons were transferred to Coastal Command and commenced TB training at Leuchars. These squadrons, 144 and 455 RAAF, had become operational within a few months but neither had suffered any operational losses in their new role before both were detached to operate from Vaenga, near Murmansk, in Russia.

The Russian Detachment

The decision to send two Hampden TB squadrons to Russia came about after the disastrous affair of Convoy PQ17 which had been ordered to scatter by the First Sea Lord in London because German naval units, including the battleship *Tirpitz*, were reported to have sailed from their base at Trondheim to attack the convoy. These German surface units were not, in the event, committed to the attack on the convoy but submarines and bombers wreaked havoc with the merchant ships which, in accordance with the orders from the Admiralty, were widely spread out and without any naval escort. The outcome of the attack on PQ17 was that 22 of the 33 merchant ships were sunk with their cargoes of war equipment that was destined for the Russians. The magnitude of the calamity was almost unbelievable, a total of 200 aircraft, 430 tanks, 3,350 vehicles and around 100,000 tons of war supplies disappearing beneath those Arctic waters, with the attendant grievous loss of the crews of the merchantmen. The total cost to the Germans was five aircraft. In spite of the massacre of PQ17, the need for more supplies to be sent to Murmansk before the onset of winter was so desperate that the decision was taken to despatch another convoy, PQ18, in September. Amongst the measures taken to try to avoid a repetition of the PQ17 catastrophe was the detachment of two Hampden torpedo – bomber squadrons to operate from aerodromes near Murmansk in the hope that their presence would at least make the Germans think twice before committing surface ships against PQ18.

Nos.144 and 455 Squadrons were ordered to prepare to fly to Murmansk and to be ready to operate from there by 7 September, at which time PQ18 would be on its way towards the Arctic waters off Northern Norway. The actual distance to be flown was 1,000 nautical miles and range trials demonstrated that a Hampden flying at 140 kts, without a torpedo, would have a safe range of 1,100 nautical miles. In order to achieve the shortest possible route to Russia, the flight was planned to begin at Sumburgh in the Shetlands, thence across the North Sea and then northwards to a point off the coast of northern Norway. From there the route was eastwards across the inhospitable terrain of Norway and the northern part of neutral Sweden, with its 6,000 – foot high mountains. The last obstacle would be German – occupied Finland with its fighter bases near Petsamo. The fact that in these latitudes there was less than four hours of darkness during this time of the year meant that the Hampdens were likely to be within range of German fighter airfields in Norway or Finland in daylight at some stage of their flight. Another problem that was appreciated was that of early morning fog over Russian airfields, and also the possibility that the Hampdens would arrive while the Russian airfields were being bombed by the Germans, whose bases were a mere fifty miles away. Vaenga, for instance, had been known to experience as many as 50 attacks in one day.

An additional complication arose from the fact that the original intention had been to use Kandalaksha as the Hampdens' first landing place and that was controlled by the Russian Army while Vaenga, which was eventually used, was controlled by the Russian Navy and it seems that the inter – service rivalry found in the capitalist world was not unknown amongst the "Comrades". It is interesting to notice that the Russian and British commands did not display any great trust towards one another despite the fact that the success of PQ18 was so essential to the general war effort. The Russians, for example, refused permission for RAF photographic reconnaissance aircraft to take photographs of the airfields at which the Hampdens were to land. The proof that this mistrust was mutual is evident in the briefing given by one RAF intelligence officer to the crew of a Catalina before it departed from Sullom Voe to fly to Murmansk. He solemnly warned the crew that they were not to disclose to the Russians the details of the IFF (Identification Friend or Foe) equipment that their Catalina carried, yet 2,000 Hurricanes had already been delivered to the Russians and all were equipped with IFF.

Further evidence of the lack of trust was forthcoming when they arrived at Murmansk and the Russians would not allow the medical staff to disembark because it was decreed that the RAF should use Russian facilities. Group Captain F L Hopper, who was to command all RAF aircraft in North Russia, arrived at Vaenga on 5 September 1942 after

Top: UB – B, P1287 of 455 Squadron, at dispersal at Vaenga. The camouflage was necessary because the base was within easy reach of Luftwaffe fighters and bombers.

Centre: The camouflage was not good enough for this Hampden which is going up in smoke after an air raid on Vaenga.

Bottom: After the smoke has cleared, disconsolate airmen, in tin hats, survey the remains of the Hampden.

flying out in a Catalina. As a sensible precaution against the possibility that ship – borne ground crew might not have arrived by the time the Hampdens landed, it had been decided that each of the 32 Hampdens should carry one ground crew tradesman so that there would at least be the nucleus of a servicing crew available at the aerodrome. Torpedoes, stores and most of the squadron ground personnel went to Murmansk by sea.

The Hampdens from the two squadrons began taking off from Sumburgh at 2025 hrs to fly separately to Afrikanda over a thousand miles away. The last of the thirty – two had disappeared to the east by 2100 on a flight that for most of them would take over eight hours and for nine of them would have an unfortunate ending. That problems had befallen the Hampdens became evident in the signal that reported their arrival:

"Twenty – seven Hampdens have arrived, three of which have crashed. One of these landed with undercarriage retracted in the Kirovsk area, another crashed in the woods between Murmansk and Polyarno and the third was shot down by Russian fighters near Vaenga; its crew are OK. Fog prevailed when the Hampdens were landing. There is no news of the remaining 5 aircraft; 24 are operational. (This figure was later reduced to 23). Contributory factors in the shooting down of AE356 were that an air raid was in progress, British aircraft were not expected N of Afrikanda and the aircraft was not in the approved corridor. The Russian pilot mistook the Hampden for a Me 110 because of its twin tail".

Of the 23 Hampdens that did land safely in Russia, twelve reported being fired upon by light or heavy flak over enemy territory or by their Russian allies and several were intercepted by Russian fighters. Some idea of the confusion that existed when the Hampdens arrived may be gathered from the reports of those that landed safely.

Squadron Leader Foster in AE231 (2) was escorted in to land at Moenchigorsk by Russian Hurricane fighters whilst the crew of X3131 (2) believed that the aircraft that had intercepted their Hampden were Russian Airacobras. They followed them in to land at what turned out to be the Russian base at Murmachi but as they were not absolutely certain of the identity of the fighters they took the precaution of eating their secret papers when they landed. Fortunately the paper was rice paper so their early breakfast caused no damage. P2126 and AE130 (3) suffered minor damage from flak as they flew around trying to find their landing place. The weather, in the form of fog and low cloud, caused problems at the destination airfield, Afrikanda. AE307 (4) had to make six circuits before it finally touched down after eight – and – a – half hours flying. L4038 (5) after twice having suffered the attention of the light flak gunners, was forced to make four attempts at landing because of the low cloud and fog before it finally succeeded in safely setting its wheels on Russian soil.

Of the nine Hampdens that were written off on the flight to Russia, six were from 144 Squadron and the remaining three from 455 RAAF. Two of the Hampdens that were written off ran out of fuel and crash – landed. AE310 (6), after being in the air for just over ten hours, bellylanded in a field not far from Afrikanda which it had been unable to find because of low visibility. Its crew were unhurt. That of P5323 (6) also escaped injury when it ran out of fuel and crashed near Kandalaksha but their arrival almost ended in a prompt departure to the happy hunting ground above as the Russian soldiers who surrounded the Hampden took a lot of convincing that they were not members of the Luftwaffe. The crew did not need a great effort of imagination to realise what their fate would have been if they had been Herrenvolk.

The third of the Hampdens that almost made it was AE356 (7) which arrived north of the area in which the Hampdens were expected. There was an air raid in progress at the time and in the general confusion, AE356 was shot down by Russian fighters which mistook it for a Bf 110. To add insult to injury, the guilty pilot was flying a Hurricane that had been supplied to the Russians. The Hampden pilot put his aircraft down close to the shore of a large inland lake and all the crew, except the lower gunner, escaped from the sinking aircraft and swam to the land. In spite of being shot at by one of the Russian fighters while still in the water and coming under rifle fire from Russian soldiers, they managed to reach the shore safely. The body of the unfortunate lower gunner, Flight Sergeant W T Tabor, RCAF, was recovered and buried by the Russians with due ceremony. It is a sad reflection on the world we live in that the name of this Canadian appears on the list of those who have no known grave.

The four men who were found in the wreckage of AT138 (8) were also buried with full military honours but post – war investigations by the Russian authorities on behalf of the Ministry of Defence failed to locate the graves and they are listed on the Runnymede Memorial. AT138 was attacked by two Bf 109s and crashed at Alakurtti near Lake Vikurij, the only survivor being the pilot, Sergeant Bray, and in his report of the event he described how his aircraft met its end. The first burst of enemy fire either killed, or incapacitated, the three members of the crew in the rear of the aircraft and a fire broke out. The navigator attempted to crawl aft to man the rear guns but the fire was too intense for him to extinguish and, as he made his way back towards the front of the Hampden, another burst of shells and bullets blasted into the fuselage and he too was killed.

The blazing aircraft fell into a spin from which the pilot realised he would not be able to regain control and he baled out. He was injured and burnt about the face and his eyesight was affected for some time afterwards; he became a prisoner of war. Two more Hampdens fell victim to the German fighters over northern Finland. P1344 (9) was shot down and crashed at Petsamo airfield, only the pilot and fifth crew member surviving. The three dead crewmen were almost certainly buried in Finland but are listed on the Runnymede Memorial because their graves have not been found. P1273 also fell to fighters and its crew must have been remarkably fortunate to escape with their lives. The navigator, Sergeant K C Smith, wrote home afterwards from a POW camp to say: "I baled out of a flaming mass of metal and landed in a swamp in Finland. Sergeant L H Bertrand must have hung on to the controls until the very last moment for all of his crew to have baled out safely; they all became prisoners of war".

Bob Anderson, navigator of AT109 (11), recalled how they had taken off from Sumburgh with a weather forecast that was far from encouraging.

The conditions encountered during the flight must have varied considerably according to the height and course maintained by individual aircraft as they crossed the Norwegian mountains. The memory of his own experience is still deeply etched in Bob's mind.

"We struck the worst possible weather conditions; the aircraft bucked so violently that I was forced to hold on with both hands to brackets above my head. Navigation was impossible and I could only pray that the pilot would hold the aircraft steady enough to keep it flying in the right general direction. The blackness of night eventually gave way to the grey light of dawn and, through the murk, the vague outline of land could be seen beneath us but with maps marked 'uncertifiable proof' and the general similarity of the hills and islets of the coastline it was some time before I realised that we were some distance north of our intended position. In the course of our stooge along the coast we had been fired at by all and sundry and Jimmy Catanach, our pilot, was convinced that fuel was being lost from holed fuel tanks so that the possibility of turning on a southerly course to fly across Varengeren Fjord would involve a near – certainty of a ditching in those desolate Arctic waters. The appearance of a strip of stunted heather adjacent to a beach ahead of us offered at least the possibility of a safe return to earth and, faced with the imminent alternative of having engines cut through lack of fuel, Jimmy gratefully put the wheels down to end the flight with an understandably rough bumpy landing".

As the crew of the Hampden attempted to set fire to their aircraft, a horde of soldiers in white – overall uniforms appeared on the high ground above the beach whilst a patrol boat arrived on the scene and commenced firing at them from a short distance out to sea. The hopelessness of their situation was apparent and in a matter of minutes the crew had become prisoners of war. Their first night was spent in a ghastly camp containing Russian slave labourers who existed under near starvation conditions. The crew were greatly relieved to be taken from there by lorry to Rovaniemi and then flown to Helsinki whence, after spending the night in the civilian gaol, they were flown next day to Frankfurt and the commencement of their prisoner of war existence at Dulag Luft. The final chapter of the AT109 story concerns its pilot, Squadron Leader Jimmy Catanach DFC, who had joined 455 RAAF Squadron in its Bomber Command days during which he was awarded the DFC before continuing as a torpedo – bomber pilot when the squadron transferred to Coastal Command. Jimmy was regarded as an exceptional pilot and at the time he became a Squadron Leader at twenty years of age was probably the youngest pilot to hold that rank outside Fighter Command. Jimmy broke out from Sagan prisoner – of – war camp in 1944 in what has become known as "The Great Escape" and was one of the fifty men who were murdered by the Gestapo after their recapture.

P5304 (12) and AE436 (12) came to grief in the mountains of northern Sweden, probably in similar circumstances, although none of the crew of the former lived to tell what happened when the Hampden crashed into the mountain side at Arvestuottar. Its five crew were buried at Overlulea but their bodies were later exhumed and moved to the Gothenburg War Cemetery. The loss of AE436 is another of the Hampden sagas as the brief report of Pilot Officer D I Evans clearly shows:

"The starboard engine failed at 0045 hrs. I flew on one engine and at 0115 hrs was reducing height to descend below freezing level at 5,000 feet. There was 5,600 feet on the altimeter when the aircraft struck the top of a mountain and burst into flames before it came to rest. The aircraft broke in half by the main spar and Corporal Sowerby, who had been sitting on the wing spar, escaped through the broken fuselage. I got out in the normal manner. I am of the opinion that the other three members of the crew were killed instantly. The Hampden burned fiercely for a long time after the crash".

The two survivors decided to head away from the bleak and desolate mountain side to seek succour and five days later staggered into the Swedish village of Kvikkjokk. A spell in hospital and plenty of food soon restored their weakened bodies while their claim that they had walked to Sweden after their aircraft crashed in Norway was investigated. There was no evidence to deny the truth of their story and the two men were repatriated a few weeks later. The grim evidence that they had crashed in Sweden did not come to light until thirty – two years later when, in 1974, two Swedish mountain climbers stumbled across the wreckage in which the other three Canadian members of the crew had been entombed. The crash site is in an area that is almost permanently covered with snow and it was only because the 1974 summer had been a particularly hot one that the Hampden had been discovered. Most of the wreckage was removed from the mountain side. Some of it arrived at the RAF Museum in Hendon where it may, at some future date, be displayed as a tribute to the Hampdens and the men who flew in them. The bodies of the three Canadians were laid to rest at Gothenburg War Cemetery where they at last joined their four Australian and one British comrades from P5304.

Some of the men who made the flight to Russia recalled their impressions of the venture. Corporal A C "Bert" Earl of 144 Squadron flew as fifth crew member on AE158 (13), the main reason for his presence being that his expertise as an engine fitter would be required when they arrived in Russia, particularly if the ship – borne main party of squadron personnel was delayed. Bert recalled that he was somewhat apprehensive about being detailed to operate the beam Vickers "K" gun, in

Top: "Heave – Ho!". Pilot Officer "Tango" Martin of 455 Squadron reveals his braces as he helps manoeuvre a torpedo into position.

Centre: Twenty – four Hampdens on an offensive patrol on 14 September 1942. They took off from Vaenga at 0820 and landed back at 1600, having seen nothing. The monoplane air tail of the torpedo and the guns in the lower position can be clearly seen on the aircraft in the foreground which appears to be UB – D (X3131) of 455 Squadron.

Bottom: Handing over to the Russians in October 1942. Wing Commander Lindeman gives Russian pilot Captain Stayanoff the low – down on Hampden controls and instruments with the help of interpreter Victor Alexandrovitch Andrushkin. (All photos: Grant Lindeman)

A Hampden of 455 Squadron at Vaenga during the detachment to North Russia. The airfield was too close to the enemy front line for comfort and suffered many air raids, hence the tree branch camouflage.
(Photo: Grant Lindeman)

the event of enemy fighters being encountered, as he had never before even fired a machine gun on the ground. Luckily for Bert, AE158 had a fast and uneventful flight even though it was over Finland in daylight. They landed at Afrikanda at 0340 after a flight of just under seven hours during which the only sign of hostility had materialised in the form of a thread of tracer shells that winged up towards them from the tree – clad hills beneath them and fortunately missed their aircraft.

The constant attention from the Luftwaffe ensured that life at the Russian aerodrome never became boring. The Russians were well versed in the art of protecting their aircraft on the ground and mainly achieved this by digging sloping pits down which they pushed their aircraft to make them less easy targets for the bombs and cannon fire of the Luftwaffe strafers. Some of these pits were provided for the Hampdens but the ground crew were reluctant to use them as they feared that once they had manoeuvred their aircraft down the slope they would not be able to drag them out again in a hurry. They had reason to regret this neglect on 29 September when, in one attack, three Hampdens were destroyed and nine more damaged. The armoury was also hit and half the Hampdens' ground equipment destroyed. Just to add a little variety to the proceedings, one of the Russian fighters crashed into the barrack block that was used by 455 Squadron but fortunately nobody was inside. In the quieter hours the traditional football matches took place between the airmen and their Russian counterparts and there was always rivalry between the 144 Squadron lads and the Aussies of 455 Squadron to keep everyone on their toes. As far as his own dealings with the Russians were concerned, Bert did not recall any of the friction, mistrust or even hostility that permeated the higher command. The skills and adaptability of the Russian Air Force mechanics with whom he came into contact made a good impression on Bert which may have been enhanced by the fact that there were quite a number of young ladies amongst them who, besides being adept at their work, also made the evening concerts in the camp dining hall considerably more interesting.

Bert's favourable opinion of the Russians was apparently shared by most of his colleagues but he could not recall that any tears were shed when they were ordered to overpaint the Royal Air Force roundels on the Hampdens with the Red Star of Russia after instructing the Russians on the maintenance and servicing routines, which was no easy matter without the aid of interpreters.

Jim Rowland, from Gateshead, a 144 Squadron airman who had made the journey to Murmansk in the American cruiser *Tuscaloosa* had memories of the frequent air raids. Usually they occurred at two – hourly intervals and a red flare would be the only warning that bombs were about to arrive from Bf 109s and Ju 88s. Jim's recollection of the actual airfield was that it was situated in a valley and consisted of levelled ground which gave a landing strip over two miles long and a quarter of a mile wide.

Flying Officer W D Barrett was the navigator of AE231 (14) and he recalled that they encountered "terrific thunder storms over Norway". They too flew over Finland in bright daylight just to the south of Lake Enare where the German and Russian armies faced each other. Russian fighters intercepted AE231 and guided it down to the aerodrome at Moenchegorsk where they landed safely. The only English – speaking Russian was an engineer officer who had learned his English from engineering books but until this occasion had never had the chance to put it to use. His talent was severely tested in the next few hours as the crew of AE231, and of P1287 (14) of 455 Squadron which had landed soon after them, were taken to the canteen and given a slap – up breakfast, accompanied by speeches of welcome from the local Commissar, the Commanding Officer and other Russian officers, to which the crews endeavoured to make a suitable reply with the aid of the overworked Engineer Officer for whom this must truly have been his moment of glory. Both aircraft took off later that morning for Vaenga where they landed safely and joined the other Hampden crews.

Flying Officer Barrett spoke well of the hospitality shown to them by the Russians and was particularly impressed by the ease with which the Russian pilots, who had been flying single – engined fighters, converted to flying Hampdens. He concluded his account thus: "I do not think that History will regard it as an epic operation. We never launched a torpedo at a German ship or shot down any of their aircraft".

"History", in general, may agree with these sentiments but the Hampden's own history has every right to be proud of this Russian detachment on its battle honours. The twenty – three Hampdens at Vaenga carried out one anti – ship patrol on 14 September when, under the leadership of Wing Commander McLoughlin, they took off at 0815 hrs for a search of the coastal waters towards Altenfjord. No ships or aircraft were sighted and all the Hampdens were safely back at Vaenga by 1550 hrs.

Convoy PQ18 battled its way through to Murmansk and came under heavy attack by German aircraft and submarines but no surface ships attacked, which may in some measure have been due to the fact that the Germans were aware of the presence of the Hampden torpedo – bombers at Vaenga. Thirteen of the merchant ships were sunk as they ran the gauntlet of the submarines and bombers on their way to Murmansk and three more were lost on the return voyage. The convoy escort group lost a tanker, a destroyer, a mine – sweeper and four aircraft; the cost to the Germans was four U – boats and 41 aircraft.

On 22 September, the convoy operation code – named "Orator" had been completed and the Russians enquired about the future plans for the Hampdens. Wing Commander Lindeman, at that time in command of 455 RAAF Squadron, recalled that he had great difficulty in persuading HQ Coastal Command that the idea of flying the Hampdens back to Britain against a headwind and using inferior grade Russian petrol would be suicidal. Tempers had become very frayed, particularly when AVM "Dusty" Durston at HQ Coastal Command had suggested that the Navy would be available to rescue any crews of aircraft that had to ditch on the return flight. The voice of commonsense eventually prevailed and on 24 September, the AOC Coastal Command wrote to the Under – Secretary of State with his comments on the problem posed by any attempt to fly the Hampdens back to Britain, as follows: "In view of the prevailing weather conditions in Norway and North Russia, and of the strong SW winds now blowing, it is felt that the risk to valuable crews is such that it would be far preferable to hand the Hampdens over to the Russians as a gift and bring the crews back by sea in H.M. ships. C – in – C stressed that on the outward flight with a following wind and with the possibility of surprising the enemy's fighter defences in Norway, only 75% of the Hampdens arrived in Russia. The return journey, making a landfall in the Shetlands with an almost certain headwind, allowed no margin at all for navigational error".

On 1 October, the Russians made an official request for the Hampdens to be handed over to them. The only alternative would have been to dismantle the Hampdens and send them back to Britain by sea and it was not thought that the Russians would co – operate in this. On 6 October the Prime Minister approved the handing over of the Hampdens and this took place ten days later to the accompaniment of complaints from the Russians that the bombsights, photographic and other detachable equipment were being removed from the aircraft and that the torpedoes

Aircrew of 455 RAAF Squadron, all of whom made the flight to Vaenga in Russia on 4/5 September 1942. They are, from left to right: Flt Lt Davenport (P1287); Sqn Ldr Catanach DFC (AT109); Wg Cdr Lindeman and his navigator Flt Lt Oliver DFC (both AD877); Sqn Ldr Holmes (L4038).
(Photo: Grant Lindeman)

were being shipped back to Britain. It was finally agreed that all equipment should remain with the Hampdens and on 22 October the detachment embarked in HMS *Argonaut* for the voyage home. They arrived back at Leuchars on 29 October for a spell of well-earned leave on the following day. The happy band of leave-takers included one officer who, with visions of the boat that would take him away from Russia, summed up the general feeling of the men in the latter days of the detachment by ending one of his reports with the wistful words "Hasta la Vista".

The 32 Hampdens (15) involved in the detachment were as follows. The nine that were written off in the course of the flight are marked with an asterisk.

L4038/455 Sqn;	P1245/455 Sqn;	P1273/455 Sqn*;	P1287/455 Sqn;
P1344/144 Sqn*;	P2095/455 Sqn;	P2122/144 Sqn;	P2126/455 Sqn;
P4415/144 Sqn;	P5304/455 Sqn*;	P5315/455 Sqn;	P5323/455 Sqn*;
X2976/144 Sqn;	X3022/144 Sqn;	X3053/144 Sqn;	X3131/455 Sqn;
AD743/455 Sqn;	AD908/455 Sqn;	AD977/455 Sqn;	AE130/144 Sqn;
AE156/455 Sqn;	AE158/144 Sqn;	AE194/455 Sqn;	AE231/144 Sqn;
AE307/455 Sqn;	AE310/144 Sqn*;	AE356/144 Sqn*;	AE436/144 Sqn*;
AN126/144 Sqn;	AT109/455 Sqn*;	AT138/144 Sqn*;	AT145/144 Sqn.

Hampdens under Russian Navy Command

Wg Cdr Grant M Lindeman had interesting memories of the Russian pilots learning to fly the Hampdens; he confirmed that they did fit dual control to a Hampden. "The instructor's position was originally in the nose but the ride proved too hair-raising for even the most steel-nerved instructor, particularly on landing when the ground seemed far too close and the speed too fast for comfort. By cutting half-way through the main spar, they managed to fit the instructor's position in behind the pilot's seat where the pupil sat. They had no qualms about weakening the aircraft in such a manner because they had no intention of performing aerobatics in it". Grant and his fellow pilots watched the efforts of the Russians with interest, thinking of those weighty Russian boots planted heavily on the rudder bar and not being surprised to see the Hampdens swinging on take-off and landing, and often ending up amongst the scrub and boulders.

There is a possibility that the Russians fitted an extra gun beneath the lower gun position as this features in a Russian drawing of the Hampden, although there is no photograph to confirm this.

It has been possible to piece together at least part of the story of the Hampdens after the RAF had departed. Robert Bock in Poznan, Poland and Hannu Valtonen and Carl Fredrik Geust in Finland were the sources of information, most of which originated from the Central Museum of the Soviet Navy in Leningrad. Seventeen Hampdens became operational with the Third Squadron of 24 MTAP (Anti-Shipping Wing); the other two squadrons were equipped with SB-2 and DB-3F twin-engined bombers. Some of the attacks the Hampdens took part in were as follows:

18 December 1942: Two torpedo-bombers, a DB-3F piloted by Captain B S Gromov, the other a Hampden with Captain C I Trunyev at the controls, sank two German cargo ships near Nordkin Hopgkuri.

14 January 1943: Two Hampdens found seven cargo ships steaming towards Vardo-Bapge. Captain Bashtyrkov's aircraft was shot down by intense and accurate flak from the convoy but his torpedo struck home and sank his intended victim. The other Hampden, piloted by Captain V N Kiselov, returned safely and also claimed to have sunk a merchantman.

15 January 1943: Two Hampdens piloted by Captains C I Trunyev and Lieutenant P N Zaychesko sank two ships near Hannibery-Xannudeps.

29 January 1943: Hampdens piloted by Captain A I Ostrowski and S A Malygin torpedoed a 12,000 ton merchant ship which sank. Ostrowski's Hampden had an engine put out of action by flak but he returned safely to base on one engine. (Note: The identity of this ship is not known; if it was of 12,000 tons, it was the largest ship to be sunk by Hampdens).

28 March 1943: Four Hampdens, piloted by Captains G D Popovich, W V Glushkov and V N Kiselyev, and Lt I I Dubinec, attacked a convoy near Sulten Fiord-Cronome and sank three ships.

25 April 1943: Hampdens attacked a convoy in Kongsfjord, Norway. Captain Vasily Nikolayevich Kiselev was killed in the attack during which he sank the biggest ship in the convoy, the *Leesee* of 2634 BRT in

With the Hampdens now in Russian hands, Soviet naval airmen manhandle a torpedo into position. The cyrillic inscription proclaims that the torpedo is "For Kisyelev". Captain W.N.Kisyelev had, posthumously, been made a Hero of the Soviet Union after he and his crew perished in an attack on a German convoy in Kongs Fjord on 25 April 1945. The aircraft had been set on fire as it made its torpedo run but Captain Kisyelev pressed home his attack and destroyed a merchant ship before crashing. (Photo: Carl-Frederik Geust)

position 70.34N:30.07E. As the Captain made his attack, his aircraft was set on fire but he pressed home his attack and either sank the ship by dropping his torpedo and then crashing his blazing aircraft into it or by actually ramming it. Accounts differ but Kiselev, who was a deputy squadron commander of 24 MTAP, was made a Hero of the Soviet Union posthumously on 24 July 1943.

20 September 1943: Six torpedo-planes, commanded by Major E W Kostykin, attacked a convoy of five ships near Makkaur-Makkayp. The convoy was escorted by twenty gunboats and 25 aircraft but the gallant Soviet pilots pressed home their attacks and sank one merchant ship and two escort vessels. In the course of the action two Hampdens, piloted by Major Kostykin and Captain A I Ostrowski, were shot down and both crews perished.

The Hampdens remained in use with 24 MTAP until the end of 1943 by which time there could only have been a few serviceable aircraft left. The Russian crews had every reason to be satisfied with the use they had made of the Hampden on operations which must often have been carried out in the most appalling conditions of Arctic weather.

Three-quarter rear view of a Hampden showing a torpedo with monoplane air tail (MAT). The MAT was fitted to stabilise the torpedo as it dropped towards the sea and broke away on impact with the water. (via Dennis Thompson)

Operations from Britain

Nos.144 and 455 Squadrons were soon back in the torpedo-bomber business once they had received replacement aircraft for the Hampdens left in Russia.

Three of 144's "Russia" crews had been lost by the end of the year. Squadron Leader Hird in P2063 (15) failed to return from a nuisance raid to Norway, the very poor weather experienced by other aircraft on the same operation possibly being the reason for the loss.

Flight Lieutenant Bowden in P1250 (16) and Pilot Officer Stilborn in P4373 (16) were shot down by flak as they attacked two escorted merchant ships on 13 December 1942. The third Hampden of the formation was also hit but returned safely after torpedoing the *Theano* which later sank. The successful pilot was Flt Sergeant J W King and his next claim to fame was that he was the last pilot to land from a 144 Squadron Hampden torpedo-bomber operations when he touched down at Lossiemouth at 0105 hrs on 19 January 1943. The squadron was taken out of the line to re-arm with Beaufighter VIs on the same day, still using Leuchars as its main base. A brief study of the early months of 144 Squadron's Beaufighter operations reveals PO J W King missing on 27 July 1943, by which date FO Fulton, FO Bond and Sqn Ldr Foster, all of the "Russia" detachment, had also failed to return from operations. The ending of 144's association with the Hampden is in itself a noteworthy event as the squadron had been equipped with the type for almost four years, which was a longer period than any other squadron could claim, and it lost more Hampdens than any other unit. Most of 144 Squadron's Hampdens were handed over to 455 RAAF Squadron which was also based at Leuchars and the code letters "OL" for 144 Squadron simply changed to "UB" for 455 Squadron once the paperwork had been attended to.

Two of 455's "Russia" crews had been involved in write-offs before the end of 1942, Warrant Officer Baker's crew having a lucky escape when the port engine of AD836 (17) cut out on take-off for operations one morning from Leuchars and they crash landed into a wood. None of the crew suffered serious injury. PO Gunton and his crew in P2078 did not have such good fortune when they failed to return from a daylight patrol of the Norwegian coast although two of the crew were later reported to be prisoners-of-war.

No.455 continued to use Hampdens as torpedo-bombers until the end of 1943 and two incidents in January of that year demonstrate how the slender thread of the Morse Code could, on occasions, be the last that was heard from an aircraft and its crew. The navigator was often so unsure of the aircraft's position that the expertise of the wireless operator had to be relied upon to obtain a QDM, i.e. a course to steer for base. On 11 January, AD792's (18) wireless operator asked for a QDM at 2213 as they neared the Scottish coast. The Leuchars ground station answered with "QDM 218". Two minutes later, another signal from AD792 ended abruptly. It was two days before the aircraft was found crashed on the north side of Hill of Warren (see 5D6).

Eight days later, X3140 (18), piloted by PO Gleeson who still had with him his same "Russia" crew, took off from Leuchars at 1330 hrs for a patrol of the Norwegian coast. At 1912 hrs, the wireless operator's request for a QDM was answered by the shore station with "QDM269" and that was the last heard of X3140. The aircraft must have gone down in the sea as three of its crew are listed in the Runnymede Memorial Register whilst the navigator's body must have either been washed ashore or picked up at sea as he is buried in the UK.

Nose aerial array of ASV (Air – to – Surface Vessel) radar fitted to Hampden P4369 in April 1943 for trials at Boscombe Down. ASV was used operationally by Nos.455 and 489 Squadrons and the cumbersome array had an adverse effect on the aircraft's performance and handling. (via Dennis Thompson)

One little – known aspect of the Hampden TB is that the aircraft was being fitted with ASV (Air – to – Surface Vessel) radar in 1943, 455 and 489 Squadrons were both being so equipped. The 455 Squadron records show that at the end of February 1943, its aircraft were being flown to St. Athan "for RDF mods", RDF being the abbreviation for "Radio Direction Finding", the alias by which radar was generally known in the early days of its use. On 20 March 1943, AT184 (19) crashed after engine failure as it was being flown back to Leuchars from St. Athan and in the section on the Air Transport Auxiliary (5J), there are accounts of the accidents to P2113 (19) and L6019 (19), both of which crashed after engine failure following which they would not maintain height due to the drag from the "Christmas Tree" aerial array. Similar problems had been encountered in Luftwaffe Bf 110 night fighters on which similar, even bulkier, aerials were mounted and which reduced the speed by 20 mph.

Some idea of the extent of the aversion to the ASV aerials may be seen in 455 Squadron's records. The entry for 17 September 1943 announces that "Two new ASV – fitted Hampdens arrived on the squadron, this makes six received in the last fortnight". A week later the sequel appears "Permission has been granted to remove the "Christmas Tree" ASV fittings from the newly – acquired ASV Hampdens". The arrival of twenty "Navigators (W)" on 455 Squadron on 8 April indicates that wireless operators had been trained in the use of radar as a navigational aid and the arrival of PO A Tait (Radar) at Station Headquarters on 11 September appears to be the first occasion on which the word "radar" was used instead of RDF. Reference to "SE Mk.II" (Search Equipment Mk.II) is first seen on 19 June 1943 in the 18 Group Appendices "Aircraft D of 455 Squadron fitted with SE Mk.II picked up the coast of Norway at 26 miles, small islands at 8 miles and the target at two – and – a – half miles from a height of 50 feet. In the return flight at 2,000 feet, the coast was picked up at 56 miles and the beacon at 47 miles, when visibility was less than two miles". The report also made it clear that "D" made a torpedo attack on the target so the radar was not merely used as a means of locating a target for other aircraft to attack.

The Hampdens of 489 Squadron were also fitted with ASV, as the report on 16 September makes clear. The mission of "R", "S" and "T" was curtailed due to weather "and radar was used on the last half – hour of the return flight". No photographs have been located showing the "Christmas Tree" aerials on the wings and fuselage of the Hampden but a photograph of the nose aerials is included in this book. The position of the operator of the set can be seen in the inset to the cutaway drawing.

Although it has not been possible to check the results of all attacks made by Hampden TBs against shipping, there are a number for which the details in the squadron ORBs may be compared with contemporary German and Norwegian records. It should be borne in mind that these comparisons, made possible through the co – operation of Sten Stenersen in Norway, only concern activity off the coast of Norway by 144, 455 and 489 Squadrons as the fourth Hampden TB squadron, 415 RCAF, operated further south. The area in which most of the successful attacks took place was off the south – west corner of Norway. The number prefixing each report is that used on the accompanying map which shows the approximate position of the incident.

1. 17 September 1942, No.489 RNZAF Squadron

Hampdens "A", "B" and "G" took off from Wick at 0917 for Norway coast patrol with orders not to approach the coast unless there was adequate

A trial installation of nose ASV (Air – to – Surface Vessel) aerials. (Photo: RAF Museum)

ASV aerial, outboard of the starboard engine. The balloon cutters can be seen on the leading edge of the wing at the extreme left and also close to the engine nacelle. (Photo: RAF Museum)

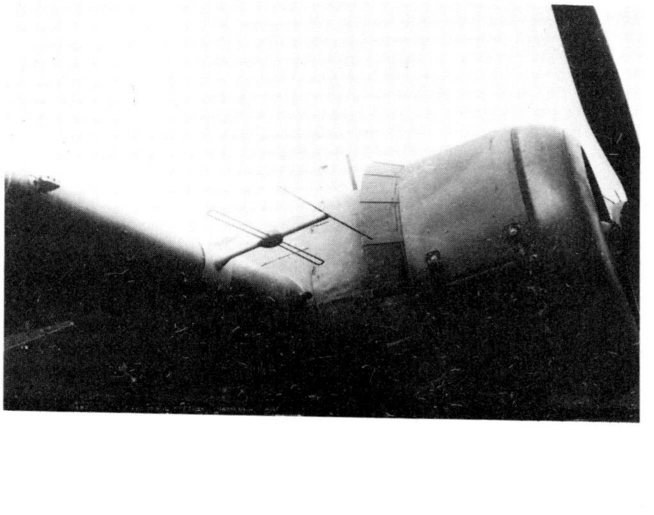

P1346 UB – T of 455 Squadron banks away after launching its torpedo against a merchant ship which is turning to starboard off Lister, Norway, on 2 August 1943. An arrow indicates the splash where the torpedo hit the water. (Photo: Ray Lassiter)

cloud cover. At 1115, two merchant vessels "one large, one small", with five escort vessels were sighted three miles west of Varhaug. F/Sgt Strain in "A" attacked first, aiming his torpedo at the larger vessel. His torpedo missed the target but by a stroke of good fortune hit the smaller one which the wireless operator of "A" observed to be on fire. Hampden "G" (FO Mottram – the "Buster" Mottram of Wimbledon fame) also attacked the larger vessel. Two explosions were seen about ten seconds apart. A large cloud of chocolate – coloured smoke appeared all over the rear part of the ship which was of the three – island type with a tall funnel, two upright masts and estimated to be of about 6,000 tons. Hampden "B" (FO Richardson) reported the two ships as being of 6,000 tons and 2,000 tons. "B" followed up "G"s attack on the larger vessel on which he had seen a torpedo strike home and saw dark grey smoke pouring out from midships to the stern. The three Hampdens turned for home and landed at Wick at 1352. Their well – merited jubilation would have been even greater had they known that both ships had sunk. They were the *Karpfanger* of 4,975 tons and the *Utlandshorn* of 4,979 tons, so the lucky hit by "A" on the "smaller" vessel had been an even bigger stroke of luck than they thought.

2. 21 October 1942, No.489 RNZAF Squadron

FO Richardson in "B" struck again on 21 October 1942 during an offensive patrol with three other Hampdens. He sighted a merchant vessel of 3,000 tons and flew in to attack it, dropping his torpedo from a height of 90 feet and range of 600 yards. He saw the torpedo strike home aft of the engine room on the stern of the ship, then climbed into the cloud layer to avoid the flak and the attentions of a Ju 88, so failing to see the outcome of the attack. The ship was the German freighter *Palatia* of 3,975 tons and it went to the sea bed near Lindesnes. According to another Norwegian source, *Palatia* was carrying many prisoners at the time and they and many of the Germans on board lost their lives.

3. 25 November 1942 No.489 RNZAF Squadron

FO Freshney took off from Wick at 1840 hrs to see what he could find off the Norwegian coast. His luck was in; off Kristiansand he sighted a 2,000 ton three – island type cargo vessel silhouetted in the moonlight about five miles distant. Lifting his Hampden from the wave tops to an attacking height of 80 feet, he launched his torpedo from 800 yards range. The rear gunner "saw the torpedo hit the vessel amidships directly below the funnel with a large flash and a shower of sparks". The crew were unfortunately unable to linger at the scene to watch the consequences of their action. It had, in fact, resulted in the *Erika Henrik Fisser* being sunk and five of its crew of 29 lost their lives.

4. 10 December 1942 No.489 RNZAF Squadron

On this occasions, a Hampden attacked a ship of about 1,500 tons. No hit was claimed on the ship which had a Norwegian flag painted on its side. Machine gun attacks were made and bullets seen to hit the superstructure. The description of the ship in the ORB report is that "It had a broad funnel amidships, a counter stern and straight bow with modern lines" A report from Norway confirmed that the *Deneb* of 937 tons had been attacked by a Hampden. The Norwegian flag was presumably a relic from the days before her seizure by the Germans when neutral shipping carried their national flags painted on the hull in an often fruitless attempt to avoid being attacked by U – boats.

5. 13 December 1942 No.144 Squadron

Three Hampdens from 144 Squadron took off at 1030 hrs from Leuchars for an offensive patrol of the south Norwegian coast. They spotted two merchant ships escorted by one naval vessel and at once went in to attack them. Two of the Hampdens, P1250 (20) and P4373 (20), were shot down by the intense and accurate flak from the ships but the third, piloted by F/Sgt J W King, although hit by flak, pressed home its attack. The torpedo hit one of the merchant ships amidships. The pilot flew his damaged aircraft back to Leuchars and landed safely. Three of the crew of P1250 became prisoners – of – war; its fourth member was never found and he is listed on the Runnymede Memorial as are all the crew of P4373. The ship was the *Theano*, a captured Dutch vessel, and it sank a few miles to the west of Kristiansand.

6. 29 January 1943 No.455 RAAF Squadron and No.489 RNZAF Squadron

Hampdens "M", "X" and "J" of 455 Squadron and "S", "W", "T" and "O" of 489 Squadron took off from Leuchars with an escort of Beaufighters. Just after 1530 hrs, they sighted a cargo ship of an estimated 6,000 tons with three islands, a raked bow and a single funnel. Six of the Hampdens claimed hits in the attack, the seventh having the frustration of not being able to get its torpedo to drop but also the satisfaction of seeing that the ship was lying over on its starboard side as he made a final circuit before rejoining the rest of the formation for the flight home. The ship, the ex – Norwegian *Kaldnes*, sank a few miles south – west of Stavanger.

7. 4 April 1943 No.489 RNZAF Squadron

Hampdens of No.489 Squadron took off at 1058 hrs for an armed reconnaissance of the south coast of Norway. Three escorted merchant vessels were sighted just to the east of Kristiansand and the Hampdens made their attack, in the course of which FO Freshney in AT259 (21) was shot down by flak. The Hampden was seen to float after it ditched but none of the crew survived; two are buried at Trondheim and two were never found. FO Latta made his attack on the largest of the merchant ships which he estimated to be about 5,000 tons and claimed his torpedo struck her just forward of the bridge. The ship was the 6,800 – ton *Altair*

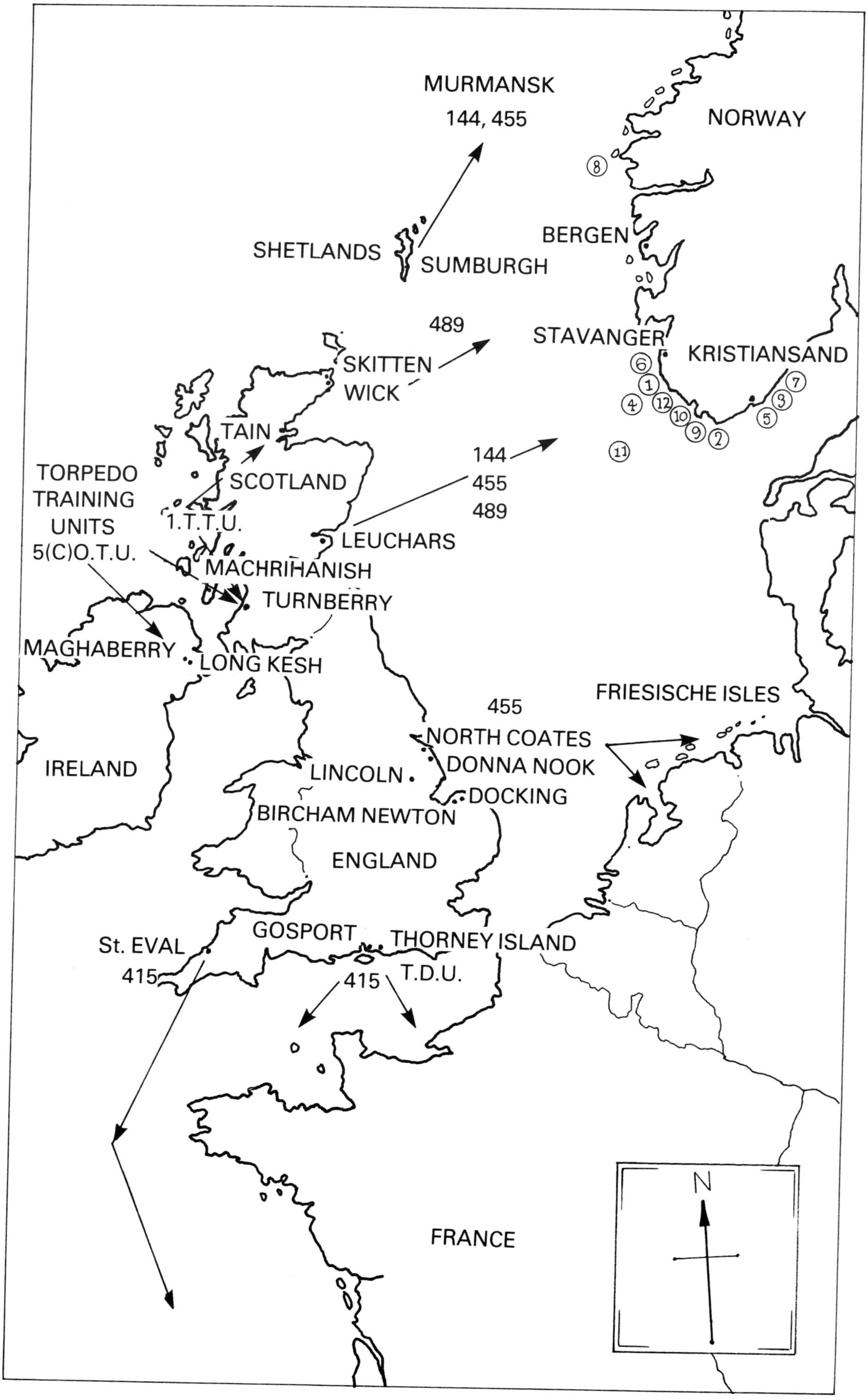

Areas of torpedo – bomber operations

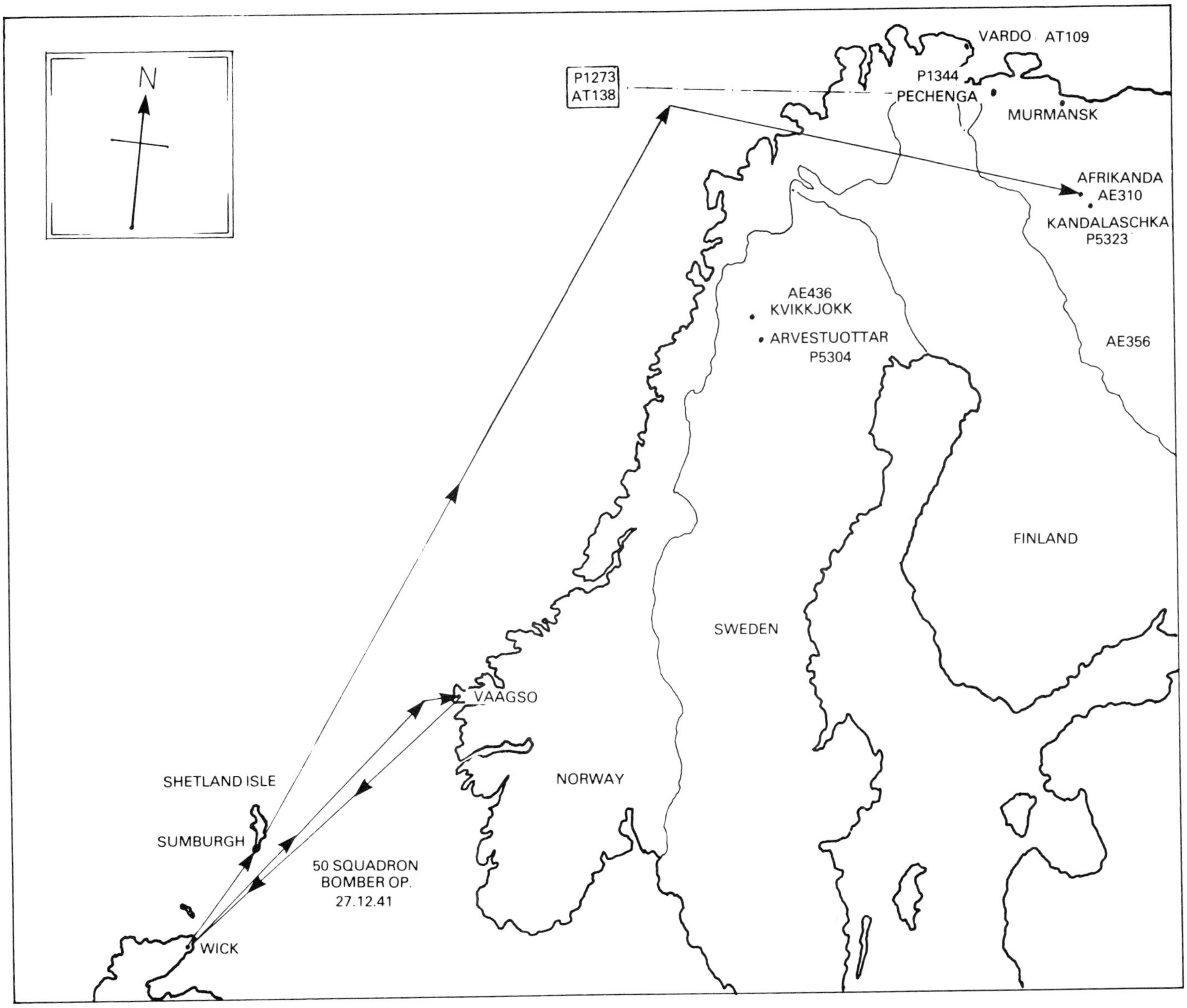

Flight to North Russia

A cartoon from 144 Squadron's news–sheet, September 1942, Vaenga, Russia.

A Hampden flies over a merchant ship which it has just attacked off Norway on 29 January 1943. (C3382)

and it sank after the attack. A German Bv 138 flying boat was also shot down. FO Latta had been the pilot of "P" on 10 December when he had unsuccessfully attacked the *Deneb*. Deneb and Altair are the names of stars used for astro-navigation but FO Latta never lived to appreciate the coincidence as his own personal star was snuffed out on 9 April 1943.

8. 9 April 1943 No.489 RNZAF Squadron

Six of 489's Hampdens took off from Wick at 1445 hrs with an escort of Beaufighters. They found three ships some 100 miles to the north of Bergen in position 6215N:0520E and commenced an attack. There was intense flak from the convoy which also had an escort which the Hampden crews identified as Ar 196s. It is almost certain that the "Ar 196"s were, in fact, the new Bf 109G-2s of 1/JG5 as two pilots from that unit, Lt Schmidt and Uffz Rodig, each claimed to have shot down a Hampden. AT258 "T" (22) in which FO Latta was flying was seen to climb steeply after the attack then dived straight into the sea and blew up. P1189 "K" (22) piloted by FO Wheeler was attacked by the fighters and was last seen heading away from the scene with an engine smoking. It must later have crashed into the sea with the loss of its crew and only the body of the Canadian wireless operator, FO L N Selthun, was found and is buried at Trondheim. The other members of P1189's crew, and all the crew of AT258, are on the Runnymede Memorial as having no known grave. A Beaufighter was also shot down.

The sadness of the returning crews at having seen two of their companion Hampdens come to grief must have been compensated to some extent by the belief that a 6,000-ton tanker had been hit and almost certainly destroyed. The crew of "P" reported a pall of black smoke which rose to over 1,000 feet and they also reported seeing FO Latta's torpedo hit a smaller vessel just before he was shot down. "Y" was the last Hampden to arrive on the scene and claimed to have seen the vessel blow up. There is, however, no confirmation from German or Norwegian sources that the tanker was actually sunk.

9. 13 May 1943 No.489 RNZAF Squadron

Four Hampdens, with an escort of Beaufighters, made an attack on a convoy of four medium-sized merchant ships with six or seven escort vessels off south-west Norway at 1505 hrs. "S" attacked a ship of about 5,000 tons which had a split superstructure and was low in the water, but did not see its torpedo run. "Y", "V" and "W" attacked a modern cargo liner with a squat funnel. Poor visibility prevented accurate observation of the results of the strike. The report from Sten Stenersen reveals that the *Klaus Howaldt* of 5,956 tons went to the bottom of the sea south of Lista.

10. 19 June 1943 No.455 RAAF Squadron

Six Hampdens from 455 Squadron took off from Leuchars at 0545 hrs for a patrol off the Norwegian coast. Attacks were made against ships of around 2,500 tons to the north of Stavanger in positions 5758N:0625E, 5849N:0530E and 5807N:0630E. In one attack F/Lt Clarke's Hampden "C" was hit by the intense fire from flak. The port main spar was damaged, the hydraulic lines severed, the starboard aileron shattered, the wireless aerial shot away and the emergency compressed air bottles exploded. The crew of "C" were obviously adept at dodging bullets as all escaped injury as did the crews of "D" and "K" which were damaged during another attack. Although there was no claim of a sinking on that day, Norwegian reports say that the submarine chaser UJ1708 *Roland* was sunk off Lista.

11. 7 July 1943 No.455 RAAF

Squadron Sqn Ldr O'Connor was on patrol just before 1900 hrs in a position about 30 miles south-west of Stavanger when he sighted two Bf 109s on his starboard beam. They were flying at a height of about 1,000 feet and converging on his Hampden. The account takes up the story;

"At 1853 both 109s attacked from directions between the starboard quarter and astern. A good burst of fire from the rear guns of the Hampden put tracers all around one aircraft which broke off its attack at once and flew away. The other 109 attacked from the Hampden's port quarter, breaking off at 100 yards. Similar attacks were twice repeated and the final attack, which was the sixth, was from the port bow. The attacks had lasted for about ten minutes until the German pilot, after apparently committing himself to another attack, seemed to change his mind and broke away and disappeared into cloud. Throughout the attacks the beam gunner gave an excellent running commentary from the astro-hatch position and kept the pilot fully informed about how the attacks were developing. The rear gunner fired 400 rounds; he claimed no hits on the second attacker but thought it probable that the first one had been damaged as it did not resume the attack after its first effort. The Hampden suffered no damage, largely due to the pilot's excellent handling of the aircraft which included violent corkscrewing manoeuvres and undulation of the height between 200 and 500 feet"

The pilot landed at Leuchars at 2055 hrs with his torpedo still on board the aircraft. German reports confirm that "A Hampden aircraft shot down Fw J Sommeregger of 1/JG5, who was flying a Bf 109". The action took place approximately 160 km south-west of Stavanger.

12. 16 September 1943 No.489 RNZAF Squadron

Three Hampdens, "H", "Z" and "D" went into action after sighting two merchant vessels with two escorts off south-west Norway in position 7807N:0632E. "H" claimed a hit on the leading merchantman which had a black hull and three islands. "D"'s torpedo missed and the result of "Z"'s attack was unobserved. Light and heavy flak was experienced from the escort and from shore batteries. A number of white cartridges were seen over the shore and were believed to have been fired from Lister airfield. Once again there was no claim in the ORB of a sinking but the *Graziella* was in fact sunk, according to the reports from Sten Stenersen.

The results of attacks on shipping by 415 RCAF Squadron are less well-known but during the months of April and May 1943, the squadron had the following ships to its credit. The locations of the attacks are evidence that 415 operated in quite a different area to the other Hampden TB squadrons at this time.

10 April 1943: The *Himalaya* of 6,240 tons, an Italian ship, damaged in the Bay of Biscay. This may have been the operation on which P1151 was lost.
14/15 April 1943: The Swedish *Tom* of 2,092 tons damaged off Borkum.
15 April 1943: The ex-Norwegian *Borga*, 4,821 tons, sunk off Ameland.
18 May 1943: *Ernst von Brieson*, 408 tons, sunk off Ameland.
Date uncertain: *M345* of 750 tons sunk off Calais.

Although other sinkings are not evident from the 415 Squadron ORB, it is obvious that there must have been some if the aircraft losses are anything to go by. No.415 Squadron flew its first Hampden TB operation on 27 April 1942. The first crew to be lost on operations was that of AT236 (23) on 29/30 May, followed less than a week later by AT240, both while attacking ships off the Friesian Isles at night. On 24/25 June, AT242 (24) failed to return from an attack on shipping between Rottum and Borkum and three nights later AT245 (24) and P1153 (24) were lost in the same area. AT235 (24) went missing on a similar operations on 1/2 July. The crew of AT241 (25) had the misfortune to have an engine set on fire as they attacked shipping in the mouth of the Gironde in the afternoon of 8 November 1942. Their colleagues watched in dismay as the blazing Hampden headed for the French shore but on this occasion the crew all became prisoners-of-war, a not very common event in a torpedo-bomber loss which usually resulted in the deaths of all concerned. Attacks off the Dutch coast claimed several of 415's Hampdens, among them AE395 (26) on 23 March 1943 and AN153 and L6055 (26), both on the night of 17/18 May 1943. AT114 (26) failed to return from a reconnaissance off Norway on 11 March 1943.

Daylight "Musketry" patrols of the Bay of Biscay, aimed at making the U-boats stay submerged, also fell to the lot of 415 Squadron and long-range German fighters may have caused the downfall of X2961 (27) on 14 June 1943, its all-Canadian crew never being found. No.415 ended its Hampden TB operations on 23 September 1943, one of its last losses being S/Ldr Peter Harris DFC, one of the flight commanders, who failed to return from a patrol off the Dutch coast on the night of 15/16 August 1943 in P1258 (28). After the Hampdens ceased operations, the squadron flew Wellingtons and Albacores, mainly against E-boats at night.

In addition to the Hampdens that failed to return, the torpedo – bomber squadrons had at least their fair share of crashes on return from operations or on training flights. The sad story of Flt Sgt McComb and his all – Canadian crew illustrates how what had, on one occasion, turned out to be something to laugh and joke about had, on another, ended in disaster. The crew returned in AT239 (29) on 1/2 July with the hydraulic system out of action. On the approach to Donna Nook an engine failed and the Hampden crashed, to become a write – off. That none of the crew were badly hurt is obvious as the same crew is together in AD762 (29) on the last night of the same month and history seemed set to repeat itself. McComb, who had been promoted to Warrant Officer by this time, returned early to Donna Nook with an engine overheating. It was just after midnight when he brought the ailing Hampden into the circuit but something went badly wrong, causing the Hampden to crash with the loss of its crew, three of whom are buried in North Coates Cemetery nearby.

On 1 November 1942, 415 Squadron was operating from St. Eval and AT193 (30) was detailed for an anti – submarine patrol. It took off at 1325 hrs and by the time it returned at 1905 hrs, it was dark and a light surface fog enshrouded the airfield. The pilot made a slight error in his approach due to the difficulty of seeing the flares and ran off the runway as he landed. In an attempt to avoid some parked Whitleys, he collided with a lorry and ended up on the fuel dump where the Hampden at once burst into flames. The wireless operator, Flt Sgt D M Coates from Ontario, made a speedy exit from the wreck but promptly returned to the blazing aircraft when he realised that the navigator was trapped and unable to escape. He helped the latter out of the aircraft and then turned his attention to the pilot who sat slumped forward in his seat motionless and possibly already dead. Displaying great heroism, he slid the pilot's canopy back and, in spite of exploding ammunition and the danger that the torpedo might be detonated by the heat, undid the pilot's safety harness and attempted to pull him from his seat. The pilot, FO J H Godfrey, was however so badly trapped in the wreckage that he was unable to be moved and the intense heat eventually forced the attempt at rescue to be abandoned. Coates staggered to safety, pausing on the way to help the injured navigator to a safer position away from the wreck. The other wireless operator, who had himself experienced difficulty in getting out of the wreck, helped him in this. Eight months later, Flt Sgt Coates was awarded the George Medal for his courageous act.

The rescue of the crew of P1157 (31) after they had gone down in the sea on 18 February 1943 is described in the Part 5B (28) but the ordeal of PO F H Mahn after he ditched AT240 (31) on 4 June 1942 seems somehow more appropriate to be included in this account of 415 Squadron. AT240 had taken off from Thorney Island at 0108 hrs for a patrol of the Friesian Isles in search of shipping. The port engine failed as they flew at 300 feet and Mahn was forced to make an emergency landing of his heavily – laden Hampden in the sea. It sank in less than a minute, taking one of the wireless operators with it. The dinghy had inflated properly and the three survivors settled down in it hopefully to await rescue; they had no food and only two quarts of water which they rationed, making the first quart last until the fifth day. For three days the weather remained calm but then it became stormy for five days; on the eighth day their hopes were raised when a Ju 88 circled them just before dusk and although a Beaufighter drove it away they were sure that help would soon be on its way. This boost to morale arrived too late to raise the spirits of the navigator, Flt Sgt Stirling, who died that evening; his body was committed to the sea.

The ninth day passed without the anticipated rescue launch arriving and the apparent hopelessness of their situation was made even more obvious when a flight of Hudsons flew past without even noticing them. On the tenth day, the wireless operator, Flt Sgt E Thomas, began drinking sea water which made him delirious; he lost consciousness and died. PO Mahn, faced with the choice of staying in the dinghy with a corpse or of putting the body over the side, understandably chose the latter course. So, after ten days Mahn was the only one left alive. He managed to catch enough rain in a piece of canvas to assuage his thirst and on the thirteenth day a seagull was foolish enough to alight on the side of the dinghy and he eagerly seized and killed it, eating as much of it as he could. After fourteen days in the dinghy, his awareness of what was happening must have been diminishing. He certainly never heard any aircraft overhead that might have spotted him but on 18 June at 1245 the sound of the engines of a rescue launch brought him to his senses. His nightmare had at last come to an end and he was soon wrapped in blankets and taken below. The remains of the seagull that probably saved his life were the only tangible evidence of his horrific ordeal in the dinghy. No trace was ever found of the other three Canadians and all are listed in the Runnymede Memorial register as having no known grave. The 415 Squadron ORB states that "PO Mahn was rescued after being adrift in a dinghy for 14 days. His condition is reported to be good". The latter part of the report appears to have been somewhat optimistic as according to one of his squadron mates, he had gangrene in both feet which subsequently had to be amputated (see 3C).

One of 415 Squadron's pilots achieved the distinction of being the only one to actually fly a single – engined Hampden. Admittedly, when he took off in AT135 (32) it did have two engines but the port engine began vibrating and fell off when they were five miles from the Isles of Scilly. The crew were picked up from their dinghy after 45 minutes.

A 144 Squadron torpedo – bomber at Leuchars in late 1942. The pilot in the cockpit is H.Woolstencroft. (Photo: H.Woolstencroft)

Another of 415's accidents, on 26 February 1943, sounds quite spectacular and this time it was the folk around Thorney Island who must have wondered what was going on. P2065 (33) caught fire as it was being refuelled in the early evening of that cold winter night and when it exploded it destroyed AE201 and damaged three other Hampdens. Very interesting – and quite amazing that there appear to have been no casualties.

Accounts from two other Canadian pilots, Bill Adams and John Enns, who flew Hampdens with 415 Squadron, give an interesting comparison of the standard of training that had been achieved in a relatively short time. Bill Adams arrived on 415 in November 1941 as a pilot, with 80 hours flying on Tiger Moths and Ansons in his log – book. For two months he learned the art of flying Beauforts and just as he was getting the hang of it the Beauforts were withdrawn and sent to the Middle East. No.415 re – equipped with Hampdens and by the time he left the squadron in July 1943, he was a Squadron Leader with a well – earned DFC and three sinkings to his credit.

Bill proved to be a great help in unravelling the mysteries of the serial numbers of the 415 Hampdens as the squadron ORB conformed to the normal Coastal Command practice of referring to aircraft by the individual aircraft letters only. This practice causes confusion to anyone researching the records because when an aircraft was written off or moved to another unit, its replacement would often be given the same letter so that even over a period of a few months, three or four aircraft might have appeared under the same individual letter.

John Enns arrived on 415 Squadron just as Bill Adams was leaving. John had 390 hours to his credit on Tiger Moths, Cessna Cranes, Ansons and Oxfords and had ended his training in Canada at 32 OTU where he trained specifically as a torpedo – bomber pilot. John was crewed – up at 32 OTU and completed torpedo training at No.1 Torpedo Training Unit at Turnberry in Ayrshire before being posted to 415 Squadron. The manoeuvrability of the Hampden impressed John and he remembered carrying out fighter affiliation exercises with a USAAF Thunderbolt squadron which found it most difficult to keep the Hampden in their sights if the pilot made a flat skidding turn. On the same theme of exercising with fighters, he also had vivid memories of going into a stabilised yaw when having fun with a Hurricane in Canada.

One interesting point about the nocturnal activities of 415's Hampdens, during John's operations on them, was the technique they used of operating only on moonlight nights so that their targets could be attacked up the moon path. He was also a bit peeved about the Lancaster dam – busters claiming that they were the first to carry out accurate low flying at night. The torpedo – bomber Hampdens in fact specialised in flying at heights of between 10 and 100 feet above the water for considerable periods of time in areas of the sea that in John's case extended from Le Havre to Heligoland.

Grant Lindeman explained a technique that 455 RAAF Squadron evolved for making a daylight torpedo attack against a major warship. Basically the plan was for two formations, each of three widely – spaced Hampdens, to make simultaneous attacks from opposite directions against the port and starboard sides of the target and then continuing flat on the water and straight across and below the warship's bows with one formation staying right down on the sea whilst the other hopped over it. Shades of "The Red Arrows"!

This tactic divided the ship's anti – aircraft fire between the six aircraft and also had the advantage that, once the torpedo had been dropped, the aircraft did not immediately turn away and present their bellies as easy targets for the ship's guns, as had previously been the practice.

Grant recalled making a practice attack on the battleship *King George V* to the west of Scapa Flow. "It was a beautiful day and she looked marvellous. What a thrill it was to fly at zero feet right under her bow with the other flight leaping over the top of you".

The general opinion of the Hampden amongst the torpedo-bomber crews seemed to be that it was indeed a good low-level strike aircraft, was a stable platform for launching an attack and had an excellent view for the pilot and navigator. It also had a better range that the Beaufort and, until ASV aerials bedecked its wings and fuselage, would usually maintain height on one engine. The main drawback of the Hampden was the narrow fuselage and consequent cramped crew positions although perhaps the slim silhouette had its advantage in providing a smaller target for enemy fighters and flak to hit.

By the end of 1943, the Hampden torpedo-bombers had all been withdrawn from service and most of them ended up at No.44 MU at Edzell in Scotland. What yarns these old warbirds would have swapped had they been able to talk as they waited to be broken up and melted down into precious ingots of aluminium with the distinct probability that these would be re-incarnated as Halifaxes or Lancasters. The serial numbers of some of these veterans were ample proof of the service they had given. At least three of them had been flying on squadron service in February 1939, L4075 with 50 Squadron, L4091 with 44 Squadron and L4205 with 61 Squadron while L4105 and P2080 (34) could boast of their participation in the "Thousand-Bomber Raid" on Cologne when they were with 16 OTU. There were even two Hampdens that began life as Herefords L6018 and L6090.

The only lasting memorials to the torpedo-bombers are the headstones in churchyards near Thorney Island, North Coates, Leuchars and other peaceful seaside locations where the bodies of many of these young men who had come to Britain from Canada, Australia and New Zealand were laid to rest. The holiday visitors who pass through the quiet churchyards have no means of knowing how these airmen came to be buried there, only that like so many more of their generation they were "killed in the war". Some idea of the perils that faced TB aircrew are obvious from the incidents described in this chapter but a 1941 review of the length of an operational tour for various flying duties provides statistical evidence. At that time the tour for a bomber crew was 200 hours and, for a crew to have an even chance of surviving a tour, the figure would need to be reduced to 160 hours. The comparable figure for a torpedo-bomber crew was 80 hours while at the other end of the scale it would be 2,000 hours on a Catalina.

To an impartial observer, the award of "gongs" to these valiant crews, who pressed home their attacks in the face of intense flak from the gunners on the ships, appears not to have been commensurate with the dangers that they faced and the results they achieved.

On 30 April 1943, Flt Sgt J S Freeth of 455 RAAF Squadron made depth charge attacks on the German U-boat *U-227* and sank it off the Faroe Islands, north-west of the Shetlands. In the late afternoon of 24 May, he and his crew in Hampden AT132 (38) were involved in a mid-air collision with
Beaufighter JL502 of 235 Squadron when practising fighter-escorted torpedo attacks five miles off Fraserburgh. Both aircraft crashed in the sea with the loss of their crews. The name of John Samuel Freeth, from Cogee in New South Wales, appears in the Runnymede Memorial, an award for the sinking of the U-boat being conspicuous by its absence. The fact that this was not an isolated instance in which an award has been overlooked may be seen in the entries of the names of "Missing" Hampden TB crews on the Runnymede Memorial. The letters DFC or DFM do not appear after any one of them, although there is proof enough in the Squadron record books that such awards were well merited.

The Hampden torpedo-bombers had their fair share of barrack room bards who composed ditties that were sung with great gusto at the various Mess functions, particularly on "Stag" nights. A few verses from one such effort from 415 RCAF Squadron provide an appropriate note on which to end this section. This particular dirge was sung to the tune of "We were only playing leapfrog" and the words of the chorus were:

"We were only flying Hampdens" (three times) ..Just to pass the time away"

Three sample verses are typical of this bit of Hampden folklore:

"We travel round the country every day and every week
From Leuchars down to Land's End via Hangman's Bloody Creek
We fly our Bloody Hampdens till they're old with Bloody age
Then order up some new ones from Handley Bloody Page"
(Chorus)
"We're not the kind to brag but we'll tell you what we've done
We love the Bloody bottle and we hate the Bloody Hun
We carry a torpedo through rain and hail and sleet
And we fly our Bloody Hampdens at a hundred Bloody feet"
(Chorus)
"Down in the Bay of Biscay, Lawrence did a Bloody trip
He stooged around for hours but he saw no Bloody ship
He lost one Bloody engine but he set her down just fine
And he broke his Bloody shoulder shooting a filthy Bloody line"

A torpedo-bomber of No.415 RCAF Squadron. Note the open hatch to the rear of the open astro-hatch; also faired-in D/F loop. (Photo: Supply & Services Canada PMR 82-11)

AE116 of 50 Squadron crashlanded in the Waddington dispersal field, having run out of fuel when attempting to land in fog. The aircraft was repaired.
(Photo: Ray Leach/Eric McCabe)

2K CRASHES ON RETURN FROM OPERATIONS

Of the 714 Hampdens lost on operations 451 failed to return, the remaining 263 were lost due to other causes. 33 of the 263 crashed on take-off or soon afterwards. 30 went down on the sea on the return flight and many of these are dealt with in the "In the sea" section. 23 were abandoned on return and a number of accounts of these are in the "Escapes by parachute" section. The total remaining, after the above causes have been taken into consideration, is 177 and this is the number of Hampdens that were written off in crashes on land in which the crews had not baled out. It has to be emphasised that many more Hampdens did crash on return from operations but were subsequently repaired. In simple terms it could be said that for every two Hampdens that failed to return from operations another one was written off on return and others damaged. Anyone who is not familiar with the background to these operations might be excused for considering the crews to be a pretty clueless lot, so a little information about the conditions that prevailed at that time might be appropriate.

The crews who began the Hampden night operations in January 1940 had little, if any, experience of flying at night in Hampdens and had to develop this skill on leaflet raids and night reconnaissance flights, including the "Security Patrols" of the bases from which the enemy He 115 mine-laying sea-planes operated. Night bombing began with the not very successful raid on Sylt on 19/20 March 1940; the minelaying campaign commenced on 13/14 April 1940 after which the night bombing operations gradually extended into Germany until by the end of August 1940 Hampdens were bombing targets as far away as Berlin. By mid-1940 most of the crews arriving on the squadrons had at least undergone training at one of the Hampden Operational Training Units which had been formed in April 1940; these were 14 OTU at Cottesmore and 16 OTU at Upper Heyford. A brief description of the night flying facilities and communications set up of the early months may help towards an understanding of the problems involved. It has to be appreciated that the general organisation on the ground, often in itself inadequate, was also like the aircrews, lacking in experience.

Some of the Officers in Charge of the Watch Offices (later known as Control Towers) were ex-1914/18 pilots who were totally unfamiliar with the requirements of returning aircraft in regard to flare path lights. There can be little doubt that the physical and mental reactions of the pilot in particular were slower, after a flight that might have lasted over ten hours, than they were when the aircraft had taken off. The aircraft might also not be in such good shape after suffering the attention of the enemy, which could have resulted in damage to fuel or hydraulic systems, or to the control surfaces of rudders ailerons or flaps. Nor did an unsuspected punctured tyre make a landing any easier.

The causes of the majority of the losses on return were adverse weather, being lost, shortage of fuel, or battle damage. Often these problems arrived in combination; sometimes they were contributed to by optimistic weather forecasts and over-zealous officers at No 5 Group. On many occasions the Hampdens were sent out in quite atrocious weather with the assurance from the meteorological forecaster that "conditions will be better over the target area, and will have improved by the time you return". These time-honoured phrases continued to be heard at briefings, and to be received with the same derisory acclaim, throughout the war. The nocturnal operations of the Luftwaffe over Britain has also to be taken into account when considering the "Crashed on Returns".

The main German bomber force was engaged in attacks on the ports and industrial areas of Britain from mid-1940 onwards, and returning RAF bombers were diverted to other aerodromes if it was considered unwise to light the flarepath at their own base because of enemy activity. By far the greatest threat to the RAF bombers and aerodromes from the Luftwaffe were the intruder aircraft, mainly Ju 88s of I/NJG2, a unit which had been formed specifically to attack the RAF bombers and their bases, and operated from July 1940 until October 1941 (see 5H).

The grass surface of the aerodromes at Hemswell and Scampton had no fixed installations of flare-path lighting although a double row of lights had been installed across the grass at Waddington for the "Lorenz" blind landing system, a system which had not achieved the anticipated success

The circumstances in which AE157 of 50 Squadron was written off on 3 September 1941 were shrouded in mystery until photographs taken by Eric McCabe, who was Engineer Officer at Waddington in 1941, came to light via Ray Leach, the Waddington and 50 Squadron historian. The movement card for AE157 confirms that it was Category E FB, which means it was written off on an operational flight on 3 September 1941. But according to 50 Squadron's Operations Record Book, it was not one of the fourteen Hampdens operating from Swinderby that night, of which five crashed on return due to fog. The ORB does reveal that AD852 had crashed into a dispersed aircraft at Waddington at 0320 but the photographs show that it was AE157 so the likely answer to the mystery would seem to be that AE157 had been reserve aircraft that night and had been operating because of unserviceability problems with AD852. The identity of the "KM"-coded 44 Squadron Hampden is not apparent from the squadron ORB nor the movement card. No Squadron Hampden is unaccounted for on 2/3 September 1941. The ORB reveals that AE254, AE152 and AE313 failed to return; AD913 was abandoned near Dorking, AD982 returned with two crew wounded after an attack by a night fighter while AD767 crashed at North Luffenham and AE192 at Coleby, both without serious injury to the crew and with repairable damage to the aircraft. Small wonder that a minor incident such as an aircraft being written off as it sat at its dispersal failed to get a mention. The only real possibility for the serial of the 44 Squadron aircraft is that it is X3025, the movement card write-off date for which is 8 September 1941. X3025 had been regularly flying on operations in July and August 1941 and last appeared in the 44 Squadron ORB on 31 August 1941. The photographs indicate that AE157 was still travelling at a fair speed when its starboard wing hit X3025's port wing at right angles to its fuselage and then demolished both its engines and the nose. In the process, its own starboard wing was ripped off, its fuselage broken aft of the lower gun position and the tail unit badly damaged. The evidence that X3025 had been peacefully picketed at dispersal can be seen in the form of the securing strop under its port wing. (Photos: Ray Leach/Eric McCabe)

mainly due to the cumbersome Morse procedure involved, but the lights occasionally came in useful. Otherwise the only lights of a permanent nature on any of the Hampden bases were the red obstruction lights on the top of hangers, wireless masts, water towers and other obstructions.

Each station had a fixed flashing beacon, sited on top of the water tower, which flashed the "peacetime characteristic" of the 'drome concerned in morse code, in red. Waddington's, for example, flashed "W". These lights were operated from the Watch Office.

The ritual of laying out a flare path took place towards dusk at aerodromes throughout the country. It began with the flare path lorry loaded with equipment and towing the "Chance" light trailer trundling out onto the grass and making its way across the aerodrome to the upwind end of the proposed flarepath, where the "Chance" light was unhitched. The "Chance" light was a powerful floodlight, powered by its own generator. It was switched on as required to illuminate the flare path for aircraft coming in to land; until the advent of aerodrome control caravans, it was the focal point of the night's activity. The flare path itself was laid out in the form of a "T" with the Chance light and "Angle of Glide" indicator at its base, the into-wind (take-off) end. The lights used on the flare path were of three types: the "Glim" lamp powered by a 2 volt accumulator, and the "Gooseneck" and "Money" flares which burned paraffin.

White "Glim" lamps, spaced 60 to 80 yards apart, formed the backbone of the flare path; an average flare path would be 600-800 yards long. Gooseneck flares were placed alongside at least every alternate Glim lamp and Money flares at four or five positions along the flare path. All three types of light were operated by hand. The Glim lamp, turned on at the flick of a switch, was the easiest to use and was adequate for aircraft taking off and for landing when conditions were reasonable.

The Gooseneck flare had the appearance of a flattened oval shaped watering can from the spout of which a thick wick protruded. The goosenecks were lit when visibility was poor; Money flares came into use if conditions such as fog materialised.

The Money flare comprised a large metal bucket with a removable flanged metal lid. The bucket contained a large wire framed wad of absorbent non-combustible material, and paraffin in which the wad was immersed. When the Money flare was required the lid was taken off the bucket and placed flange upwards on the grass. The wad was then placed on the lid and lit by means of a burning gooseneck. The Money flare's supply of paraffin was topped up from time to time by pouring more fuel into the lid. It was quite a common occurrence for the flare path party to have to extinguish the lights in a hurry. The Glim lamps went out at the touch of a switch, the goosenecks were doused by pushing the wick back down the spout; a blazing Money flare needed strong arms and resolute attention. A hooked metal bar had to be inserted into a wire loop on the top of the flare, and the flare, burning more fiercely than ever as it was lifted at arm's length, was dropped back into the bucket. The bucket lid, still too hot to be handled with bare hands, had then to be placed back on top of the bucket before the flare was finally extinguished.

Mobile flashing beacons known as "Pundits" and the more powerful white lights of the "Occult" navigational beacons were sited around the countryside. Their positions and characteristics were changed regularly to make it more difficult for the Luftwaffe to make use of them; this also made them equally useless to a navigator in a RAF bomber if he mislaid his list of their positions.

During night flying operations the aerodrome control pilot controlled take-offs and landings from his position by the Chance light; a field telephone linked him with the Watch Office. An aircraft wishing to take off or land signalled his aircraft letter to the aerodrome control pilot; if permission was granted his letter would be flashed back in green on an Aldis lamp, if refused the letter came back in red. A red signal cartridge, fired from a Very pistol, was the most emphatic refusal of permission to land; it was also a signal when fired from an aircraft that a pilot had problems and needed to land as soon as possible. Signal rockets or a signal mortar that fired a 3-inch white signal flare up to a height of 1000 feet could be used from the Watch Office to attract wayward aircraft back to base; not unnaturally there was a marked reluctance to poop these off if the enemy was around.

Other lights that the returning bombers might see, in addition to the fires started by the Luftwaffe, were the "Q" site dummy flare paths and the "Starfish" sites on which fires burned convincingly in the hope of attracting the bombs of the enemy.

Many lessons were learned from the experiences of the first few months of the RAF bomber effort. Flare path lighting was improved until the "Drem" system, with its leading-in funnel lights and ring of lights around the airfield became standard. By the end of 1942 the grass surface of the aerodromes was being replaced by concrete runways and perimeter tracks with a built-in lighting system that could be controlled from a console in

AD983 of 44 Squadron crashed into the Girls' High School. Lindum Hill, Lincoln, when returning from minelaying on 22/23 July 1941. The four crew and a lady from the school lost their lives. (Photo: Lincolnshire Echo)

the control tower at the flick of a few switches after the appropriate runway had been selected. Army searchlights were incorporated into a "pointer" organisation in which they swung their beams in the direction in which an aircraft should fly to reach safety. The searchlights near the bomber bases trained their lights to form a signpost in the form of a cone, the apex of which was over the aerodrome.

Fog was the main weather hazard that faced the returning bombers, often causing diversions to other aerodromes because Lincolnshire was affected by fog. There were occasions when every airfield within range was blotted out by fog; one such night was 2/3 September 1941 when 51 Hampdens were operating against Berlin and other targets. Five of them were lost over enemy territory and because of dense fog on their return eight more were written off in crash landings and a further five seriously damaged. The attacks on Berlin were really beyond the safe range of the Hampden and allowed no time for a returning aircraft to search around for a landing place. In fact, as can be seen in Part 5B, a number of them went down in the sea on their return flight because they had insufficient fuel to even reach the English coast.

General bad weather conditions, with low cloud covering high ground, also presented its problems, particularly if the aircraft was out of wireless contact; a fearsome dilemma faced the crew of an aircraft under such circumstances. It was quite likely that they would not have been able to fix their position with any accuracy since they took off and static electricity or lightning may have given them cause to doubt the accuracy of their compasses. The wireless operator was usually the saviour on these occasions, but if for any reason he was unable to obtain the vital bearings or courses to steer for base, then the navigator would only be able to make an inspired guess as to their whereabouts. After wandering around the night sky for seven or eight hours under such circumstances without seeing a sign of the earth beneath them apart for an unwelcome anti-aircraft shell burst as they flew above dense cloud over enemy territory, the navigator could not be sure of their position to within fifty miles. As the needles on the fuel gauges flickered towards the "Empty" mark the pilot had to make a decision. Should he order his crew to bale out before the engines failed? Or should he descend slowly and attempt a crash-landing, probably with only the light from the aircraft's landing light to warn him, at the very last moment, of what lay beneath him?

The pilot would probably offer the crew the choice of baling out or staying with the aircraft if he decided to try for the crash landing option, but while there was no certainty that they were over land the grim prospect that if they baled out they might land in the sea would persuade the crew that the better choice would be to remain in the aircraft. Some examples of the problems that faced aircraft as they returned from night operations may be seen in the accounts in the Squadron Operations Record Books:

L4153 (1) of 44 Squadron took off from Waddington at 1825 hours on 26 February 1940 for a Security Patrol of the Norderney-Borkum-Sylt area. It returned safely but at 0250 encountered the balloon barrage at Grimsby. With low cloud and mist about and being almost out of fuel the pilot crash-landed three miles south of Grimsby. The crew were unhurt, but this good fortune did not stay with them, as they had all, with the exception of the pilot, been killed by the Autumn in different aircraft.

L4111 (2) of 61 Squadron took off from Hemswell at 2018 for a Security Patrol which was apparently carried out successfully. "Owing to adverse weather conditions and difficulty in contacting the aircraft by W/T it did not return to base. It eventually located Digby airfield at 0530 and crashed nearby after a flight of 9 hours 15 minutes. Flying Officer Clinkard had made right hand circuits at Digby and appeared incapable of turning the aircraft to the left. Examination of the scant remains of the aircraft indicate that the wireless operator LAC Wood had been severely wounded by enemy action; several bullets had passed right through him."

The Station Signals Officer reported that "It was an amateur and not a trained operator working the set ... FO Clinkard and his crew made a determined and almost successful attempt to get down, flying for three hours up and down Yorkshire and Lincolnshire. Enemy action, coupled with mist over the whole of England, proved too much for them but the full extent of their difficulties can only be surmised. Was the wireless operator revived sufficiently to make a stumbling and faulty effort to get in touch with base? Did the second pilot Sergeant Hobbs try to work the set, the wireless operator being already dead?" (It is at least as likely that the observer/rear gunner, Sergeant Glasson tried to use the wireless.) "How many of the crew were injured before the crash? How controllable was the aircraft? The answers to these questions will never be known and consequently no recommendations for posthumous decorations were made". The account in the ORB was written by Pilot Officer Pascoe who was later "Missing" in P4341 (3). FO Clinkard, who was a New Zealander from Wanganui, had only recently married a local girl, and he was buried in the churchyard near Hemswell at 1130 on 12 March 1940, by a sad twist of fate at almost exactly the same time the 61 Squadron Anson crashed 7 miles from Brize Norton whilst on a training flight with the loss of the five men on board.

The crash of L4063 (4) of 50 Squadron on 16/17 March 1940 as it returned from a Security Patrol of the Heligoland Bight was the direct result of a fault in the direction finding equipment at its base at Waddington. "Owing to a failure in the sense of the goniometer at base

X2997 of 44 Squadron, 16/17 October 1940. Thick cloud at Waddington on the return of the aircraft at 0130 caused the pilot to attempt a landing using Lorenz "ZZ" system. The aircraft crashed on to the top of an old wooden-roofed hangar, bounced off and landed on its belly close to the petrol dump which is visible in the background. P4414 struck the same hangar two hours later; it was repaired but X2997 was written off. There were no fatalities.
(Photo: Harry Moyle)

DF station the aircraft was given incorrect QDMs (Courses to steer) when trying to reach base and crashed into the hills at Jedburgh (over 100 miles from Waddington). The aircraft had been flying at 13,000 feet and when the pilot was thought to be only a few miles from Waddington he was instructed to descend to 2,000 feet. The hills in the region of Jedburgh are 3,000 feet high." The Hampden had taken off from Waddington at 1915 and crashed at Cocklaw Foot at 0300 hours. All four crew were killed. An unusual aspect of this crash is that although it happened on land, two of the crew are listed as being missing and their names appear on the Runnymede Register.

L4043 (5) of 49 Squadron returned from a minelaying operation to the West Baltic on 14/15 April 1940; its pilot Squadron Leader Lowe DFC reported: "We were lost but eventually found an aerodrome beacon flashing red letter "D". I circled around the beacon flashing the letters "SOS" but got no reply. I then fired a red Very light and searchlights came on to the north of us, about 15 miles away and I headed towards them. One of the engines stopped and I told the crew to prepare for a forced landing when the good engine began to come on in jerks. I asked if any of the crew wished to bale out, but they all agreed to remain. I made my forced landing approach up the coast, intending to land on the western edge of the beach. At 500 feet I put my landing light on, my wheels and flaps were up. As I was holding off to land, the aircraft suddenly swung violently towards the cliffs and the rocks below them. I managed to correct this slightly but not enough to clear a bunch of rocks running into the sea. When I was holding off, and practically stalled, I pulled the nose up and landed slightly on a tail slide, with very little forward speed. The crew had been instructed to take up crash positions amidships but owing to Pilot Officer Bryan Smith, the lower gunner, being only half-way through the rear door at the moment of impact, he was killed instantly. The rest of us were only slightly injured."

The crash was at Ryhope, south of Sunderland. The mystery beacon flashing red "D" could have been Dishforth or Driffield. In this case and in that of Flying Officer Clinkard at Digby there appears to have been a shameful lack of interest from those who were safe on the ground to the predicament of their comrades in distress in the sky above them.

Other reports of crashes on return are typical of so many that occurred later, with just the bare details of what had happened because no one survived to tell the full story. L4119 (6) of 61 Squadron on the night of 30 April/1 May 1940: "After carrying out a bombing attack on Aalborg in Denmark L4119 was contacted by wireless,... a plot from its IFF showed the aircraft to be searching for a gap in low cloud. At 0440 it was reported that it had crashed at Croxton Kerrial, Leics, where the cloud was down to the ground. The crew were all killed in the crash which occurred after L4119 had been in the air for over nine hours."

The loss of L4069 (7) of 83 Squadron on 14/15 May 1940 came about in most mysterious circumstances. The aircraft had taken off from Scampton at 2229 hours and was back over Lincolnshire at 0210, less that four hours later, so it appears to have been returning with some sort of problem, possibly engine trouble. "At 0210 L4069 was challenged by an anti-aircraft battery. The correct answer was given after which the aircraft suddenly changed course, made a sharp circle, gained speed and dived into the ground and blew up near Louth, Lincs." The crew of L4069 were all killed. A word of explanation about the "challenge" from the anti-aircraft battery. There was a code of recognition signals, which were made in the form of single letters flashed in morse; it was indeed a sort of aerial password. In answer to the code challenge letter the aircraft replied with the correct reply letter. All very simple.

L4066 (8) of 83 Squadron returned from Frankfurt on 7/8 July 1940 and crashed at Clacton, Essex, with the loss of its crew. P4376 (9) of 83 Squadron took off for operations from Scampton at 2145 hours on 2 August 1940. It returned less than two hours later with an engine out of action and crashed 4 miles east of Scampton at 2340. The pilot and wireless operator were injured, the navigator and lower gunner lost their lives in the crash.

The crash of L4055 (10) of 83 Squadron on 22/23 May 1940 with the loss of the crew, came about when they "returned from bombing operation and crashed into a hill at Holmfirth near Huddersfield in Yorkshire at 0400 hours, in fog."

The problem that high ground presented to returning aircraft, particularly in conditions of low cloud or fog, is apparent from the following fatal crashes; the location of some of the incidents also underlines the navigational failings of the Hampden era. X2920 (11) crashed on the Yorkshire moors near Leeming, L4178 near Warmsworth in Yorkshire. X3124 (12) piloted by Wing Commander W W Stainthorpe AFC who, in pre-war days had been an instructor at the Central Flying School, crashed at Derrington Cross near Stafford. Dartmoor claimed its victims too, amongst them AD748 (13) and according to one report X3054, although the 49 Squadron records say that it crashed at Moreton-in-Marsh. AD730 (14) had overflown fog-bound England and reached Southern Ireland on its return flight from Berlin when it crashed into the high hills near Blessington in County Wicklow. In most of these cases of aircraft crashing into high ground the need to find somewhere to land had been made a matter of urgency due to fuel running low.

Bad weather in the form of heavy rain, strong winds and thunderstorms were regularly encountered by the crews; the winters of the first three war years had more than their fair share of severe conditions which in early 1942 in particular were quite Arctic with heavy snowfalls. On 16/17 October 1940 X2979 (15) had been airborne for nearly nine hours when it crashed in bad weather whilst attempting to force-land near Swaffham in Norfolk. Shortage of fuel must have been the initial cause but, had the weather been more favourable, the pilot may have made a successful crash-landing and he and his crew may have survived instead of all being killed. Bad weather was definitely the cause of four Hampdens being written off on 1/2 March when they returned from bombing operations.

Out of eight Hemswell aircraft detailed to attack Cologne three came to grief, AD723 (16) at Caistor, Lincs, X3147 (16) near Bircham Newton and P1253 (16) near Stradishall. The only survivors from the three crashes were the seriously injured pilot and injured lower gunner from P1253.

X2984 (16) of 50 Squadron crashed near Driffield on the same night in the same bad weather; in this instance the Hampden was lost, running out of fuel and flew into high ground as it tried to find somewhere to land. In a number of crashes on return the cause was difficult to ascertain, because none of the crew survived and there had been no report from the aircraft that it had a problem before it crashed. AD785 (17) for instance "exploded in the air and crashed near the Whitby to Scarborough road". X2917 (18) "Broke up in the air in bad weather and crashed in a wood near Marham in Suffolk". P5392 (19) hurtled out of the night sky and crashed with such violence that identification of the individual crew members was well nigh impossible. AD983 (20) "crashed into the Lindum Girl's High School at Lincoln and a civilian, Miss Fowler, who lived at the school died of shock".

There are other crashes in which, although none of the crew survived, the aircraft had reported by wireless that it had problems. L4185 (21) had sent out distress calls as it flew homewards, it eventually crashed in Half Acre Creek in the Thames estuary. P4405 (21) sent out an SOS

VN-A, L4062 of 50 Squadron, 26/27 September 1940. The aircraft hit trees on approach to Lindholme in poor visibility with no serious injuries to the crew. (Photo: George Leonard)

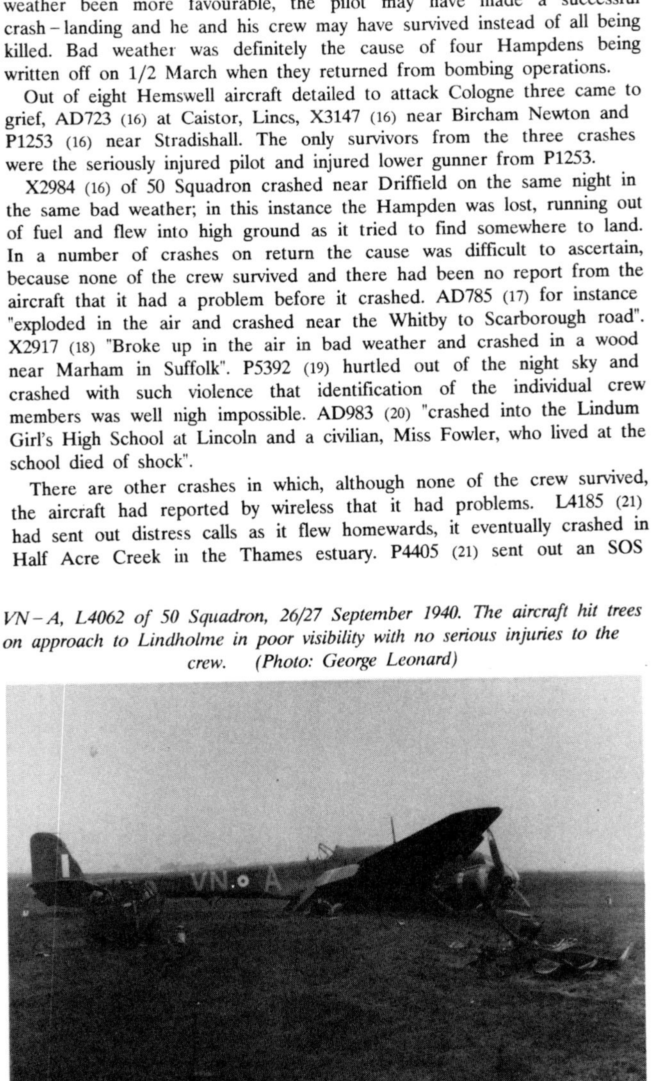

when 60 miles out to sea on its way back from Wilhelmshaven, it crashed near Oulton Park in Norfolk. AD783 (22) came back early after developing engine trouble and crashed near the village of Eagle whilst in the circuit of Wigsley its home aerodrome. Many of the other Hampdens that crashed on return were attributed to "pilot error" in undershoots, overshoots and undercarriage collapses due to a particularly heavy landing. Fortunately in most of these cases the crews escaped with nothing more that a severe shaking up and the aircraft were subsequently repaired.

The drama of a crash caused by a mechanical defect is typified in the report about X2998 (23) of 144 Squadron which took off from Hemswell at 1715 hours and returned just after midnight on 26 October 1940. "The aircraft crashed at 0010 hours at Upton near Hemswell, due to the throttle levers jamming. The pilot, Sergeant R J Curtis, was thrown thirty yards from the aircraft when it crashed. He returned to drag the unconscious wireless operator, Sergeant E H Woods, and the navigator, Sergeant A Goulder, from the blazing wreck and undoubtedly saved their lives". A local farmer, Ernie Till, later received a letter of commendation from the Commanding Officer of Hemswell for his assistance with the rescue of the crew.

The personal accounts of Hampden aircrew who survived a crash on return amplify the details of such an experience. On 7/8 June 1940 Pilot Officer Peter Tunstall was a pilot flying as navigator to Pilot Officer Alec Webster in P4349 (24). The 61 Squadron crew, after having made one practice night cross country flight together, had set out on their first operational flight to bomb Hannover; it was on the return flight that their troubles began. Peter estimated that they ought to have been back over England by 0300 hours but by that time no sign of a coastline had materialised. As the darkness gave way to the greyness of dawn it was not sea or land that appeared beneath them but an endless milky white blanket of fog; this aviator's nightmare enshrouded everything. The wireless operator was having problems with his set, so no help was forthcoming from that direction; their only hope lay in stooging around in the hope that the fog would lift before their fuel ran out. The possibility that they might have to bale out or attempt a blind crash landing in the fog were equally unattractive alternatives, particularly as they were unsure whether they were over land or sea. The anxious eyes of all on board stared down at the murk beneath them. The fuel gauges were already flickering near the "Empty" mark when Alec and Peter both spotted a darker patch in the whiteness, as they flew over it they saw a vague pattern of ploughed furrows. Alec needed no prompting to keep this link with Mother Earth in sight and warned the crew to take up their crash positions as he circled to make his approach. Peter moved up from the nose of the Hampden and sat on the main wing spar behind the pilot. From this vantage point he watched Alec's every move with a critical eye. The nose of the aircraft dipped towards the hole in the fog and they glided down towards the furrows. Peter noticed approvingly that Alec switched off the fuel, ignition and electrical master switch as they neared the ground. Their arrival back on earth was a short, chaotic, bucking affair accompanied by rending and grinding noises and amidst flying clods of earth and bits of the Hampden which ended in a drunken sideways slither. The peace and quiet of the sudden silence that followed seemed almost deafening. Only a faint crackling noise of the hot exhaust collector rings around the engines as they cooled disturbed the silence. None of the crew, who were all uninjured, required this reminder that the aircraft could still catch fire; they did not remain in the Hampden to contemplate their good fortune but departed as though from a ticking time bomb. Their arrival had not gone un-noticed, the local soldiery were soon on the scene to guard the wreck whilst the crew were whisked away to the local stately home where they were entertained as though they were Royalty until the following afternoon when transport arrived to take them back to the reality of Hemswell and the war.

Clear memories of another foggy flirtation with fate came from George Leonard who was flying in 50 Squadron L4062 (25) on 26/27 September 1940. George was a Pilot Officer direct entry commissioned air gunner who had previously served in the Army; he had been posted to 50 Squadron at Lindholme on 17 August 1940 after completing three months training at No.1 Air Armament School, Manby. On 26 September, George occupied the lower gunner's position in L4062; his pilot with whom he had already made six operational flights was 21-year-old Pilot Officer Micky Mulligan, described by George as "a lovely New Zealander from Wellington". The other two crew members were navigator Pilot Officer Kilner and wireless operator, 19-year-old Welshman Sergeant Cyril J Smith from Ystrad, Glamorgan. They took off from Lindholme at 1923 hours to bomb shipping at Calais and made their attack successfully, being fired at by heavy flak as they did so. After 4 hours 45 minutes flying they arrived back over Lindholme to find it covered in a shroud of mist some 30 feet thick. To quote George Leonard, "The met forecast was the usual joke". As they flew directly over the aerodrome it was possible to see the flare path flares and Micky decided to have a go at landing. He made an approach to where he thought the flare path would be, but unfortunately as he dipped down into the mist he hit some trees on the aerodrome boundary and L4062 ended up as a mangled wreck on the 'drome. The pilot and navigator were injured and George, who against all the rules was still sitting in the lower gun position when the Hampden crashed, suffered a bruised backside as the aircraft hurtled along the ground on its belly before coming to rest. George did another op with Micky Mulligan in October of 1940 and was then crewed with another New Zealander, Geoff Weston, for the rest of his Hampden tour, which was a stroke of good luck as Micky Mulligan and his crew, which included his Welsh wireless operator from L4062, failed to return from a raid on Dusseldorf in X3004 on 7/8 December 1940 and were presumably lost over the North Sea as no trace was found of aircraft or crew.

On 27/28 February 1942 Flight Sergeant Doug L Barrie, the wireless operator in AE218 (26), was no doubt congratulating himself on having obtained the vital QDMs that had been instrumental in bringing the aircraft back to its base after they had become lost on the way back from Kiel. Doug's self satisfaction changed to dismay when an engine began misfiring through fuel shortage and finally cut out as the aircraft made its final approach for landing. The Hampden stalled and went into a flat spin from which the pilot was unable to recover, and they crashed onto a field near the aerodrome. The ambulance was soon on the scene to collect the crew who were all concussed and in need of hospital treatment for assorted cuts, sprains and minor injuries. Doug Barrie wryly recalled how, even at four o'clock in the morning the smell of alcohol on the Medical Officer's breath had been as good as an anaesthetic. The Officers' Mess at Skellingthorpe obviously held some good parties.

A closer study of the squadron record books reveals how the fortunes of individual crew members changed as they moved from one crew to another. Sergeant Oakley of 61 Squadron piloted P4398 (27) on 29/30 July 1940 with an observer, Sergeant A A Horn as his navigator. On their return flight from a Baltic minelaying op they encountered a severe storm and, owing to a 30 degree error in the pilot's compass, made a landfall over the north coast of France. Turning northwards they arrived over the English coast and with petrol almost exhausted attempted to land at Shoreham airfield at 0620 hours by which time it was daylight. Both engines failed on the final approach and Sergeant Oakley retracted the undercarriage and made a successful crash landing on the side of a hill. None of the crew were injured and P4398 was subsequently repaired.

Sergeant Oakley's encounter with three Ju 88s on 23 November 1940 is described in the "Intruder" section (see 5H11); on this occasion Sergeants Durtnall and Clelland were manning the rear gun positions. On 2/3 January 1941 Sergeants Horn, Durtnall and Clelland were Flight Lieutenant Powdrell's crew in X3126 (28). Fuel shortage at the very last moment again proved to be a problem. Both engines failed as they came in to land at Hemswell at 0736 hours and they crashed 500 yards behind the Officers' Mess. The pilot and Sergeant Horn were injured, the two in the rear escaped unhurt. On 5/6 July 1941, now Flight Sergeant Horn and flying with a different crew, he failed to return in AD806 (29), the crew of which are all listed in the Runnymede Memorial Register. The luck of wireless operator air gunner R R B Durtnall, also now a Flight Sergeant, ran out on the night of 30/31 July 1941 when he flew in AE266 (30). The pilot of AE266 was making his first operational flight and crashed on return as he made his approach to land at Upwood. Only the navigator, Flight Sergeant Scott, survived the crash. Efforts to follow individual histories are often frustrated by poorly kept Squadron Record Books and confusion is increased when there was more than one sergeant of the same name on a Squadron. On 50 Squadron, for instance, in the latter half of 1940 the following "Smiths" were occupants of one of the rear positions in a write-off. Smith C J in L4062, Smith A H in L4078, Smith E S in X2896 and Smith G G in X3141.

Communications

Communication between aircraft and base was by Morse code wireless transmissions, the equipment in use on the Hampden being the Type T1083 transmitter and the Type R1082 receiver. Both receiver and transmitter had to be tuned in manually and errors in frequency were common, particularly amongst inexperienced operators, with the result that there could be a total lack of communication. The introduction of the TR9 VHF radio greatly improved matters, even though its range was only 20 miles under normal circumstances. The TR9 enabled the pilot to talk to base and, when the "Darky" distress organisation came into being, provided the vital link through which many aircraft were guided to safety. Wireless operators maintained radio silence on the outward flight until they sent a brief "Target attacked" signal on completion of their task; often, owing to the distance involved, the signal failed to reach base. When the navigator believed that they were within 90 miles of the English coast the aircraft's identification signal was transmitted and the IFF transmitter switched on. These initials stood for Identification Friend or Foe, the signal enabling the British coastal radar to identify a returning aircraft as friendly and most aircraft had been fitted with IFF by early 1940.

It is worthy of note that the RAF bomber crews were completely unaware of the existence of British radar, the conspicuous aerials at places such as Stenigot on the Lincolnshire Wolds being referred to as RDF stations i.e. Radio Direction Finding stations, which sounded a reasonable enough explanation. British radar by the beginning of 1940 covered the eastern and southern coasts of Britain and it could detect incoming aircraft flying at 10,000 feet when 100 miles away. This state of ignorance of their own radar defences extended to a blissful unawareness of the existence of German radar which at that time would seem to have been at least as good as that of Britain.

Emergency situations arising through engine trouble, fuel shortage etc put the skill of the wireless operator to the test. The slender thread of dots and dashes of his Morse code messages sometimes ended happily, with the aircraft and its crew reaching safety; at other times the emergency ended in "SOS" "SOS" "SOS" followed by an uninterrupted signal as the operator clamped the signal key down so that the shore stations could fix the aircraft's position before it crashed into the sea. That last signal was on many occasions the final link between the aircraft and its crew and the land of the living. The wireless operator also operated the Direction Finding loop aerial with which he sought to obtain bearings of British and German radio stations for use by the navigator; the fixes obtained from these bearings at best only gave a general indication of the aircraft's position.

A Hampden torpedo-bomber, possibly P1346 of 455 RAAF Squadron, returns from an attack on shipping off the Norwegian coast on 30 January 1943. Note the damage to the rudder and that both rear gun positions are manned. The extra depth of the bomb-bay is evident just forward of the lower gun position. (Photo: IWM C.3389)

2L CRASHES DUE TO BAD WEATHER

At least a hundred Hampdens literally disappeared on operations in the sense that no trace of the aircraft was ever found and the enemy had made no claim in regard to the loss. Other Hampdens that crashed on enemy soil were not attributed to enemy action and one cannot help wondering what caused these unexplained losses. An appraisal of Hampden crashes in the United Kingdom reveals several instances of apparent structural failure, mid – air explosions, losses due to pilots attempting to change places, loss of control and the "stabilised yaw" phenomenon. It would be reasonable to expect a proportion of the mystery losses to be attributable to such causes and also to assume that others, that were out of wireless contact, crashed into the North Sea after engine failure.

There are known incidents of minelaying or torpedo – bomber Hampdens descending too low and hitting the sea; others are likely to have suffered the same fate, particularly when mist or low cloud obscured the surface of the sea. Some of the losses, off the French coast in particular, may have been due to enemy action as it is difficult to ascertain German flak and fighter claims for that area. However, even when a generous allowance is made for the aforementioned causes, there still remains a substantial number of losses that are inexplicable. The likelihood that bad weather was responsible for many of these is borne out by the accounts of crews who did return after hair raising experiences, although often the only mention in the record book is that "bombs were jettisoned due to bad weather". Weather forecasts in the early war years were even less reliable than they were later on when more information became available from the Meteorological Flight aircraft, and the Met officer at briefing, who optimistically forecast that "Cloud will clear by the time you reach the target area", learned to ignore the derisory cheers of the veteran crews who had experienced other "clearing weather" operations.

"Bomber" Harris, when Air Officer Commanding No.5 Group, is reported to have boasted that his Hampdens were sent on operations on nights when bombers of other groups stayed on the ground because the weather was too bad. Reading through the reports from the Hampden crews about their experiences one wonders whether many of his crews really shared his enthusiasm for these foul weather operations which in many cases were unproductive. Some of the losses due to weather are dealt with in "Crashes on return" (see 2K). Others, including X3000 and AE394 which were due to icing, are included in the "Escapes by Parachute" section (see 5A). The problems that wintry conditions presented became apparent as soon as the Hampdens began night operations in January 1940 on leaflet raids and "Security Patrols". Scampton aerodrome was closed completely from 25 January to 13 February whilst Waddington Hemswell and Finningley also suffered periods of enforced inactivity.

In addition to heavy snowfalls the temperature was unusually cold, one Waddington aircraft operating on a leaflet raid on 26 February 1940 took off with the air temperature down to minus 10 degrees Centigrade. Over Hamburg at 15,000 feet it was minus 40 and on this and other similar occasions frost – bite became a danger, especially to the lower gunner who was at the far end of the ducted warm air of the heating system. The chaos that the heavy falls of snow brought about was soon counteracted with "Snow Plans" at each aerodrome. Lorries became makeshift snow ploughs and gangs of airmen often augmented by soldiers from local units laboured round the clock to keep a landing strip clear across the grass surface of the 'drome. The direction of the cleared strip would afford the longest possible take – off run towards the most likely wind direction. The wind was not always sufficiently obliging, so that it was not unusual for aircraft to have to take off with a strong cross – wind blowing and on 7 February 1942, L4125 (1) of 49 Squadron developed a swing as it took off from Scampton and hit the bank of snow at the side of the cleared strip and crashed, luckily without injury to the crew.

Just a few weeks earlier AT129 (1) crashed as it was taking off from Scampton for a night bombing attack on Brest and its crew were all killed; the report of the crash gives a clue to the cause "there was icing at ground level and intermittent snow at the time". Besides the fact that aircraft iced up and went down out of control, the intense cold had a habit of freezing things. Anyone who has experienced a frozen lock on a car door will appreciate the problem that Sergeant Farrow faced when the "power bolt" inside the cockpit froze and could not be moved to bring the hydraulic system into operation so that the bomb doors could be opened and the bombs released. The incident is described more fully in the "Operational Incident" section. (See 2F19)

The freezing up of latches or hinges on escape exits may explain why crew members sometimes failed to get out of their aircraft in time to parachute to safety. The terse accounts in the squadron record books give some idea of the conditions encountered on these bad weather nights which of course were not necessarily in winter. Eight Hampdens took off from Scampton on 25/26 April 1940 to lay mines in Kiel Bay; owing to very bad weather conditions only one aircraft claimed to have dropped its mine successfully and L4040 (2), L4092 (2) and P1319 (2) all of 49

Heater covers fitted to a 83 Squadron Hampden at Lossiemouth in March 1940. (Photo: D.Garton via Peter Green)

Squadron failed to return. There were no survivors from any of the crews; a night fighter claimed one Hampden but the cause of loss of the other two in unknown.

P4322 (3) of 49 Squadron failed to return from a Brest minelaying operation on 5/6 January 1941 with newly – commissioned Pilot Officer G E Price DFM at the controls. The aircraft is believed to have gone down in the sea some ten miles south of Sidmouth as it was being diverted to St.Eval. No trace of the aircraft or its crew was ever found and although in this case flak damage may have been a contributory factor the sinister words "severe icing encountered" appear in reports of those that did return.

No.144 Squadron lost AE252 (4) and AD784 (4) on 30/31 July 1941 and AD903 (4) and AE140 (4) a week later; there were no survivors from any of the crews, two of which are entirely missing. One crew is buried at Cambrai and from the fourth three men were missing and one is buried at Bergen in Holland after being washed ashore. There are no German claims in respect of the four Hampdens and "other aircraft had reported icing and electrical storms".

AT126 (4) of 49 Squadron, the crew of which are listed in the Runnymede Memorial Register, was similarly lost on the night of 6/7 April 1942. No.61 Squadron had five Hampdens detailed to bomb Hamburg on 13/14 November 1940. One crashed on take – off, one came back after experiencing icing trouble, another could not find the target due to appalling weather. P4396 (5) failed to return and is believed to have crashed into the sea twenty miles north west of Heligoland after being hit by flak over Hamburg; the fifth crashed at Leconfield after being damaged in an attack by Spitfires.

The crews of the ten 50 Squadron aircraft that were operating against Munster on the night of 28/29 January 1942 had a far from enjoyable night out which ended with all the aircraft returning without a single one dropping its bombs on the target due to the atrocious weather conditions. The squadron commander Wing Commander Oxley is on record as having "abandoned his task as it was snowing so hard in the cockpit that he was unable to see his instruments". The Wing Co's report about the general conditions would have been sufficiently colourful to convince the Intelligence Office at de – briefing that the weather really had been bad.

Further proof would have been forthcoming when Pilot Officer Milner telephoned base from Norwich were he and his crew escaped without serious injury after a crash – landing that ended a truly nerve shattering flight in AE373 (6). Their troubles began over enemy territory in a snowstorm in which all the instruments except the turn and bank indicator had failed due to icing which also caused the aircraft to go down out of control on two occasions. Because of the persistent thick cloud the bombs were eventually released from 14,000 feet and the Hampden turned to head homewards. On return to England they were unable to see the coast and neither the altimeter nor air speed indicator were working. The Hampden iced up again and the pilot ordered the crew to abandon the aircraft, but due to the intercom failing the order was not heard by any of the crew. After putting both feet on the instrument panel and heaving with both arms and all his strength on the control column he pulled the aircraft out of the dive. He could only have been about fifty feet above the ground as he levelled out and found to his horror that he was heading for some wireless masts, which he just managed to avoid. Another problem arose when he found that the Hampden was refusing to gain height, probably because of ice having formed in the engine air intakes; the only solution was to crash – land and this he did successfully, not realising that he had missed some electricity grid cables until he had passed beneath them.

This extremely fortunate escape may have been the pilot's second brush with fate as a Pilot Officer Milner crashed on take – off in AE124 four months earlier. Two of his crew on the AE373 crew were navigator

Sergeant Hudson and wireless operator Sergeant Mills and there were two men similarly named who baled out of AT173 six weeks later, so they too had Good Fortune on their side in their Hampdens days at least. One requirement vital to night operations was that of ensuring that the bombers did not arrive over enemy territory whilst it was still daylight and that they were well on their way back across the North Sea before dawn. The calculation of the time by which the enemy coast could be crossed under the cover of darkness on 25/26 August 1941 seems to have been somewhat optimistic as some of the Hampdens despatched to bomb Mannheim that night reported that it was still almost light as they crossed the Belgian coast. AD918 (7) of 144 Squadron failed to return on this occasion and the fact that two of its crew are buried near Brussels suggests that a German fighter pilot or flak gunner may have had the Hampden in his sights in the half light of dusk and shot it down.

The account by Squadron Leader Freddie Newall, an 83 Squadron flight commander, of his experience in AE155 (8) on 18/19 August 1941 highlights the state of confusion that an encounter with cumulo – nimbus cloud (thunder – storm cloud) could create. Turbulence in the cloud became so violent that the Hampden was thrown about as though it were a toy, being turned onto its back and losing height from 16,000 feet to 1,600 feet before, after a series of desperate manoeuvres, Freddie was able to regain control. The immense feeling of relief at being back in command of the aircraft turned to one of bewilderment when, after calling up his crew members on the intercom, he received no reply from the navigator. The reason for this lack of response became apparent after the wireless operator had crawled up to the nose of the aircraft to find that there was no sign of him.

Exactly what caused the navigator to leave the aircraft will never be known. It is possible that his intercom became disconnected as Freddie battled to control the Hampden so that when he called up his pilot he received no reply and being under the impression that the other crew members had already baled out the navigator followed suit. The circumstances of his departure from the aircraft may be something of a mystery; the outcome was that the unfortunate navigator, Pilot Officer James Gould, who was himself a pilot, lies in Kiel War Cemetery. Whether he actually had his parachute on, if he did whether it failed to open, or whether he baled out normally and then landed in the sea and drowned are other unanswered questions.

An incident to one of 420 RCAF Squadron's Hampdens on 13 March 1942 might almost be added as a postscript to Freddie Newall's affair. The aircraft, AE260 (9) had completed its mission but ran into severe icing at 8,000 feet over England on return. As the pilot struggled to control the aircraft he ordered the crew to "Prepare to abandon aircraft". After losing 5,000 feet in height he once again had the Hampden flying normally but was dismayed to find that the state of normalcy did not extend to the crew, as his lower gunner, Sergeant Wilkinson, was no longer in the aircraft having mistaken the order and baled out. All ended happily as the sergeant landed safely and the crew were soon flying together again. There is at least one known occasion when a Hampden pilot lost control in a storm over England and ordered his crew to bale out before doing so himself. Sadly the pilot was the only one to get out of the aircraft.

"There but for the Grace of God go I"

There can be little doubt about the fact that, even when all the aforementioned possibilities have been taken into account, a substantial number of unexplained losses will remain. It is obvious that many of the weather – related losses described in this section could so easily have ended in complete disaster for the aircraft and their crews, so it is a safe assumption that others were not so fortunate.

The entry "Severe ice accretion rendered the aircraft unmanageable" appears in the margin of the 44 Squadron Operations Record Book on the night of 7/8 August 1940 against P4374. The memory of that night is still vivid in the mind of its navigator, Sergeant Harry Moyle.

We took off from Waddington just as it was getting dark with a load of four 500 – lb semi – armour – piercing bombs for delivery on German battleships at Kiel. It was raining steadily and we were in cloud before the flaps had been raised; an hour later we had climbed to 12,000 feet and were still in cloud.

Ginger Bell, our wireless operator, reported in his broad Belfast accent that "There's a funny sort of light on the aerial mast, Sir". Joe Crossley, our dour Canadian pilot, was unimpressed. "It's static electricity, St. Elmo's Fire, nothing to worry about". Our Hampden was soon a mass of eerie light; trails of vapour that formed at the propeller tips danced away in a vortex of white sparks; this awesomely beautiful display intensified and it was almost a disappointment when, after about ten minutes, the spectacle began to fade until it died away.

On returning to the task of sorting out the charts and checking the course for the final part of the outward flight, I received a reminder of the voltage involved in the display we had just witnessed in the form of a sharp electric shock when I grasped the "Course and Speed Calculator" that hung on a length of wire by the side of the chart table.

We were now about half – way across the North Sea, still flying in cloud, and it was obvious from the occasional slight jerks on the rudder that Joe had put "George", the automatic pilot, into operation. This relatively peaceful progression through the murky night sky became bumpier and Joe took over the controls as violent turbulence began to throw the aircraft about. The worst of the shaking up lasted about twenty minutes, accompanied by the vicious glare of lightning flashes. Tension mounted at this fearsome display of the death – dealing forces of nature that surrounded us and we were all immensely relieved when we eventually flew out of this thundery area and George was once more put in control of P4374.

The mass of cloud showed no sign of breaking; my concentration on what lay ahead of us was broken by a most unwelcome sound, a "Splat, splat" against the side of the fuselage. Joe heard it too; "Reckon that was ice coming off the props", spoken in a manner that suggested that he hoped he would be proved wrong.

We considered reducing height to below icing level but decided to press on at our present height of 12,000 feet, because if we descended to a safer height we would still have to climb back up again as our semi – armour piercing bombs would fail to penetrate armour if dropped from a lower altitude. There was also the chance that the Met man would be proved correct and we would simple emerge from the cloud into clear sky. The rattle of ice from the props against the sides of the fuselage was a continuing reminder of our increasing peril. Ginger and Wattie at the rear of the aircraft added to the tale of woe by reporting that they could see ice forming on the wireless aerial and on their guns.

From my own position in the nose I was usually able to see, on either side of me, the comforting red glow of the exhaust collector ring round each engine. That glow, seen through the ice that had formed on the perspex windows, seemed dimmer now and there was no sign of rain or snow leaking in through the front gun hatch. Thick ice was forming; we definitely had problems. Joe was at the controls now, and in an attempt to break up some of the ice he began putting into operation a routine that he had practised on a previous occasion, which basically consisted of operating everything that could be operated.

Bomb doors were opened and closed, flaps lowered and raised, undercarriage lowered and retracted and the controls worked vigorously to prevent the ice rendering them completely useless. Joe gave a running commentary as he carried out each manoeuvre, so that the fusillade of ice from the changing of pitch of the propellers or the backfiring of the engines as he choked them to try to clear the ice from the air intakes, would not alarm his three crewmen unduly. Our faithful P4374 responded to this unaccustomed treatment by lurching, dropping, climbing and general staggering about the sky in protest.

I was sitting on the small round navigator's seat with my back resting against the webbing support that extended across the 3 – foot wide fuselage when Joe announced for the umpteenth time "I'm opening the bomb doors". It seemed as though he had opened the doors to Kingdom Come. The ice had formed to such an extent that our poor old Hampden simply stopped flying and fell out of the sky. There was nothing gradual about it. From a sitting position, my head was forced down onto my knees and I was hurled back under the pilot's seat to slam against the main wing spar that ran through the fuselage. The invisible force of gravity clamped me against the spar as we plunged down out of control in a vertical dive.

The initial mental reaction was for my brain to tell my body "Put your parachute on" but it was impossible even to move my hands towards the parachute, let alone clip it on. Frustrated at this lack of physical co – operation, the next flash of inspiration materialised: "Don't just die, do something, for Christ's sake" then realising the true nature of the crisis more sombre sentiments emerged: "You've had it .. no date with Edna for you tonight .. you are going for a Burton". The messages ended in profanity as I waited for infinity, more angry at being so helpless than frightened.

Joe, of course, did not want to die either. Thoughts of his wife waiting for him back in Lincoln spurred his desperate efforts to regain control of P4374 as it dived down at ever – increasing speed. The unseen force of gravity that seemed to have welded me to the wing spar began to ease slightly. My ears pained me and I gulped and gasped in disbelief. Then I moved my arm a little. Joe had done it. We were coming out of the dive. My intercom plug had disconnected in the upheaval; I reconnected it just in time to hear Joe ask "Is everyone alright?". Our sighs of relief assured him that we were. Joe had jettisoned the bombs as we began to go down and it was extremely fortunate for us that the bomb doors had been open at the time. There was no point in continuing towards Kiel so I gave an approximate course for base and set about recovering my charts and navigation instruments. A glance at my altimeter showed that it read 400 feet. That was much too close to the sea for comfort.

Joe climbed to 3,000 feet and we landed back at Waddington just six hours after we had taken off. After we had gathered our gear together and removed our parachute harnesses, our crew gathered together near the tailplane to combine in relieving ourselves in a sort of ritual Urinatory Celebration of our safe return – besides which we were usually desperate for a run out. On this occasion, the usual comments of "Christ I needed that" and "Who's for eggs and bacon?" were eclipsed by the thought that Ginger uttered: "To be sure I thought I'd had my last piss a couple of hours ago". No one disagreed with him.

Aircrew of 408 Squadron RCAF; from left to right:
Front row: Sgt A.T.McMillan (P1212), Windsor, Ont.; Sgt A.R.Hunter, Fort William, Ont; Sgt J.K.Fraser, Ottawa, Ont.; Sgt H.E.Marshall (AE437), Edmonton, Alberta; Sgt.A.W.Wood, Toronto, Ont; Sgt W.M.Fraser (AT219), Vancouver, BC
Back row: Sgt.D.M.Ross, Toronto, Ont.; Sgt H.D.Murray, Vancouver, BC; Sqn Ldr W.J.Burnett DFC (AE393), Fredericton, NB; Wg Cdr H.W.Timmerman DSO, DFC, Kingston, Ont; Flt Lt N.Janet, Montreal, Quebec; Sgt.R.M.Gifford (P1218), Sherbrook, Quebec; Sgt K.W.McGrail, Montreal, Quebec; Sgt D.L.Henderson (AE286), Winnipeg, Manitoba
(Photo: National Archives of Canada PL – 4704)

PART THREE: THE CREWS

PART 3A THE HAMPDEN CREWS

The Hampden was designed to carry a crew of four which is listed in the Aircraft Tables in the following order. Pilot; Navigator; Wireless operator/upper gunner; lower rear gunner. A fifth crew member who manned the beam gun position was carried on a few occasions such as the defensive night patrols in 1940 over areas of Britain that were being bombed by the Luftwaffe and on some daylight operations from 1941 onwards.

The Pilots

At the outbreak of war there was a mixture of older Regular Officer pilots, NCO tradesmen/pilots and an increasing proportion of Short Service Commission and Direct Entry pilots. The pilots of the RAF Volunteer Reserve, most of whom had learned the rudiments of flying in their spare time before the war, began arriving on the Hampden squadrons by the end of 1939.

The Navigator/Bomb Aimer

The controversy about who was to occupy this position will be seen later, but basically he was either a second pilot or an Air Observer, some of whom were the older tradesmen/observers, a number of whom sensibly accepted the option of reverting to their ground trade soon after the outbreak of war. Direct Entry NCO Air Observers began arriving on the Hampden squadrons in June 1939.

The Wireless Operator/Upper rear gunner

Almost all of the W/Op AGs at the outbreak of war were relatively young, had been trained at Cranwell and few had even attained the rank of Corporal. They had a justifiable grievance when Acting Sergeant Air Observers began arriving on the Hampden squadrons after being transformed from inexperienced civilians into Observers in a matter of five months. Besides the advantage of much better pay these "wet behind the ears" observers, with their shiny new stripes, enjoyed the better facilities of the Sergeants' Mess. Fortunately this anomaly disappeared in May – June 1940.

The Lower Rear Gunner

The tradesmen/air gunners, sporting the brass winged bullet badge on their left arm, constituted the majority of the "straight" Air Gunners. In

A 50 Squadron crew in front of AE422. The two bullet holes in the prop were by courtesy of a Ju 88 over the Frisians on the night of 22/23 January 1942. The aircraft was diverted to Mildenhall where it landed on a snow-covered airfield with both main wheel tyres burst from bullet damage. Crew, from left to right, were: Nairn ("Robbie") Robertson, W.Op/AG; Terry Taerum, navigator, who later navigated for Guy Gibson VC on 617 Squadron; Joe Graves, lower gunner; Norman Goldsmith from Rhodesia, pilot, who was awarded a DFC for his part in the attack on German warships off the Dutch coast on 12 February 1942. (Photo: Nairn Robertson)

mid-1940 there was an influx of Air Gunner Officers, some of whom were either not medically fit enough, or were too old to be accepted as pilots or observers; they included one or two veterans of the previous war.

Air Observers often flew as lower gunners, but a second wireless operator eventually became the more likely occupant of that uncomfortable position. In the early Hampden days there was considerable controversy on the subject of whether the navigator's seat should be occupied by a second pilot or by an Air Observer. This dilemma was resolved at a meeting at the Air Ministry on 8 January 1940 which approved the deletion of the Air Observer from the crew which was henceforth to be composed of a first pilot/captain, a second pilot/navigator bomb aimer and two wireless operator/air gunners, one of whom was to man the lower rear gun position. The Air Observers were "to be distributed among the Battle and Blenheim squadrons..", but fortunately for them the postings never materialised and they remained on the Hampden squadrons, being utilised mainly as lower gunners until mid-1940 when many of them became navigator/bomb aimers.

There were also some Air Gunner Officers who, after teaching themselves the rudiments of navigation flew on operations in that capacity, as did Julian Badcock (5A52). Despite the decision of 8 January about Air Observers, HQ Bomber Command soon afterwards came up with the idea that aircraft crews should be re-organised when the larger bombers materialised, with the Air Observer as Captain of the aircraft. In the early months of the war there was almost a "caste" system on at least one Hampden squadron (144) whereby the Commanding Officer decreed that where there were Officer and NCO pilots on the squadron the Officer should fly the aircraft and the NCO pilot do the navigating ... not a popular ruling amongst the NCO pilots who were in many cases the more experienced pilots.

In early 1941 the Air Member for Training, Air Vice Marshal A G R Garrod,complained of the wastefulness of using two pilots on Hampdens. "The second pilot is trained initially chiefly as a bomb-aimer/navigator in a period of six weeks, and when he is due to become a Captain he returns to the OTU for a further period of four weeks training. During the last half of the second period he will be crewed up with other members and go forward to an operational squadron as captain of a complete crew." In March 1941 it was agreed that three Hampden squadrons could be reduced to one pilot aircraft with a crew of pilot, observer, and two wireless op/AGs and that 16 OTU should turn out complete crews, whilst 14 OTU fed the squadrons which remained on a two pilot basis.

The following incidents demonstrate the two sides of the one pilot or two pilot dilemma. The difficulty experienced by a second pilot in taking over the controls was described by Robin Hooper, who practised the change-over drill with Len Ratcliff at 16 OTU in 1941. "After letting down the hinged back of the pilot's seat, the first pilot lay back whilst the second pilot took hold of the control column, then, sitting astride the prostrate body of his comrade, carried out the trickiest part of the exercise – that of moving the first pilot's feet from the rudder pedals and inserting his own in their place. The final act was the removal of the original pilot by dragging him out from under the posterior of his replacement." Robin seemed quite satisfied that they had carried out the operation with a loss of height of only 5,000 feet, and remembered that the exercise was later removed from the syllabus because it was too dangerous. Proof of the problems involved may be seen in the incidents in which L4155 and P1327 were written off. The loss of L4155 (1) on 25 May 1939 was believed to be due to a practice change-over of pilots going dreadfully wrong, but none of the five men on board survived when the aircraft spun down out of control and crashed five miles SE of Newark.

The experience of Squadron Leader Oxley in P1327 (2) as he returned from operations in the early hours of 1 August 1940 underlines the danger of practising the change-over. P1327 was within 50 miles of the English coast when the pilot apparently decided that it would be good practice for his navigator/pilot to change places with him. They were in the process of doing this when the Hampden went down out of control in a near vertical dive, from which it was extremely fortunate that one of the pilots managed to regain control of the aircraft; they then flew back to base. The effect of this terrifying experience on the crew can only be guessed at, but proof of the stress imposed on their aircraft by this "abnormal evolution" is the fact that the wings had been strained to such an extent that the aircraft had to be written off. The "Crash" card for P1327 states that the incident in which the aircraft was written off occurred 40 miles east of Flamborough Head and this is given in some accounts as the crash site. By a remarkable coincidence P4383 of the same squadron, on the same date and at almost the exact time, was last heard of 32 miles east of Flamborough Head just before it crashed into the sea with the loss of its four crew. (5B25)

The loss of AD825 (3) on 18 April 1941 when the crew all baled out safely (see 5A13) poses a question as to whether the navigator was in fact an Air Observer, who could not have been expected to be able to land the aircraft in the emergency which had arisen. Had another pilot been flying as navigator there may easily have been a different ending to that particular story. Evidence to prove that it was possible to carry out a change of pilots may be seen in two remarkable incidents involving Hampdens from Waddington within ten weeks of one another when pilots flying as navigators took over control from their wounded pilot, both incidents occurred at night and at low level.

On the night of 4/5 May 1940 Squadron Leader Good of 50 Squadron flew to Norway in L4062 (4) to lay a mine in Oslo harbour. As he flew in low to drop the mine the light flak batteries hammered away at the Hampden, and unknown to any of the crew the pilot suffered wounds in both arms. Pilot Officer E G Gardiner, a pilot flying as bomb aimer navigator, released the mine successfully, and becoming apprehensive about the manner in which the aircraft was being flown he made a tactful enquiry of the pilot as to what he had in mind. When he received no reply the navigator scrambled under the pilot's seat and stood up to see for himself what was happening. He at once realised that the Squadron Leader had been wounded, the blood which ran from cuts on his face being more dramatically evident than the injuries to the arms. The erstwhile navigator took hold of the control column and eased the Hampden into a gentle climb then, when he had attained a safer height, enlisted the aid of the wireless operator to help him haul the wounded pilot from the seat so that he could take over control.

Sergeant Wallace, the wireless operator, then attended to his Squadron Leader's wounds; the sergeant had been with Squadron Leader Good on 12 April in the attack on Christiansand (see 2C) and had been awarded the Distinguished Flying Medal for his part in the action. Pilot Officer

Air Observers under training at No.7 Elementary & Reserve Flying Training School at Desford in April 1939. Before the end of 1939, Harry Moyle, in centre of photograph, had survived being shot down in L4089; Len Thom, on his right, had been killed in L4203; Charlie Williams on his left had perished in L4132. Lawrie Wheatley, extreme left, was taken off flying because of persistent air sickness; Chas Thomas, extreme right, did not complete his training. (Photo: Harry Moyle)

Gardiner, who survived the war and retired as a Group Captain, and Squadron Leader Good each received the Distinguished Flying Cross for this event; the Hampden was not seriously damaged and landed safely. The gallant squadron leader recovered from his wounds and returned to flying duties with 50 Squadron later on. It would be nice to end on that "happy ever after" note, but this cannot be so, as less than a year later Squadron Leader Good DFC was lost with his crew in AD728 on another mine – laying operating, this time to La Rochelle. There was not even the consolation of the crew being buried together, as the bodies of three of them lie in different graveyards in the La Rochelle area, whilst that of Squadron Leader Good was never found and his only memorial is his name on the Runnymede Memorial.

The second of the Waddington change – over incidents happened on the night of 18/19 July 1940. As 44 Squadron Pilot Officer Walker made his bombing run over Eschwege airfield at 3000 feet he was forced to take evasive action to dodge the stream of tracer shells being pumped up at the Hampden from the ground defences. In the nose of the aircraft the navigator was another pilot, Canadian Pilot Officer David Romans, and Dave sensed that something was amiss when the Hampden began to dive earthwards and his pilot failed to reply on the intercom. Dave crawled hurriedly back under the pilot's seat and up onto the wing spar from where he could see his friend slumped over the controls, obviously unconscious. Realising the peril of the situation Dave set about lowering the back of the pilot's seat so that he could take over the controls. He must have been mightily relieved when he eventually had the control column in his hands and his feet on the rudder pedals, even though he was still sitting astride the body of his stricken comrade.

The wireless operator had been completely unaware of the drama that had taken place in the front end of the Hampden until he happened to glance through the window in the bulkhead door, but once he realised the situation he hastened to assist Dave move the limp body out of the pilot's seat. After handing up the charts from the navigator's position so that Dave could estimate a course for base the wireless op gave his luckless pilot a shot of morphine from the first aid kit after doing what he could to attend to the wound on his head.

Dave flew the Hampden back to Waddington and landed safely but unhappily the flight ended with the officer who had set out in the pilot's seat being transferred to hospital where he died from his wounds. Dave Romans was awarded a DFC for his prompt action which had at least saved the lives of his other two crew mates and brought the aircraft back safely. Before he left 44 Squadron Dave had led an "interesting" life, in the course of which he put two Hampdens down on the sea in less that a week. (See 5B6 – 7 which also describes how Dave was killed on operations with 90 Squadron when flying in one of the RAF's first "Flying Fortresses".)

Although the question as to whether there should be a second pilot in the navigator's position was a somewhat contentious one, there was no disputing the fact that the utilisation of a second wireless operator in the rear lower position made good sense as it enabled the newly trained wireless ops to have valuable experience of operational flying before they in turn became the first operator and moved to the upper position.

The crew of a 415 Squadron torpedo – bomber on return to base after a running battle with six Ju 88s whilst on anti – submarine patrol over the Bay of Biscay. When the Hampden finally escaped into cloud, it had 200 holes in it, the only casualty being the pigeon which had its feathers grazed by a cannon shell. From left to right: FO Frank Brockwell RAF, the navigator from Waterford, Herts; PO Warren ("Doc") Savage RCAF, the pilot from Portage La Prairie, Manitoba; wireless operator/air gunners Sgt Steve Babyn RCAF, from Toronto and Sgt Earl ("Shorty") Mason RCAF from Ottawa. Note the light sea grey camouflage of the upper fuselage. (Photo: Public Archives of Canada PL – 22086)

No.144 Squadron at Leuchars in June 1942, having been transferred from 5 Group, Bomber Command, to 18 Group, Coastal Command, on 21 April 1942 to become a torpedo – bomber unit. (RAF Museum P8089)

A group of erks being talked to by the High Commissioner for Rhodesia when he visited No.44 (Rhodesia) Squadron at Waddington. The title was bestowed from 12 September 1942. (Photo via Ray Leach)

Wireless Operator/Air Gunners and Air Gunners of 44 Squadron pose with a 500 – lb bomb at Waddington in the early summer of 1940. From left top right they are:Bill Street from New Zealand; Jim Taylor, who became a prisoner of war during his tour on Manchesters; Ginger Bell (P2087); Gerry Preston (X2986); Pete McLaren from New Zealand (L4154). (Photo: Jim Taylor)

A Rhodesian pilot at the controls of a Hampden of 44 Squadron at Waddington. (Photo: IWM CH 3488)

A 83 Squadron W.Op/AG behind the upper position twin Vickers K guns. The purpose of the shaped rail below the guns was the lessen the chance of bullets being fired into the aircraft's tailplane. The "ring and bead" gunsight can be seen over the centre of the gun barrels and the more sophisticated "reflector" sight is at the rear of the guns to which ammunition pans are not fitted. (Photo via Peter Green)

Flying Officer Ford of 50 Squadron poses with his aircrew and groundcrew in front of "K for Chris" (sic!) (Photo via C.G.Jefford)

A 50 Squadron crew poses with a couple of 250 – lb bombs with their ground crew. (Photo via C.G.Jefford)

An official posed photograph of aircrew at Scampton watching one of their aircraft return in February 1941. Hampden "D" on the ground is "OL – D" of 83 Squadron. (Photo: IWM CH 271)

PART 3B: CREW CASUALTY TOTALS AND SOURCES OF INFORMATION

The total number of Hampden crew casualties is as follows: –

Operational:
Killed: 1,077 Missing: 739 POW: 349
Interned in Eire: 4 Injured: 149
Non – operational:
Killed: 570 + 16 non – aircrew
Missing: 38* Interned in Eire: 4
Injured: 118 + 2 Air Cadets
Totals
Killed: 1647 + 16 non – aircrew
Missing: 777*
POW: 349
Interned 8
Injured 267 + 2 Air Cadets
* Totals include 18 men Missing at 32 OTU in Canada who are not listed in the Runnymede Memorial Register.

Abbreviations

The following abbreviations appear in the Aircraft Tables to describe crew casualties: –
(I) Injured; (P) Prisoner of War; (K) or b. Killed or b. buried, e.g. b.Kiel = buried Kiel War Cemetery; (M) Missing Body never found, name recorded in Runnymede Memorial Register.

Sources of information on crew fates

The Squadron Record books provided the initial information on casualties; they occasionally included International Red Cross reports of the fates of crews that had failed to return. In the early years of the war most squadron record books were, to say the least of it, not well kept; many were written up in a longhand that would have been better suited to a doctor's prescription and even when typed out clearly names were often mis – spelt and information omitted. The actual serial numbers of the aircraft were often entered incorrectly, with the letter of the serial number being given wrongly e.g. P instead of X or L etc, while the advent of the AD, AE, AJ, AN and AT prefixed numbers further complicated matters. More confusion arose when Hampden squadrons began operating with Coastal Command and adopted the standard practice of using the aircraft code letter only in the record books.

It was also not unusual for the loss of a Coastal Command aircraft and its crew not to be processed until two days after it had been lost. A good example of this appears in the 415 Squadron record for 17/18 May 1943 when two aircraft "B" and "D" and their crews were lost, the serial numbers of the aircraft proving difficult to track down because the write – off dates were different from the date of loss. It was also apparently considered by some squadrons that the omission of missing aircraft and crews from the Record Book caused anyone who later inspected the Book to be less aware of the losses being sustained, a case of "What the eye didn't see, the heart didn't grieve over" perhaps? There were also times when the Forms 541 were completed from the crew reports; as there were no reports from missing crews, nothing was recorded.

The Record Books became more standardised as the war progressed with those of 455 Squadron RAAF, 408 Squadron RCAF and 32 OTU in Canada being the best of the Hampden records. The whole task would have been so much easier had the Hampden squadrons adopted the same procedure as that used by one of the Blenheim squadrons in 1940, in which the serial number and squadron letter of each aircraft and the name, initials and service number of each crew member were all neatly typed out for each operational sortie.

The entries in the 49 Squadron Operation Record book for 12 February 1942 highlight the inadequacy of information, it states "20 aircraft detailed to attack the *Scharnhorst* and *Gneisenau* at sea. Four aircraft and their crews were lost." There is no record of the crews, or serial numbers for the aircraft lost, in the "Detail of Work Carried Out" (Form 541), although the aircraft and crews that returned safely are given. Next day sixteen aircrew are posted missing, and from the aircraft "Movement cards", it is apparent that AE396 (1), P5324 (1), AE132 (1) and AE240 (1) were lost in battle by 49 Squadron on that date, so the sixteen missing aircrew had to be shared out between the four Hampdens, hopefully in something approaching the correct crew composition.

By checking out previous crew lists it was possible to find two crews which had previously flown together, and the remaining eight names were formed into two crews of a pilot, navigator, wireless op/AG and air gunner. The only officer among the sixteen names was a Pilot Officer S W A Way who was, not unnaturally, assumed to be a pilot until it was noticed that he was a wireless operator air gunner who had been appointed to a commission on 4 February 1942, just a week previously. One could not help wondering whether 21 – year – old Pilot Officer Stanley W A Way, from Shoreham – by – Sea, ever had a chance to wear his new uniform before he failed to return from that calamitous operation. (See 2H)

On the training units the Unit or Station Record Book did not list the crews involved on training flying, although the "Thousand Bomber" raid participation of 14 and 16 OTU is proudly recorded in the approved manner. Accidents on training flights were usually entered in the daily "Summary of Events" which in many cases appear to have been oriented towards the interest of the person who wrote it. The arrival of Section Officer Fanny Adams might be considered worthy of note by a fellow WAAF officer who would fail to mention that, on the same day, one of their Hampdens had crashed with the loss of its crew. It is interesting to see how these Training Unit records vary; in some entertainment or sporting activity predominate, in others every message about Air Raid warnings is meticulously recorded. Visits to the unit by inspecting officers receive their share of space, as do the occasions when rats (presumably four-legged ones) had been seen in the WAAF huts.

One training unit Station Commander appears to have been obsessed with the idea that a Station Band was a prime necessity which, once formed, enabled the station "to hold its first full parade". It conjured up a picture of the Commanding Officer being one of the "Old Brigade" who desperately sought to turn the clock back to the good old pre-war days. He would, indeed, have had an uphill struggle in his attempts to instil the old "bull" disciplines into his trainee aircrew, many of whom were "Colonials" who did not enthuse about being used as Parade Ground toy soldiers. The use of the word "Colonial" to one of the lads from Canada, Australia or New Zealand was, incidentally, a certain means of starting a fight unless that particular chap was already a firm buddy.

Other sources including personal accounts from ex-Hampden men; information from overseas researchers; Air Historical Branch; Commonwealth War Graves Commission; Public Record Office where, besides record books, etc., the 5 Group Roll of Honour and a 1945 list of Officers and NCOs heldin German Prisoner of War Camps have been researched.

AT112, EA-M of 49 Squadron, 14/15 February 1942. Having run out of fuel after being airborne for over nine hours it crashed at Glossmore Farm, on the Ponders Bridge to March road, near Upwood. The crew escaped without serious injury. (Photo: R.N.Hamer)

PART 3C: THE INJURED

The abbreviation (I) appears after the names of over 250 Hampden aircrew in the Aircraft Tables. It signifies that the man was injured in the incidents. Many aircrew who became Prisoners of War were quite seriously injured in the process, as is evident in the accounts in Part 3D. The injuries to some of the men who survived a crash were so severe that even when skilled medical attention was immediately available they died with a short time; in such instances the casualties are shown as having been killed in the actual crash.

Most of the aircrew who are shown as having been injured were left with a permanent reminder of their Hampden days. Sergeant Ken Lodge (1), for instance, was hit over the eyebrow by an incendiary bullet when L4090 of 44 Squadron was shot down, a fragment of phosphorous entered the eye and he subsequently lost the sight of his right eye. John Day (2) had been navigator of a 50 Squadron aircraft on the Vaagso daylight operation of 27 December 1941; from his home in South Africa John recalled that day. "I was aiming the gun in the nose at a flak gun when the shot hit me. I felt no pain at all but had no us in my leg. I lay back on the floor and Robbie, the wireless operator, gave me morphia and fixed a tourniquet. After I lost my leg I became a Staff Officer at HQ Bomber Command and later served on 139 (Pathfinder) Squadron, then 439 Squadron and finally with the Bombing Development Unit at Feltwell."

The consequences of the horrific injuries suffered by Sergeant Robert R L Adams (3), wireless operator in P2099 on the afternoon of 17 May 1941, were revealed when enquiries were made into the cause of his death almost forty years later. After P2099 had developed engine trouble the pilot decided to make a forced landing on a field at Uffingham, near Wittering but stalled in the attempt; the aircraft crashed and burst into flames. A farm worker who had seen the crash rushed towards the blazing wreck and valiantly dragged Bob Adams away from the flames. The injuries that Bob sustained were truly terrible; the fierce heat had burnt away his facial skin, his nose, eyelids and fingers being almost unrecognisable. The violence of the impact had shattered both of Bob's kneecaps and it had also killed the pilot and the other two men in the aircraft. For two years Bob endured the pain of skin grafts and operations at East Grinstead, where Sir Archibald McIndoe and his team worked miracles of surgery to reconstruct his face.

After Bob left East Grinstead, his ex-WAAF wife bravely battled by his side to help him achieve some sort of happiness over the next forty years. Arthritis added to Bob's problems and eventually he decided that life was more than he could bear and, after a final drink at his local pub, he went home, took an overdose of drugs and went to bed to wait for the pain to end. Bob Adams was 59 years of age when he died; at the inquest into his death the Coroner commented "I regard this death as if he had died in battle at the age of 20, except that he had to experience more pain and suffering."

Bill Johnstone (4) of 50 Squadron had been observer/navigator on AE375 when it crash landed, out of fuel, near Patrington in Yorkshire on 11 October 1941; the crew escaped injury and AE375 was repaired. Six weeks later Bill was in P1202 in another crash landing at Swinethorpe near Skellingthorpe and although there is nothing in the records about him being injured he was unfortunate enough to lose the sight of his left eye as a result of the crash. The pilot and lower gunner in the AE375 and P1202 incidents, Sergeants Williams and Caunce were together again in AE250 on 11 January 1942 when it crashed on return as it was trying to land at the aerodrome at Crosby in Cumberland. Sergeant Caunce survived the crash but was the only to do so, and was posted "non - effective sick", which covered a multitude of possibilities.

Another example of misleading information may be seen in the 415 RCAF Squadron Record Book on 18 June 1942, when Pilot Officer Mahn was reported as having been "rescued after being adrift in a dinghy for 14 days. His condition was reported as being good." A subsequent report from one of his squadron colleagues told a different story. Pilot Officer Mahn, a well built six - footer who tipped the scales at twelve and a half stone, had weighed just on six stone when picked up. He spent eighteen months in hospital after his awesome ordeal and had both legs amputated below the knee because gangrene has set in; he nevertheless returned to flying duties.

The scarred faces, sightless eyes and missing limbs were the visible manifestations of the permanent injuries sustained by the unfortunate aircrew who have just been mentioned, but it should be remembered that the bomber crews suffered not only physical injury; there were also the injuries to the minds of many of them.

The subject of these mental injuries is usually surrounded by a ritual "taboo", but it has to go on record that they existed and evidence that their treatment was causing some soul - searching among the more enlightened Medical Officers is to be seen in a letter dated 31 December 1941 (AIR15 - 105) which reads: "Medical Officers are not often in a position to observe early the sign of flying strain. The majority have been qualified for under 12 months and few of them have that practical knowledge of medicine in its widest sense, as distinct from the knowledge of medicine obtained from book - work. They do not realise they are dealing with essentially physiological individuals who are exposed to a pathological environment to which the majority have acquired only a minor degree of immunity. The inexperienced MOs are inclined to think that every case of indigestion must be due to a peptic ulcer, whereas the majority of cases of indigestion are due to alien psychological factors such as anxiety in regard to risk of life, etc. Accidents are in many cases caused by utter carelessness. The essence of crashes is thought to be the inability to think clearly and act accordingly when frightened by a difficult situation. Fear drives out all logical thought. He who can be afraid, yet think clearly and do the right thing is of the essence in the aircraft, it is pointed out. Many experienced pilots agree. With regard to the Neuropsychiatry Unit at Matlock, it is pointed out that the aircrew themselves have no faith whatever in the Neuropsychiatric treatment, and the "loony bins", surely a basis such as this dooms psychological treatment from the start."

In other words, most RAF Medical Officers, operating from the comfortable world of their Station Sick Quarters, may have been capable of dealing with the bodily wounds and ailments of their patients, but they had absolutely no comprehension of the mental strain that wartime flying in general, and operational flying in particular, was imposing on the aircrew concerned. Any suggestion that a member of an aircrew could actually lose consciousness through any cause other than a defective oxygen supply, or that one of these young men in the prime of life could experience a complete failure of co - ordination of mind and body, was utterly unacceptable to most Medical Officers.

An air observer/navigator at Waddington recalled the treatment meted out to him after he had been advised to see the MO because after almost a year on operational flying he was experiencing loss of consciousness and suffering delusions of weightlessness, amongst other things, when flying in or near cloud. These troubles had mainly originated three operational flights prior to him reporting to MO, in an icing up incident described at the end of Section 2L, the observer concerned having become progressively more useless when flying anywhere near cloud and it was ultimately apparent that he had become a liability to his crew mates. The initial reaction of the Medical Officer concerned was to diagnose sinusitis and despatch the patient to the RAF hospital at Rauceby for treatment, which consisted of a thrice daily session during which the patient sat, with his head covered under a large towel and inhaled the steam from a jug of boiling water to which menthol crystals had been added. Not surprisingly this did nothing at all to soothe the patient's nerves; if anything the enforced idleness of hospital routine aggravated his fears, so that by the time he returned to Waddington even the prospect of flying reduced him to a dithering wreck, and his squadron commander once more pointed the lad in the direction of Station Sick Quarters.

The Medical Officer, who had arranged to have this NCO's complaint cured, was most indignant to hear that the patient actually believed that he was still ailing, and offered his personal opinion in no uncertain manner. "If I had my way I'd have you shot for cowardice, like they did in the last war" was his solution to the problem. The sergeant concerned had, some eight months previously when still only eighteen years old, almost had his life ended by British bullets when he had been shot down into the sea by Spitfires. In his present state of nervous exhaustion the prospect of facing more British bullets from a firing squad caused no alarm; it would at least have released him from the mental torture that he was enduring. The outcome of it all was that he was put on non - flying duties for six months then posted to non - operational flying for a year before being taken off flying altogether and being re - mustered to flying control duties. During the intervening two years there was no further interest from the medical side at all even though the iniquitous "Lack of Moral Fibre" label was being attached to some of his less fortunate colleagues who had suffered similar problems. A study of the whole business of the occurrence of nervous troubles in wartime aircrew, and its consequences, is long overdue; it would make interesting reading.

Nursing staff at RAF Hospital, Rauceby, Lincs. Left to right: Sister Cross; VAD Nurse Higgs; Sister Cranston. Rauceby, which had been a pre - war civilian mental hospital, was taken over by the RAF on the outbreak of war. (Photo: Harry Moyle)

Hospital Blues at RAF Hospital, Rauceby. Harry Moyle and W.Op/AG Brooks in late August 1940. (Photo: Harry Moyle)

"Clipped Wings". Kriegies at Oflag IVC Spangenburg. All are identified as Hampden pilots from Hemswell: Back row left to right: "Spooky" Barrett (P4346); "Digger" Young (X2967); "Gonga" Coste (L4132); Ken Jones (P4344); fifth man unknown. Middle row left: Peter Tunstall (P4324); right: "Skid" Morley (L4112). Front row left: Bill Hartop (X3129); centre unknown; right Aiden Skehill (P4291). (Photo: Peter Tunstall)

PART 3D: HAMPDEN PRISONERS OF WAR - THE KRIEGIES

The symbol (P) appears in the serial number appendix against some 350 names of Hampden aircrew who failed to return from an operational flight; it indicates that they became Prisoners of War, the "Kriegsgefangenen" or "Kriegies".

The first Hampden "Kriegies" were four members of different crews from the five 144 Squadron Hampdens shot down on 29 September 1939 after a battle with over twenty Bf 109s north of Wilhelmshaven. Each of the four was the sole survivor from his aircraft after it had crashed in flames into the sea and disintegrated; all owed their lives to their "Mae West" lifejackets keeping them afloat until German patrol boats picked them up. The next arrivals were three crew members from L4083 (1), one of the six Hampdens that were lost from Waddington on 12 April 1940 in an attack on Kristiansand when Norway was invaded. The seven who became prisoners of war, some of whom were badly injured, were the only survivors from the 44 aircrew in the eleven Hampdens shot down in these two daylight encounters with enemy fighters.

The Hampdens commenced night operations in January 1940 and were the spearhead of the minelaying campaign from April 1940, playing a full part in the night bombing operations of the early war years. It was from a minelaying operation, on 20/21 April 1940, that those aboard L4088 (1) became the first complete crew of a Hampden to become prisoners, after an encounter with flak ships resulted in damage that brought about a crash landing in Denmark. The crew of torpedo-bomber X2898 (1) became the last Hampden crew to go behind the barbed wire on 19/20 August 1943.

Space does not permit a description of all the incidents during the intervening period of over three years that resulted in Hampden aircrew joining the "Kriegie Club" but the following accounts illustrate the diverse circumstances of their arrivals, which also qualified many for membership of the "Caterpillar", "Goldfish" or "Guinea Pig" clubs, as they arrived by parachute, dinghy, or escaped from a blazing wreck and suffered burns that subsequently required skin grafts. A number of incidents are included in other sections of the book, such as the "Wilhelmshaven", "Escape by Parachute", "Dinghy" and "Minelaying" chapters.

NAVIGATION ERRORS

The least physically-painful way of becoming a "Kriegie", was, no doubt, to make a forced landing when you were lost and out of fuel, but nevertheless believed that you had landed in some part of the Mother Country. What an unpleasant surprise it must have been when German soldiers arrived on the scene to do the "For you the war is over" routine, as happened in the cases of L4146, P2070 and at least two other Hampdens.

L4146 (2) had been sent to attack railway targets in Western Germany and it landed in the grey light of dawn on a patch of clear ground in an otherwise wooded area. The crew were so certain that they had landed in Scotland that the pilot sent the navigator to the village nearby to telephone the nearest aerodrome for RAF transport to come and pick them up. The grim realisation that they were down in Germany, in the Black Forest region between Horb and Rotweil, arrived too late for them to do much about it; even had the pilot thought that taking off again might be a good idea, the wheels of the Hampden had already sunk into the soft ground, thereby making it impossible.

Len Walker, the wireless operator of L4146, believed that their troubles stemmed from a faulty gyro-compass, but whatever the cause of this and other similar incidents they were not only happening to the Hampden crews at this stage of the war, with its "By guess and by God" navigation. Len also recalled that his observer/lower gunner crewmate, Sgt E R Corrigan, contracted peritonitis whilst a prisoner of war and died in hospital, in spite of the efforts of the hospital staff to save his life.

P2070 (3): A German radio broadcast on 26 August 1940, revealed another slight error of navigation on the part of a 50 Squadron crew in P2070. Reading it now, over forty years later it sounds quite amusing, although it is doubtful whether the unfortunate crew thought it so at the time. The broadcast described how the inhabitants of Worms, in southern Germany, heard the drone of an aircraft flying at low height over the town and anti-aircraft batteries firing. Shortly afterwards a twin-engined bomber (P2070) landed at Lautersheim. To the astonishment of persons who surrounded the aircraft, the crew calmly alighted and its Officer pilot addressed them in English. To the several German Air Force Officers who arrived on the scene, he announced that he thought that they were in Scotland. When asked if they were not surprised at being shot at by AA guns, he replied, "Oh no. We're used to that sort of thing in England". The Hampden's carrier pigeons were still in their baskets; if released they would have had a long flight home!

AD768, AT226 (4,5) At least two other Hampdens were involved in similar incidents in enemy territory, whilst the crew of AD768 ended up being interned in Eire after baling out over County Donegal. Proof that

Top: L4146 "QR – P" of 61 Squadron landed out of fuel in a grass – covered clearing in the Black Forest on 24 May 1940. Note the absence of modifications to the perspex side panel of the navigator's position. Other views show clearly the codes and tail stripes of red, white and blue of even width. Len Walker, who was the wireless operator, believed that the aircraft was well on fire when the crew left it but it is possible that smoke from the Very cartridges which had been fired into the fuselage had created an impressive amount of smoke without causing much damage. (Photo: T.E.Willis/Heinz Nowarra)

Centre: P4324, "QR – P" of 61 Squadron, landed out of fuel on the Vliehors sandbank on the Dutch island of Vlieland in the early dawn light of 27 August 1940. Note the modification to the perspex side panels of the navigator's position and the wide white gap between the red and blue stripes of the tail – fin marking. (Photo: Dirk Bruin)

Bottom: AT109 "UB – C" of 455 Squadron salvaged by German troops after it had force – landed on a beach near Vadsoe in northern Norway. The aircraft had been damaged by flak as it flew towards Vaenga on 4/5 September 1942. (Photo: Randolph Kugler via Franz Selinger)

Top: An unidentified Hampden in Luftwaffe Beutepark 5 in the Nanterre district of Paris. Another view of the aircraft shows that it bore the individual code letter "M" but the squadron code is obscured. AE265 of 144 Squadron, which made an emergency landing when out of fuel at Ypenburg airfield near The Hague on 25/26 August 1941, would seem to be the most likely identity of this aircraft. The presence of B – 17 wreckage and also the balloon cable cutter on the port wing of the Hampden, close to the fuselage, suggest a date nearer to 1942 than to 1940. (Photo: T.E.Willis)

Centre: A captured Hampden taking off from Rechlin, the Luftwaffe equivalent of Boscombe Down. The identity of the aircraft is not known for certain but L4146 seems to be likely with P2070 or P4324 as possibilities. The last – mentioned was flown from Vlieland a few days after it had landed there but one unconfirmed report says that, although it was escorted by German fighters, it was shot down by German flak. (Photo: T.E.Willis)

Bottom: Although this photograph is of poor quality, the markings on the tailplane and fuselage appear to be different from those on the Hampden taking off from Rechlin. The perspex side panel of the navigator's position is unmodified, which suggests that the aircraft is either L4146 or P2070. (Photo: Archiv Schliephake)

Sergeant Phil Draper, sole survivor of AD835 of 83 Squadron which crashed on Schiermonnikoog after being shot down by a night fighter on 25/26 July 1941. (Photo: Gerrie Zwanenburg)

the "navigational errors" still existed in 1942 can be seen in the case of AT226, which made a crash landing when lost and out of fuel. This crew were so certain that they were in enemy territory that they set fire to their aircraft in the approved manner, only to discover a few minutes later that they had come down at Grendon Underwood in Buckinghamshire. England may still have seemed like a foreign country to three of that crew; the pilot was from Australia, the observer/navigator from Southern Rhodesia and the lower gunner from New Zealand. The crew did not last long enough to be too embarrassed by their chums with wisecracks such as "Set fire to any Hampdens lately?", as they were all in AT218 when it failed to return a week later, and only the wireless operator Sergeant Rickard survived as a "Kriegie". His three crew mates who had travelled so far to fly together in a Hampden lie at rest in Hannover War Cemetery.

CRASH LANDINGS

Many Hampden "Kriegies" arrived on enemy soil in a crash landing, often as the result of enemy action, sometimes after engine failure or through running out of fuel. It would serve no useful purpose to give an account of all such incidents, nor to go too deeply into the manner of their arrival, as the only difference between these crash landings, and the Hampdens that crashed in the United Kingdom, lies in the unhappy consequence for the crew, i.e. several years behind the wire of a prison camp.

X3052 (6): The crash landing of X3052 was described by its navigator, Des Dunphy, a sergeant who had qualified as a pilot in the RAF(VR) before the war and after a period of training on Hampdens, had arrived on 49 Squadron at Scampton in September 1940, at a time when the awards of Victoria Crosses to Flight Lieutenant Learoyd and Sergeant Hannah were still a main topic of interest. Des was rightly proud to remember them, particularly as he had known John Hannah and even slept in the next bed to him at South Carlton. Although loud in the praise of others, Des did not mention that his own name appears in a file at the Public Record Office about Stalag XIB Prisoners of War, as being recommended at the end of the war "for valuable services".

None of the crew of X3052 had previously flown together. The man behind the controls was Sergeant Phillips, another "VR" pilot who, in pre-war days, "had something to do with a brewing firm in Birmingham"; the wireless operator was nineteen-year-old Sergeant Terpey, from Newcastle. The fourth member of the crew, Aircraftman Second Class Hoskins, seemed somewhat out of place, as by this time all aircrew flying on operations had been given the rank of Sergeant, at least. Des supplied the answer to that mystery. AC2 Hoskins, if Des remembered correctly, was around 30 years of age and had managed a pub in Leatherhead before enlisting in the RAF after war had broken out. Our Mr.Hoskins, however, had not been in the Service long enough to grasp a very basic rule - "Never volunteer for anything". He had, no doubt, found life as a member of the Aerodrome Defence Guard somewhat tedious and this, coupled perhaps with envy of the dashing young sergeants who led such different exciting lives, prompted him to enquire of his Flight Commander whether he could be considered for an Air Gunner's course. "You have fired a Vickers G.O. machine gun, I suppose, in the training for your present job?" enquired the Flight Commander, duly impressed with this eager erk. "Yes Sir", came the reply. The magnanimous "Sir", looking at the crew list for the night's operations, noticed that X3052's crew was lower gunner-less, and made the sort of decision that only an inspired leader could make. "Tell you what, how would you like to go for a trip tonight to see what it's all about?"

Hoskins staggered from the Flight Office, hardly believing his luck; what a fine chap "Sir" was, letting him have his first flight in an aeroplane, and on an operation too. "Sir" entered the name of AC2 Hoskins as the fourth man on the crew, delighted at having been spared the trouble of phoning around for a real Air Gunner, Observer or Wireless Op to fill the vacant space, unmindful of the fact that AC2 Hoskins could, on his first flight of any kind, end up as a "Kriegie", which must, when it happened, have qualified him for some sort of entry in the "Guinness Book of Records"; he certainly deserves this mention in "The Hampden File"! X3052 met its end when it had the misfortune to be still over enemy territory in the early dawn light of 25 November 1940. It was even more unfortunate to be intercepted by three Bf109s from III/JG54; Leutnant Wubke received the credit for shooting the Hampden down.

For Des Dunphy it was a most unwelcome present for his 22nd birthday, as it coincided practically to the minute with the anniversary of his arrival on this earth on 25 November 1918. The Hampden had been set on fire in the attack and the pilot was forced to crash land near Den Helder, in Holland, and AC2 Hoskins received injuries to a foot and ankle but recovered fully in hospital. Des paid this tribute to Sergeant Phillips "The fact that we were not all killed is due entirely to the skill and coolness under fire of Sergeant Phillips, to whom we all owe our lives. The Hampden was completely burned out; the only fatal casualties were our pigeons"

X3049 (7): One of 44 Squadron's crews all became prisoners on 10/11 December 1940, after a successful minelaying operation in the Gironde estuary, on the French Bay of Biscay coast, from which the U-boats ventured forth to attack the British Atlantic convoys. The last message from X3049 came by W/T at 0901 hours; it read "OK, down in France", and the answer to the question as to just how it came to be transmitted was supplied by Peter Tebbutt, who had flown as second wireless operator in X3049. Flak had damaged the fuel tanks of the Hampden and, when they were still many miles from home, the engines began to falter. It had already become light enough for the pilot to attempt an undercarriage-up crash-landing, which he did without injury to the crew who scurried away from their battered Hampden at a commendable rate of knots, anticipating that it would catch fire. X3049 refused to oblige voluntarily however, so the crew returned to the aircraft to offer such assistance as was available in the fire raising line. After a couple of Very cartridges had been fired into the fuselage the Hampden began to go up in smoke, so Peter decided it was time to depart. As he made to leave the aircraft he noticed that the wireless transmitter was switched on. So, from that far away field in France, he tapped out on the Morse key "OK, down in France". Peter then climbed out over the side from the rear upper position and made another fast time that must have made him a strong contender for the "Sprint from a Burning Hampden" Trophy. It must be acknowledged that the prospect of fuel and/or bombs exploding was a great incentive towards achieving a personal best for the hundred yards sprint - one lad even managed it with both ankles broken.

ARRIVALS BY PARACHUTE

It has not been possible to ascertain exactly how many Hampden "Kriegies" simultaneously qualified for membership of the "Caterpillar Club" on the occasion of their arrival. Some are, in any case, included in the "Escape by Parachute" section but a few examples are included here to demonstrate the differing circumstances. The varying fortunes of five squadron leader pilots who ordered their crews to "Jump" show that even with the advantage of the "seat-type" parachute they could still be unlucky.

P4375 (8): Squadron Leader J G Macintyre of 44 Squadron ordered his crew to bale out after his Hampden had collided with a balloon cable 8,000 feet over Hamburg. His three crew landed safely but the Squadron Leader's body was found near the crashed Hampden; his parachute had opened but, possibly due to damage from flak fragments, had not functioned properly. The navigator of P4375, Andrew Coveyduck, had been surprised that German balloons were operating at such a height, as the Intelligence Officers believed their maximum operating height was around 3,000 feet, which was what the British balloons reached at that time. Bomber crews who reported having seen balloons flying at a much greater height over Germany were airily advised by the "experts" - "Must have been smoke from a shell burst, old chap". The Germans had in fact developed a High Altitude Balloon capable of operating at over 10,000 feet. P4375 is the only confirmed case of a Hampden being brought down by an enemy balloon, although "friendly" balloons accounted for several more.

P4287 (9): Squadron Leader F A Willan of 50 Squadron also had problems on a Hamburg op, the exact nature of which are unclear. Whatever the trouble was, his wireless operator was able to tap out the dramatic words "Jumping for it" which were received back at base, before baling out. As all four men reached the ground safely, one gets the impression that it was a fairly orderly affair.

AD912 (10): Squadron Leader E P Willcox DFC of 83 Squadron came to grief over Holland; F/Sgt Jack Lawrence, the wireless operator, vividly

Colditz Castle, to which persistent escapers and trouble – makers were sent from other POW camps. The German adjutant of Colditz, Rheinhard Eggers, still remembered Peter Tunstall as being "the worst of the naughty boys" when he wrote about Colditz forty years later. Peter clocked up over 400 days in solitary confinement during his Kriegie days. (Photo: Stadler, Colditz)

remembers what happened.

"None of us even saw the night fighter that shot us down. We were at 14,000 feet and it was just after midnight. The initial burst of fire demolished the port wing and set the engine on fire. I am not sure whether the pilot was hit but one of the bullets removed the earpiece from my own flying helmet. The Hampden went into a spin and the pilot gave the order to bale out". Jack, the navigator and the second wireless operator, landed safely by parachute, although Jack's landing place, actually within the perimeter of a labour camp, did not allow any thoughts of evading capture to even begin to form as he was captured immediately. Squadron Leader Willcox may have been wounded in the attack and was killed, apparently still in the Hampden when it crashed. Both he and Jack Lawrence were on their second tour of operations and were "showing the ropes" to the navigator, Pilot Officer Organ, and the other wireless operator, Sergeant Gabbitas, the latter having arrived at Scampton three days earlier; the new boys were on their first and only operational flight.

In the course of his Kriegie days Jack, who had been with 61 Squadron at Hemswell from 1937 and done his first tour from there, met up with a number of his pals from Hemswell and Scampton. He also changed identities with an Army private for a while and went out on working parties in the hope that an opportunity to escape would arise and on one occasion whilst labouring at an aerodrome at Goerlitz, he was told that a complete Hampden was stored in one of the hangars there, but never had an opportunity to look at it.

AD900 (11): Another Hampden that suffered at the hands of a night fighter was AD900 of 144 Squadron, again over Holland; its pilot, Squadron Leader Colin Rawlins, subsequently gave this terse account.

"After a short fight with a night fighter and some minutes flying on one engine, the aircraft stalled. I recovered from the stall, but the Hampden began to burn. We eventually baled out at about 1,000 feet, and the aircraft crashed and burned up."

The crash site is at Hoogkarspel. Two of the crew, who may have been mortally wounded in the attack, failed to escape from the Hampden and are buried at Bergen in Holland. At the time when Colin Rawlins told his story some forty years after the event, he was the Director of Regents Park Zoo in London. Did the sight of the animals in their enclosures ever remind him of the days when he was himself caged up for four years?

AD916 (12): A fifth Squadron Leader, New Zealander Neil Svendsen DFC of 83 Squadron, was a veteran of many operations, one of which is described in the "In The Sea" chapter (see 5B2). Neil's "Kriegie" connection came about when he was flying in AD916; this was not his usual aircraft and its performance was not at all to his satisfaction, which was probably why they had been the last aircraft to arrive over Dusseldorf, where they received the undivided attention of the anti – aircraft gunners. Neil described what happened:

"One shell burst underneath us, putting the starboard engine out of action and, as this old crate had difficulty flying on two engines, you can imagine how it went on one. We went into a spin from which I managed to recover and I promptly ordered the crew to bale out, as I did not want to make a single – engine crash landing in the dark. After a minute or so I called up each member of the crew on the intercom and, receiving no answers, I assumed they were all out so I then baled out myself. What happened to poor Irving I will never know."

Pilot Officer Irving, the navigator, was killed and is buried at Duren. The other three joined the "Kriegie" and "Caterpillar" clubs and became guests of the Third Reich for the duration. When Neil wrote from Papakura, New Zealand, he recalled many of his comrades from his Hampden and Prisoner of War days.

P4346 (13): Neil Svendsen's sadness at the loss of his navigator in AD916, is echoed in accounts from other pilots who lost one or more members of their crew. Flight Lieutenant Barrett had suffered the attention of the Du sseldorf heavy anti – aircraft batteries in P4346 a year earlier, when a shell had exploded so close to the Hampden that the tail assembly was either blasted off completely to such an extent that there was no response at all to the controls, so that the aircraft simply plummeted earthwards. The pilot had no alternative but to order his crew to jump and make good his own escape before the aircraft crashed. Sadly, in the case of P4346, no – one else managed to get out.

AD788 (14): The contents of a letter that Flight Sergeant Hind wrote home from Stalag VIIIB, six months after he became a Prisoner of War, survive in a file in the Public Record Office, apparently because the Intelligence Branch thought that the phrase "Going for a Burton" and "missing a bacon and egg breakfast" had some hidden meaning. Other Hampdens operating over Kiel that night saw an aircraft shot down over the target and this seems likely to have been AD788.

Flight Sergeant Hind's letter was to his mother and father; it began by saying how pleased he was to receive his promotion to Flight Sergeant and to hear that he had been awarded a Distinguished Flying Medal. He explained some of the misfortune that had overtaken him and his crew.

"It was a terrible time with the fires, but I stuck trying to help Hancock and Co (his crew was Sergeants Ray, Whiting and Hancock) until I was exhausted. Then God stepped in, and with the tail gone it was hopeless, but it was duty."

The proof of that "terrible time" shows up later in the letter; almost apologising for being a nuisance, he wrote:

"So sorry about details of my injury. You have heard all now, apart from slight head burns. My nose is somewhat scarred but my face is coming on well. Left leg and hand are well scarred too. My trousers were burned right off me and I was a sorry mess when I reached earth."

Flight Sergeant S N Hind DFM was another of the Volunteer Reserve pilots, as his service number 745404 shows. The contribution of these "VR" pilots, in 1941 in particular, would not be overstated if one said that the majority of the Hampden pilots at this period had arrived from that source, and the "VR" contribution to the Observers, Air Gunners and Wireless Operator/Air Gunners should also be borne in mind.

In the same file that contains Flight Sergeant Hind's letter, there is a message from a "secret source in Denmark". It accompanies a photograph of three RAF survivors (not identifiable as Hampden lads) taken by an illicit photographer on Kolding railway station with the explanation "There were great demonstrations of sympathy. People rushed the station and deluged them with cigarettes; sweets and apples, whilst the Germans vainly tried to hold them back, and were furious". Not relevant to the Hampden File, maybe, but it demonstrates the high regard in which the Danes held the RAF lads, who were at that time Britain's only real means of taking the war to the enemy.

P2087 (15): That the name of P/O Vollmer features in the "Kriegie" list and not the 44 Squadron Roll of Honour seems to be due to a miracle. John Vollmer, a pilot, recently arrived on the Squadron and making his first operational flight as navigator of P2087, sat in the nose of the Hampden. The bombs had been dropped on Krefeld and they were near Munster when disaster arrived. A heavy AA shell scored an almost direct hit, set the starboard engine and wing on fire and gave the pilot no option but to order his crew to bale out. Before John had managed to clip on his parachute, P2087 lurched drunkenly to starboard and plunged down in a spiralling dive of rapidly increasing velocity. Trapped in the smoke – filled fuselage and aware of the raging inferno of the starboard wing, John awaited the imminent crash that would end his days on earth; his only thoughts were of the sorrow that his death would cause to the folks at home. The sound of a terrific bang, followed by a shattering tumult of disintegrating metal and perspex, climaxed in a final, roaring, fiery explosion that convinced John that he was about to solve the mystery of whether there really was a life hereafter.

The sudden awareness that he was no longer inside the fuselage of the stricken Hampden but outside it and falling rapidly towards the earth galvanised John into action. He had, somehow, still held on to his parachute which he frantically attempted to snap into position on the two clip hooks of his harness on his chest apparently without success. John knew that the ground could not be far below him. In a last desperate gamble he held on to the parachute with one hand and pulled the ripcord with the other. The parachute opened with a surprising jolt, was wrenched from his grasp, breaking two of his fingers in the process and for a few brief moments John stared at the face of Eternity. It was dispelled by a sudden jerk on his parachute harness and a rapid decline in his rate of descent until, a few moments later, he made a safe but muddy landing in a peat bog. In his frantic fumbling earlier he had managed to make one of the clip hooks secure, maybe with the help of his Guardian Angel who had certainly worked hard that night. The pilot had also escaped safely but John's two companions were out of luck and, trapped inside the blazing tangled mass of metal that had been P2087, lost their lives when it smashed into the ground.

AD732 (16): The "Angel in Charge" of 61 Squadron's aircrew had snatched PO Davis (in P4343 – see 2E3) from the very Jaws of Death over Wilhelmshaven in July 1940. On 18 April 1941, she hovered over Cherbourg as AD732 dropped down from the cover of cloud, in daylight, and roared across a nearby Luftwaffe airfield, unloaded its bombs across the fighters on the ground at their dispersals then headed back up into the grey refuge of the stratus above. Just as the Hampden crew believed that they were out of danger, a well-directed cannon shell from the ground defences transformed the port engine into a mass of flame and smoke. After operating the engine fire extinguisher which appeared to have no effect, the pilot, F/Lt Aldridge, realising that the blaze was feeding on fuel from damaged tanks and that AD732 was doomed, gave his crew the option of baling out or remaining in the Hampden as he attempted to make a crash landing.

Ken Downing, in the lower gun position, decided to get out, clipped on his parachute and pulled the emergency release on his escape hatch; precious seconds had to be wasted in kicking away the door which refused to jettison. When the door did disappear Ken quickly followed it out into the slipstream, pulled his ripcord, felt his parachute jerk open, then remembered nothing until he realised that he was on the ground. He was so close to the burning wreck of AD732 that the flames had already set fire to the canopy of his own parachute; it was obvious that no one else had survived the crash. Ken dragged himself away from the flames and exploding ammunition. Unable to walk he half-crawled, half-clawed his way across the ground until German airmen running to the scene of the crash found him and an ambulance took him to the Naval hospital at Cherbourg. Those on the ground who had witnessed Ken's escape estimated that it had been from a height of some 40 metres. He was congratulated on his amazing escape next day by the Luftwaffe Commander at Cherbourg. In the course of his stay in hospital recovering from the broken right thigh and leg, injuries which he had sustained as he hit the ground, and later whilst a Kriegie, Ken had ample time to reflect on his lucky escape and offer thanks to the Guardian Angel of No.61 Squadron.

AD962 (17): A few months later, three of the crew of AD962 baled out near Brest when their Hampden was set on fire after being attacked by a Bf 110; again it was a daylight op, this time against the battlecruisers at Brest. The navigator was the unfortunate crew member who failed to get clear of the aircraft, and he is listed in the Runnymede Memorial Register.

AE300 (18): The wireless message "SOS 50 miles from Sylt. Starboard engine cut", was the last that was heard from AE300 after it had taken off from Coningsby at 2140 for a bombing raid on Rostock. Reports in the Civil Defence records at Copenhagen, forwarded by Birger Hansen, show how the various members of the crew were eventually tracked down and taken prisoner; the details of all the identification marks on parachutes, harnesses, "swimming jackets" and flying helmets are meticulously recorded. The police at Horsens certainly seemed to have been kept busy, and one of the dogs used in the search no doubt earned himself a tasty piece of liverwurst for finding "the helmet of an airman", which was marked 758247 Banister (yet another VR pilot). One entry certainly highlighted how one of the crew had been left with (or without) a permanent reminder of the end of his operational flying; it read: "A torn-off finger and a door from an aircraft were found near Daugaard. The finger belongs to the wireless operator, who is in Hornsyld Hospital". The door would almost certainly have been the rear lower door through which the Air Gunner, Sergeant Davies, and the Wireless Operator, Sergeant Philson, would have escaped. The air gunner should have been the first to leave the aircraft but if the "Emergency Release" on the door had been difficult to operate, Sergeant Philson may have offered to assist in releasing it. The door was jettisoned by pulling hard on a wire cable that disengaged the hinge pins at the top of the door to allow it to fall free; Sergeant Philson's finger was probably trapped behind the wire cable so that, when the door went, the finger went with it.

UNCONVENTIONAL ARRIVALS

The most frustrating way to become a Kriegie was to arrive by parachute and then learn later that your aircraft had got back to England, as happened in the following cases.

L4067 (19): The pilot of L4067 ordered his crew to bale out after flak had caused severe damage to the Hampden's rudder and ailerons. Pilot Officer H O Jones and Corporal W G Smith landed safely and became Kriegies, the third man perished when his parachute failed to open. The pilot, Pilot Officer Coton, then nursed L4067 back to England, where after attracting the attention of our own defences, he abandoned the aircraft, which crashed at Rainham in Essex.

P1355 (20): When P1355 was hit by flak and set on fire as it went in to attack invasion barges at Antwerp, a chain of events was set in motion that was to result in the Wireless Operator, Sergeant John Hannah, being awarded the Victoria Cross for his action in extinguishing the fire and the pilot, Canadian Pilot Officer C A Connor the DFC. The rear gunner, Sergeant George James, had no option but to bale out from the inferno that his position had become, and the navigator, Sergeant Hayhurst, for

Australian Alex Gould (P4358) and South African Peter Tunstall (P4324) at Spangenburg. The photo was taken just before Peter made what he considered to be his best escape through the gate in the background. Dressed as a German officer, he was saluted by the guard as he walked out. Peter sent the photo back to England, with a secret message concealed in the split photo, to MI 9. He asked for the return of the head-and-shoulders, suitably modified, for use on a forged "Ausweiss" (Identity Card). The modified photo that MI 9 returned is shown in the small inset. He received it while in Colditz but never had a chance to make use of it. A typical forged "Ausweiss" is shown on the right.
(Photos: Peter Tunstall)

Four photographs of the burial, with full military honours of John Shaw of 49 Squadron who became a POW when he crash – landed P4404 in northern France on 6/7 December 1940. John attempted to escape from Stalag Luft a few weeks later and was shot by one of the guards as he stood with his hands held up to show he knew he had been seen trying to get through the barbed wire. This infamous deed aroused great anger among his fellow – prisoners and in an attempt to placate their ire, the Germans accorded John Shaw's body a burial with full military honours. Clockwise, the photos show: (1) fellow – prisoners falling in on parade. (2) At the graveside, PO B.H.Evans, John's navigator (extreme left) with Wing Commander "Wings" Day on his left, salute. Sergeants David Young and W.K.O'Leary, also from John's crew, wait to lay wreaths. It was a tragic irony that PO Evans was later himself to die by a German bullet. He was one of the fifty officers murdered by the Germans when recaptured after "The Great Escape". (3) The coffin is lowered into the grave. (4) The German Guard of Honour fire a volley of shots in final tribute. (Photos from David Young)

whom this was supposed to have been the end of his operational tour, heard the order to prepare to abandon the aircraft and soon afterwards made his own escape, believing the Hampden to be doomed. The pilot flew P1355 back to Scampton (see 2F2); the two sergeants became Kriegies. P4392 and L4051 (21): The "Guardian Angel" at Scampton may have made a good job of directing two Victoria Crosses towards Scampton within the space of just over a month. In addition to John Hannah, Flight Lieutenant R A B Learoyd received the award for an attack on the Dortmund – Ems canal on 12 August, but this same Angel seems to have had a weird sense of humour. Two members of different 83 Squadron crews were convinced that either the order to bale out had been given or that their Hampden was so out of control that it was time to leave. Pilot Officer J C Bowman baled out of P4392 over Kiel on 21 July 1940, and on 8 August Sergeant T F Newton departed from L4051. The mention of the word "parachute" when the night sky was filled with bursting anti – aircraft shells and the aircraft was cavorting madly about the sky, is reported to have been the cause of one of the departures. The word "parachute" had been used by the pilot in a request for a parachute flare to be dropped; the misunderstanding resulted in a long spell behind the barbed wire for the unfortunate individual.

ARRIVALS BY DINGHY

Compared with the number of Hampden Kriegies who arrived in crash landings or by parachute, the number who came in from the sea in their dinghy, or were picked up at sea in a dinghy or when supported by their "Mae West", was comparatively small.

P4350 (22): This crew were probably the first complete Hampden crew to be picked up from the sea and taken prisoner after attacking a target near Stettin and coming down in the sea near Calais, almost certainly because they were out of fuel.

AE193 (23): AE193 was 40 miles north west of Texel when both engines cut. It was obvious that there was no chance of surviving if they baled out so far from land and in the dark, so the pilot ordered the crew to prepared to "ditch" (i.e. go down on the sea). The aircraft hit the sea with an almighty wallop and the crew only just managed to scramble clear and get into their dinghy seconds before the Hampden sank.

In the course of the next two days the four sergeants endured rough seas, cold, seasickness and a very heavy rain storm; then the weather changed and they had two long, cloudless, hot days. Sleep was impossible and they had no food whatsoever but, using their flying jackets as sails and making paddles from their parachute harnesses, they tried to make headway in the direction of the Dutch coast. Ships and aircraft came into view from time to time but none saw their frantic signals as they waved from the dinghy. The wireless operator became too ill to help with the paddling and the navigator became delirious. All seemed lost when a big wave hit the dinghy and tipped the pilot into the sea; the next wave capsized the dinghy and all four men found themselves in the water. Their immediate frantic attempt to right the dinghy ended in shouts of disbelief, as each in turn realised that his feet could touch the bottom; the water was only chest deep! Tired and weak as they were they waded towards the coast, towing the dinghy behind them. The water became shallow enough for them to pick up mussels from the sea bed; they ate them greedily, and trudged on. There seemed to be no end to the vast expanse of shallow water and at last fatigue overcame them so, rigging a make – shift anchor for the dinghy, they clambered aboard and tried to get some sleep. The night passed, daylight arrived, and the sun shone down on them with a dazzling, blinding brilliance, its warmth aggravating the raging pangs of thirst that had been intensified by the saltiness of the mussels.

Just after lunch time on the fourth day, a small fishing boat appeared on the horizon and after much desperate shouting and waving they attracted its attention. The crew took them on board and made them as comfortable as possible, at the same time trying to explain that they would not be able to take them back to Zoutcamp until the tide had flooded for a few more hours. The physical condition of the four airmen made it impossible for them to attempt to evade capture and they agreed that they would on no account put the lives of their Dutch fishermen friends, and their families, at risk by asking for further assistance. The Germans were informed of their presence after the Dutch had plied the airmen with hot coffee and cigarettes; their next stop was Groningen where they received medical attention and more food and drink before moving to the prison in Amsterdam, and eventually to the ranks of the Kriegsgefangenen.

AT257 (24): An article, printed in German of course, appeared in the journal of Luftflotte Nord, the German Air Force in Norway, in February 1943. It is headed "Four from the other side", and as it relates to a New Zealand Hampden crew, the "other side" could be interpreted as "from the other side of the world".

The German war correspondent, Walter Henkels, describes how the 489 RNZAF Squadron Hampden torpedo-bomber came to grief after flying too close to a small German merchant vessel, which proved to be armed with light anti-aircraft guns manned by gunners who knew how to use them. The encounter proved fatal for the Hampden, putting its starboard engine out of action, so that soon afterwards it had to put down on the sea where the crew took to their rubber dinghy. The Hampden remained afloat for only a very short time; the words used to depict its disappearance are worth recording "... the sea licked inquisitively at the machine and then opened its jaws like a monster and with a gargling, yawning slurp, swallowed the Hampden." The crew were picked up by a Norwegian fishing boat and landed ashore where the Germans took them into custody. Walter was present soon after the men arrived. He expressed the sentiments of the captors thus:

"Guy Gaskill, the radio operator, sat smiling at us and said "Yes, for us the war is over...we're at least all still alive but New Zealand is a long way away". Suddenly we all realised that these four men, sitting there quietly, had come thousands of miles from the other side of the world, to show that they cared for a world that was not even theirs, and one that they did not know." So, the crew of AT257 joined the "Kriegies". This story has a happy ending, as, with the assistance of the New Zealand "Opotiki News", it was found that all the crew had survived and had returned to their native land and, in 1984, were all alive and well. Bruce McKenzie, AT257's navigator, had "pleasant memories of the Hampden as a very reliable aircraft. With a good range, as a platform for launching torpedoes it was excellent. Its only drawback was not much comfort on patrols of up to eight hours duration". Guy Gaskill wrote to say that he had returned to the same dairy farm, at Opotiki, to resume the life that his service in the Royal New Zealand Air Force had interrupted.

AE359 (25): The problems that beset a crew of an aircraft that went down on the sea, and the drama of the struggle for survival, are evident in the brief letter that Sergeant Laing, the navigator of AE359, wrote from captivity. The Hampden had gone down in the sea after bombing Brest on the night of 31 January/1 February 1942; the weather at the time was bitterly cold. Sergeant Laing wrote:

"I am safe and unhurt; we were picked up from the sea after about twenty hours in our dinghy. The two wireless operators are safe here, but our pilot Sergeant Le Queux is almost certainly drowned, as he never made it to the dinghy. We could not get near him and lost him in the darkness. The sea became rough next day, we were capsized once, but got back into the dinghy safely. We were pretty cold when we were picked up but were given dry clothes and plenty of hot coffee, which did us a power of good."

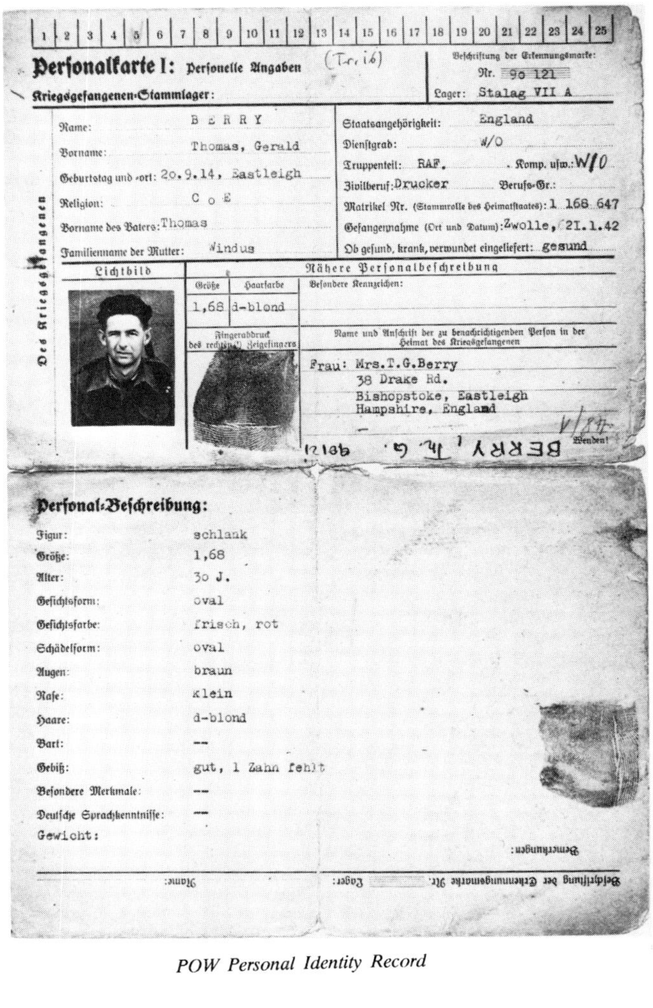
1 2 3 4 5 6 7 8 9 10 11 12 13 14 15 16 17 18 19 20 21 22 23 24 25

Personalkarte I: Personelle Angaben (T.r. i6)

Kriegsgefangenen-Stammlager:

Beschriftung der Erkennungsmarke: Nr. 90 121

Lager: Stalag VII A

Name: BERRY

Vorname: Thomas, Gerald

Geburtstag und -ort: 20.9.14, Eastleigh

Religion: C o E

Vorname des Vaters: Thomas

Familienname der Mutter: Windus

Staatsangehörigkeit: England

Dienstgrad: W/O

Truppenteil: RAF. Komp. usw.: W/O

Zivilberuf: Drucker Berufs-Gr.:

Matrikel Nr. (Stammrolle des Heimatstaates): 1 168 647

Gefangennahme (Ort und Datum): Zwolle, 21.1.42

Ob gesund, krank, verwundet eingeliefert: gesund

Des Kriegsgefangenen

Lichtbild

Nähere Personalbeschreibung

Größe	Haarfarbe	Besondere Kennzeichen:
1,68	d-blond	

Fingerabdruck des rechten (!) Zeigefingers

Name und Anschrift der zu benachrichtigenden Person in der Heimat des Kriegsgefangenen

Frau: Mrs.T.G.Berry
38 Drake Rd.
Bishopstoke, Eastleigh
Hampshire, England

BERRY, Th. G. 90121

Wenden!

Personal-Beschreibung:

Figur: schlank
Größe: 1,68
Alter: 30 J.
Gesichtsform: oval
Gesichtsfarbe: frisch, rot
Schädelform: oval
Augen: braun
Nase: klein
Haare: d-blond
Bart: --
Gebiß: gut, 1 Zahn fehlt
Besondere Merkmale: --
Deutsche Sprachkenntnisse: --
Gewicht:

Bemerkungen:

Beschriftung der Erkennungsmarke Nr. Lager: Name:

POW Personal Identity Record

The unfortunate pilot had, obviously, failed to get into the dinghy with the other three before the Hampden sank and, even had he only been a few feet away from it, the wind would soon have carried the dinghy away from him. At that time of the year the coldness of the water would have brought about his death from hypothermia in a matter of an hour or so. The three who did survive had more than a fair share of luck in that they all managed to get back on board the dinghy when it capsized as, even under practice conditions, this was quite a difficult operation.

The Commonwealth War Graves Commission Register "France 1196-1265" contains the details of all British servicemen who are buried in the Cherbourg peninsula area. One of these is twenty-year-old Sergeant (Pilot) Peter Lawrence Edward Le Queux from Barnstaple, Devon, who lies at rest in the churchyard at Coutances on the west coast of the Cherbourg peninsula. Peter is the only British serviceman to be buried at Coutances and his body must either have been washed ashore or picked up at sea. No doubt Peter's three Kriegie crew mates had ample time to reflect on their own relative good fortune in not having suffered a similar fate on that cold winter night.

P4404: A CRASH LANDING AND ITS CONSEQUENCES

On the night of 7/8 December 1940, P4404 (26) of 49 Squadron took off from Scampton; its mission was to seek out Luftwaffe training airfields to the south of Paris and attempt to emulate the success that the German "Intruder" aircraft of NJG/2 had achieved over England. The temperature was well below freezing and light snow was falling as they headed south from Scampton and with a weather forecast that included snow, ice and low cloud the crew could not have been over-optimistic about the chance of their mission achieving anything. Sergeant John Shaw, the pilot of P4404 must have thought that it was his lucky night when he actually found a German airfield, all nicely lit up and with aircraft in the circuit. No doubt with visions of a DFM floating before him, he entered the circuit with the object of shooting one of the Jerries down and then unloading his bombs on the hangars and buildings of the airfield.

The inconsiderate Germans, however, had realised that there was a fox after their chickens. Their radar had probably tracked P4404 in from the coast, so when the aircraft headed in towards the airfield and came within range of the defending light flak guns, it received the full treatment. Streams of tracer poured up towards the Hampden and many shells struck the aircraft, fortunately without wounding any of the crew. When he was clear of the hail of shells, John took stock of the situation and deciding that the Hampden was too badly mauled to justify returning for more punishment he turned on to a northerly course in the general direction of England. Both engines began losing power; it seemed probable that the fuel tanks had been damaged. which could result in a ditching in the cold English Channel from which their chances of survival would be almost nil. John decided that a wiser plan would be to turn south and head towards the Spanish border, so that when they crash landed or baled out, they would have a better chance of evading capture and reaching Gibraltar and, eventually, good old England.

Just as the visions of a DFM had earlier come to nothing, the prospect of a visit to the south of France began to fade as the two stricken Pegasus engines continued to lose power. John Shaw, realising that they were already too close to the ground for the crew to escape safely by parachute, decided that a crash landing was the only possible course of action and ordered his crew to take up their crash positions. He eased P4404 down towards the ground which materialised in the form of a forest, some fifty miles south of Paris. The Hampden crashed down into the trees, the wings were ripped off but the fuselage continued gamely onwards through the trees until it bounced out of the wood and came to rest in a field, which as David Young, one of the two wireless operators in the crew, remembered thankfully "was covered in deep snow". A French family from a nearby farm offered aid but, as some of the crew needed hospital treatment, the offer was declined, particularly as it was known that the Germans dealt harshly with anyone helping RAF aircrew to hide or escape. The crew asked that the Germans be informed of their presence and they were soon in the care of the enemy.

After they had recovered from their injuries the crew were moved to Stalag Luft, on the Baltic coast. David Young had no complaint about the manner in which they were treated at this time as although rations were meagre the Germans never resorted to physical punishment or torture. The attempts at escape from the camp usually ended in recapture and were regarded by their captors as part of the prison camp routine, for which the punishment was a ritual spell of solitary confinement.

Wing Commander "Wings" Day, the Senior British Officer at Stalag Luft, had established an escape committee and in due course John Shaw came up with a simple escape plan, which was to camouflage himself with a white sheet and then do an "Invisible Man" act to crawl across the snow to the wire fences surrounding the compound. He would then cut his way through the wire. The plan received the blessing of the Escape Committee and when it was put into operation all went quite well until a vigilant sentry, operating a searchlight on one of the look-out towers, spotted John cutting away at the wire. The sentry called out to one of the patrolling guards, who challenged John, who, realising that his bid for freedom had ended, raised his hands in surrender and turned to face the guard, who promptly shot him dead.

Whether the shooting had been accidental, maybe in a moment of panic, or as some feared, deliberate, was a matter of great concern to all the prisoners at Stalag Luft. The Germans too, seemed to be aware of the gravity of the incident, as they instituted a full enquiry and decreed that John Shaw's body should be buried with full military honours; David Young was present at the grave-side and sorrowfully remarked that, on this occasion, the rifles of the Luftwaffe Guard of Honour fired blank ammunition. He paid tribute to him thus, "John was a good pilot, and a very brave man. He died a hero's death, doing his duty for his country. I hope that his family have heard of the circumstances of his death."

David lost track of Pilot Officer B H Evans, the navigator of P4404, during his subsequent years of captivity. He was appalled to learn that he too had been shot by the Germans in an act of cold-blooded murder after what has become know as "The Great Escape".

HAMPDEN KRIEGIES INVOLVED IN THE GREAT ESCAPE

Fifty officers were, in fact, executed on the direct order of Hitler, having been recaptured after escaping through a tunnel from Stalag Luft III at Sagan. This infamous act of mass murder has been the subject of official enquiries, books have been written and a ludicrous film made about it; it is generally known as "The Great Escape". The nominal roll of those who made their way out of that tunnel lists seventy-nine men and besides Pilot Officer B H Evans' name on the list of the fifty men who were murdered after recapture, there is Squadron Leader J Catanach DFC, an Australian pilot who had become a prisoner of war when his torpedo-bomber Hampden, AT109, force landed on a beach in northern Norway after being damaged by light flak on its flight to Russia (see 2J11).

Two more Hampden men were recaptured after the "Great Escape" and eventually returned to captivity. Pilot Officer "Pop" Green, a 44 Squadron Air Gunner Officer, who had been awarded a Military Cross in the 1914-18 war, became a Kriegie in July 1940 (see L4087) and was spared the murder bullet because he was relatively old. Flight Lieutenant R S A Churchill had been a Pilot Officer with 144 Squadron and was the pilot of P4370, one of the first Hampdens to be shot down by a night fighter. He is believed to have been spared because his name was Churchill and if, as the Germans thought, he might be related to Winston Churchill, he was worth more alive than dead. One more Hampden Kriegie was actually arrested at the exit from the tunnel and remained at Sagan; had he been captured further afield he would doubtless have been a candidate for the murder squads. This last officer's name was PO L Reavell-Carter, an Air Gunner Officer and lower gunner on P4305 who had been shot down by flak in a minelaying operation on the Kaiser Wilhelm Canal (The Kiel Canal).

Alan Bryett, himself a Stalag Luft III Kriegie and sole survivor from a Halifax crew, kindly assisted with information about the "Great Escape".

COLDITZ

The Colditz Story has become a legend through books, film and television. The very word "Colditz" conjures up a vision of a grim-looking medieval German fortress, a maximum security camp. It was to this camp, where the guards outnumbered the prisoners, that officer prisoners who had attained notoriety as escapers were transferred and in fact practised their expertise at escaping to such good effect that more successful "home run" escapes were made from Colditz than from any other camp.

Several Hampden Kriegie veterans graduated to Colditz including FO Middleton and FO "Weasel" Donaldson, survivors from L4083, shot down in the daylight attack of 12 April 1940, and Peter Tunstall of 61 Squadron. Peter had arrived in the Fatherland after an unfortunate "error of navigation" when his Hampden P4324, staggering along on its last dregs of fuel after over ten hours in the air, had coughed and spluttered its way down to a landing on a beach that had appeared as the crew prepared to alight in the sea. They might, as far as the crew was concerned, have been anywhere; the observer/lower gunner, Sgt Joyce (a nephew of Lord Haw-Haw, the pro-German who broadcast propaganda on the German radio) confidently asserted that they were in Ireland. The proof that they were not came when, a few minutes later, they looked in through the window of a beach hut and saw to their dismay a popular Dutch magazine on a chair; they had come down on the Dutch island of Vlieland.

Peter met up with several other Hampden Kriegies in the course of his days behind barbed wire prior to his sojourn at Colditz. Among them were "Abdul" Abbott (AE251), "Spooky" Barrett (P4346) and "Bacchus" Baughan and "Gonga" Coste, both veteran Kriegies from the 144 Squadron debacle on 29 September 1939. The misdeeds that resulted in Peter being despatched to Colditz included several escapes, on two of which he attempted to steal military aircraft, the "Luftwaffe" uniform he sported on these occasions having been manufactured in the Kriegie tailor's shop. Although Peter did not succeed in escaping from Colditz after his arrival there in 1942, his activities resulted in his being court martialled five times by the Germans as a result of which he spent a total of 145 days in solitary confinement; he was only saved from serving a further sentence of nine months in solitary by the fact that the war came to an end.

Peter Tunstall and "Weasel" Donaldson, so nicknamed because of his small stature, feature in a number of war books of which P R Reid's "The Colditz Story" is probably the best known. Another account, this time from the German point of view is given in "Colditz Recaptured" by Reinhold Eggars who was the German officer responsible for security at Colditz. Reinhold "regarded Dick Howe as the main thorn in his flesh; the others he regarded as good prisoners, with the exception of Peter Tunstall who, by virtue of his speciality of creating noisy diversions, spent the best part of 400 days in solitary confinement." In the same book, Captain Muller, an assistant to the Commandant and Adjutant's deputy at Colditz, recalled his responsibility for arranging Courts Martial. He was obviously well acquainted with Peter Tunstall whom he remembers thus: "Flight Lieutenant Tunstall was accused in a number of trials and of all the "bad boys" I remember him best". Unsolicited testimonials, indeed. After over 25 years Peter Tunstall was not forgotten by his German gaolers.

Peter's persistence in creating problems for the Germans no doubt made him unpopular with certain Senior British Officers, one of whom subsequently decided that Peter's efforts should go unrewarded. It is ironic that Sergeant M J Joyce, Peter's observer/lower gunner, should have been awarded the Military Medal for his services whilst a prisoner of war. The award was cancelled when British Intelligence discovered after the war that Sergeant Joyce had in fact co-operated with the Germans by obtaining information from newly-arrived aircrew prisoners about their squadrons etc. Sergeant Joyce fared better than his uncle "Lord Haw Haw" who was hanged as a traitor for broadcasting propaganda for the Germans, an act of treason that was really less harmful to the Allied cause than Sergeant Joyce's treachery. If it is any consolation to Peter Tunstall, he does not appear to have been the only Kriegie whose efforts went unrewarded which seems all the more surprising when one realises that, in addition to his passion for making life difficult for his gaolers, his inventive mind developed a method of concealing messages in Kriegie photographic negatives that were posted home.

John Dominey's book "The Sergeant Escapers" contains references to Hampden sergeants, Johnnie Shaw (P4404), Ferdie Farrands (L4087), Neil Prendergrast (P4358) and others. There is also a description of the Hampden as "that God-awful monstrosity of the air". Readers of "The Hampden File" are, as is John Dominey, entitled to a personal opinion; it is to be hoped that this book has presented the Hampden in a light that will merit it a more favourable epithet. The manner in which Pilot Officer Robert Kee, pilot of AD915, became a Kriegie is described in the "Minelaying" section (see 2D10). In post-war years, Robert became an eminent author, historian and radio and television personality. In his book "A Crowd is not Company", he describes his war-time experiences and life as a Kriegie (29).

BITTER ENDINGS

As will have been seen in this chapter, not all those who became prisoners of war survived to see the end of the war. A few died from "natural causes", if such words are applicable when the very circumstances that they endured for several years must have eroded the will to live. Three Hampden "Kriegies" were shot by the Germans during and after escape attempts and at least one Hampden lad, Geoff Watson of L4088, a prisoner since 1940, was amongst those tragically killed and others wounded when, during the very last days of the war, "friendly" fighters strafed their straggling column of prisoners as it shuffled along a road in Eastern Germany, on the move from one camp to another as the Russian Army advanced westwards. The bodies of those who perished in this unfortunate incident are now buried in the Berlin War Cemetery, amongst them at least twenty aircrew and a number of soldiers, mostly from Scottish Regiments. The only evidence to suggest that they all lost their lives in the same lamentable accident is the same date on the headstones, 19 April 1945, almost five years since Geoff Watson had begun his Kriegie existence on 21 April 1940 (30).

EVADERS

There were at least five Hampden aircrew who came down in enemy territory but managed to evade capture and escape from enemy territory. They were FO J A Whitecross DFC, AD834; FO B J A Rennie, AD924; Sgt D McCallum, X3132; F/Lt G C Fisher and FO Van Den Bok DFC, both of AE197. The escape route for all of these was, as far as can be ascertained, via Spain and Gibraltar. Within weeks of returning to 50 Squadron, FO Whitecross was lost in P4408, crashing into the North Sea whilst on an Air Sea Rescue mission, and FO Rennie died in a non-flying accident at a Hampden Operational Training Unit (31).

ESCAPERS

Of all the Hampden aircrew who became Kriegies only two seem to have actually escaped from imprisonment; these were Flying Officer L M Hodges and Sergeant J H Wyatt. FO Hodges, in P1347 (31), had been airborne for nine hours and was on his way back from Stettin when, in the cold light of a September dawn, he sighted a coastline. Being, as the saying goes "uncertain of their position" the crew optimistically decided that the coastline could only be North Cornwall, and that the aerodrome which presently materialised was St.Eval; no doubt with thoughts of Cornish Pasties in mind, FO Hodges reduced height and prepared to land.

As they entered the circuit they received a most unpleasant surprise as the light flak defences opened up to assure them that it was not a good place to land at all. Flying Officer Hodges wasted no time in getting out of range of the hostile gunners but it was not many minutes later, when at a height of 2,000 feet, that the coughing and spluttering of his engines reminded him that he had a fuel problem. He ordered his crew to bale out but by the time he had held the Hampden steady long enough for his crew to escape, he decided that the height was too low to permit his own safe escape so he made a crash landing in a field which turned out to be near St. Brieuc in Brittany. As he went to get out of his cockpit, he was most surprised to see Sergeant Wyatt still in the aircraft. The sergeant had not heard the order to jump, as his intercom socket was unserviceable.

Flying Officer Hodges and his crew all became prisoners of war and while in a prisoner of war camp at Toulouse in southern France Bob Hodges and Sergeant Wyatt managed to escape and returned to England via Gibraltar on 31 July 1941. In November 1942, Bob Hodges, now a Wing Commander, joined 161 Squadron at Tempsford to fly on secret operations, at first using Halifaxes to drop supplies to resistance groups in occupied countries. Bob became the Commanding Officer of the squadron and soon graduated to actually landing in occupied France in Hudsons and Lysanders to set down, and pick up, agents and other important people who were flown back to England.

At least three other ex-Hampden pilots became involved in these hazardous missions, all former colleagues of Bob Hodges. They were Squadron Leader G de G Sells, Len Ratcliffe and Flight Lieutenant Robin Hooper. Robin had the misfortune to have his Lysander become so badly bogged down when he landed in France that he had to set fire to it and go into hiding with the French Resistance for a month. Wing Commander Hodges was the pilot of the Lysander that flew into pick up Robin and the story of these two intrepid pilots had a happy ending as both survived the war and were granted Knighthoods in later years. Hugh Verity's book "We Landed by Moonlight" gives a graphic account of these clandestine operations, including many in which Bob Hodges participated.

The flag and flower bedecked coffin of AC.1 Allan Wilson is guarded by a member of the Koningklijke Marine (Royal Dutch Navy) and a soldier of the Landmacht (Dutch Army). Allan's body was washed ashore on the Dutch island of Schiermonnikoog on 24 October 1939. He was buried with full military honours by the Dutch who were still neutral at that time. Allan Wilson was the W.Op/AG of L4127 which was one of the five 144 Squadron Hampdens shot down on 29 September 1939. (Photo: Wyb Jan Groendijk via Dirk Bruin)

PART 3E: THE DEAD

The inscription on the cross reads "Here lie McKinglay and his three comrades who died on 14.5.40". "McKinglay" was, in fact, Sgt F.W.McKinlay who was observer/lower gunner in P4286 of 44 Squadron which failed to return from an attack on Breda on 14 May 1940. The aircraft crashed at Oosterhout, Netherlands, where they are buried; the cause of the loss is not known. The placing of the tail fin of the Hampden at the rear of the graves was a fitting tribute to the crew. (Photo: Gerrie Zwanenburg MBE)

A total of 1647 aircrew and 16 non – aircrew were killed while flying in Hampdens. The overall casualty figures, when approximated, show that for every five men killed two more were missing and one became a prisoner of war. The symbol (K) or b. appears in the Aircraft Tables against the names of the men who lost their lives and who have a known grave. In the case of natives of the United Kingdom the next of kin of those who died in crashes in the United Kingdom were, whenever possible, offered the opportunity of having their loved one laid to rest in a grave in his local churchyard. The bodies of their comrades from the Commonwealth countries were buried in the United Kingdom, often in churchyards adjacent to their own aerodrome; many of these graves are situated in War Graves Plots, each of which has a Cross of Sacrifice overlooking the neat rows of headstones, the care of which is supervised by The Commonwealth War Graves Commission whose headquarters is at Maidenhead in Berkshire.

The burial of the dead was carried out with varying degrees of ceremonial. On the occasions when the burial was in the man's home churchyard the local RAF station would, if the next of kin so desired, provide a bearer party for the coffin; at aerodromes such as Castle Bromwich on the outskirts of Birmingham, the handful of aircrew based there became quite familiar with the sombre routine of pall bearer duty with its attendant cluster of heart – broken women and grieving menfolk around the grave – side. The dead who were laid to rest in the churchyard near their own aerodrome, on the other hand, were treated to such ceremonial as the circumstances permitted, with the volley of rifle shots and the sounding of the Last Post being mandatory on such occasions. One such funeral took place when three of the four men killed when L4046 crashed at Corby on 18 July 1939 were buried at Waddington.

Each wreath and flower – covered coffin lay on its own trailer unit of the articulated transport affectionately known to the RAF as a "Queen Mary"; behind them a long column of hundreds of airmen tailed back from the guardroom gates. The RAF band from Cranwell headed the procession and, on a shouted order from the Station Warrant Officer, began blasting forth, sending the plaintive strain of the "Dead March in Saul" soaring up into the cloudless blue sky.

The initial progression was far from smooth, as the drivers sought desperately to keep their charges moving at the very slow speed required to match the measured slow marching of the men, but by the time the column came in sight of the churchyard entrance, some semblance of order prevailed.

Observer Sergeant E W Jones was one of the men being buried; he had arrived on 44 Squadron less than a month earlier, direct from training school; six of the lads with whom he had trained, now acting as his pall bearers, flanked the flower – bedecked coffin. One of them, Harry Moyle, remembered reflecting sadly on the unreality of it all as he slow – marched beside the body of his erstwhile friend.

"Bees and butterflies, busy in flowers around the bier,
Urged on by warmth of sun in sky so clear,
All seemed at peace, as though to say
"Death has no place on such a day"."

At the entrance to the graveyard the cortege halted, pall bearers lifting the coffins from the trailers and carrying them through the shuffling, sobbing group of mourners that had gathered to bid farewell to their loved ones.

"The coffins lowered in the ground,
Rifles crack and bugles sound,
The ranks reform, their duty done,
And then quick march away as one"

The march "Colonel Bogey" will always be remembered by servicemen as having associations with defiance and vulgarity. The choice of that particular melody for the jaunty quick march back to camp had the suggestion of being a gesture of defiance towards the Almighty, who had ordained that these men should die. He was to have a busy time during the next six years.

The Grim Reaper harvested his crop of Hampden dead in many different ways, sometimes singling out one man, more often scything away a complete crew. The manner of their departures varied too, a few had time to contemplate the outcome of that fatal journey, some never even realised that they were making it. It may be of interest to hear the recollections of the lucky ones who, already convinced that they were about to die, escaped from the very jaws of death by a matter of seconds. None recalled screaming in terror, nor did panic arise, and that oft – publicised filling of trousers failed to materialise – maybe things happened too quickly for that particular gut reaction to develop anyway.

Not one man had visions of his past life flashing before his eyes, nor did the Angelic Host and the Pearly Gates spring to mind. Once self preservation had ceased to be a possibility the thoughts that occupied most men's minds were of anger and frustration at being about to die, and of regret that their demise would cause anguish to their kinfolk in the form of a mother's Christmas being spoiled, a girl friend being left waiting at a rendezvous, a beloved wife becoming a widow. The less fortunate ones, who kept their tryst with death, must have perished with similar thoughts in mind. Death treated some more kindly that others; sometimes a well – aimed bullet, a broken neck or fractured skull left a man's body whole, on other occasions death arrived with such a shattering impact that the aircraft and its human content were smashed beyond recognition.

The 408 Squadron Record Book describes the manner in which P5292 (1) met its end as it returned from Cherbourg in the small hours of 15 December 1941: –

"This aircraft crashed at Newfarm, Longdown, near Christchurch and was entirely burnt out. Only three bodies were recovered from the debris, and are so far unidentifiable. The fourth member of the crew has not yet been recovered and instructions were issued that a communal burial should take place. Two bodies were eventually identified, all four were buried at RAF Calshot." (They are buried in the churchyard of Fawley All Saints Church.) The demands of the distraught father of one of the unfortunate men, that he be allowed one last look at the mortal remains of his beloved son, were understandably refused.

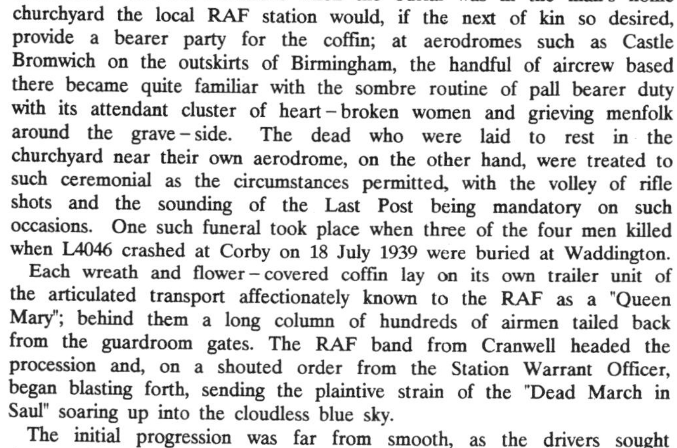

AE250 VN – B of 50 Squadron and crew at Swinderby. This crew subsequently perished in AE367 on 12/13 October 1941 and are buried together in Reichswald War Cemetery. From left to right: Sgt N.R.(Rolf) Lawson, W.Op/AG, PO W.E.(Wally) Hinchcliffe, Second pilot/navigator – bomb aimer, PO J.M.(Jock) Waddell RNZAF, pilot, from Auckland and Sgt R.W.(Ronny) Curtis, W.Op/AG. (Photo: Jim Mitchell)

Harpswell (St.Chad) churchyard near Hemswell is typical of many other adjacent to wartime airfields and thirteen Hampden aircrew are among the twenty servicemen buried there. Crashes on take – off accounted for two men from L4163, one from X3006 and one from AD720. One man from AD725 and another from L4137 were killed in training; the other seven were killed in crashes on return from operations, the four crew of P4345 which crashed at Felixstowe after hitting a balloon cable and one man from each of the crews of L4119, P4365 and X2987. (Photo: Kenneth Green)

John Enns of 415 RCAF Squadron, on the other hand, having seen how a fellow Canadian pilot had crashed to his death was well aware how easy it had been for the remains of his late friend's lofty frame to have been contained in a coffin four feet long. There were other instances of crews being well nigh obliterated in crashes and when L4063 (2) flew into a hillside near Jedburgh (See 4K4) the destruction to aircraft and crew was so great that two of the crew were apparently not found and their names appear in the Runnymede Memorial Register as having no known grave. The crew of AD753 (3) at least received a military funeral when they flew into a hillside near Blessington in County Wicklow, Eire.

"Crashed and caught fire" are words that appear with almost a monotonous regularity in the write – offs, particularly at the OTUs where the trainee Hampden pilots, practising to acquire their flying skills, were particularly vulnerable. The prospect that death might arrive in such a manner was without doubt the one that aircrew dreaded most and the grounds for such fears is obvious in the horrendous crashes in which Hampdens crashed and caught fire with members of the crew trapped inside the wreckage. The utter ghastliness of such incidents is apparent in crashes described elsewhere in this book e.g. AT193 (see 4J30), X2959 (3C21), P2099 (5C3), P4358 (4E2). In the early days of their flying careers most of these young men had an unshakeable belief in their own invincibility i.e. "Jack would be all right". Accidents would only happen to the other crews, and they at least began their operational careers with the same unshakeable belief that "Jack" would not be the one that failed to return. Even in the mythical "Jack" crews it would have been an interesting experiment to have fitted some form of pulsometer to measure their heart – beat.

It is one hundred per cent certain that on take – offs in general, and on operational take – offs in particular when the engines had to drag the aircraft into the air with its bomb load and a load of three tons of 100 octane fuel on board, that as the throttles moved forward and engine revs increased a commensurate rise in heart – beat would have been indicated even in the least sensitive of men, and that it would not have returned to normal until the undercarriage and flaps were raised and a safe height attained.

Death came to at least three Hampden men in another unpleasant form when they baled out and fell to their deaths with parachutes unopened, but is is possible that one at least may have been rendered unconscious through hitting his head on the tailplane as he jumped. The entire crew of L4054 perished after they had had the extreme misfortune of descending by parachute and landing in the sea (See 5A5); others took to their dinghies and died before rescue arrived (See Part 5B).

On the occasions when crew members were being kept afloat by their Mae West life – jackets their hopes of staying alive were largely governed by the temperature of the sea. In winter they would be lucky to last for an hour; in high summer it would be longer. Hypothermia was the medical term for this form of death and it was basically caused by the man's body temperature being gradually reduced, this in turn acted upon the brain and numbed it so that the victim became unconscious and died just as peacefully as though he were asleep. Of all the ways of dying, death from hypothermia would seem to have been the least unpleasant according to accounts from men who were picked up when they had already begun to succumb to its effects and had arrived at a state of semi – consciousness which, although they did not realise it at the time was the harbinger of death. Several men drowned either because their Mae Wests had failed to inflate or because they were not wearing one, and this "normal" sort of drowning would have been no different from the tragic accidents that happen around the peacetime waters of Britain. The rapidity with which death could arrive as the result of hypothermia is described in Part 5 which relates to the loss of AE428.

The Hampden men who lost their lives and are buried in the United Kingdom and Canada represent only a part of the casualty total; many more lie in graves in Europe and again it is the Commonwealth War Graves Commission (CWGC) that attends to their care. Some of the burials are in separate War Cemeteries, others are in church – yards, sometimes in separate War Graves plots overlooked by a Cross of Sacrifice. The earthly remains of others lie in smaller groups whilst here and there one finds a lone airman occupying a grave in a village church – yard. Sergeant Peter Le Queux of 144 Squadron, pilot of AE359 (4) is an example of a solitary burial, he lies in the churchyard of Creances on the west coast of the Cherbourg peninsula. His three crew mates survived as prisoners of war and the sad circumstances in which Peter Le Queux was lost are described in Part 3D.

The loss of AD727 (5) and its crew is a sad instance of how the members of a crew could perish together yet be buried apart. Three of the crew had been involved together in X3005 on 23/24 March 1941 when they had struggled back from Kiel after the port engine had seized up after being hit by anti – aircraft fire; they made landfall on the Yorkshire coast and baled out safely near Driffield. On 10/11 June 1941 AD727 with 19 – year old Flying Officer Peter Pritchard at the controls, took off from Hemswell to lay a mine in Kiel Bay. The aircraft failed to return and is understood to have crashed into the sea at Buelk near Kiel although the cause of loss is not known. The body of Peter Pritchard from Marylebone was washed ashore on Maglehoj beach on Lolland Island, Denmark on the 12 August 1941, two months after the crash and he was buried in Kappel churchyard on 15 August. The names of Peter's navigator Pilot Officer Caunter – Jackson and his wireless operator, Flight Sergeant Breene, are both listed on the Runnymede Memorial Register as having no known graves. The fourth member of the crew, the lower gunner, wireless operator air gunner 21 – year – old Sergeant John Bestwick from West Bridgeford near Nottingham is buried in Kiel War Cemetery. A letter printed in The Times in 1946 shows the depth of feeling and esteem in which our friends in Denmark held Peter Pritchard. The letter, which had been send to Peter's mother, Mrs Edith M Pritchard, reads: – "Dear unknown relations of the English airman Pritchard, With the deepest feelings of compassion I send you, who are unknown to me, a last message from the airman Pritchard ... Pritchard's body was washed to our beautiful shore one fine summer's day in 1941. Gently he was, by Danish hands, put on the grass and soon the ground all around him was decorated with flowers – red ones, white ones, blue ones – for we knew indeed that he was a friend of ours, although unknown to us. We should have liked to bury him ourselves in Danish earth. But the Germans came to take charge of the funeral ... We were not allowed to decorate the tomb with wreaths and flowers, because in this way we showed our sympathy with the English. Nevertheless flowers and wreaths appeared there, more that the Germans could take away, and at last they gave up ... Now the tomb is decorated by friendly hands – as if it were one of our

F/Sgts B.A.Dakin, D.Pearce and V.Whelpley are buried on Schiermonnikoog. All were Canadians from 415 Squadron's AT245 which was shot down by flak 3 km north of the island at 0011 hrs on 28 June 1942. The fourth member of the crew, pilot WO I.N.Garfin, is buried at Sage War Cemetery, West Germany. (Photo: Wyb Jan Groendijk)

own, especially at Church festivals and every Saturday evening throughout the summer – as is the custom in this country ... Pritchard fell for his country in the fight against Nazi tyranny, but he also fell for our freedom. So we are thankful to him, and to you, that are unknown to us."

The names of Hampden crew members appear on War Cemetery headstones from the north of Norway, southwards through Denmark to Germany and eastward in Germany as far as Berlin, thence westward through Holland, Belgium and France to the Bay of Biscay. Finland and Russia also have their Hampden dead from the Russia detachment of September 1942 (See Part 2J) but, although these men were accorded a military funeral at the time, the whereabouts of their graves is not now known and they are listed in the Runnymede Memorial register as having no known grave. Burials in neutral Eire and Sweden are in graves with Commonwealth War Graves Commission headstones.

Eire honours RAF dead. The crew of AD730 are buried with full military honours at Blessington, Co.Wicklow. The Hampden from 50 Squadron had flown into a nearby hillside as it returned from Berlin on the night of 17/18 April 1941. (Photo: Irish Times via Damian Smyth)

The Kiel War Cemetery Register

A study of the Kiel War Cemetery register provides an interesting insight into the manner in which the fortunes of war attended the Hampden crews. The bare statistics at the beginning of the Kiel register show that of the 897 burials, 803 are "Air Force", comprising 581 RAF, 147 RCAF, 33 RAAF, 33 RNZAF and 9 Polish Air Force; there are also 62 unidentified Air Force burials. A closer study of the 803 Air Force names reveals that 68 of the men were from Hampden crews, amongst them are representatives of every Hampden squadron except 408 RCAF, and a check of squadron records show that they came from 28 Hampden losses on operations which covered every facet of the aircraft's activity. Sixteen of the 28 aircraft came to grief on mine – laying operations or on bombing attacks on Kiel itself, where the docks and warships under construction, being fitted out or repaired were regular targets.

Hampdens destined for, or returning from, bombing operations against Christiansand, Berlin, Hamburg, Bremen, Munster and Dusseldorf added another 8 aircraft to the total; the remaining four were lost on Hampden torpedo – bomber operations. The aircrew categories of the 68 men were: – 20 pilots, 9 observers, 37 wireless operator/air gunners and 2 air gunners. Twelve officers were among the casualties, all of them pilots. These figures do not give a true picture of the casualties for each of the four crew positions in the Hampden as pilots sometimes flew as navigators, observers as navigators or lower gunners, and whilst a wireless operator always manned the rear upper gun position there was frequently another in the lower position (See Part 5A). The casualties materialised from the following Hampdens: – Ten complete crews from P1171, P2082, P4323, P4361, AD796, AD827, AD984, AE236, AE256, AE430; four crews from which only the pilot survived as a prisoner of war: AD788, AE202, AE299, AE389. The lower gunner of X3127 and the wireless operator of L4112, their crew colleagues became prisoners of war. The wireless operator of AT121, whose lower gunner crew mate is buried in Holland whilst the other two crew became prisoners of war. The navigator of AE155 which, with the other three crew members returned safely to Scampton. (See 5A25) From the remaining ten Hampdens none of the crew had survived, but only one member is actually buried at Kiel; from L4064, P1319, X3127, AD727, AE189, AE239 and AN153 he is the lower gunner; from L4112 and P1153 the wireless operator and from L6055 the pilot. The circumstances in which one member of the crew of L4064 and two from AD795 came to be buried at Kiel are something of a mystery. The crew of L4064 were seen to be safely in their dinghy after going down in the sea on 12 April 1940 (see 5B22), and although the dinghy was re – located

Kiel War Cemetery on the Baltic coast of West Germany is the last resting place of 888 UK and Commonwealth servicemen and nine Polish airmen. 803 are aircrew and of these 68 are from Hampden crews. (Photo: Commonwealth War Graves Commission)

The body of Peter Prichard of 61 Squadron, pilot of AD727, awaits burial at Kappel Cemetery, Denmark. (Photo: Birger Hansen)

Burial of the crew of AD969 at Bevern. The group on the right are probably from nearby POW camps; most appear to be soldiers and sailors. (Photo: Peter Soltau)

next day it was empty. The most likely explanation is that the dinghy had been overturned during the night by a wave and that the four men had been unable to clamber back on board. The bodies of three of the crew were never found but that of the lower gunner is buried at Kiel some 200 miles away from where he and his crew mates had taken to their dinghy.

AD795 was believed to have ditched 60 miles east of Rattray Head off NE Scotland on 14/15 September 1942, yet two members of its all New Zealand crew are buried at Kiel; the other two Kiwis were never found. It is possible that the reported position of the ditching was inaccurate, but if it were correct the two men who are buried at Kiel had drifted over 300 miles. As happened in other tragic cases in which at least one member of the crew lived to tell the tale, as for example happened to Pilot Officer Mahn in AT240 (See 2J31) all or most of the crew of AD795 may have taken to their dinghy successfully and drifted across the North Sea until all succumbed to the effects of exposure and hunger. The pilot and navigator may have perished first and been buried at sea by their comrades who also later died and whose bodies may have remained in the dinghy until they were either washed ashore or picked up at sea by a patrolling seaplane or passing ship and taken to Kiel for burial. The explanation of these, and other, arrivals at Kiel Cemetery can only be surmised; the deeds of valour and stories of self sacrifice that attended their deaths will never be known.

The majority of the Hampden aircrew who lie at rest at Kiel were around twenty years old; a few were in their late twenties. One noticeable exception is the navigator of AE236, Sergeant Pilot Eric Durham who, at thirty-eight years of age, was old enough to have been the father of the nineteen-year-old wireless operator on that crew, Patrick Duffy; they are buried together with the other crew members of AE236 in Collective Grave number 9-12 in Row C of Plot Two.

The Hampden men at Kiel originated from every corner of the United Kingdom and of the Commonwealth; the various English, Welsh, Scottish and Irish accents would have complemented those of their overseas comrades who had come from Victoria, BC and Saskatchewan in Canada, from Auckland, Dunedin in New Zealand, the Cook Islands in the Pacific, Victoria in Australia, Cape Town in South Africa and Madras in India.

Most of the Hampden fliers had made their last take-off from the aerodromes around Lincoln, and the dates of their arrivals at Kiel Cemetery began on 12 April 1940 and ended in May 1943, with those after mid-1942 being from the Coastal Command torpedo-bomber Hampden squadrons. It would obviously not be true to say of these overseas burials that "There is some corner of a foreign field that is forever England" but the sentiment is appropriate and the stickler for truth may mentally substitute "Wales" "Canada" etc as required for the word "England". Peter Pritchard would have been the first to agree that the tribute to him, from Denmark, that:- "He fell for his country in the fight against Nazi tyranny, but he also fell for our freedom", was one that should be shared by all his aircrew comrades who gave their lives in those dark days of war.

The graves of the crew of AD969 of 83 Squadron, which crashed at Bevern, 45 km NNE of Bremen on the night of 22/23 June 1941 when apparently dazzled by a searchlight. The crew are buried at Bevern burial ground. The cross on the right bears the name "R.J.Heavens, English airman"; the other three are marked "Unknown English airmen". After the war the bodies were identified and moved to Becklingen War Cemetery. (Photo: Peter Soltau)

The Runnymede Memorial commemorates the members of the Air Forces of the British Commonwealth who while serving with, or in association with, the Royal Air Force and operating from bases in the United Kingdom and Europe in France, The Netherlands, Belgium, Germany, Denmark, Norway, Finland, Luxembourg, Czechoslovakia and Russia, lost their lives and have no known grave. The names of 20,436 men and women of the British Commonwealth are commemorated on the Memorial, 759 of who were from Hampden crews. The overall total is made up as follows:

Royal Air Force	*15,318*	*Women's Auxiliary Air Force*	*14*	*Royal Canadian Air Force*	*3,050*
RAF personnel from Newfoundland	*25*	*Royal Australian Air Force*	*1,397*	*South African Air Force*	*17*
Royal New Zealand Air Force	*576*	*Royal Indian Air Force*	*7*	*Air Transport Auxiliary*	*10*
Air Training Corps	*4*	*Ferry Command*	*9*	*War Correspondents*	*2*
		British Overseas Airways Corporation	*7*		

(Photo: Commonwealth War Graves Commission)

PART 3F: THE MISSING

The Runnymede Memorial commemorates 20,436 members of the Air Forces of the British Commonwealth who, whilst operating from bases in the United Kingdom or Europe, lost their lives and have no known grave. Sited on Cooper's Hill, within the Urban District of Egham, the Memorial overlooks the valley of the Thames, and the names of the airmen commemorated are inscribed on stone panels on the cloisters. In the centre of the cloister rests the Stone of Remembrance and in the beautifully engraved glass of the great window of the shrine, two angels hold a scroll on which appears the following verses from Psalm CXXIX:

"If I climb up into Heaven, Thou art there: If I go down to Hell, Thou art there also. If I take the wings of the morning: and remain in the uttermost parts of the sea. Even there also shall Thy hand lead me: And Thy right hand shall hold me."

The names of 759 Hampden men are recorded on the memorial, most of them having been lost on operational flights. A further 18 men were similarly missing on Hampden training flights at 32 OTU in Canada, but being outside the area covered by the Memorial, their names are not included upon it. Thus the overall total of Hampden men without a known grave is 777, the abbreviation (M) being used in the table of individual aircraft to indicate this. Many of the men who are listed as "Missing" are known to have perished in aircraft that crashed in the sea after being shot down by fighters in such incidents as the early daylight operations against Wilhelmshaven on 29 September 1939, Kristiansand on 12 April 1940, and the "Channel Dash" which, with its associated minelaying operations, accounted for 46 men missing in February 1942. Night fighters accounted for something like a hundred Hampdens, many of which were shot down over the North Sea with the loss of the entire crew; in some cases bodies were recovered, in others they were lost without trace.

Engine failure, sometimes attributable to damage from flak or fighters, accounted for other aircraft which vanished whilst limping back across the North Sea under circumstances such as described in the chapter on ditchings. The chance of an aircraft making a successful ditching at night when conditions were good was slim; on a dark and stormy night they were non – existent and although some of the reports of Hampdens which went down in the sea out of fuel may seem to indicate otherwise, the likelihood that anyone who did survive a ditching would be picked up was, particularly in the early war years, very slender to say the least. Many unidentified airmen were washed ashore on the coast of Europe and representatives from Hampden crews are certain to be amongst them, but even when allowances are made for this and the possible causes of losses previously mentioned have been taken into account, it is apparent from the aircraft tables that at least seventy – five Hampdens disappeared completely. Entries in the Squadron record books on these sad occasions vary from a blank space or a dash in the "Time Landed" column to a brief "Failed to Return", "Missing" or "Nothing was heard from this aircraft after it took off".

In the course of subsequent years, the manner in which at least two crews came to grief has been brought to light. Both were aircraft that had crashed on remote mountain sides, AE436 in Sweden and AN132 in Western Canada. Time alone will tell whether the reasons for any more of these unexplained losses will come to light (1).

John McWilliams during his W.Op/AG training in Canada at Calgary in 1941 with his girl friend Doris. Before John (who was always known as Jack) left for England, Doris gave him a cigarette case with the inscription "From Doris to Jack 14.6.41". The discovery by Dirk Bruin of Vlieland of a 1946 report in his local police archives referred to a cigarette case with the same markings having been found in the wreckage of an unidentified aircraft that had been discovered by fishermen on a sandbank. The wreck has still not been positively identified, its remains having sunk beneath the sand but Dirk and Harry Moyle are convinced that it is AT236 of 415 Squadron, which was missing on 29/30 May 1942 and it is hoped that a dig to recover identifiable wreckage will prove their contention.

(Photo: Richard McWilliams, brother of Jack McWilliams)

The seventy-five Hampdens shown as totally missing do not represent the whole of the unfortunate story as that total does not include other aircraft that were lost through undetermined causes but from which at least one crew member's body was subsequently either washed ashore or picked up at sea.

One Hampden that comes into this category is AT236, from which the body of wireless operator Flight Sergeant Conway washed ashore near Wilhelmshaven; he is now buried at Sage War Cemetery. It now appears likely that a hitherto-unidentified aircraft, discovered on a sandbank on the Dutch island of Vlieland in 1946, will turn out to be AT236.

Dirk Bruin, who lives on Vlieland, found the report during his research in the local Police records; it describes how fishermen had come across the wreckage which contained "burned human remains". The only positive clue to its identity was a cigarette case bearing the initials "J.M." on the outside and the inscription "From Doris to Jack 14.6.41" on the inside. Dirk asked Harry Moyle for assistance in identifying the wreck.

A combination of good fortune and detective work, in which he was ably assisted by Eric Grove in Canada, eventually led Harry to suggest that "J.M." was John McWilliams, the wireless operator in the all-Canadian crew of 415 RCAF Torpedo-Bomber Squadron Hampden AT236, which had failed to return from an attack on a convoy some 60 km east of Vlieland on the night of 29/30 May 1942.

This notion was based on the fact that John McWilliams' brother Richard, from Cornwall, Ontario, confirmed that, although John was christened as such, he was always called Jack. Richard also provided a photograph, taken at Calgary in 1941, of Jack and Doris.

Doug Stallard, who had served with Jack on 415 Squadron, recalled that Jack's nickname had been "Smokey" because he was a heavy cigarette smoker, so the cigarette case was an appropriate gift for Doris to have given Jack before he left for England.

Dirk Bruin, after talking to a fisherman who had actually seen the wreck in 1946, worked out its likely location and eventually persuaded Captain Petterson of the Royal Netherlands Air Force recovery unit to visit Vlieland with his team of experts to carry out a search. Their detection equipment located metal objects 3 to 4 metres beneath the sand and, although the depth was too great for an immediate attempt to be made to recover any part of the wreckage for identification purposes, Dirk intends to make such an attempt in the late summer of 1989.

The search for Doris has now been extended to Australia, as it is believed that she may have emigrated there after the war as the bride of a Royal Australian Air Force lad, whom she probably met whilst he was under training at Calgary.

The loss of so many aircraft and men in unknown circumstances provides food for thought as to just how they met their end. From the accounts already given of aircraft managing to return after suffering engine failure or being damaged by enemy action, it would seem fair to assume that a proportion of the missing aircraft were lost through similar causes and that circumstances had prevented an SOS or any indication of the aircraft's plight being transmitted back to base.

The absence of such a message is not difficult to explain. Enemy action may have put the wireless operator, and/or his equipment, out of action as happened to L4111 (2). Failure of the aircraft's transmitter or an error by its operator when tuning manually to the required frequency could result in his desperate out-pouring of dots and dashes being unheard by the listening ears at base. Another fact that has to be taken into consideration was that the range over which communication could be established varies according to the height at which the aircraft was flying. For example, at 12,000 feet a range of 200-300 miles was normal whereas at 100 feet on a torpedo bomber mission, the operational range would be greatly reduced. Similarly, an aircraft flying at 500 feet on a minelaying operation over the Baltic would have little chance of its message being heard if it suffered damage or engine failure, whilst its chance of climbing to a reasonable height for effective W/T communication would be minimal.

Confusion about the content of a message can also lead to disaster as apparently happened to AE437 (3) which ditched 300 miles west of Brest; a failure of the ground station equipment at Waddington led to L4063 and its crew crashing into a hillside 100 miles from base. (See 2K4)

Misinterpretation of the Morse Code dots and dashes was easy enough in the comfort of a classroom on the ground; in the noisy, uncomfortable wireless operator's position in the Hampden, it was obviously much more likely, particularly at the end of a flight which may have lasted for over eight hours when mental fatigue would be a contributory factor. Calamity could arise, for instance, from the morse code figure "2", which is "..- - -", being mistaken for the figure "1", which is ".- - - -", with the consequence that the course given to the pilot might be "190" instead of "290". The wireless operator was not, however, the main source of such problems; more often than not it was his expertise that resulted in a safe return to base.

Accounts given elsewhere in this book demonstrate only too well the navigational problems experienced by night bomber crews and it seems reasonable to presume that many of the unexplained Hampden losses were due to such causes. A navigator misreading a chart, miscalculating a course, using an incorrect wind speed or direction or failing to make allowance for the effect of height on the air speed indicator reading, or a pilot setting a wrong course on his compass were all ingredients for disaster. There are cases on record of pilots setting a reciprocal course on their compasses i.e. setting "red on black" instead of "red on red". On one such occasion, a Hampden pilot took off on a dark and stormy night and circled over base as he climbed through the cloud layers before, as he thought, setting course for the Fatherland. An hour later the cloud cleared to reveal to the incredulous eyes of the navigator a well-defined estuary instead of the blank expanse of the North Sea. The estuary was soon identified as the River Mersey and the pilot's embarrassment at discovering that he had been flying westwards instead of eastwards can be well imagined.

Two incidents involving Hampdens carrying "Imp" impact land-mines which were basically sea mines fitted with a fuse that operated when the mine hit the ground, may well point to the cause of other losses. Photographs of the remains of X2996 (4) which had set out from Waddington to drop an "Imp" mine on Berlin on the night of 14/15 November 1940, suggest most strongly that the aircraft had blown up in the air. How else is it possible to explain the fact that houses in the vicinity of the wreckage have escaped without their windows being blown in?

It is possible, of course, that X2996 suffered a flak hit that detonated the mine but the fact that on 20 December 1940 it had been found necessary for X2981 (4) of 61 Squadron to carry out a drop trial of an "Imp" mine on the coastal bombing range at Theddlethorpe strengthens the suspicion that the mine-detonating fuse mechanism may have been under investigation. All three men in the crew of X2981 were killed when it blew up as the mine detonated prematurely. Other missing Hampdens may have been carrying Imp mines when they were lost, and over a dozen Hampdens on sea-mining operations disappeared without trace. It is conceivable that magnetic mines could have been affected by a build-up of static electricity on the aircraft and the possibility of a mine-layer descending too low and actually flying into the sea to be lost without trace is another likely answer to a few of the unsolved disappearances.

The fact that several Hampdens fell victim to our own balloon barrages at night, points to the possibility that there may have been others, and the fact that RAF fighters shot Hampdens down in daylight suggests that their night fighter colleagues may have made similar mistakes.

ABBREVIATIONS

A&AEE	Aeroplane & Armament Experimental Establishment
AAS	Air Armament School
ATGp	Atlantic Transport Group
AAU	Aircraft Assembly Unit
A/C	Aircraft
AC 1	Aircraftman First Class
A/F	Airfield
AFDU	Air Fighting Development Unit
AG	Air Gunner
AGS	Air Gunners School
AOS	Air Observers School
APC	Armament Practice Camp
ASR	Air – Sea Rescue
ASV	Air – to – Surface Vessel (Radar)
ATA	Air Transport Auxiliary
ATC	Air Training Corps
b	Buried
Bbg	Bombing
BDU	Bombing Development Unit
Bf	Bayerische Flugzeugwerke (Messerschmitt)
BGS	Bombing & Gunnery School
(C)OTU	(Coastal) Operational Training Unit
CFS	Central Flying School
CGS	Central Gunnery School
Cr	Crashed
Cv	Converted to
DBF	Destroyed by fire
Do	Dornier
E/A	Enemy Aircraft
EF	Engine failure
F/Sgt	Flight Sergeant
Flt	Flight
F/Lt	Flight Lieutenant
FO	Flying Officer
FPP	Ferry Pilots Pool
FTR	Failed to return
Gp	Group
HP	Handley Page
HQ	Headquarters
(I)	Injured
Ju	Junkers
LAC	Leading Aircraftman
(M)	Missing
Met	Meteorological
M/L	Minelaying
MU	Maintenance Unit
N/F	Night fighter
NFD	No further details
NonOP	Non – operational
Nr	Near
OAD/PU	Overseas Aircraft Delivery/Preparation Unit
Op	Operation
OTU	Operational Training Unit
(P)	Prisoner – of – war
PO	Pilot Officer
RAAF	Royal Australian Air Force
RAE	Royal Aircraft Establishment
RCAF	Royal Canadian Air Force
Rhod	Rhodesian
RLG	Reserve Landing Ground
RNZAF	Royal New Zealand Air Force
Sgt	Sergeant
SOC	Struck off charge
S of TT	School of Technical Training
Sqn	Squadron
S/Ldr	Squadron Leader
TB	Torpedo – bomber
TDU	Torpedo Development Unit
TO	Take – off
Trg	Training
TTU	Torpedo Training Unit
U/C	Undercarriage
Wg Cdr	Wing Commander
WO	Warrant Officer
WopAG	Wireless operator/air gunner

K4240, the first prototype Hampden, displays its angular lines on a test flight over Bridgemarch Island on the River Crouch in Essex (RAF Museum 5732–15)

PART FOUR: AIRCRAFT TABLES

180 Handley Page Hampden Is delivered between August 1938 and June 1939 by Handley Page, Cricklewood

Serial	Units	Date	Cause	Crew	Duty	See
L4032	AAEE & RAE	31. 5.43	SOC; to 2711M 5.41			5K2 1–6
L4033	AAEE & RAE	31. 5.43	SOC; to 2782M 5.41			2F1 1–6
L4034	CFS/49	23.11.39	Crashed into hangar at Waddington in bad weather	S/Ldr P McG Watt; Cpl T Keating; AC1 S Taylor; AC2 F Talbot; AC1 L McGarvie; LAC W Kelly; (All K)	Training	
L4035	AAEE/16 OTU	11.11.41	Stalled on take–off, Croughton	Sgt A Stack RCAF; Sgt J Yeo RCAF; Sgt S Carberry (All K)	Training	6B15
L4036	49(EA–R)	11/12. 8.40	Shot down by flak during raid on Dortmund	PO R F Gower; PO E A Harris; Sgt A J Baum; Sgt P W N Jelley (All buried at Reichswald)	Bombing	
L4037	AAEE/TDU/Cv TB/RAE/144/489	13. 2.43	Engine caught fire; crashed in sea near Wick	Sgt H Wrench RAAF; Sgt R E Burt; LAC A Richardson RAAF (All missing)	Air Test	
L4038	49/106/25 OTU/ 14 OTU/Cv TB/455	12.10.42	Handed over to Russian Navy	Flown out by S/Ldr Holmes, PO Dick, Sgt Whitson, Sgt Olcorn, F/Sgt Raebell	Left in Russia	2J5 2J15
L4039	49	1. 8.39	Overshot landing at Scampton	F/Lt G F Lerwill (unhurt)	Night training	6A1
L4040	49	25/26. 4.40	Missing, cause not known	FO D White; Sgt N L Jones; LAC N Hornsby; Sgt P J Hurst (All buried at Hamburg)	Minelaying in Kiel Bay	2L2
L4041	49/7/16 OTU/ 14 OTU/Cv TB/ 415/	27. 2.44	SOC			

Serial	Units	Date	Cause	Crew	Duty	See
L4042	49/7/16 OTU/ 44/106/408/ 1401 Flt/521	23.12.43	SOC			
L4043	49	14/15. 4.40	Crashed on coast at Ryhope on return	S/Ldr G Lowe DFC (I); PO Beauchamp (I); Cpl Appleton (I); PO A Bryan Smith (K)	Minelaying West Baltic	2K5
L4044	49	8/ 9. 6.40	Engine failed on take-off, Scampton	PO E D Parker; PO Lockhead; Sgt Lloyd; Sgt Robinson (All injured)	Bombing Amiens	
L4045	49(EA–Q)	11/12. 1.41	Crashed in bad weather on return, Northorpe, Lincs. on return	PO H Newhouse; Sgt H Irving; Sgt L Jackson; Sgt P C Prosser (All killed)	Bombing Wilhelmshaven	5A7
L4046	49/44	18. 7.39	Pilot lost control in cloud; crashed, Corby Lincs.	FO D I Jobson; Sgt J A Hawes; Sgt E W Jones; W/Op (name not known) (All killed)	Training	5F6 6A1
L4047	49	17. 1.39	Crashed in bad weather 10m N of Nottingham	PO B A Mitchell (unhurt) First Hampden write-off	Training	6A1
L4048	83/14 OTU	18. 8.41	Stalled and crashed, Barrow near Cottesmore, cause not known	Sgt J Lee; Sgt A D W Clay; PO W Ellis DFM; Sgt J Webber; Sgt F Appleby (All five killed)	Training	
L4049	83	23/24. 9.40	Hit by flak; crashed near Bethen	S/Ldr A Bridgeman DFC(P); PO F Watson DFC*; Sgt J Goodwood DFM*; Sgt A Blatch RNZAF* *buried at Becklingen; (P) = POW	Bombing Berlin	
L4050	83/16 OTU	11. 9.41	Overshot landing at Upper Heyford	Sgt J Ruddock and crew unhurt	Night training	6C9
L4051	83 (OL–M)	14/15.11.40	Crashed at Neuruppin	PO B Arnold; Sgt J C Threlfall; Sgt E Taylor; PO B A de Bressey (All buried Berlin)	Bombing Berlin	5A46
L4052	83	28. 2.39	Overshot Northolt	FO A O Bridgeman unhurt	Training	6A1
L4053	49/83(OL–N)	8/ 9. 8.40	Crashed near Mannheim	PO I M Muir (P); Sgt W Barber (P–wounded) Sgt J Leakey (P); Sgt F Stone (P–wounded)	Bombing Ludwigshafen	
L4054	83	6/ 7. 4.40	Abandoned over Whitley Bay out of fuel	PO W Roberts (K); PO K Brooke-Taylor (K) LAC D W Sharp (M); Sgt A McNichol (K)	Security patrol	5A3 5A5
L4055	83	22/23. 5.40	Flew into hill in fog 4m S of Holmfirth, Yorks., on return	Sgt S W Jenkins; Sgt P W Josse; AC1 W Thornton; Sgt A Marsh (All killed)	Bombing communications	2K10
L4056	83	11. 4.39	Engine cut on take-off, Northolt	PO A B Harrison unhurt; one crew injured	Training	6A1
L4057	83(OL–F)	15/16.11.40	Missing, cause not known	Sgt R Bucknell; Sgt P H McLoughlin; Sgt S Dale; Sgt H S Haste (All missing)	Bombing Berlin	
L4058	83/16 OTU	18. 8.41	Engine cut on take-off, Upper Heyford	Sgt E W McBarnet (Rhod); Sgt D E Davies; Sgt L G Wright; Sgt W B Sykes (All K)	Night training	6B7
L4059	83	18. 5.40	Overshot Newton	Sgt Prichard (I); three crew unhurt	Training	
L4060	49/5 BGS	17. 2.44	SOC			
L4061	83	7. 8.39	Engine cut on take-off, Scampton	Sgt S W Jenkins unhurt	Training	6A1
L4062	50	26/27. 9.40	Crashlanded in fog on return, Lindholme	PO P T Mulligan (I); PO Kilner (I); Sgt C.J.Smith; PO G Leonard (both unhurt)	Bombing Calais	3A4 2K25
L4063	50	16/17. 3.40	Flew into hill, Cocklaw Foot, Cheviots	FO V H Ayres (K); PO P A F Addie (K); LAC A V Wallace (M); Sgt G A Rowling (M)	Security Patrol	3E2 2K4
L4064	50	12. 4.40	Damaged by Bf 109s and ditched in North Sea	PO J B Bull (M); PO A D Pilcher (M); Cpl G Fawcett (M); Sgt W S Nevinson (Lastnamed buried at Kiel)	Day bombing Kristiansand	5B22 2C13
L4065	50	13/14. 4.40	Believed crashed off Skegness on return	F/Lt R J Cosgrove (M); Sgt E W Everatt (M) Cpl J Doran (M); Sgt F W Batchelor (M)	Minelaying Great Belt	2C19 5B23
L4066	49/83	7/ 8. 7.40	Crashed on return, Clacton, Essex	PO O H Launders (K); Sgt R C Hallett (K) Sgt L Howard (K); Sgt B Kinton (K)	Bombing Frankfurt	2K8
L4067	49/144	21/22. 5.40	Damaged by flak; pilot gave order to abandon	PO E H Coton (flew aircraft back); PO H O Jones (P); Cpl W G Smith (P); Sgt C J Shewry (K)	Bombing Krefeld	5A3 3D19 5A12
L4068	49	11/12. 5.40	Crashed near Le Treport after engine failure	PO Drakes; PO Fox; AC1 Watson; Sgt Fennell. All unhurt and returned home	Bombing M-Gladbach	
L4069	83	14/15. 5.40	Dived into ground and blew up near Louth	FO N Johnson (K); PO E A Vagg (K); AC1 S G Pickard (K); Sgt D Wolstenholme(K)	Bombing Ruhr	2K7
L4070	83/14 OTU/ 16 OTU/14 OTU/ 1403 Flt	3. 6.43	Overshot landing and stalled, Gosport	Sgt W Courtney Cox (K); one other injured	Training	5A42
L4071	83	30.10.39	Retracted undercarriage prematurely, Scampton	PO M G Lewis unhurt	Training	
L4072	49	21.12.39	Crashed on approach, Acklington, probably fuel shortage	Sgt Marshall (I); PO Irvine (?); AC1 E Humphrey (K); Sgt K Potts (K)	Reconnaissance off Norway	2C7
L4073	50	12. 4.40	Shot down by flak or Bf 109s, Kristiansand	Sgt G M Wild (M); Sgt D W A Chambers (M); LAC S Kershaw (M); Sgt H Bratton (M)	Day bombing Kristiansand	2C14 6E11
L4074	50/44/25 OTU/ 5 OTU	17. 2.40	SOC			2C11
L4075	50/16 OTU/ 14 OTU/Cv TB/ 455	7. 3.44	SOC			2J34
L4076	50/14 OTU/Cv TB/ 5 OTU/415	6. 9.43	Crashed in sea in circuit, Thorney Island	PO R E Ritscher RCAF(K); PO A K Main RCAF (K); Sgt H J Hanson RCAF (K)	Night training	
L4077	50/49	20/21. 7.40	Crashed out of fuel on return, Hunstanton	PO Michie; Sgt Lowe; AC1 Bellamy; Sgt Corbett (All unhurt)	Minelaying, Baltic Sound	6E11

The blunt nose of K4240 is evident in this view taken at Martlesham Heath during trials (Photo: A&AEE MH5734)

Serial	Units	Date	Cause	Crew	Duty	See
L4078	50	26/27. 6.40	Shot down by flak, Hannover	PO A S Luxmoore; Sgt W C Harding; Sgt G Nichols; Sgt A H Smith (All buried at Hannover)	Bombing Langenhagen airfield	
L4079	50	30/31. 8.40	Crashed at Angerlo, NL; cause obscure	PO K R K Smettem; Sgt A A Horsfall; Sgt E Skinner; Sgt H Best (All buried at Groesbeek)	Bombing M – Gladbach	
L4080	50	17.10.39	Brakes failed while taxying and hit hangar, Waddington	PO R Lloyd unhurt	Training	
L4081	50	12. 4.40	Shot down by flak/ Bf 109s, Kristiansand	PO M Thomas (M); Sgt S Vart (M); Cpl G Barrass (M); Sgt J R Jopling (M)	Day bombing Kristiansand	2C15
L4082	50	20. 3.39	Crashed out of control, Boultham, Lincoln	PO R A McAlister; Sgt W H Freestone; AC1 N Newsham; AC2 F R Greenshill All k.	Training	5F5 6A1
L4083	50	12. 4.40	Shot down by Bf 109s on island off SW Norway	FO M W Donaldson (P); PO F Middleton (P); Cpl A C Henry (P – inj); Sgt J Greet (K)	Day bombing Kristiansand	2C16 3D28 2C20
L4084	50/25 OTU/16 OTU/ Cv TB/5 OTU/415	3. 4.43	Spun into ground, Marsh Farm, Sussex, on fighter affiliation exercise	Sgt D Montgomery RCAF (K); Sgt W Mackay RCAF (K); Sgt J D Scott RCAF (K); Sgt J W Whipple (K)	Training	
L4085	44	1. 8.40	Ditched out of fuel off Aberystwyth on return	Sgt E D Farmer (K); Sgt R D Hobbs*; Sgt K Wood (K); Sgt D Seager* * both picked up by rescue boat	Bombing Misburg	5B35
L4086	44/14 OTU/49/420/ 408/Cv TB/1 TTU	20. 1.44	SOC			
L4087	44	19/20. 7.40	Hit by flak and crashed in sea off Jutland	Sgt E L Farrands (P); Sgt P Nixon, Sgt R Miller (Both b.Skagen); PO B Green MC(P)	Minelaying Frederiks – haven	3D29 3D27 2D4
L4088	44	21/22. 4.40	Hit by flakships; crashed near Kiel	F/Lt F.G.Dutton (P); Sgt J Slowey (P); Cpl W Watson (P); Cpl W Davidson (P)	Minelaying Baltic Sound	3D1 3D30 2D3
L4089	44	21.12.39	Shot down by Spitfires off North Berwick	PO R J Sansom; Sgt E L Farrands; AC1 J Lyttle; Sgt H R Moyle All picked up by fishing boat	Day bombing off Norwegian coast	3D29 2D4 2C8
L4090	44	21.12.39	Shot down by Spitfires off North Berwick	PO P F Dingwall*; Sgt J A M Reid*; LAC T Gibbin (K); Sgt W K Lodge (I)* *Three picked up by boat	Day bombing off Norway	2C9 3C1
L4091	44/16 OTU/ Cv TB/455	11. 5.44	SOC			2J34 2C18
L4092	49	25/26. 7.40	Missing, presumed in sea	FO D P W Rowan – Robinson (M); PO C A Weeks (M); LAC T Hogarth (M); Sgt A F James b.Becklingen	Minelaying Kiel Bay	2L2

Serial	Units	Date	Cause	Crew	Duty	See
L4093	83(OL – J)	3/ 4.11.40	Damaged by flak, Kiel; ditched off Humber	PO C A Connor DFC (K); Sgt G Stubbings (M) Sgt J W C Gibson (M); Sgt R Norris (K)	Bombing Kiel	2F2 3D20
L4094	83(OL – R)	25/26. 7.40	Missing; cause not known but probably flak	PO W O D Tweddell DFC*; Sgt G W Jones*; Sgt P Tointon (P); Sgt D Townsend* *Buried at Reichswald	Bombing Dortmund – Ems Canal	
L4095	83(OL – R)	10/11.11.40	Crashed in sea 25m off Dutch coast	FO C F Kendall; Sgt N Tyson; Sgt A Woodburn; Sgt G R Wilson (All missing)	Bombing Mannheim	
L4096	50	31.10.39	Crashed after take – off, Waddington	Sgt R S W Cordle; AC1 J F Burge; AC2 L J Pentley (All killed)	Night training	
L4097	50	10/11. 9.40	Shot down by flak off Ostend	F/Lt R J Reed*; PO P W Roylance*; Sgt R Halls (M); Sgt W A Johnson (M) *both buried at Ostend	Bombing Ostend	2F3
L4098	44	2. 5.40	Landed with undercarriage unlocked, Waddington	PO P F Dingwall unhurt	Training	
L4099	50	12. 4.40	Shot down by flak/ Bf 109s, Kristiansand	FO W G Taylor; Sgt J R Tonkiss; Cpl H Brown; Sgt W Ison (All missing)	Day bombing Kristiansand	2C10
L4100	44/106/14 OTU	8.11.42	Undershot landing, Saltby	Sgt East and three crew. No serious injuries	Training	5E1
L4101	44	8. 5.39	Undercarriage collapsed on landing, Waddington	PO J D Haskins unhurt	Training	6A1
L4102	44	25/26. 3.40	Crashed in sea on return flight	Sgt S A Williams; Sgt A R Jones; AC1 A C Hamman; Sgt R Cross (All missing)	Night rec. Elbe Esty.	
L4103	61/106	7.12.40	Engine failed in circuit, Finningley	Sgt J D Shilling; Sgt Thompson; both killed. Pilot from No.50 Squadron	Night training	
L4104	61/83	18/19.10.40	Abandoned near Malton, Yorks., out of fuel	Sgt G Loveluck; Sgt J Dall; Sgt F G Constable; Sgt W K Thompson All landed safely	Bombing Hamburg	5A3
L4105	61/16 OTU/ 14 OTU/Cv TB/ 455	2. 8.43	Damaged by flak; crash – landed at Leuchars	FO I H Masson; FO F J Skidmore; FO J B Mullan RNZAF; FO G A Lennox RCAF. All unhurt	Day Torpedo bombing off Norway	2J34
L4106	61/83	18. 8.40	Engine cut on approach, Scampton	FO D W F Barker unhurt	Training	
L4107	61	6. 6.39	Stalled and crashed in sea on air firing exercise, Leuchars	PO A G Newton; Sgt G H Fawcett; AC1 R G Thomas (All killed)	Training	6A1
L4108	61/144/14 OTU	27. 3.42	Hit ground in poor visibility, Saltby	Sgt A R Fraser RAAF; Sgt R F King Both killed	Night training	
L4109	61/14 OTU	13/14. 9.42	Shot down by night fighter 15m SE of Groningen	F/Sgt K Grabek RCAF*; Sgt G Coldron RCAF*; Sgt R Carlson (P); Sgt F Needham (P) *buried at Stadskanaal NL	Bombing Bremen	2A1
L4110	61/14 OTU	8. 3.42	Stalled landing at Saltby	Sgt J F McPhail RCAF (K); Sgt A H Hanson RAAF (K); two others injured	Training	
L4111	61	7/ 8. 3.40	Fuel ran out; crashed on approach, Digby	FO D C G Clinkard; Sgt C C Hobbs; LAC W K Wood; Sgt R P Glasson.All killed	Security patrol, Sylt	2F2 2K2
L4112	61	27/28. 6.40	Hit by flak; crashed in Kiel Bay	PO S B Morley (P); PO N E Hore (P); Sgt W F Anderson (b.Kiel); Sgt B D McPherson (P)	Minelaying Kiel Bay	
L4113	61	14/15. 4.40	Missing, cause not known	Sgt V Emanuel; Sgt R Etheridge; All AC1 P G Graham; Sgt J Atkinson missing	Minelaying Elbe Esty	
L4114	61	14. 6.39	Crashed on landing, Leuchars	PO K A Porter and crew unhurt	Training	6A1
L4115	61/44/16 OTU/ Cv TB/5 OTU	1. 5.43	Swung on landing and undercarriage collapsed, Maghaberry	FO Sandberg (I)	Training	
L4116	61	17. 4.40	Overshot flarepath, Hemswell	PO S B Morley; LAC J N Prendergrast; Both unhurt	Night training	
L4117	61/14 OTU	31.7/1.8.42	Damaged by flak and night fighter; crashed near Loddiswell, Devon	PO D J Curtin (unhurt); PO S J Walker (I); Sgt R Hawkes (I); PO H G Clarke (K)	Bombing Dusseldorf	
L4118	455/Cv TB/ 5 OTU/1 TTU	20. 1.44	SOC			
L4119	61	30.4/1.5.40	Crashed in bad visibility, Croxton Kerrial, Leics on return	FO C W Glover (K); PO J E A Friend (K); AC1 J B Greenhall (K); Sgt C Frost (K)	Bombing Aalborg	2K6
L4120	61/106	16.12.40	Crashed after engine failure, Finningley	Sgt I P Mapp of 61 Sqn unhurt	Training	6D13 5A20
L4121	144	29. 9.39	Shot down by Bf 109s N of Wilhelmshaven	FO J T B Sadler (M); Sgt R L Galloway (P) AC1 J W Cummings (M); Sgt W H Stephens (M)	Day bombing	2C2
L4122	144	3. 7.39	Undercarriage collapsed on landing, Hemswell	PO T E Sanders unhurt	Training	6A1
L4123	CFS/16 OTU	12. 1.41	Undershot Upper Heyford	Sgt W M Davies; Sgt C J Smith (I)	Training	6C8
L4124	144/83/25 OTU	17. 5.41	Control lost on take – off; dived vertically into ground, Finningley	Sgt R S K Hickmott; Sgt D G Crichton; Sgt E D Foster All killed	Night training	
L4125	144/49	7. 2.42	Hit bank of snow on take – off, Scampton	Sgt Reid RNZAF and crew unhurt	Minelaying	2L1
L4126	144	29. 9.39	As for L4121	FO R D Baughan (P); Sgt N M Kirkus (M); AC1 J H Anthony (b.Hamburg); Sgt H H Turner (M)	Day bombing	2C3
L4127	144	29. 9.39	As for L4121	FO N C Beck (b.Sage); PO R Turner (M); AC1 A Wilson (b.Schiermonnikoog); Sgt P E B Sproston (b.Sage)	Day bombing	2C4

L7271, the second prototype Hampden, in natural metal finish. It was later converted to take Napier Dagger engines and become the Hereford prototype.

(Photo via Don Hannah)

Serial	Units	Date	Cause	Crew	Duty	See
L4128	144	2. 6.39	Crashed from 400 ft on approach, Hemswell	PO H S James; Sgt H Kinlock; AC1 D E Shimmin All killed	Training	6A1 5F2
L4129	144/49	16/17.10.40	Shot down by flak	Sgt D S Imber; Sgt R F Rose; Sgt K C Friend; Sgt F Corbett All buried 7m NE of Bordeaux	Minelaying Bordeaux	
L4130	144	24. 8.39	Undercarriage collapsed, Hemswell Became 1647M	PO A P B Hordern and crew unhurt	Training	6A1
L4131	144/14 OTU (PL – A)	10/11. 9.42	Missing, cause not known	Sgt J C Kerr; Sgt H D Chapman; Sgt W O Anderson; Sgt W D Deatherage All RCAF; buried at Reichswald	Bombing Dusseldorf	2F12
L4132	144	29. 9.39	As for L4121	PO R M Coste (P); F/Sgt P Williams*; AC1 S Isherwood*; Sgt C G Williams* *all buried at Sage	Day bombing	2C5
L4133	144/83/14 OTU	6. 6.42	Spun into ground near Saltby	PO Broadley; Sgt Thurgar; Sgt Norris; Sgt Henning All killed	Training	
L4134	144	29. 9.39	As for L4121	W/Cdr J C Cunningham (M); Sgt R E Herd*; AC1 H Liggett (P); Sgt A E Povey (M) *buried at Sage	Day bombing	2C2
L4135	144	27. 5.40	Broke back when taxying with full bomb load, Hemswell	PO A G Trueman; Cpl H Dickenson; LAC A W Bennett; Sgt P Robert All unhurt	Training	
L4136	144	20. 4.39	Stalled and crashed attempting forced landing, Sunk Island, near Hull	PO C E Jones (K)	Training	6A1
L4137	76/144	20. 3.40	Dived into ground near Scunthorpe on navex	FO R H Bullard – Davies; Sgt A Woodward; LAC Dowty All killed	Night training	
L4138	76/7/16 OTU	13. 6.40	Collided with P4339 of 61 Sqn on take – off, Upper Heyford	PO Rowe (I); PO Blythe (I); LAC Evitt (unhurt)	Training	5C5 2F2
L4139	76/106/7/16 OTU	26. 7.40	Undercarriage collapsed on landing, Finningley Became 2170M	PO K R K Smetterm unhurt	Night training	
L4140	76/7/16 OTU/ 408	27/28. 3.42	Shot down by night fighter N of Terschelling	PO G F Beaver RNZAF; WO G P McLean RCAF; Sgt S F Bentley; Sgt R N Beale (All M)	Minelaying Weser Esty	5G3
L4141	76/144/1 AAS/ Cv TB/415	4. 6.44	SOC			
L4142	76/16 OTU/32 OTU	30. 4.44	SOC in Canada			5K8;5K13
L4143	76/144	22. 4.40	Crashed near Kirton – in – Lindsey	PO Robson; AC1 Elcoat. Both unhurt	Air test	
L4144	76/16 OTU/ Cv TB/489	4. 6.44	SOC			5C20
L4145	76/16 OTU/Cv TB/ 489/32 OTU	29. 2.44	SOC in Canada			5K8
L4146	76/61	23/24. 5.40	Forcelanded in Germany out of fuel	F/Sgt T G Ross (P); Sgt J W Wootton (P); Cpl L Walker (P); Sgt E R Corrigan (P*) *died while POW	Bombing rail targets	3D2

Serial	Units	Date	Cause	Crew	Duty	See
L4147	76	24.11.39	Hit trees while night flying and forcelanded at Bicester	PO T C Hynes unhurt	Night training	
L4148	76/16 OTU	22. 4.40	Lost control on test flight; crashed at Fritwell, Oxon	F/Lt B P Jones (Ser I); LAC Holt (K); LAC J Fisher – Dobson (K)	Air test	5A27
L4149	76/106/50	10/11.11.40	Missing, probably in sea	Sgt T Thomas (M); Sgt R C Hyland (M); Sgt W Cave (M); Sgt Hamilton (P)	Bombing Mannheim	
L4150	76/106/50/ 16 OTU	27. 2.44	SOC			5F3
L4151	76/16 OTU	6. 3.41	Stalled on approach to Brackley LG	PO G D G Hudson and crew unhurt	Training	
L4152	76/83	14/15. 4.40	SOS received as aircraft homed on Manston; ditched	FO K R H Sylvester; Sgt E R Clarke; LAC J H Edwards; Sgt G C Perry All missing	Minelaying West Baltic	
L4153	44	26/27. 2.40	Forcelanded short of fuel 3m S of Grimsby; became 1888M	S/Ldr Norris; Sgt Smythe; AC1 Preston; Sgt Nixon. All unhurt	Security Patrol Borkum/Sylt	2K1
L4154	44	20/21.10.40	Forcelanded short of fuel in bad weather near Colchester	Sgt A R Atkins; Sgt Hobbs; Sgt Muir; Sgt McLaren All unhurt	Bombing Berlin	2C18
L4155	7	23. 5.39	Control lost while practicing change of pilot; spun in 5m E of Newark, Notts.	FO J G McGrail; PO G Lloyd; Sgt H K Trencham; AC2 E B Bretnall; AC1 R Lumley	Training	3A1 6A1
L4156	7/16 OTU	26. 5.40	Collided with L4158 over Bristol Channel	PO N G Dryburgh (K); LAC T Baird (M) LAC R J Aitken (M)	Air firing training	5C4
L4157	7/16 OTU Cv TB/32 OTU	30. 4.44	SOC in Canada			5K8
L4158	7/16 OTU	26. 5.40	Collided with L4156 over Bristol Channel	PO R C Frost (K); PO T A Nixon (K); LAC H Sharpe (M); LAC J Whyte (K)	Air firing training	5C4
L4159	7	6.11.39	Hit tree on approach, Upper Heyford	PO A C Manaton unhurt	Training	
L4160	7	29. 3.40	Forcelanded at Brackley LG after engine failure	PO J E Newton – Clare and crew unhurt	Training	
L4161	7	5. 9.39	Dived into ground during turn, Cockwood Farm, near Finningley	PO A R Playfair (K)	Training	
L4162	7/14 OTU	2. 9.42	Crashed after getting OK to land, Thistleton, near Cottesmore	Sgt Wilson (K); Sgt J K Johnston RCAF (K); F/Sgt J J P Wardrop RCAF (K); Sgt P V Norris RCAF (K)	Training	
L4163	7/144	17. 4.40	Hit cottage on take – off, Hemswell	Sgt J Windsor; Sgt R F Carter; AC1 D Burnside; Sgt W Cuthbert. All killed	Minelaying West Baltic	
L4164	7/50/1 AAS	9.10.41	Crashed in bad weather on approach, Manby	Sgt J Cechrzycki. Reported as "One of crew killed"	Training	
L4165	7/144/25 OTU/ Cv TB/415/489/ 5 OTU/1403 Flt/ 1404 Flt	27. 4.44	SOC			
L4166	7/144	19.11.39	Crashed in heavy landing, Doncaster	Sgt H Shields unhurt	Training	
L4167	7/144	30.10.39	Overshot Doncaster	PO W F Tudhope and crew unhurt	Training	
L4168	7/44/50/44/ 16 OTU	31. 8.41	Overshot landing at Odiham	Sgt R N McDougall and crew unhurt	Night training	
L4169	7/16 OTU	30. 7.40	Controls became locked; abandoned; crashed at Bishops Itchington, Wks	PO G E Weston baled out safely	Air test	5A28
L4170	7/16 OTU/ Cv TB/415	6. 1.44	SOC			
L4171	44	23/24. 5.40	Shot down by flak near Aachen	S/Ldr C E Johnson; PO D C Barker; Cpl W Crook; Sgt H Collins All buried at Rheinburg	Bombing rail targets	2C17
L4172	144	25/26. 5.40	Shot down by flak	Sgt A E Jones (P); Sgt J A Smyth*; Cpl T Baybutt*; Sgt C Woodhouse* *Buried at Rheinburg	Bombing rail targets	
L4173	144/14 OTU	30/31. 5.42	Crashed on return, Horsham St Faith	F/Lt W L Cameron; Sgt J E Sheridan; F/Sgt M F Porter; Sgt E H East (All killed)	Bombing Cologne	
L4174	106	31. 5.40	Spun in 2m W of Finningley	Sgt Wilkes, AC1 Everson both unhurt	Night training	
L4175	106	24.10.39	Undershot and hit trees, Finningley	PO A C Dier (I) only occupant	Night training	
L4176	106/7/16 OTU/ 25 OTU	9.11.41	Hit trees on approach, Finningley	Sgt H A Ogden (I); one crew (I)	Night training	
L4177	106	24. 8.39	Undercarriage collapsed, Lossiemouth; to 1883M	PO L J Ashfield unhurt	Training	6A1
L4178	106/44/144	26/27. 2.42	Ran out of fuel and crashed, Hexthorpe, Lincs. on return	Sgt A R Bland; FO H E Aspey; Sgt J A Eves; F/Sgt J B Gray RCAF All killed	Bombing Kiel	2K11
L4179	106	19. 8.39	Crashed into plantation after engine failure 4m SE of Inverness	Sgt R Orbell (I); others unhurt	Training	6A1 6D7
L4180	106	29/30.10.40	In sea 20m E of Spurn Head on return	PO J J Hill; PO J V F Gibbs; Sgt G L Howe Sgt D S Bazeley. All missing	Minelaying Baltic Sound	5B20

L7271 after conversion to a Hereford on test at Martlesham Heath. Note lack of dihedral on outer wing section, the sliding rear upper position and peace–time silver colour scheme with serial on rudder. (Photo: A&AEE MH6438)

Serial	Units	Date	Cause	Crew	Duty	See
L4181	106	6. 7.40	Overshot and hit tent, Finningley	PO G A C Potts and crew unhurt Three killed, two injured in tent	Night training	
L4182	106/1 AAS/ Cv TB/TDU	3.10.44	SOC			
L4183	106	29.11.40	Crashed soon after take–off, Finningley	FO W K Burr–Thomas; Sgt J E Sowden; Sgt J C Wiggall; Sgt Flynn. No serious injuries	Night training	
L4184	106	13.10.40	Swung on take–off and undercarriage collapsed, Finningley	Sgt R Grainger and crew unhurt	Night training	
L4185	106	4/ 5. 7.41	Crashed at Half Acre Creek, Thames estuary on return	PO G F E Greenhalgh; PO G F Wadeson; Sgt R A Gilbert; Sgt G G Hutson All killed; bodies recovered	Bombing Dortmund	5B17 2K21
L4186	106	11.11.39	Stalled on low level bombing practice and crashed, Haxey, Lincs.	PO A B Loew; Sgt W Styles (I)	Training	
L4187	106	7. 8.40	Dived into ground, Owston Ferry, Lincs. Dazzled by searchlight?	Sgt R F Wilson; Sgt A S Dalgress; Sgt W C Goldsmith; Sgt D K Mills All killed	Night training	
L4188	106	1. 9.40	Blew up in air near Halton, Bucks	Sgt A Duncan; Sgt L E Ashforth; Sgt E Knight; Sgt R Capeling. All killed	Night training	
L4189	106	30. 9.40	Flew into high ground near Buxton while lost	Sgt J G Gow; Sgt C O Cook; Sgt L J Burt; (all killed); Sgt N Powell (ser.inj)	Night training	
L4190	7/16 OTU	24. 7.40	Overshot and hit fuel tank, Upper Heyford; burnt out	Sgt L R Adams (I); PO G Hall; PO S C Wise; Sgt F Jowett Last three unhurt	Night training	
L4191	185	28. 9.39	Landed with undercarriage jammed in midway position, Perth	PO Jacklin and crew unhurt	Training	
L4192	185	18. 8.39	Dived into ground from 300 ft after take–off 2m S of Thornaby	PO R R Harper on first Hampden solo killed	Training	6A1
L4193	185/14 OTU/ 1401 Met Flt	23. 7.43	Port engine cut soon after take–off; crash–landed, Bircham Newton	F/Sgt R Wilson; F/Sgt R Buddle; Sgt J Dale; F/Sgt Workman All unhurt	Met flight	
L4194	195/106	22/23.11.40	Hit by flak and ditched off Lorient	Sgt J M Ward; Sgt T W Glenn; Sgt A W Newton; Sgt N W Hull. All buried at Vannes	Minelaying Lorient	
L4195	185/49 (EA–H)	16/17.10.40	Ran out of fuel and crashed, Lenham, Kent	PO Evans; Sgt Molyneux (I); Sgt Potter(K) LAC Glover. One killed, one injured	Minelaying Bordeaux	
L4196	185/14 OTU/ Cv TB/5 OTU	24. 1.44	SOC			
L4197	185/14 OTU	15. 4.40	Undercarriage failed on landing, Cottesmore	PO R R Sandford unhurt	Training	
L4198	185	24. 2.40	Undercarriage failed on landing, Newton Down	PO R P J Radbourne unhurt	Training	
L4199	185/10 BGS/ CGS	23. 6.41	Rolled on to back and dived into sea off Isle of Man	F/Sgt Cannell and three crew All missing	Training	5F7

Serial	Units	Date	Cause	Crew	Duty	See
L4200	185/14 OTU	21. 4.40	Overshot night landing, Cottesmore	PO T A Freke unhurt	Night training	
L4201	185/14 OTU/ Cv TB/5 OTU	10. 5.44	SOC			
L4202	185/16 OTU/ 6 OTU/1401 Flt/521	10.10.43	Lost at sea; presumed ditched after engine failure	No crew details known	Met flight	6B2
L4203	185	30.11.39	Dived into ground in poor visibility, Hughenden, High Wycombe	PO J Musgrave; S/Ldr N Hope; AC2 O'Regan; Sgt Thom All killed	Training	
L4204	185/408/14 OTU/ Cv TB/5 OTU/ 1401 Flt/521	10.10.43	Hit gunpost on take-off in bad visibility, Docking	PO W D Cooper (K); Sgt C Curd (K); Sgt J Allaway (I)' FO J Maxwell (I)	Met flight	5G3
L4205	185	1. 1.40	Stalled on overshoot, Cottesmore	PO E A C Hill (K)	Training	
L4206	185/5 AOS	31.12.42	To GI airframe 2166M at No.3 S of TT			
L4207	No RAF service	.6.39	To English Electric as pattern aircraft			
L4208	No RAF service	.6.39	To Canada as pattern aircraft			5K1
L4209	No RAF service	.6.39	As for L4208			5K1
L4210	Canada/Retd UK/ Cv TB	6. 6.44	SOC. Originally pattern aircraft.			5K1
L4211	Canada/Retd UK/ 1403 Flt/1404 Flt	29. 5.44	SOC. Originally pattern aircraft			5K1

* * * * * * * * * * * * * * * * *

100 Handley Page Hereford Is delivered between August 1939 and June 1940

The Herefords were all built by Short and Harland at Belfast. The first batch of 100 (L6002 to L6101) had been delivered by 8 June 1940. The first production aircraft, L6002, arrived at the Aeroplane and Armament Experimental Establishment, Martlesham Heath, for performance trials on 23 August 1939. The fifty N-series Herefords were completed between June and September 1940. Of the 150 Herefords, twenty-three were converted to Hampdens, thirty were written off in accidents and the remaining ninety-seven were stuck off charge. No Herefords served with operational units.

Serial	Units	Date	Cause	Crew	Duty	See
L6002	Mkrs then 3023M	31. 5.43	SOC			
L6003	AAEE/Napiers then 2800M	26.11.41	SOC			
L6004	CFS/1 STT as 2712M	31. 5.43	SOC			
L6005	185	22. 9.39	Forcelanded after engine failure, Melton Mowbray	PO R Hodgingham unhurt	Training	
L6006	AAEE	2. 7.40	Undercarriage collapsed on landing, Boscombe Down; to 2164M	S/Ldr R G Slade unhurt	Testing	
L6007	185	28.10.39	Undercarriage collapsed on landing, Cottesmore; to 2163M	F/Lt G G Avis unhurt on first Hereford solo	Training	
L6008	185/14 OTU	15. 6.40	Forcelanded after engine failure, Pickworth, Lincs	Sgt D S Matthews unhurt	Training	
L6009	–	12. 4.45	SOC as 3024M at 11 STT			
L6010	AAEE	17. 3.42	SOC			
L6011	See under Hampden conversions below					
L6012	185/14 OTU/5 BGS	20.10.41	SOC and reduced to spares			
L6013	5 BGS	6.12.41	SOC and reduced to spares			
L6014	14 OTU/5 BGS	23.12.41	SOC			
L6015	–	8. 2.45	SOC as 3028M at 12 STT			
L6016	185/14 OTU	9. 5.40	Crashed after engine failure, Tugby, Leics.	Sgt D J Fielder unhurt. LAC D Sharpe (I), LAC J L Brooks, LAC A C Wight baled out	Training	5A33 6D1
L6017	–	26.11.41	SOC as 2801M at Hereford			
L6018–6020	See under Hampden conversions below					
L6021	–	31. 5.43	SOC as 2716M at Hereford			
L6022	–	25. 8.43	SOC as 3020M at 2 STT			
L6023	5 MU	19. 7.40	Engine failed on take-off, Kemble	PO W J W Kingston unhurt	Air test	
L6024	16 OTU	23. 7.40	Crashed on landing, Upper Heyford	PO R G Goode unhurt	Training	
L6025	185/16 OTU	21. 1.44	SOC as 3130M at 7 SS			
L6026	16 OTU	19. 7.40	Crashlanded after engine failure, Weston-on-the-Green; to 2968M	Sgt W W Huggins unhurt	Training	
L6027	16 OTU	13. 1.41	Port engine failed on take-off; swung and hit P2141, Upper Heyford	Sgt P C T Forster unhurt	Training	5C9 6B4
L6028	–	8. 2.45	SOC as 3030M at 12 STT			
L6029	AAEE/TDU	3. 7.41	Undercarriage collapsed on landing, Gosport	FO J D Willis unhurt	Trials	
L6030	5 BGS	22.12.41	SOC and reduced to spares			
L6031	5 BGS	2.11.41	SOC and reduced to spares			
L6032	–	.41	Became 3031M			

L6002, the first production Hereford, on the same piece of taxiway as the previous photograph at Martlesham Heath. This aircraft has dihedral on the outer wing and an improved rear upper gun position canopy. (Photo: A&AEE MH6439)

Serial	Units	Date	Cause	Crew	Duty	See
L6033	24 MU	17. 5.41	Engine failed on ferry flight; crashlanded near Audlem, Cheshire	PO R F Forbes unhurt	Ferry flight	
L6034	–	30. 9.43	SOC as 3018M at 1 STT			
L6035	14 OTU/5 BGS	19. 2.42	SOC at Tollerton			
L6036	14 OTU	30. 9.40	Crashed after engine failure 4m NW of Llanelly	F/Lt N W Timmerman unhurt	Training	
L6037	14 OTU/5 BGS	2.11.41	SOC and reduced to spares			
L6038	–	31. 3.43	SOC as 3095M at 4 STT			
L6039	–	31. 5.43	SOC as 2713M at 4 STT			
L6040	16 OTU/5 BGS	12.12.41	SOC			
L6041	5 BGS	19.11.41	SOC			
L6042	–	8. 2.45	SOC as 3029M at 12 STT			
L6043	–	31. 5.43	SOC as 2802M at Hereford			
L6044	–	12.11.43	SOC as 3021M at 10 STT			
L6045	–	12. 4.45	SOC as 3025M at 11 STT			
L6046	–	9. 2.44	SOC as 3026M at 11 STT			
L6047	14 OTU	17. 5.41	Engine caught fire; bellylanded at Cottesmore	F/Lt R J Dunlop – Mackenzie; LAC Mountain both unhurt	Training	
L6048	14 OTU	12. 4.45	SOC as 2803M at Melksham			
L6049	See under Hampden conversions below					
L6050	–	30. 6.43	Broken up as 2715M			
L6051	–	31.12.43	SOC as 3022M at 4 STT			
L6052	185/14 OTU	9. 8.40	Crashed on take – off, Cottesmore	Sgt Hughes (I); Sgt Grant (K)	Training	
L6053	16 OTU	6. 8.40	Stalled on approach, Upper Heyford	Sgt A J Fetherston killed on first Hereford solo	Training	2F4
L6054	16 OTU	5.10.41	SOC as 2717M			
L6055	16 OTU	21. 2.41	Engine failed on take – off and undercarriage torn off. Converted to Hampden (see below)	FO R W Reynolds unhurt	Training	
L6056	16 OTU/5 BGS	19.11.41	SOC			
L6057	16 OTU	20. 8.41	Engine failed on take – off, St.Angelo	PO W R Kelly	Training	
L6058	–	1. 6.41	SOC			
L6059	16 OTU	5.10.41	SOC			
L6060	–	.41	SOC as 3032M			
L6061	14 OTU/5 BGS	10.12.41	SOC and reduced to spares			
L6062	14 OTU/25 OTU	26.11.41	SOC			
L6063	14 OTU	12. 9.41	SOC as 2714M			
L6064	–	31. 5.43	SOC as 2162M			
L6065	14 OTU	13. 5.40	Crashed on approach, Aston Down; to 2969M	No details	Training	
L6066	16 OTU/5 BGS	31. 3.42	SOC and reduced to spares			

Serial	Units	Date	Cause	Crew	Duty	See
L6067	16 OTU	1. 8.40	Hit ground in vertically banked condition, probably after engine failure, Souldern near Bicester, Oxon.	Sgt O G Day; FO G H Thevenard; Sgt S N Baker; AC2 G W Rhodes All killed	Training	5F4 6B12 6D2
L6068	16 OTU	24. 6.40	Starboard leg failed to lock; crashed on landing, Upper Heyford; became 3033M	PO F R W Palmer unhurt	Training	6C1
L6069	See under Hampden conversions below					
L6070	185/6 FPP	5. 1.41	Engine failed; belly-landed, Thornton-le-Moor, Lincs.	First Office E E Beville unhurt	Ferry flight	6D4
L6071	16 OTU/5 BGS	21.10.41	SOC			
L6072	16 OTU/5 BGS	21.10.41	SOC for spares			
L6073	14 OTU/25 OTU	26.11.41	SOC			
L6074	14 OTU/5 BGS	30. 4.42	SOC as GI airframe at 7 STT			
L6075	14 OTU/5 BGS	27. 4.41	Engine caught fire; abandoned and crashed 2m S of Jurby	Sgt Astbury and crew unhurt	Training	6D1
L6076	See under Hampden conversions below					
L6077	–	1. 4.45	SOC as 2964M at 6 SS			
L6078	16 OTU/5 BGS	19. 2.42	SOC			
L6079	16 OTU	28.11.41	SOC as spares			
L6080	16 OTU/25 OTU	26.11.41	SOC as spares			
L6081	16 OTU/5 BGS	28. 1.42	SOC as spares			
L6082	–	9.12.41	SOC as 2831M			
L6083	–	10.12.41	SOC as spares			
L6084	–	3.11.41	SOC as spares			
L6085	See under Hampden conversions below					
L6086	–	28.11.41	SOC as spares			
L6087	–	28.11.41	SOC as spares			
L6088	14 OTU/5 BGS	20.10.41	SOC as spares			
L6089	See under Hampden conversions below					
L6090	See under Hampden conversions below					
L6091	–	31. 7.43	SOC as 3027M at 12 STT			
L6092	SDF Christchurch	21. 1.44	SOC as 2967M at 7 SS			
L6093	–	27.10.41	SOC as spares			
L6094	–	30. 6.43	SOC as 2914M at 1 SS			
L6095	–	9.12.41	SOC as 2832M			
L6096	See under Hampden conversions below					
L6097	–	17.10.41	SOC for spares			
L6098	1 AAS	15.10.41	SOC as 2753M at 1 STT			
L6099	–	31.12.43	SOC as 2963M at 6 SS			
L6100	8 FPP	29. 1.42	Engine failed in circuit, Cranwell; crashed 1m S of Wellingore	First Officer S E Mitchell ATA unhurt	Ferry flight	6D5
L6101	1 AAS	8.10.41 22. 5.43	To 2737M at 1 STT SOC at 6 STT			

* * * * * * * * * * * * * * * * *

Prototype H.P.53 Hereford

Serial	Units	Date	Cause	Crew	Duty	See
L7271	RAE/S&H/AAEE	7. 7.40	To 2057M			

* * * * * * * * * * * * * * * * * * * *

Thirteen Handley Page Hereford Is converted to Hampden Is from above batch

Serial	Units	Date	Cause	Crew	Duty	See
L6011	14 OTU/5 AOS	26. 5.42	Hit wires on approach to Hellaby Q-site in error and crashed	Sgt J W G Hodgkinson (I); Sgt Parsons (I); LAC Hughes (I); LAC Alter (I); LAC Griffiths (K)	Night training	5E11
L6018	5 AOS/455	4. 6.44	SOC			
L6019	5 AOS/Cv TB/ 7 FPP	25. 7.43	Crashlanded after engine failure near Bamburgh, Northumberland SOC 3.3.44	First Officer R J Pass ATA unhurt	Ferry flight	2J35 6D15 5J2 2J19
L6020	14 OTU	15. 2.42	Engine caught fire; crashed at Risegate, Lincs.	PO P Chinn and crew baled out; all unhurt	Night training	
L6049	5 OTU/BTU	16. 4.44	SOC			
L6055	16 OTU/14 OTU/ 415 (GX-D)	17/18. 5.43	Shot down by ship's flak off Dutch coast	PO K Wathen RAAF (b.Kiel); FO J M Crawford (M); F/Sgt I M Sykes RNZAF (M); Sgt J Steward (M)	Night torpedo bombing	2J26
L6069	5 AGS/1 AAS/ 32 OTU	30. 4.44	SOC in Canada			5K8

Hereford L7271 on the ramp at Queen's Island, Belfast, where a production line of Herefords was laid down in 1939. (Photo: A&AEE MH6437)

Serial	Units	Date	Cause	Crew	Duty	See
L6076	CGS/ATA	20. 4.44	SOC			
L6085	5 OTU/1404 Flt	20.12.43	SOC			
L6089	5 BGS/Cv TB/ 415	25. 4.44	SOC			
L6090	5 BGS/Cv TB/ 455	11. 8.43	Overshot landing at Tain	Sgt G E Batchelor unhurt	Training	2J34
L6096	14 OTU	23.12.41	Stalled on take-off, Cottesmore	Sgt I R Beldam (K); Sgt F Lambert (K)	Training	6B11

* * * * * * * * * * * * * * * * *

Fifty Handley Page Hereford Is delivered in June and July 1940 by Short and Harland, Belfast

Serial	Units	Date	Cause	Crew	Duty	See
N9055	–	26.11.41	SOC as 2804M at 12 STT			
N9056	–	14. 1.42	SOC			
N9057	20 MU	31. 7.40	Undercarriage collapsed on take-off, Aston Down; to 2192M	Not known	Air test	
N9058	AAEE	26.11.41	SOC as 2805M at Melksham			
N9059	–	27.10.41	SOC for spares			
N9060	–	8.11.41	SOC for spares			
N9061	–	27.10.41	SOC for spares			
N9062	See list of Hampden conversions below					
N9063	–	28.11.41	SOC for spares			
N9064	See list of Hampden conversions below					
N9065	See list of Hampden conversions below					
N9066	5 BGS/5 AOS	17.12.41	SOC for spares			
N9067	5 BGS/5 AOS	26. 1.42	SOC			
N9068	–	21. 5.43	SOC as 2913M at 1 SS			
N9069	5 BGS/5 AOS	1.12.41	SOC for spares			
N9070	See list of Hampden conversions below					
N9071	5 BGS/5 AOS	10.12.41	SOC for spares			
N9072	5 BGS	9.12.41	SOC for spares			
N9073	5 BGS/5 AOS	10.12.41	SOC			
N9074	5 BGS	17. 3.41	Dived into sea after port engine caught fire at 6,000 ft.off Point of Ayre, Isle of Man	F/Lt Austin and five crew. Two baled out and four seen in water but all missing	Training	6D3
N9075	AAEE/RAE	12. 4.45	SOC as 2830M at 12 STT			
N9076	5 BGS/5 AOS	19.11.41	SOC			
N9077	5 BGS	25. 4.41	Engine failed; crashed in forced landing SE of Jurby	Sgt L Molyneux and crew. One injured	Training	
N9078	5 BGS	10.12.41	SOC			
N9079	–	14. 7.42	SOC as 2914M at 7 SS			
N9080	See list of Hampden conversions below					

Serial	Units	Date	Cause	Crew	Duty	See
N9081	23 MU	11. 6.41	Undershot landing at St.Angelo; DBF	PO C W C Farrow unhurt	Ferry flight	
N9084	–	15. 5.42	SOC as GI airframe at 5 SS			
N9085	–	27.10.41	SOC for spares			
N9086	See list of Hampden conversions below					
N9087	–	27.10.41	SOC for spares			
N9088	–	3.11.41	SOC for spares			
N9089	–	21. 5.43	SOC as 2019M at 2 STT			
N9090	See list of Hampden conversions below					
N9091	5 MU	23. 9.41	Crashed after engine failure, Lodminton, Glos	F/Lt R Ferguson unhurt	Ferry flight	6D6
N9092	8 FPP	8. 4.42	Undercarriage collapsed on landing, Staverton	First Officer L O Mills ATA unhurt	Ferry flight	
N9093	1 AAS	15.10.41	SOC as 2754M at 1 STT			
N9094	5 BGS	11. 3.41	Undercarriage collapsed on landing, Jurby; to GI airframe	F/Lt F Austin unhurt	Training	
N9095	5 BGS	29. 7.40	Undercarriage collapsed on landing, Jurby	S/Ldr J A Harris unhurt	Training	
N9096	See list of Hampden conversions below					
N9097	TDU	15.11.41	SOC			
N9098	1 AAS	6.11.41	SOC for spares			
N9099	–	9.12.41	SOC for spares			
N9100	–	16. 2.42	SOC			
N9101	See list of Hampden conversions below					
N9102	5 BGS	20.10.41	To 2767M			
N9103	5 BGS	20.12.41	SOC for spares			
N9104	1 AAS	22. 5.43	SOC as 2781M at 1 STT			
N9105	See list of Hampden conversions below					
N9106	See list of Hampden conversions below					

* * * * * * * * * * * * * * * * *

Eleven Handley Page Hereford Is converted to Hampden Is

Serial	Units	Date	Cause	Crew	Duty	See
N9062	14 OTU	31.7/1.8.42	Shot down by night fighter, Moergestel, Netherlands	Sgt E Collier (P?); Sgt L Lobb (b.Tilburg NL); Sgt G Russell RNZAF (P); Sgt M S Whiting RAAF (P)	Bombing Dusseldorf	
N9064	CGS/Cv TB	7. 2.44	SOC			
N9065	16 OTU	28. 2.42	Caught fire after heavy landing, Upper Heyford	Sgt J H Stillborn RCAF and crew unhurt	Night training	6C14
N9070	1 AAS	7. 6.42	Spun into sea from 12,000 ft. 3m E of Saltfleet, Lincs	Sgt D R Williams (K); PO S W A Gibbs (K)	Bombing training	
N9080	1 AAS/BTU	21.12.43	SOC			
N9086	144	14. 1.42	SOC			
N9090	5 BGS/5 AOS	2. 3.42	Stalled and crashed off Jurby Head, Isle of Man	Sgt Swain (K); LAC Robson (K) LAC Aisbitt (K)	Weather test	
N9096	1 AAS/5 AOS/ Cv TB/415(GX – Y)	22. 6.43	Swung off runway landing in fog and hit X3061, St.Eval	Sgt A F Hildebrandt; F/Sgt W MacCausland; Sgt N S Hastings; Sgt W K Mailman All unhurt	Anti – sub patrol	5C20
N9101	5 AOS/1 AAS	21.12.43	SOC			
N9105	1 AAS	14. 1.44	SOC			
N9106	16 OTU/Cv TB/ 415 (GX – N)	20/21.11.42	Missing from shipping attack in bad weather off Ijmuiden	Sgt A J Tippett RAAF (M); PO J J Lynn*; Sgt D L G Miller*; Sgt R J Sleep* *buried at Pihen – les – Guines, Calais	Night torpedo bombing	

* * * * * * * * * * * * * * * * *

200 Handley Page Hampden Is built by Handley Page, Cricklewood, and delivered between June 1939 and February 1940

Serial	Units	Date	Cause	Crew	Duty	See
P1145	16 OTU/Cv TB/ 455/5 OTU	10. 1.43	Swung on take – off, Maghaberry; SOC 3.1.44	Sgt D H Mann and crew unhurt	Training	
P1146	16 OTU/Cv TB	9. 6.44	SOC			
P1147	16 OTU/Cv TB/ 5 OTU/1404 Flt	2. 5.43	Stalled on landing and undercarriage collapsed SOC 15.5.44	Sgt M Green and crew unhurt	Training	
P1148	16 OTU	29. 6.41	Abandoned in bad weather and crashed, Heath Farm, Brackley, Oxon	PO C D Magyer; Sgt A S Chantrell Both baled out safely	Night training	5A31
P1149	16 OTU	7. 2.41	On take – off, collided with P1302 which was landing, Upper Heyford	Sgt W Walker (I); Sgt H James (K)	Training	5C6
P1150	Cv TB/455/ 144/415	10. 4.44	SOC			
P1151	144/14 OTU/ Cv TB/415 (GX – P)	10/11. 4.43	Shot down by flak during dusk attack off, Brittany	F/Sgt G A Cline RCAF (M); PO W B Keown (M) WO A M Kelly RCAF (M); Sgt J M Ryan RNZAF (M)	Torpedo bombing	

P1296 at the Torpedo Development Unit, Gosport, where it was based at from 6 February to 13 August 1943 for torpedo trials before going to 415 Squadron. (Photo: Frank Rushworth)

Serial	Units	Date	Cause	Crew	Duty	See
P1152	44/50	15/16.11.41	Flew into high ground on return, Goldsborough, Nr.Whitby	Sgt Young (I); Sgt Symes (I); Sgt Bernard (I); Sgt Neale (K)	Minelaying, Frisian Is	2H6
P1153	49/455/ Cv TB/415 (GX – V)	27/28. 6.42	Shot down by flak in attack on convoy off Borkum	WO J McCallum (B.Sage); Sgt H F Hasleden (M); F/Sgt R M McNeil (b.Kiel); F/Sgt L H Latimer (b.Sage).Crew all RCAF	Night torpedo bombing	2J24
P1154	16 OTU/5 OTU	28. 6.44	SOC			
P1155	14 OTU	27.11.41	Dived into ground soon after take – off, Saltby	Sgt R S Carson; Sgt J A Tulley (Rhod); Sgt N J Griffiths; Sgt B W Bucknell RAAF; All killed	Training	6B10
P1156	50/455 (UB – F)	12. 2.42	Missing from attack on German warships off Dutch coast	F/Lt W E Perrin; PO A R Abbott RNZAF; FO E G de T Symons; Sgt R E Tomlinson All missing	Day bombing warships at sea	2H2
P1157	14 OTU/Cv TB/ 415 (GX – O)	18/19. 2.43	Hit by flak off Dutch coast and ditched 2257 hrs 18.2.43	FO A B Brenner; F/Sgt E L Rowe; Sgt A Glass; Sgt E Vautier All picked up by ASR Walrus	Night torpedo bombing	5B28 2J31
P1158	14 OTU/Cv TB/ 455	28.12.43	SOC			
P1159	16 OTU	23.12.41	Forcelanded in bad weather; Denham Golf Course, Bucks	Sgt T Reynolds and crew unhurt	Training	
P1160	CGS/Cv TB/ 415/16 FPP	29. 2.44	Collided with Beaufighter NE590 while taxying, Kirkbride	First Officer E R Kershaw ATA unhurt	Ferry flight	
P1161	1 AAS	16. 8.42	Crashed in sea after engine failure on bombing practice off Mablethorpe, Lincs	Sgt J A Dent and three crew all killed	Training	6D8
P1162	CGS	27. 7.41	Stalled after take – off, Castle Kennedy	Sgt H G Turner, Sgt G E Appleton and three other crew members all killed	Training	
P1163	CGS	17. 8.42	Hit trees at Terrington St.Clements, Norfolk	Sgt S L Vinten on first Hampden solo killed	Training	
P1164	144/Cv TB/ 455/415	17. 2.44	SOC			
P1165	408 (EQ – B)	28/29.12.41	Shot down by night fighter, Winterswyk, Netherlands	PO S B Brackenbury (P); Sgt D W Thane RNZAF*; Sgt D Williams*; Sgt J Isaac* *buried at Winterswyk	Bombing Huls	
P1166	144/50/144/ 408/Cv TB/455	23.12.43	SOC			2F23 5K11
P1167	16 OTU/14 OTU/ 32 OTU	30. 4.44	SOC in Canada			5K8
P1168	14 OTU	9.12.41	Stalled and crashed 4m S of Saltby	PO Fisher RAAF (K); Sgt K F Thornton RAAF (K)	Night training	
P1169	RAE/AAEE/RAE/ AAEE/RAE/ Cv TB/415/489	17. 2.44	SOC			
P1170	61	14.10.39	Crashed on landing, Doncaster	Pilot unhurt; name not known	Night training	
P1171	61/83 (OL – P)	1/ 2. 7.40	Shot down by flak over Kiel	PO D Redmayne; Sgt C Lee; Sgt O S Gander; Sgt G F Little. All buried at Kiel	Bombing Kiel	

Serial	Units	Date	Cause	Crew	Duty	See
P1172	144	5/ 6. 9.40	Missing; presumed crashed in sea	PO J E Newton Clare; Sgt S O Clarke; Sgt W Thompson; Sgt W L Powell. All missing	Bombing Hamburg	
P1173	44	12. 4.40	Shot down by Bf 109 into sea off Norway	FO H Robson (b.Frederikshaven); PO K Johnstone (M); Cpl J Evans (M); Cpl W Evans (M)	Day bombing Kristiansand	2C12
P1174	49/16 OTU/ 5 OTU	16. 8.42	Forcelanded while lost on Isle of Skye	Sgt Forbes (I); Sgt Harding (I); Sgt Rock, Sgt Simrock both unhurt	Training	
P1175	49	21/22. 4.40	Overshot landing at Scampton on return	FO D White; Sgt Jones; LAC Hornsby; Sgt Hurst. All unhurt	Minelaying W Baltic	
P1176	49/14 OTU/ 1404 Flt	4. 6.44	SOC			
P1177	49/25 OTU/ 16 OTU	16. 2.42	Failed to level off when landing, Weston–on–the–Green	Sgt J McKenzie and crew unhurt	Night training	
P1178	83	3/ 4. 6.40	Shot down by night fighter; location not known	FO F J Haydon; PO C E Greenwell; Sgt D Thomas; Sgt J Perry All buried at Rheinburg	Bombing Emmerich	
P1179	76/16 OTU	2. 6.40	Hit tree on take–off, Upper Heyford	PO B Simpson killed on first Hampden solo	Training	6B16 5B31
P1180	76/16 OTU	27. 7.40	Hit tree on take–off, Brackley LG	PO D M Marsh (K)	Night training	6B17
P1181	76/16 OTU	24. 5.40	Engine cut on take–off, Upper Heyford	Wg Cdr K H R Elliott; F/Sgt E G Crouch Reported as 'one injured'.	Training	6B5
P1182	76/16 OTU	23.12.40	Engine failed in circuit; crashed near Bicester	Sgt Allison and crew unhurt	Training	
P1183	RAE/83 (OL–K)	18/19. 9.40	Shot down by flak, Le Havre	S/Ldr J A Pitcairn Hill DSO, DFC (K); PO A P Linsdell (K); Sgt C G McCarthy (M); PO V Rendall (K)	Bombing Le Havre	2F3
P1184	16 OTU	13. 9.41	Hit tree on undershoot, Croughton	Sgt T E Barnett and crew unhurt	Night training	6C8
P1185	16 OTU/ 14 OTU	31.7/1.8.42	Believed shot down by fighter near Ostend	Sgt N E McGovan RCAF (P); PO D G McLean RCAF*; Sgt S Shailer RAAF*; F/Sgt G Bloomer*. *buried at Ostend	Bombing Dusseldorf	5A32
P1186	14 OTU	26. 1.42	Dived into ground near Spalding, Lincs. Cause obscure	PO J M Marok RCAF; PO W H Lynch RAAF; PO S S White RAAF; Sgt E O'Connell RAAF All killed	Bombing practice	
P1187	44/420 (PT–X)	4/ 5. 5.42	Damaged by flak and night fighter; crashed at Great Bentley, Essex	Sgt F S Hiley (I); Sgt Germain unhurt; Sgt Johnson unhurt; Sgt J F Halward (K)	Bombing Stuttgart	2F14
P1188	14 OTU/Cv TB/ 489 XA–A	3. 8.43	Damaged by Fw 190s and bellylanded at Wick. SOC 21.12.43	WO J Strain; FO E Fordham; WO H Smith all unhurt; F/Sgt J F Marks (I)	Day torpedo bombing off Norway	
P1189	16 OTU/Cv TB/ 489 (XA–K)	9. 4.43	Shot down by fighters off Norway	FO D Wheeler (M); Sgt G Smith (M); FO L Selthun RCAF (b.Trondheim); F/Sgt A B McDonald RCAF (M)	Day torpedo bombing off Norway	2J22
P1194	25 OTU	27. 9.41	Engine cut on take–off, Balderton	PO L S Coulter and crew Reported two injured	Training	
P1195	14 OTU	3. 2.44	SOC			
P1196	14 OTU/5 OTU/ 1402 Flt	12. 4.44	SOC			
P1197	16 OTU	25. 3.42	Overshot landing and hit hut, Croughton; caught fire	Sgt A C Hulme unhurt	Night training	
P1198	106/144/Cv TB/ 5 OTU/1 TTU	29. 4.43	Misjudged practice torpedo drop and hit sea; ditched off Lady Island, Firth of Clyde	Sgt Batchelor and crew picked up unhurt	Training	6E15
P1199	16 OTU/5 OTU	1. 9.42	Collided with Beaufort on landing, Turnberry	WO J E Jones RNZAF (K); Sgt J C Newman RCAF (K). Two killed, one injured in Beaufort	Training	
P1200	16 OTU/Cv TB/ 32 OTU	18. 1.44	Failed to return from night navex, presumed due to icing	Sgt R F I Watson; PO W R Dishman; Sgt E G Quigley all missing. No trace found of aircraft or crew. They were the last Hampden fatal casualties	Night training	5K8 5K9
P1201	455 (UB–P)	7/ 8.11.41	Missing, cause not known	PO J A C Gordon RAAF (P); PO Cordwell (P); Sgt Stokes (P); Sgt Holt (P–ser.inj.)	Bombing Cologne	
P1202	50	30.11/1.12.41	Crashed out of fuel 2m E of Skellingthorpe on return	Sgt Williams; Sgt W J Johnston (I); Sgt Smith; Sgt Caunce. Three unhurt	Bombing Hamburg	3C4
P1203	455 (UB–T)	6/ 7. 4.42	Shot down by night fighter SW of Zwolle, Netherlands	Sgt A Wincott; FO T Roberts RAAF; F/Sgt C Gammie RAAF; Sgt K McIlrath All buried at Amersfoort, Netherlands	Bombing Essen	
P1204	25 OTU/16 OTU/ 1404 Flt	16. 7.43	Hit by Whitley BD276 while parked, St.Eval	Nil	Parked	5C19
P1205	14 OTU	31. 8.42	Yawed and struck ground, Grimsthorpe bombing range, Lincs	Navigator PO D F Foy died from burns. Other crew members, names not known, suffered shock	Training	
P1206	49 (EA–K)	8/ 9.11.41	Shot down by night fighter 20 km E of Alkmaar, Netherlands	WO C Saunders DFM*; Sgt J D'Arcy*; Sgt S G Mullenger (M); Sgt J E Kehoe (M) *buried at Amsterdam	Bombing Bocholt	
P1207	Cv TB/AAEE/ 144/455 (UB–P)	27.10.43	Ditched out of fuel off Aberdeen	FO T G Davis RAAF; FO A O Peters RAAF; F/Sgt F Mechan RCAF; F/Sgt D Vitarelli RCAF. All unhurt	Air–sea rescue	

Hereford N9097 at the Torpedo Development Unit, Gosport, in late 1941. A torpedo loading trolley can be seen beneath the bomb bay. The oil – bespattered fuselage and propeller, plus the fact that the engine panels have been taken off, indicate problems with the port Napier Dagger engine. (Photo: Frank Rushworth)

Serial	Units	Date	Cause	Crew	Duty	See
P1208	16 OTU/Cv TB/ 455 (UB – C)	7.10.43	Went into stabilised yaw soon after take – off from Wick	F/Sgt A E Jones; FO R Lassiter; F/Sgt J H Taylor RAAF; F/Sgt K F McDonald RAAF All unhurt	Anti – sub patrol	5F17
P1209	16 OTU/14 OTU/ 1406 Flt	8.12.43	SOC			
P1210	16 OTU	19. 6.41	Overshot and crashed, Balderton	PO R R Michell (Rhod)(I); Sgt E A M Grange RCAF (I); PO A E Sinclair (K); Sgt R A Connel (K)	Training	
P1211	14 OTU	18. 6.41	Developed yaw on approach and dived into ground, Cottesmore	PO A E Cooke (K); Sgt Dent (K)	Training	
P1212	408 (EQ – T)	20/21.10.41	Stalled on approach to Coningsby on return	Sgt D G Bradley RNZAF (K); Sgt A T McMillan RCAF (K); Sgt E L Coles (I); Sgt R A Stansfield (K)	Bombing Bremen	
P1213	14 OTU	16. 3.41	Control lost in turn at 800 ft near Cottesmore	PO Ball; PO Priest; Sgt Cheetham (I); Sgt Hendra, Sgt Liddell	Training	
P1214	16 OTU/Cv TB/ 489 (XA – X)	7. 4.44	SOC			
P1215	14 OTU/CGS/ATA	9. 6.44	SOC			
P1216	16 OTU/14 OTU/ BTU	18. 1.44	Stalled and dived into Braid Fell, Wigtown	PO A Coumbis (K); LAC E T A Pottinger (K) One other crew member, name not known (K)	Training	6E16
P1217	14 OTU/1402 Flt/517	14. 1.44	Crashed on landing, Tollerton, en route for scrapping	F/Sgt J Smith; F/Sgt E Bacon; Sgt P Evatt; Sgt K Neill All unhurt	Ferry flight	6C4
P1218	408 (EB – Q)	22/23.10.41	Missing, cause not known	F/Sgt F A Titcombe*; Sgt Gifford (P); Sgt P A G Ragg*; F/Sgt R R Walker* *buried at Durnbach	Bombing Mannheim	
P1219	16 OTU/Cv TB/ 489 (XA – W)	22. 7.42	Crashed in sea 2m S of Selsey Bill, Sussex	FO P E Daniels unhurt; PO Lawrence (I); F/Sgt Warde (M); LAC Riches (M); LAC Gaitens (K)	Training	
P1220	25 OTU	30. 5.41	Overshot landing at dusk, Finningley	Sgt K Richardson (I)	Training	
P1221	25 OTU/16 OTU	26. 3.42	Flew into ground 4m S of Polebrook	Sgt Dobie; Sgt N G Boyce; Sgt R J O'Leary Sgt D C Lindsay. All RCAF. All killed	Training	
P1222	25 OTU/16 OTU/ 5 OTU	18. 1.43	Hit sea and ditched 1m W of Dubh Artach lighthouse, Argyll	PO J S Newcombe RNZAF; PO J A C Watts; Sgt K F Ray RCAF; Sgt J Munro RCAF. All killed	Torpedo attack training	6E14
P1223	50/1404 Flt	25. 3.43	Crashed out of fuel, Trevose Head, Cornwall in bad weather	Sgt Bartram; FO McLaughlin; Sgt Eaton; Sgt Jackson Reported minor injuries	Met flight	
P1224	RAE/TDU	22.12.43	SOC			
P1225	16 OTU	9.11.41	Heavy landing in fog at night, Croughton. Caught fire.	Sgt Stipe RCAF and crew unhurt	Night training	6C13
P1226	49	17. 3.42	Flew into hill in fog, Branscombe, Devon	PO D A Cook (K); PO R E Manders (K)	Collecting aircraft	
P1227	16 OTU	14.10.41	Overshot landing at Croughton	Sgt Halcro, Sgt H Kendall unhurt	Night training	
P1228	50/106 (ZN – L)	30.11/1.12.41	Missing; presumed ditched in North Sea	Sgt A J Moore; Sgt A J W Parker; Sgt T F Gowan; Sgt D D Jones. All missing	Bombing Hamburg	

Serial	Units	Date	Cause	Crew	Duty	See
P1229	16 OTU/Cv TB	24. 1.44	SOC			
P1230	16 OTU/14 OTU/ 32 OTU	.44	SOC in Canada			5G5 5K8
P1233	25 OTU	18/19. 5.41	Missing on navex; believed crashed in North Sea	PO P Dixon – Spain; Sgt B Halbert; Sgt E K Dawson; Sgt T Morgan Crew all missing	Night training	
P1234	25 OTU	25.10.41	Hit balloon cable in bad visibility, Weybridge while off course	PO J W Atkin and three crew killed	Training	
P1235	14 OTU/Cv TB	21.12.43	SOC			
P1236	25 OTU/Cv TB/ 455 (UB – G)	3. 2.44	SOC			
P1237	16 OTU/Cv TB/ 144/489/455	3. 1.44	SOC			
P1238	14 OTU/Cv TB/ RAE/TDU	3.12.43	SOC			
P1239	50/420 (PT – Y)	12/13. 4.42	Shot down by night fighter N of Terschelling	F/Sgt R Johnson; Sgt C J E Butler; F/Sgt R H Black; Sgt J Salmon All missing	Bombing Essen	
P1240	14 OTU	8. 4.41	Crashed after engine failure, Little Bytham, Lincs.	PO E Crump and crew unhurt	Night training	
P1241	14 OTU	15. 1.41	Crashed after engine failure near Cottesmore	Sgt F Rowney on first Hampden solo unhurt	Training	
P1242	14 OTU/1 AAS	16.12.43	SOC			
P1243	14 OTU/5 OTU/ 455/5 OTU/ 1 TTU	26. 1.44	SOC			
P1244	16 OTU/455/ 408 (EQ – Y)	27/29. 8.42	Missing, cause uncertain	F/Sgt A F Jennings; WO L G Chaston RCAF; Sgt C H Thompson RAAF; Sgt J W Todd All buried at Hannover	Bombing Kassel	
P1245	25 OTU/Cv TB/ 455 (UB – J)	12.10.42	Handed over to Russian Navy	Flown out by Sgt W S Martin; F/Sgt Waites; Sgt Ackroyd; Sgt J A Miller; LAC Walkley	Left in Russia	2J15
P1246	16 OTU/Cv TB/ 455/RAE/455	17.11.43	Crashed after hydraulic failure near Leuchars	FO L W Carr RAAF unhurt	Training	5F18
P1247	16 OTU	16. 4.42	Stalled on overshoot, Croughton	Sgt W H Phillips RCAF (K); Sgt M A Bent RCAF (I)	Night training	6C18
P1248	25 OTU	19. 4.41	Hit balloon cable and crashed in Concord Park, Rotherham Road, Sheffield	PO R A P Allsebrook (I); PO J B Ransom (K)	Training	5E7
P1249	14 OTU/Cv TB	28. 6.44	SOC			
P1250	76/16 OTU/CGS/ Cv TB/144(PL – M)	13.12.42	Shot down by flak off south – west Norway	F/Lt H J W Bowden (P); FO L J Mennill RCAF (P); Sgt J Vosper (M) Sgt C D Elsmore (P)	Day torpedo attack off Norway	2J20 2J16
P1251	76/16 OTU	5. 5.40	Port engine failed on take – off, Upper Heyford. Crashed at Fritwell	S/Ldr Sheehan (I); Sgt Stilwell (I) Three other crew injured	Training	6B9
P1252	1 AAS	24.11.41	To 10 STT as GI airframe No record of accident			
P1253	106/61	1/ 2. 3.41	Ran out of fuel and crashed near Stradishall	PO Noble (I); Sgt McKinnon (K); Sgt J Fletcher (K); Sgt Healin (I)	Bombing Cologne	2K16
P1254	106/14 OTU	20. 2.44	SOC			5K7
P1255	106/BTU	18. 9.44	SOC			
P1256	106	27. 9.40	Crashed at Misson, near Finningley	PO J E Lowe (of 44 Sqn K); Sgt L McLeod (K)	Night training	
P1257	16 OTU/455/ 420/Cv TB/489	13. 4.43	Missing from shipping reconnaissance off Norway	FO A E B Barnard (M); WO F M Holobow RCAF* F/Sgt W J MacErlich RNZAF*; Sgt I Griffiths RNZAF*. *buried Haugesund	Day torpedo bombing	
P1258	106/25 OTU/ Cv TB/489/ 455/415	15/16. 8.43	Shot down by night fighter off Dutch coast	S/Ldr P N Harris DFC (M); PO R V Procter (B.Texel); WO A G Innes (M); F/Lt J O Bostock (b.Bergen)	Night torpedo bombing	2J28
P1259	106	18/19. 9.40	Shot down by flak, Elbe Estuary	PO E T Watkin; Sgt C Kenmure; Sgt J McLean Fraser; Sgt J B Keats. All missing	Minelaying Elbe Esty	
P1260	7	1. 1.40	Flew into Snaefell in cloud, Isle of Man	PO H M McGregor (K); Sgt Dennis (K); Cpl P Brightmore (I); Sgt R J Bailey (K)	Training	5D1
P1261	7/16 OTU/4 OTU	17. 9.43	SOC			
P1265	7/16 OTU/14 OTU/1402 Flt	8. 9.43	Missing; last heard of 100m NE of Londonderry	F/Sgt H K Bartlett; FO R Woodford; WO W Coleman; Sgt A W Bentley. All missing	Met flight	
P1266	7/16 OTU	1. 8.40	Crashlanded after engine failure, Babdown Farm	PO R F W Palmer and crew unhurt	Night training	
P1267	76	11.12.39	Hit tree in forced landing in poor visibility, Princes Risborough	PO C D Stevenson first Hampden solo at dusk killed	Training	
P1268	76/16 OTU	21. 6.40	Undershot and hit bowser, Upper Heyford	Sgt H Woolstencroft unhurt	Night training	6C8
P1269	76/16 OTU	24. 7.41	Overshot and hit hangar, Upper Heyford	Sgt E W Phillips (I)	Solo training	6C10
P1270	76	4. 3.40	Stalled on take – off, Upper Heyford	PO W G C Edmonds (badly burned)	Night training	6C12
P1271	76/16 OTU	8. 10.40	Broke up in air 3m W of Gloucester	PO A F Jefferys, Sgt J D Shields, Sgt R T Bowtell, Sgt R A Keeling (all K)	Training	
P1272	TDU/14 OTU/ 455	30.11/ 1.12.41	Missing	Sgt J Shannon RAAF; PO A G Sands, Sgt A C Shorey, Sgt V Towers (all b.Hamburg)	Bombing Hamburg	

P1316, GL – P of No.14 OTU, on the ground at Cottesmore in July 1940. (CH711)

Serial	Units	Date	Cause	Crew	Duty	See
P1273	14 OTU/144	4/ 5. 9.42	Set on fire by fighters; abandoned and crashed in swamp near Petsamo	Sgt L H Bertrand (P); Sgt K C Smith RCAF RCAF(P); Sgt K S Piper RCAF (P); Sgt E D Bind RCAF (P); Cpl E R Desforges (P)	In transit to Russia	2J10 2J15
P1274	185/14 OTU	6. 5.40	Control lost on over – shoot; crashed 2m NNE of Cottesmore	Sgt O Beck (K); PO K A Ramsay (K) Two others (K), names not known	Night training	6C15
P1275	185/14 OTU	19. 5.40	Hit trees on approach, Cottesmore	FO A B Lennie "one injured"	Night training	
P1276	185/14 OTU	7.12.40	Control lost after engine failure, Cottesmore	Sgt Lovejoy (I); Sgt Clarke (I)	Night training	
P1277	14 OTU	2. 5.42	Overshot landing, Saltby	PO J H Hood unhurt	Night training	
P1278	14 OTU	17. 7.41	Stalled after engine failure on take – off, Cottesmore	PO Bowker (K); Sgt J W Rowe (K)	Training	6B6
P1279	185	24. 2.40	Undercarriage collapsed, Cottesmore; to 1857M	S/Ldr D B Field unhurt	Training	
P1280	14 OTU	12. 6.40	Undercarriage collapsed, Cottesmore	PO E B Liddell unhurt	Night training	
P1281	14 OTU	1. 6.40	Stalled on take – off, Cottesmore	PO G V Eason unhurt	Night training	
P1282	14 OTU/Cv TB/ 415/TDU	25. 6.43	Undercarriage collapsed at Gosport; SOC 17.2.44	PO L R Lord unhurt	Night training	
P1283	14 OTU	3. 3.41	Failed to become air – borne; hit paint store, Cottesmore	PO Kerridge; Sgt Hendra; Sgt Rowney; Sgt Baker; all unhurt	Night training	
P1284	16 OTU/Cv TB/ 5 OTU	28.10.42	Forcelanded while lost and out of fuel, Letterkenny, Eire	Sgt D Montgomerie and crew unhurt; interned	Training	
P1285	16 OTU	1. 9.41	Overshot landing and hit lorry, Upper Heyford	PO J D Bridge; Sgt A G Harbottle	Training	6C9
P1286	14 OTU/415	17. 2.44	SOC			
P1287	25 OTU/Cv TB/ 455 (UB – B)	12.10.42	Handed over to Russian Navy	Flown out by F/Lt Davenport; PO Bowman; Sgt E Smith, Sgt Harrison, LAC McKay	Left in Russia	2J14 2J15
P1288	16 OTU	25. 2.41	Crashed near Kidlington	PO H M Macrossan and crew unhurt	Night training	
P1289	14 OTU	15. 7.41	Crashed into Anson R3386 at Cottesmore	Crew unhurt Names not on record	Training	5C11 6C20
P1290	106	7/ 8.11.41	Shot down by flak, Oslofjord	PO D C Firth*; Sgt W King*; Sgt C J Walton (P); Sgt A Archer (P) *b.Oslo	Mine – laying	5J7
P1291	25 OTU/16 OTU/ 14 OTU/1401 Flt	18. 7.43	Ditched in North Sea after engine failure	PO H Parkinson; PO T A Fettes; F/Sgt J Battis; Sgt G Lindsay; all picked up	Met flight	

Serial	Units	Date	Cause	Crew	Duty	See
P1292	14 OTU	23. 2.41	Failed to take – off; hit hedge, Cottesmore	PO Keartland; Sgt Hammond; Sgt Banham; Sgt Everett. All unhurt	Night training	
P1293	14 OTU/1404 Flt	23. 7.43	Crashed near St.Merryn in bad weather	PO A E Trebbek and three crew all killed	Met flight	
P1294	14 OTU	30.10.41	Came out of cloud in vertical dive near Oswestry, Salop	PO G D Kerr; Sgt I M Williams; Sgt D Tatton; Sgt H Playforth. All killed	Training	
P1295	144 (PL – L)	28/29.12.41	Missing	Sgt J W Woodall (M); Sgt E Whitaker RNZAF*; Sgt A Collier (M); Sgt J Skingsley* * b. Cherbourg	Bombing Huls	
P1296	16 OTU/Cv TB/ TDU/415	24. 5.44	SOC			
P1297	25 OTU/16 OTU	20. 4.42	Crashed in flat spin at Fox Covert, Trismore, Oxon	F/Lt J E Morgan; F/Sgt D J Taynan; Both RCAF and killed	Training	
P1298	14 OTU	25. 3.42	Overshot and hit lorry, Cottesmore	Sgt D E Randall RAAF unhurt	Training	6C9
P1299	25 OTU/16 OTU	6. 4.42	Crashed in flat spin from 1,000 ft, Locking	Sgt A B McKeith RCAF; Sgt W B Wilson RCAF; Sgt F A Dann; Sgt H G Stalker. All killed	Training	
P1300	14 OTU/1 AAS/ 5 AOS/1 AAS	15.12.43	SOC			
P1301	14 OTU	12. 9.41	Collided with P4303 near Saltby	Sgt A P Overall DFM; Sgt V H Griffiths RNZAF; Sgt D G K Smith; Sgt E F Chamberlain (All killed)	Training	5C7
P1302	16 OTU	7. 2.41	Collided on landing with P1149, Upper Heyford	FO S Moxham; Sgt Fenton Both killed	Training	5C6
P1303	106/5 Gp TF/ 25 OTU/14 OTU	29. 4.42	Dived into ground on single – engined practice 5m N of Cottesmore	Sgt W J Nichols RAAF; Sgt J M Stewart Both killed	Training	
P1304	106	21.12.40	Hit trees on approach to Finningley, Mount Pleasant, near Bawtry	Sgt P C Wells; Sgt Richardson Both unhurt	Night training	
P1305	14 OTU	19. 8.40	Dived vertically out of cloud, Sharnbrook, Beds.	PO W A King; Sgt S Britnor; Sgt J Bishop; Sgt T A Jackson; all killed		
P1309	14 OTU	7. 7.40	Overshot and hit pillbox, Cottesmore	PO C O Dunkels unhurt	Training	
P1310	185/14 OTU/49/ Cv TB/415/455	27.12.43	SOC			
P1311	106/32 OTU	30. 4.44	SOC in Canada			5K8
P1312	185/14 OTU/Cv TB/ 5 OTU/415	4. 6.44	SOC			
P1313	CGS/Cv TB/415	26. 6.43	Ran out of fuel and crashed, St.Merryn SOC 17.2.44	Sgt K T Ashfield and crew unhurt	Anti – ship sweep	
P1314	CGS/49/420/ 408/Cv TB/ 1402 Flight	25. 3.43	Swung off runway on landing and broke back, Aldergrove	Sgt Collins unhurt	Training	6C7 2F14
P1315	CGS	13. 5.41	Destroyed in air raid, Sutton Bridge		On ground	
P1316	14 OTU	14. 2.41	Set on fire by heater lamp, Saltby		On ground	
P1317	50	26/27. 8.40	Damaged by flak; engine failure on approach, Hemswell;	Sgt Abbott; Sgt Bryce; AC1 Green; Sgt Harney. All unhurt	Bombing Leipzig	
P1318	49	25/26. 5.40	Missing from attack on rail targets between Krefeld and Aachen. Flak	FO P S Butler (M); PO J K Bennett*; LAC G R J Parsons*; Sgt J Harrison (M) * b.Zeebrugge	Bombing rail targets	
P1319	49	25/26. 4.40	Believed shot down by night fighter near Sylt	PO A H Benson (M); PO A P B Hordern (M); LAC J D Openshaw (M); Sgt R I L McKenzie (b.Kiel)	Minelaying Kiel Bay	2L2
P1320	106	25.11.40	Hit ground in spiral dive 4m NW of Stamford, Lincs.	Sgt Bagnall (Ser.I); Sgt E F Carthew (K); Sgt M C Mair (K); Sgt Canham (I)	Night training	
P1321	106/50	25/26. 7.40	Damaged by flak; crash – landed on beach near Happisburgh, Norfolk	PO W A C Mulloy; Sgt Watt; Sgt Stewart; PO T H Taylor. All killed	Bombing Castrop – Rauxel	5B34
P1322	106/44/25 OTU/ 16 OTU/14 OTU	7. 2.44	SOC			
P1323	61/49/16 OTU	12. 9.41	Dived into ground 3m S of Upper Heyford	Sgt S Graaf (K); Sgt J H Bendon (K)	Night training	6C17
P1324	44	1/ 2.10.40	Undercarriage collapsed in forced landing, Brackley LG; to 2261M	S/Ldr Broad and crew unhurt	Minelaying Bordeaux	
P1325	44	11/12. 6.40	Missing	Sgt W Jeffrey; F/Sgt C L Sumpster; Sgt J Simpson; Sgt J F Sandall. All b.Lestrem	Bombing rail targets	
P1326	144	11/12. 5.40	Hit by flak over target and abandoned, Belgium	W/Cdr A N Luxmoore (b.Finnevaux, Belgium); PO R Allitt; Cpl Jolly; Sgt H Wathey safe	Bombing M – Gladbach	5A3 5A11
P1327	50	31.7/1.8.40	Overstressed over North Sea; to GI airframe	S/Ldr R J Oxley and crew unhurt; landed safely at base and SOC	Minelaying	3A2
P1328	144	1. 2.41	Crashed after control lost in snowstorm, Willingham, Lincs.,	Sgt J F Thurlbeck (K); Sgt J T Hawthorne (K)	Night training circuits and landings	

Three Herefords of No.14 OTU in echelon to starboard. GL – A2 is L6070. (Photo via Dennis Thompson)

Serial	Units	Date	Cause	Crew	Duty	See
P1329	50	26/27. 6.40	Missing; radioed 'going down 15m from Dutch coast'	W/Cdr N D Crockart (K); F/Sgt W S Southey (M); Sgt A Ingram (K); Sgt E Turner (M)	Bombing Hannover	5B21
P1330	50/ SF Farnborough/ AAEE/ 5 AOS	17. 2.44	SOC			
P1331	44	23/24. 4.40	Damaged undercarriage on take – off; crash – landed, Waddington	PO L J Ashfield and crew unhurt	Security patrol	
P1332	Hendon/16 OTU	4. 2.41	Hit ground in snowstorm, Conington, Hunts	PO A Smith; Sgt E L Clayton; Sgt A Cunningham; Sgt H M Pattinson. All killed	Training	
P1333	49	16/17. 8.40	Crashlanded near Breda, Netherlands	Sgt M G P Stetton (P); Sgt C H Butcher (P); Sgt K N Farrow (P); AC1 Lindsay (P)	Bombing Merseburg	
P1334	83	29/30. 8.40	Returned with engine failure and undershot, Scampton	S/Ldr Collier DFC; PO Turner; Sgt Johnson; PO de Bressey. No serious injuries	Bombing Gelsenkirchen	
P1335	Hendon/14 OTU/ Cv TB/489	25. 3.44	SOC			
P1336	106	24. 5.40	Hit balloon cable over Coventry and crashed	PO J M D Irvine; Sgt J R Collingham; AC1 Firth. All killed	Night training	5E5
P1337	106/5 Gp TF/ 455/489 (XA – M)	11/12. 8.42	Missing from attack on *Lutzow* off Dutch coast	S/Ldr A C Tiptree (M); F/Lt J J Hurley (M); Sgt J O Wellard (M); Sgt G A Jones RNZAF (M)	Torpedo strike	
P1338	44	11/12. 9.40	Missing from attack on liners *Bremen* and *Europa*	F/Lt T L S Smythe DFC; PO S C Wise; Sgt W A Jones; FO W F E Coombes. All b.Becklingen	Bombing Bremerhaven	2F3
P1339	44/1 AAS	17. 1.44	SOC			
P1340	44	3/ 4. 6.40	Hit balloon cable and crashed in River Orwell near Harwich	Sgt E J Spencer (baled out); PO R W Roots; Sgt A J Kendrew; Sgt S P Connell (all M)	Bombing Emmerich	5E5 5A25
P1341	16 OTU/106	15/16. 1.42	Shot down by flak near Esburge	Sgt W S Dashwood RNZAF; Sgt C Rousseau RCAF; Sgt W S Horseman; Sgt Moore (All P)	Bombing Hamburg	5H6
P1342	185/14 OTU	9. 2.41	Stalled on landing, Cottesmore	Sgt Austin, Sgt Rodney, Sgt Power all unhurt	Training	
P1343	185/14 OTU	22.10.41	Hit trees forcelanded in fog, Wittering	Sgt Irving, Sgt Eastwood, Sgt Robertson all unhurt. Sgt Williams (I)	Training	
P1344	14 OTU/Cv TB/ 144	4/ 5. 9.42	Crashed near Petsamo airfield, Finland	PO E H E Perry (P); F/Sgt G E Miller RCAF (M); Sgt J M Robertson (M); Sgt D C Garrity (M)	Ferry flight to Russia	2J9
P1345	16 OTU/14 OTU/ Cv TB/5 OTU	25. 5.44	SOC			

P1346	185/14 OTU/ Cv TB/455	5. 5.44	SOC			
P1347	49 (EA – D)	4/ 5. 9.40	Hit by flak and crashed, Brittany	FO L M Hodges (P*); Sgt S J Hitchings (P); Sgt J H Wyatt (P*); Sgt L C Turnbull (P) *escaped and returned to UK	Bombing Stettin	3D31
P1348	83	5/ 6. 6.40	Hit by flak and crashed, Barnkrug near Stade	S/Ldr D B D Field; Sgt E H Paul; Sgt R Eadie; Sgt L L Kerr. All b.Becklingen	Bombing Hamburg	
P1349	1 AAS	14. 1.44	SOC			
P1350	RAE/16 OTU	15. 5.41	Port engine failed; stalled and dived into ground, Fritwell	Sgt M J Hopkins; Sgt W K Clarke RCAF; Sgt J L Newton; Sgt H Crawford; AC2 R Robson. All killed	Training	
P1351	RAE/16 OTU	5. 5.42	Control lost on take – off, Saltby	Sgt Peters and two crew all injured	Training	
P1352	14 OTU/16 OTU/ 5 OTU	5. 3.43	Hit rising ground on take – off 3m SW of Long Kesh	Sgt Dyer (K); Sgt Henry (K)	Night training	
P1353	14 OTU/1401 Flt/521	23.12.43	SOC			
P1354	144/83 (OL – Y)	25/26. 8.40	Ran out of fuel and ditched in The Wash	FO N H Svendson DFC; Sgt Threlfall; Sgt Gear; Sgt Dale. All picked up unhurt	Bombing Berlin	5B2
P1355	83	15/16. 9.40	Hit by flak and set on fire over Antwerp; RTB	PO C A Connor; Sgt D A E Hayhurst (P); Sgt Hannah; Sgt G James (P)	Bombing Antwerp	3D20
P1355	83/5 BGS/ 1401 Flt/521	22.12.43	SOC			
P1356	83/50/16 OTU/ Cv TB/144/ 489 (XA – D)	26. 3.44	SOC			

* * * * * * * * * * * * * * * * *

75 Hampden Is built by English Electric, Preston, and delivered between February and July 1940

Serial	Units	Date	Cause	Crew	Duty	See
P2062	AAEE/14 OTU	31. 3.41	Control lost on approach, Woolfox Lodge	Sgt Holden; PO M J West; PO J E Richmond; (three pilots); Sgt Jessop. All killed	Night training	
P2063	49/144	22/23.11.42	Missing on nuisance raid	S/Ldr J R D Hird; Sgt D W Hepplewhite; Sgt R H Alvey; Sgt R J Coles. All missing	Bombing Norway	2J16
P2064	7/14 OTU/408/ Cv TB/415	11. 4.44	SOC			
P2065	16 OTU/455/ Cv TB/415	26. 2.43	Caught fire refuelling at Thorney Island		On ground	2J33
P2066	16 OTU	27. 3.41	Stalled landing with failed engine, Wroxton Heath near Banbury, Oxon	Sgt D V Bidlake (K); Sgt C H D Frost (K)	Training	
P2067	16 OTU/14 OTU/ 32 OTU	30. 4.44	SOC in Canada			5K8
P2068	49	19. 6.41	Engine cut on take – off, Scampton	Sgt J L Alden and two crew injured	Training	
P2069	4 FPP	7. 8.40	Engine failed, crashed on approach, U. Heyford	PO E C Shand unhurt	Ferry	6C3
P2070	50	25/26. 8.40	Forcelanded out of fuel, Lautersheim, Germany	PO Wawn; Sgt J Scholfield; Sgt R Smith; Sgt Bushell. All POW	Bombing Berlin	3D3
P2071	106	23.12.40	Abandoned while lost on night flying test near Nantwich, Cheshire	Sgt J E Sowden; Sgt Davies. Both baled out safely	Night test	
P2072	14 OTU	5.10.40	Hit hill in bad visibility near Richmond, Yorks	PO Fielder; Sgt Hawes; Sgt White; Sgt Gray. Two injured	Night training	5D3
P2073	106/408/ Cv TB/5 OTU	20.11.43	To GI airframe 4348M at Aberporth			
P2074	14 OTU/Cv TB/ 5 OTU	20.12.43	SOC			
P2075	14 OTU/455/ Cv TB/5 OTU/ 1 TTU	16.12.43	SOC			
P2076	14 OTU	15.12.41	Throttle lever broke off; undershot landing on one engine, Cottesmore	Sgt Roy; Sgt Hugall (I)	Night training	6C8
P2077	44	13/14. 8.40	Damaged by flak; crash – landed S of Ouisthuizen, Netherlands	PO H P Clarke; Sgt F M Wilkes; Sgt Brooks; Sgt G E Harris. All POW	Bombing Bernburg	
P2078	16 OTU/Cv TB/ 455 (UB – P)	11.12.42	Missing from Rover patrol off Norway	PO G I Gunton (P); FO T A Edgoose (M); Sgt D F Thomas (P); Sgt A E Crossley (M) All RAAF	Torpedo bombing (day)	2J17
P2079	144	8/ 9.11.40	Abandoned out of fuel near Hemswell	Sgt R H Pearman; Sgt S R Streeter; Sgt H E Turner; Sgt L W Jones. All safe	Bombing Munich	5A3 5A9 5A17 2J34
P2080	144/16 OTU/ Cv TB/489/455	20. 2.44	SOC			
P2081	144/25 OTU/ 16 OTU	14. 2.42	Hit trees on approach, Chipping Warden	Sgt J R Viau RCAF(K); Sgt W J Hodge RCAF (K); two W/ops injured	Night training	

AT225 of No.1404 (Meteorological) Flight, which later became No.517 Squadron, in January 1943 wearing Coastal Command white colour scheme.
(Photo: RNAS Yeovilton 213)

Serial	Units	Date	Cause	Crew	Duty	See
P2082	61	6/ 7.11.40	Missing, Kaiser Wilhelm Canal	S/Ldr J H R Oldfield; Sgt R S E Hamilton; Sgt J I Moyes; Sgt A Robertson. All b.Kiel	Minelaying in Kiel Canal	3E6
P2083	106/5 Gp TF/ 106	26/27. 5.41	Crashed in forced landing, Wellesbourne Mountford	Sgt L A Forty (K);Sgt A Wood (I); Sgt N G Lines (I); Sgt R Thompson (I)	Minelaying Brest	
P2084	CGS/Cv TB/ 144/489 (XA – O)	19/20. 5.43	Missing from shipping reconnaissance, Oberstadt to The Naze	F/Sgt R W Easton (M); Sgt E M Whitsed (M); WO McEachern RCAF (M); Sgt Keeping RCAF b. Stavanger	Torpedo bombing	
P2085	16 OTU/455	10/11. 3.42	Undercarriage collapsed on landing, Wigsley	Sgt Busfield; Sgt Waites; Sgt Gower; Sgt Biggin	Bombing Essen	
P2086	4 FPP	7. 8.40	Engine cut on approach, Honeybourne	PO H Malcolm (I)	Ferry	6C3
P2087	44 (KM – M)	6/ 7. 9.40	Shot down by flak	PO S R Taunton (P); PO H J Vollmer (P); Sgt W Bell*; Sgt J Bracegirdle* * b. Reichswald	Bombing Krefeld	5A4 3D15
P2088	61/14 OTU	17. 6.41	Stalled on approach, Cottesmore	F/Sgt H G A Reed (K); Sgt A M Greenaway (K)	Night training	
P2089	61	19/20. 8.40	Crashlanded low on fuel near Lyme Regis	PO P M de Mestre; PO I H Stewart; Sgt J I Doyle; Sgt L Hodgson. All unhurt	Bombing Bordeaux	
P2090	61	26/27. 9.40	Damaged by flak and crashed, Oresund, Sweden	PO R P Earl; PO W D Hermon; Sgt D Sinclair; Sgt J C Williams. All b.Kviborg	Bombing Kiel	
P2091	16 OTU/Cv TB/ 489/5 OTU	22.12.42	Dived into sea m W of Rillane Castle, Ayrshire	gt Hoskins and two other crew killed Sgt Maul picked up slightly injured	Bombing training	6E13
P2092	14 OTU	8. 4.41	Crashed near Little Bytham, Lincs. Believed shot down by intruder	Sgt Holborrow (K); PO W E Cochrane (K) Sgt W J Cunningham (K)	Night training	
P2093	50/1 AAS	23. 7.43	Overshot landing at Manby; SOC 17.2.44	Sgt Rembecki unhurt	Training	
P2094	144/50/420	20. 5.42	Hit by Lancaster L7581 at dispersal, Waddington		On ground	5C16
P2095	49/25 OTU/455/ Cv TB/455	12.10.42	Handed over to Russian Navy	Flown out by PO Humphrey, Sgt Robertson, Sgt Miller, Sgt Perry, Cpl Taylor	Left in Russia	2J15
P2096	83	27. 9.40	Crashed on approach, Scampton	Sgt T H Whitehead unhurt	Training	
P2097	83 (OL – S)	27/28.12.40	Crashed at Appleford Bridge near Abingdon	F/Lt L J Rooney; PO D J Fielder; Sgt J D Hunt; Sgt H B Whalen. All killed	Bombing Lorient	
P2098	106	27/28.12.40	Missing, presumed ditched	Sgt C E Sidebotham; Sgt W Chambers; Sgt W M Eaton; Sgt W R N Beavers. All missing	Minelaying Brest	
P2099	106 (ZN – K)	17. 5.41	Crashed in forced landing, Uffington, Rutland	PO S J Harvey (K); Sgt R R Adams (I); two passengers Sgt W A Campbell and AC2 T Kane both killed	Training	3C3
P2100	16 OTU/455/ 5 OTU	19. 7.42	Flew into hill 2m SW of Maybole, Ayrshire	Sgt W E Rayner (K)	Training	5F18
P2110	16 OTU	7. 8.40	Swung on take – off, Upper Heyford	Sgt W W Huggins and crew unhurt	Training	6B1

Serial	Units	Date	Cause	Crew	Duty	See
P2111	49/25 OTU	12. 6.41	Missing from navex over sea	Sgt W Allison(M); Sgt J A Tyrell (M); Sgt G T Corbett*; Sgt H W White*. * bodies recovered from sea at Burnham-on-Sea	Training	
P2112	49/14 OTU	8.12.41	Stalled on approach, Saltby	Sgt W G S Martin unhurt	Training	
* P2113	16 OTU/25 OTU/ Cv TB/16 FPP	20.10.43	Crashed on Tod Brow. Cumberland, after engine failure; had ASV aerials	Flt Captain J E Martens, ATA	Ferry	6D15 5J1 2J19
P2114	16 OTU	5. 1.41	Crashed in bad weather due to icing 8m W of Abingdon	Sgt R A Bridgart (K); Sgt C A Jones (I); Pilot on first night solo on Hampden	Night training	
P2115	16 OTU	17. 4.41	Wing sheared off in dive, Otmoor Bombing Range	Sgt P C Forster DFM; Sgt W W Walker; Sgt K L Clarke; Sgt F E Hubank; Sgt B Lodge All killed	Bombing practice	
P2116	14 OTU	30/31. 5.42	Shot down by night fighter 3 km N of Deventer, Netherlands	PO T E P Ramsay (P); PO W H Gorton*; PO V E Woolnough DFM*; Sgt F H Falk* *b. Diepenveen, Netherlands	Bombing Cologne	5A40
P2117	144	23/24. 8.40	Crashed after engine trouble, Boscombe Down; to 2270M	F/Lt Boylan; Sgt Wathey; Sgt Elcoat; PO J K Drewry. All unhurt	Bombing Brest	5A11
P2118	14 OTU/1406 Flt/519	25/26. 8.43	Hit mountainside in bad weather, Ben Loyal, Sutherland	F/Lt H R Puplett DFC (K); PO C Ritchie (K); FO C A Faulks (I); Sgt T R Hudson-Bell (K)	Air-sea rescue	5D6
P2119	14 OTU/Cv TB/ 415	3. 2.44	SOC			
P2120	14 OTU/16 OTU	16. 5.42	Hit ground in spin near Saxilby, Lincs.	PO W H Parr; Sgt D C McNab; Sgt K G McKee all killed. Rear gunner, name not known, baled out	Night training	5A37
P2121	44	17/18. 9.40	Shot down by flak, Antwerp	Sgt T V Henderson (K); PO R G Goode (K); Sgt J Angus (b.Antwerp); Sgt J Sugden (K)	Bombing barges	2F3
P2122	44/Cv TB/ 144	12.10.42	Handed over to Russian Navy	Flown out by F/Sgt Fendt, Sgt Jones, Sgt Westwood, Sgt Muirhead, LAC Holt	Left in Russia	2J15
P2123	44	31.8/1.9.40	Ditched out of fuel off Cromer, Norfolk	PO Romans DFC; PO Stewart; Sgt Logan; Sgt Mandale. All came ashore in dinghy	Bombing Berlin	5B6
P2124	50	25/26. 8.40	Ditched out of fuel off Scarborough, Yorks	PO Potts; PO Hook; Sgt Wood; Sgt Ashmore; All picked up unhurt by trawler	Bombing Berlin	5B3 2F23
P2125	83 (OL-L)	26/27.11.40	Swung on take-off, Scampton	Sgt Baxter; Sgt Powell, Sgt Robinson; Sgt Hunt. All unhurt	Bombing Cologne	
P2126	83/Cv TB/ 455 (UB-S)	12.10.42	To Russian Navy; damaged in air raid 27.9.42	Flown out by F/Lt O Connor; PO Lindeman; Sgt Walters, Sgt Collins, LAC Forbes	Left in Russia	2J3 2J15
P2127	14 OTU/12 MU	20.10.41	Crashed on approach, Kirkbride	Sgt A Raw (K); one non-RAF person killed	Air Test	
P2128	14 OTU	31. 7.41	Landed too fast and undercarriage collapsed, Scampton	Sgt D Calvert and crew unhurt	Training	
P2129	106/16 OTU/ 14 OTU	1. 8.42	Flare ignited in aircraft; crashed near Cottesmore	Sgt Coates and three crew. No serious injuries	Night training	
P2130	AAEE/1407 Flt	20. 2.44	SOC			
P2131	16 OTU	8. 8.40	Engine failed; overshot Upper Heyford; to 2177M	Sgt J H Morphett and crew unhurt	Training	
P2132	16 OTU	23. 8.40	Overshot and hit house, Upper Heyford	Sgt S L Bennett unhurt	Night training	6C8
P2133	16 OTU/Cv TB/ 489/32 OTU	27. 4.44	SOC in Canada			5K8
P2134	49	29. 9.40	Hit high ground on cross-country, Haigh near Wakefield, Yorks	Sgt E N Catley; Sgt Nicol; Sgt B V Hastie All killed	Night training	
P2135	49	31. 8.40	Stalled on climbing turn after take-off, Scampton	Sgt Roberts; AC1 T Robinson; AC1 G W Snoden; AC1 S Hall. All killed	Transit to satellite	
P2136	44/1 AAS	7. 2.43	Control lost on approach, Manby	Sgt R C Poole unhurt	Training	
P2137	44	20/21.10.40	Shot down by flak and crashed near Berlin	Sgt C W Hartop; Sgt C V Stubbs; Sgt R G Bennett; Sgt R L Milbourne. All b.Berlin	Bombing Berlin	
P2138	83/16 OTU/ 14 OTU/1402 Flt/BTU	18. 9.44	SOC			2F16
P2139	16 OTU/25 OTU/ 16 OTU/14 OTU/ 1402 Flt	29.11.43	SOC			
P2140	16 OTU	12.10.40	Undercarriage collapsed in fast landing, Upper Heyford	Sgt R Hanmer and crew unhurt	Training	
P2141	16 OTU	14. 1.41	Hit by Hereford L6027	No crew on board	On ground	5C9
P2142	44/16 OTU	10. 2.42	Abandoned after engine failure, Tusmore Park, Oxford	Sgt E A Robinson and crew landed safely by parachute	Night training	5A35 6D9
P2143	49	16/17.10.40	Crashed at Andover, Hants, presumed due to fuel shortage	FO C D Pitman; FO R C Parker; Sgt J Brown; Sgt P Flanagan. All killed	Bombing Bordeaux	
P2144	61	2/ 3. 6.41	Engine cut; crashlanded at Bircham Newton	Sgt Sleight; Sgt Woods; Sgt Burrows; Sgt Lodington. All unhurt	Bombing Dusseldorf	
P2145	49/16 OTU/ 455	16. 7.42	Spun into ground after engine failure, Crail	Sgt R D Wells RAAF (K); Sgt F Dawburn RAAF (I)	Training	

Two Hampdens of 408 Squadron flying above cloud. The bomb racks under the wings, extended flare damper exhaust and faired–in direction finding aerial show up clearly on P1166, the nearest aircraft. (Photo via Dennis Thompson)

120 Hampden Is built by English Electric, Preston, and delivered between February and August 1940

Serial	Units	Date	Cause	Crew	Duty	See
P4285	50/44 (KM–U)	8/ 9. 9.41	Engine failed; hit trees on approach to Coningsby at New York	PO Anekstein; Sgt Armstrong (I); Sgt Walker (I); Sgt Campbell	Bombing Kassel	
P4286	50/44	15/16. 5.40	Crashed at Oosterhout, Netherland; cause not known	PO L J Ashfield; PO C B Crawley; Cpl F Preston; Sgt F W McKinlay All b. Oosterhout	Bombing Breda	
P4287	50	8/ 9. 9.40	Last message said "Jumping for it" Cause not known	S/Ldr F A Willan; Sgt Wright; Sgt A B Cox; Sgt Brook. All POW	Bombing Hamburg	3D9
P4288	50	9. 7.40	Engine cut; crashed 2m E of Waddington	PO W A C Malloy unhurt	Training	
P4289	50 (VN–X)	8. 6.40	Crashed after engine failure, Coleby Grange	PO K R K Smettem unhurt	Training	
P4290	44 (KM–B)	5/ 6. 9.40	Ditched out of fuel off Lowestoft, Suffolk	PO Romans, PO Kerr, Sgt Logan, Sgt Simons All picked up unhurt by patrol craft	Bombing Stettin	2C17 5B7
P4291	144	16/17. 8.40	Missing; cause not known	PO H A T Skehill (P); Sgt W A Hughes*; Sgt D Perritt (P); Sgt R L Wake* *b, Reichswald	Bombing Merseburg	
P4292	16 OTU	7.12.40	Engine failed; aircraft rolled over on to its back and dived into ground from 300 ft, Weston–on–the–Green	Sgt A J Coad; S/Ldr H D Beck; Sgt M R Dards; Sgt Hadfield; Sgt Binks All killed	Night training	6D11 5F15
P4293	AAEE	23. 3.40	Crashlanded after engine failure, Andover; SOC 7.6.40	S/Ldr J W McGuire AFC unhurt	Training	
P4294	76/16 OTU	5. 8.40	Stalled on approach to Upper Heyford, Tusmore Park	PO G G Llewellyn (K)	Night training	
P4295	76/16 OTU	4. 6.40	Stalled landing at Brackley LG	Sgt G E Cowan unhurt	Night training	2F8
P4296	76/16 OTU/1 AAS	3. 2.44	SOC			
P4297	76/16 OTU	13. 6.40	Stalled while circling school, Iwerne Minster, Blandford, Dorset.	FO J E S Macalister; PO E A Sedgeley; AC2 J Lennon. All killed	Training	
P4298	61/25 OTU	30. 8.41	Stalled on approach, Finningley	Sgt C J H Blunt. Shown as '1 killed, 1 injured' in records	Training	

Serial	Units	Date	Cause	Crew	Duty	See
P4299	49	4/ 5. 2.41	Missing, cause not known	Sgt W J Baird; Sgt R I Eastwood; Sgt L W Homard; Sgt T R H Hawkes. All POW	Bombing Dusseldorf	
P4300	14 OTU	30. 6.41	Missing on night navex presumed ditched	PO J S Curl; PO J Mares; Sgt C J Shadbolt; Sgt A R Tour. RNZAF;. All missing	Night training	
P4301	16 OTU	23.10.41	Crashed at Charlton, Oxon., while seeking Croughton runway	Sgt J C Hill (Rhod); Sgt H Weir; Sgt R C D Jones (Rhod); Sgt R H Brickett. All killed	Night training	
P4302	106/25 OTU	16. 9.41	Dived into ground on overshoot, Balderton	PO D D Christie (K). One injured	Night training	
P4303	14 OTU	12. 9.41	Collided with P1301 near Saltby	Sgt D Bartlett; Sgt J E Wall RNZAF; Sgt G R West; Sgt H R Tuff. All killed	Training	5C7
P4304	49/Cv TB/ 489 (XA–X)	14. 5.43	Missing off Norway. Last seen with port engine on fire	PO J Dubbury; F/Sgt A H Jones RNZAF; F/Sgt D A Brier RNZAF; F/Sgt G R Burt RNZAF All POW	Torpedo bombing (day)	
P4305	49	26/27. 6.40	Shot down by flak, Kaiser Wilhelm Canal	F/Lt B A Mitchell (P); Sgt W A Spickett*; Sgt R Dutton*; PO L Reavall–Carter (P) *b. Kiel	Minelaying Kiel Canal	3D27
P4306	14 OTU/420/ Cv TB/415	28. 4.44	SOC			
P4307	7/14 OTU	4. 5.41	Engine cut on take–off, Cottesmore	PO Chase, Sgt Cowell unhurt	Training	6B3
P4308	14 OTU	27. 9.40	Crashed after control lost, Oakham	PO J Lufton (K).	Training	
P4309	7/14 OTU	13. 6.40	Overshot landing at Cottesmore	Sgt Linsdell unhurt	Training	
P4310	44	12/13. 6.41	Missing, cause not known	F/Sgt C T Mercer; F/Sgt R K Main; F/Sgt A Kossick; Sgt J R Park. All b.Reichswald	Bombing Soest	
P4311	7/14 OTU	17. 9.40	Stalled in forced landing, Kidwelly, Carm.	PO Souter; Sgt Brown; Sgt Blair; Sgt Cooper. All killed	Air firing exercise	
P4312	7/14 OTU/Cv TB/ 144/489/455	7. 2.44	SOC			6B3
P4313	7/14 OTU	27. 3.41	Flew into hill in bad visibility near Laxey, Isle of Man	Sgt D R Barclay (49 Sqn)(K); Sgt Case (K); Sgt Hawkins (K); Sgt Harper (I)	Training	5D1
P4314	7/14 OTU/106	3. 1.41	Engine cut; crashed 1m S of Finningley	Sgt P Mapp (61 Sqn) unhurt	Night training	5A20 6D13
P4315	7/14 OTU/Cv TB/ 415/489	8. 4.44	SOC			
P4316	TDU/14 OTU	21. 6.41	Dived into ground after take–off, Cottesmore	Sgt N McLafferty (K); PO I M Begbie (K); Sgt Leach (K); Sgt Thompson (I)	Training	
P4317	25 OTU/14 OTU/ 489	18. 4.44	SOC			
P4318	106/25 OTU/ 16 OTU/14 OTU	16. 8.42	Flew into hill in cloud, Arkengathdale Moor, Yorks	Sgt Henry (I); Sgt Longair; Sgt Phillips (I)	Night training	
P4319	16 OTU	10.10.41	Spun into ground soon after take–off 1m SW of Upper Heyford	Sgt C P Rose–Innes (Rhod); Sgt C van Deemter (SAAF); PO B Miller RNZAF; Sgt P W Harris. All killed	Night training	5A34 5F18 6B14
P4320	16 OTU	1. 9.41	Control lost on approach, Upper Heyford	PO W Lines (I)	Training	
P4321	49/1 AAS	13. 4.43	Bellylanded at Manby	Sgt W Wasiak unhurt	Training	
P4322	49 (EA–N)	5/ 6. 1.41	Missing, believed in sea 10m S of Sidmouth	PO G E Price DFM; Sgt W B Winning; Sgt C A Small; Sgt T H Chamberlain. All M	Minelaying Brest	2L3 5B16
P4323	16 OTU/106	23/24. 2.42	Shot down by naval flak over Heligoland Bight	Sgt S A Kent; Sgt R S Anderson; Sgt S R McLeod; Sgt A W Blake. All b.Kiel	Minelaying Heligoland	
P4324	61	26/27. 8.40	Forcelanded out of fuel, Vlieland, Netherlands	FO R P D Tunstall; Sgt A E Murdock; Sgt W J Brook; Sgt M J Joyce. All POW	Bombing Merseburg	3D28 2F24
P4335	61	12/13. 8.40	Missing, presumed ditched	PO T C Cundhill; PO D E Gardner; Sgt W B Taylor; Sgt D R Aldom. All missing	Bombing Salzbergen	
P4336	61	9/10. 6.40	Shot down by flak near Krefeld–Verdingen	S/Ldr L S Lawrence; FO J R Guthrie; Sgt J Fraser; Sgt H Davies. All b.Reichswald	Bombing Ruhr railways	
P4337	61/5 OTU	17. 9.43	SOC			
P4338	61/144/AAEE	21. 4.44	SOC			5E2
P4339	61	13. 6.40	Collided with L4138, Cottesmore	PO D A Helsby (I); PO G A Atkinson (K); Sgt Hutchinson (I); Sgt Golding (K)	Training	5C5
P4340	83	12/13. 8.40	Shot down by flak near Mesum, 5m SE of Rheine	F/Lt A R Mulligan (P); Sgt S Hill*; Sgt R Abel*; Sgt W E Younger (P).*b.Reichswald	Bombing D–Ems Canal	2F1
P4341	61	29/30. 6.40	Shot down by flak near Amsterdam, prob.in sea	PO A G Pascoe; PO A C Ennis; Sgt F Pursglove; Sgt J W Greig. All missing	Bombing Geestacht	2K3
P4342	61/25 OTU	6. 8.41	Crashed on air firing practice off Bridlington, Yorks	PO D S Jackman and three crew picked up from dinghy unhurt after 40 minutes	Training	
P4343	61	20/21. 7.40	Shot down by flak during low level attack	PO D H Davis (P); PO M R Tagg*; Sgt F S Waltho*; Sgt A M Cockburn*. *b.Sage	Minelaying Wilhelmshaven	5A4 2E4
P4344	61	20/21. 7.40	Hit by flak and crashed near Jever airfield	PO K Jones; Sgt Wright; Sgt D D Cain; Sgt R Bonson. All POW	Minelaying Wilhelmshaven	2E4
P4345	144	12/13. 6.40	Hit balloon cable and crashed into flour mill, Felixstowe	W/Cdr J J Watts DSO; PO J F E Andrews; Sgt R Jolly DFM; Sgt A Winstanley. All killed	Bombing railways	5E6
P4346	61	21/22. 6.40	Flak destroyed tail–plane; fell out of control	F/Lt D M Barrett (P); PO J G Mitchell*; Sgt C N Webster*; Sgt R Houghton–Brown* *b. Reichswald	Bombing Schwerte	3D13

Three Hampdens of 44 Squadron over Lincolnshire. (CH3482)

Serial	Units	Date	Cause	Crew	Duty	See
* P4347	144/Cv TB/ 489/16 FPP	2.12.43	Taxying accident at Kirkbride	First Officer M A Murtagh ATA unhurt	On ground	
P4348	144/16 OTU	1. 3.42	Spun into ground, Upper Heyford	PO R A Ray (K) on solo practice flying	Training	
P4349	61	7/ 8. 6.40	Crashlanded short of fuel in fog near Kings Lynn, Norfolk; to 2157M	PO A Webster; PO P Tunstall; Sgt R Jones; Sgt W Bissett. All unhurt	Bombing Hannover	2K24
P4350	49 (EA – L)	5/ 6. 9.40	Ditched off Calais, probably out of fuel	F/Lt J D Haskins; Sgt W L Evans; Sgt R Cartwright; PO J V Silverston. All POW	Bombing Stettin	3D22
P4351	49	4/ 5. 8.40	Ditched 10m E of Skegness out of fuel	Sgt J Unsworth; Sgt Imber; Sgt McDonald; LAC Allwood. Picked up from dinghy unhurt	Bombing Kiel	5B1
P4352	44	3/ 4. 7.40	Ditched in North Sea when returning with failed engine	PO D A Todd (M); Sgt E T Apperson (M) Sgt A Baird (M); PO W S Bull (washed ashore and b. Denmark)	Minelaying Great Belt	5B24
P4353	44	25. 8.40	Engine cut on take – off, Waddington	FO F G Quick. Two reported as injured	Air test	
P4354	AAEE	22.10.40	Collided with Buffalo AS430 12m SW of Salisbury, Wilts.	FO W J Carr; F/Lt S Wroath (both baled out safely). Mr J Parsons, Mr E Oxley of A&AEE both killed	Training	5A41
P4355	61	20/21. 6.40	Shot down by flak near Rheine	FO A S Smith; PO C Matheson; Sgt W S Rowcliffe; Sgt R S Powell. All b.Reichswald	Bombing railways	
P4356	61	29/30. 6.40	Shot down by flak near Rotterdam	FO B M Wyatt*; PO L S Adams (P); Sgt H Hill (P); Sgt R Burnett*. *b.Becklingen	Bombing Geestacht	
P4357	61	5. 8.40	Engine cut; ditched off Skipsea, Yorks.	Sgt H S Blakeman (I); Sgt J L Shardlow (I); Sgt G H McGrory drowned	Air firing practice	
P4358	61	20/21. 7.40	Shot down by flak at low level	PO A H Gould (P); Sgt J F Cowan (P); Sgt J N Prendergrast (P); PO D S Carnegie (b.Sage)	Minelaying Wilhelmshaven	3D29 2E2
P4359	144	8/ 9. 2.41	Engine cut in bad weather; abandoned, Taversham, Norfolk	Sgt R H Pearman; Sgt H Masters; Sgt H E Turner; Sgt A Simmons. All safe	Bombing Mannheim	5A3 5A9 5A18
P4360	144	21/22. 8.40	Missing; barge lift on Mitteland Canal	PO A Robson DFC; PO D L Wingate; Sgt A A Wilmot; Sgt H Chambers. All missing	Bombing Rothensee	2F3
P4361	144	4/ 5. 7.40	Shot down by flak, Kiel	FO E B Lancaster; PO J H Gilling; Sgt P F Bailey; Sgt A I J Nichol. All b.Kiel	Bombing Kiel	
P4362	144	5/ 6.10.40	Missing, presumed in sea	Sgt R J Neale; Sgt H Bennett; Sgt G Wood; Sgt P N Yorath. All missing	Minelaying Elbe	
P4363	144	23/24. 6.40	Missing, cause not known	PO B A Taylor*; PO C D Frost*; Sgt H G Cumpstey (P); Sgt D H Dunt (P).*b.Sage	Minelaying Heligoland	
P4364	144	18/19. 6.40	Crashed after engine failure near Scampton	PO B A Taylor; PO C D Frost; Sgt Riddell; Sgt H G Cumpstey. All unhurt. See P4363	Transit flight	
P4365	144	16/17. 8.40	Crashed on return, Hemswell; cause not known	S/Ldr P H Rebbeck; PO D Hutchins; Sgt H Dickenson; Sgt B H Foster. All killed	Bombing Merseburg	

Serial	Units	Date	Cause	Crew	Duty	See
P4366	144	11/12. 7.40	Shot down by flak, Kessel	PO I M Hossack; Sgt E France; Sgt E Leamy; Sgt C Rose. All b.Jonkerbos, Netherlands	Bombing Wanne – Eickel	
P4367	144	20/21. 7.40	Damaged by flak and ditched 90m E of Skegness, Lincs	F/Lt D H Edwards; Sgt B N Dingle; Sgt W Wylie; Sgt A J Giles. All missing	Minelaying Wilhelmshaven	2E1 2F11
P4368	144	9/10. 8.40	Crashed in Ysselmeer, cause not known	PO N F Tudhope (b.Kampen); Sgt L S Belton DFM (M); Sgt D McKay (M); Sgt A J Griffiths (M)	Bombing Homberg	2E5
P4369	144/25 OTU/ Cv TB/TDU	17. 1.44	SOC			
P4370	144	2/ 3. 9.40	Shot down by night fighter, Wynandsrade	PO R S A Churchill (P); PO C Kirk (P/I); Sgt A E Walker (K); Sgt A C H Edmeads (K)	Bombing Ludwigshaven	3D27
P4371	44	10/11. 9.40	Missing from bombing invasion barges	F/Lt D J Rogers (M); Sgt R J Brading (M); Sgt B N Westhorp (K); Sgt H G Lacey (M)	Bombing Calais	2F3
P4372	44	29/30. 8.40	Shot down by flak	PO T G Hynes (P); PO C O Dunkels (b.Rhein – berg); Sgt Wicks (P); Sgt Wainwright (P)	Bombing Gelsenkirchen	
P4373	44/25 OTU/ 16 OTU/Cv TB/ 144 (PL – R)	13.12.42	Shot down by ships' flak off SW Norway	PO H J Stillborn RCAF; PO G W Sproule RCAF; Sgt H F Dell; Sgt G H Foster. All missing	Torpedo bombing (day)	2J16 2J20
P4374	44 (KM – J)	2. 9.40	Undercarriage collapsed on landing, Radlett	PO C P D Price unhurt	Ferry	
P4375	44	27/28. 7.40	Hit balloon cable 8,000 ft over Hamburg	S/Ldr J G Macintyre (b.Hamburg); Sgt A Coveyduck; Sgt Cross; Sgt J H R Edgar. Last three POW	Bombing Hamburg	5A1 3D8
P4376	83 (OL – E)	2/ 3. 8.40	Returned early with engine failure and crashed 4m E of Scampton	Sgt Harpham (I); Sgt R Beales (K); Sgt Jackson (I); Sgt E J Clarke (K)	Bombing Hannover	2K9
P4377	106/49 (EA – K)	6/ 7. 8.40	Missing, cause not known	Sgt R J Jennings DFM; Sgt R F Batho; Sgt J F Lindsay; Sgt J Jones. All missing	Minelaying Frederikshaven	
P4378	144	5/ 6. 9.40	Stalled on overshoot and dived into ground, Hemswell	S/Ldr Lerwill (I); Sgt J W Caiter (K); Sgt W I Mackay(K); PO F J Drake Carrell(K)	Bombing Hamburg	
P4379	61	12/13. 8.40	Missing, cause not known	PO H C Sheldon; F/Sgt W Ward; Sgt R C Morrison; Sgt C E Thrower. All b.Hamburg	Bombing Salzbergen	
P4380	83 (OL – Z)	25/26. 8.40	Ditched out of fuel off Grimsby, Lincs.	PO A J G Mills; PO Mundy; Sgt Groves; Sgt T Highton. All picked up in dinghy	Bombing Berlin	5B4
P4381	83	3.11.40	Developed stabilised yaw at 600 ft and hit tree 1m E of Fiskerton	Sgt Garwell and crew unhurt	Training	
P4382	50	10/11. 8.40	Missing on practice operation flight to Vlieland	PO G G K Nolan; Sgt F C Shook; Sgt J M Beattie; Sgt E S Quinn All missing	Night training	
P4383	50	31.7/1.8.40	Ditched 32m E of Flamborough Head returning on one engine	PO B S Bell (b.Bergen – op – Zoom); Sgt W E Miller (M); Sgt C D Evans (M); Sgt D Tailford (M)	Minelaying off Fehmarn	525
P4384	49	27/28.12.40	Crashed near Abingdon; cause not known	Crew all killed. Names not recorded	Non – op'l	
P4389	50	18. 6.41	Lost height after take – off, Lindholme	Sgt L R Hinde; Sgt A Campbell; Cpl M E Balkin; LAC H Barton; AC2 P C Kemp. All K	Air test	6E8
P4390	61	7/ 8. 7.40	Shot down by flak, Nordhorn	F/Lt E C Fewtrell (P); PO J Eadie (b.Reichswald); Sgt E V Gawith (P); Sgt K B Wood (P)	Bombing Dortmund – Ems Canal	
P4391	144/14 OTU	28. 8.41	Blew up 1m NE of Akeman Street RLG	Sgt G H Bradley; Sgt F Grove; F/Sgt W H Morley; Sgt H S Davies. All killed	Night training	
P4392	83	21/22. 7.40	Returned without navigator	PO J C Bowman (P) baled out over Kiel	Bombing	5A47 3D21
P4392	83 (OL – P)	27/28. 9.40	Abandoned over Lincoln while lost in bad weather	PO D D Snooke (K); PO Turner, Sgt McKenzie, PO Brighouse baled out safely	Bombing Lorient	5B8 5A2
P4393	44	9. 7.40	Crashed on landing, Waddington	F/Sgt J F Clayton unhurt	Ferry flight	6C2
P4394	144	1/ 2. 3.41	Control lost in cloud after engine failure; crashed, Wainfleet	PO D H Skinner (I); Sgt F E Darnell (K); Sgt J F Shutt (K); Sgt R J Jones (K)		5A3 5A24
P4395	50/14 OTU/Cv TB/5 OTU/1 TTU	16.12.43	SOC			
P4396	61	13/14.11.40	Damaged by flak; believed ditched 20m NW of Heligoland	PO J E Langford; PO C D Chapman; Sgt R Sinclair; Sgt C E Russell. All missing	Bombing Hamburg	2L5
P4397	61	24. 9.40	Hit by X2911 on ground; possibly repaired and later to 14 OTU		On ground	
P4398	61/83/106	28/29. 1.42	Missing in icing conditions	Sgt H Sporny (K); Sgt J R Anyan*; Sgt A J Dickerson*; Sgt R M Young*.*b.Reichswald	Bombing Munster	2K27
P4399	61	30/31. 7.41	Control lost in thunder – storm; crashed near Dartford, Kent	Sgt Baker+; F/Sgt R J W Geater(K); Sgt W Chambers(K); Sgt W Butler(K)	Bombing Cologne	5A3 5A22
P4400	61/25 OTU/ 44/420 (PT – J)	12. 2.42	Shot down by fighters or flak	PO J R Topping DFM; Sgt F W Ashfield; Sgt T H Mate; FO E G Fowler. All missing	Bombing B/cruisers	2H3 2F9
P4401	61/Cv TB	17. 2.44	SOC			
P4402	83 (OL – D)	8/ 9.11.40	Abandoned 2 m NNW of Louth out of fuel	PO J A Mundy, Sgt Constable, Sgt Thomas, Sgt Bale all baled out safely	Bombing Munich	5A3 2F1

Formation of No.44 Squadron Hampdens in echelon: AE257 (KM – X), AE202 (KM – K) and AD981 (KM – A). Note bomb markings indicating raids on AE257 just aft of the nose perspex. These three aircraft were all subsequently lost on operations. (CH3481)

Serial	Units	Date	Cause	Crew	Duty	See
P4403	49 (EA – M)	3/ 4. 4.41	Overshot landing and hit hedge, St.Eval	Sgt Ball, Sgt Batchelor, Sgt Bentley, Sgt Calvert all unhurt	Minelaying Lorient	2F2
P4404	49 (EA – R)	6/ 7.12.40	Crashlanded in France after flak damage	Sgt J C Shaw; PO B H Evans; Sgt W K O'Leary; Sgt D Young. All POW	Bombing airfields	3D26
P4405	61	9/10. 2.41	Crashed at Oulton, Norfolk after SOS	F/Lt W E Frutiger; Sgt R A Chanin; Sgt J E Hill; F/Sgt R A Gapp. All killed	Bombing Wilhelmshaven	2K21
P4406	44	27/28. 7.41	Shot down by flak, Lorient	Sgt P C Gammon (M); Sgt B Winchester (M); F/Sgt G Slater (M); Sgt R Whittaker (b.near Lorient)	Minelaying Lorient	
P4407	144	29/30.11.40	Abandoned near Wittering after engine/fuel trouble	FO Page, Sgt Brain, Sgt Fraser, Sgt D Laing all baled out safely	Bombing Le Havre	5A3
P4408	50	15. 8.41	Crashed in North Sea	FO J A Whitecross DFC (M); Sgt M O Fisher (M); F/Sgt W Gray DFM*; Sgt R Taylor* *b.Sage	Air – sea rescue	3D31 5A26
P4409	49/50	1. 4.41	Swung on take – off and undercarriage collapsed, Lindholme	Sgt J Ratcliffe unhurt	Training	
P4410	83 (OL – H)	12/13. 8.40	Hit by flak, presumed ditched on return	FO E H Ross DFC; Sgt J H Dewhurst; Sgt K A Sayer DFM; Sgt J R Buckley. All Msg.	Bombing D – Ems Canal	2F1
P4411	50	30.9/1.10.40	Overshot single – engined landing, Docking	PO Powell, Sgt Martin, Sgt Drake, Sgt Green all unhurt	Bombing Berlin	
P4412	83/25 OTU	29. 8.41	Lost control and crash – landed, Balderton	Sgt Groves unhurt	Training	
P4413	106 (ZN – J)	15/16. 9.41	Engine cut; overshot landing, Pocklington	PO Loftus and three crew all injured	Bombing Hamburg	
P4414	44/106	21/22. 2.42	Missing from night intruder, Cologne – Koblenz presumed ditched	Sgt P D Bishop; Sgt R F W Hirst; Sgt N D Wright; Sgt F H Denton All missing	Bombing	
P4415	44/1 AAS/ 144	12.10.42	Handed over to Russian Navy	Flown out by Sgt Stilborn, Sgt Sproule, Sgt Dell, Sgt Foster, AC1 Fretwell	Left in Russia	2J15
P4416	49 (EA – L)	25/26. 8.40	Missing, believed ditched on return	PO N B Fawcett; Sgt J C Clarke; Sgt J Baker; AC1 G Reay. All missing	Bombing Berlin	
P4417	50	5/ 6.10.40	Missing, probably ditched	Sgt F J Brooker; PO A S Boak; Sgt J D Marshall; Sgt C H Wise. All missing	Bombing Cologne	
P4418	61/14 OTU/ Cv TB	24. 1.44	SOC			

80 Hampden Is built by Canadian Associated Aircraft and delivered between September 1940 and August 1941
Up to P5337 assembled by Quebec Group and remainder by Ontario Group

Serial	Units	Date	Cause	Crew	Duty	See
P5298	7 AAU/5 BGS/ 5 AOS/32 OTU	2. 8.44	SOC in Canada			5K3 5K6 5K8
P5299	7 AAU	23. 4.42	SOC at 7 AAU			5K4
P5300	7 AAU/5 BGS/1 AAS	14. 1.44	SOC in UK			
P5301	7 AAU/5 BGS/Cv TB/415 (GX – D)	26. 7.43	Missing on shipping patrol	F/Sgt R Grainger (M); Sgt A Millican (M); Sgt R B Mackenzie*; F/Sgt T A Ivey*. *b.Calais	Torpedo bombing (day)	
P5302	7 AAU/5 AOS/Cv TB/455 (UB – S)	3. 2.44	SOC in UK			
P5303	7 AAU/16 OTU	1.11.41	Spun into ground near Leamington	Sgt Butterworth; Sgt A R Enderby; Sgt T G Williams; Sgt R C Hollingsworth RCAF. All killed	Training navex	5A32
P5304	7 AAU/16 OTU/Cv TB/455 (UB – H)	4/ 5. 9.42	Flew into mountain, Arvastuotter, Sweden	Sgt E J Smart RAAF; F/Sgt T G Nichols RAAF; F/Sgt L A Biggin; Sgt J N O Harris RAAF; Cpl D H Nelson RAAF. All b.Kviborg	In transit to Russia	2J12 2J15
P5305	7 AAU/16 OTU	22. 7.41	Undershot and hit trees, Brackley LG	PO W M Major (Rhod) (K); Sgt Butler (I); Sgt D F Stephen, Sgt A C Willis unhurt	Night training	
P5306	7 AAU	23. 4.42	SOC at 7 AAU			5K4
P5307	7 AAU	23. 4.42	SOC at 7 AAU			5K4
P5308	7 AAU/16 OTU	29. 9.41	Flew into hill, Corble Farm, Brill, Oxon	Sgt E M White; Sgt J K Howe; Sgt K D Greenwood; Sgt J Amott. All killed	Bombing practice	5D2
P5309	7 AAU/16 OTU/CV TB/5 OTU/415	27. 2.44	To GI airframe 3075M			
P5310	7 AAU/16 OTU	17.10.41	Control lost after prop fell off; crashed 5m SE of Aberayron, Card.	PO J Scroggie (K); Sgt R Young (K); Sgt J Bremmer (I); Sgt Hart (I)	Training	5G4
P5311	7 AAU/25 OTU/ 14 OTU/Cv TB/ 1404 Flt	20.12.43	SOC in UK			
P5312	7 AAU/25 OTU/ 14 OTU	25/26. 6.42	Missing from "1,000 bomber raid"	F/Lt Count T S Salazar DFC; Sgt T S Gaffney; F/Sgt R J Horlock; Sgt S H Cusden. All b.Sage	Bombing Bremen	
P5313	7 AAU/16 OTU/ 5 OTU	24. 7.42	Swung on take – off, Turnberry	Sgt K Coates and crew unhurt	Training	2F16
P5314	7 AAU/16 OTU	20. 9.41	Shot down by intruder 4m NE of Upper Heyford	Sgt N van der Merve; Sgt D A Paine; Sgt J H Ixer; Sgt R Leagas. All killed	Night training	5H5
P5315	7 AAU/16 OTU/ Cv TB/415 (UB – K)	12.10.42	Handed over to Russian Navy	Flown out by Sgt Storry, FO Lovell, Sgt Sheedy, Sgt D Collins, LAC Cryer	See Russia	2J15
P5316	7 AAU	24. 7.41	Collided with Tiger Moth N5447 near Hooton Park	Captain Bailey, test pilot, and two passengers killed plus two in 19 EFTS a/c	Test flight	5C22 5K4
P5317	7 AAU/16 OTU/RAE	18. 5.44	SOC in UK			
P5318	7 AAU/16 OTU	22.10.41	Flew into hill near Wigginton Church, Bloxham, Oxon.	Sgt E Robinson; Sgt R Gordon; Sgt S K Holland. All killed	Night training	5D2
P5319	7 AAU/16 OTU	26. 2.42	Control lost after engine failure; crashed near Bourton – on – the – Water, Glos.	PO J Maura (baled out); Sgt J W Connell RCAF (K); Sgt T G Tallentyre (K); Sgt K Twidale (K)	Night training	5A36 6D10
P5320	7 AAU/14 OTU/455	10. 3.44	SOC in UK			5K7
P5321	7 AAU/44/408/ 14 OTU	30/31. 5.42	Collided with Halifax W1013 near March, Cambs.	S/Ldr D B Falconer unhurt; F/Sgt C Holmes (K); Sgt J H Knowling(K); PO H S Little(K)	Bombing Cologne	5A39
P5322	7 AAU/83/ 14 OTU	31.7/1.8.42	Missing, cause not known	PO R E Guthrie*; Sgt C A B McMullin (M); Sgt E T Hunking*; Sgt J H Morrison* *b. Groesbeek	Bombing Dusseldorf	
P5323	7 AAU/106/Cv TB/455 (UB – L)	4/ 5. 9.42	Ran out of fuel and crashed, Kandalashka	PO Patrick, Sgt McIver, Sgt Darnell, Sgt Clokessy, LAC Bryans all unhurt	See Russia	2J6 2J15
P5324	7 AAU/83/49 (EA – T)	12. 2.42	Shot down by flak or fighters	Sgt T K Downes*; Sgt T H Wood (M); Sgt B Hunter (K); Sgt D Poxon*. *b.The Hague	Bombing B/cruisers	2H4 3B1
P5325	7 AAU/455 (UB – V)	2/ 3. 4.42	Missing, believed in sea	PO J E Maloney; Sgt C F Woodburn; FO H N Young; Sgt H E Rowley, all RAAF. All Msg	Minelaying Lorient	
P5326	7 AAU/455 (UB – V)	12. 3.42	Engine cut on take – off; dived in, Wigsley	Sgt Andrew (K); Sgt Miflin (I)	Training	
P5327	7 AAU/455/Cv TB	1. 4.44	SOC in UK			
P5328	7 AAU/455 (UB – Q)	2/ 3. 1.42	Crashed at Folly Farm, Haddenham, Bucks. Cause not known	PO C Ludwig; FO J L S Willox RAAF; Sgt Stansfield; PO R G Christie RCAF. All killed	Minelaying La Rochelle	
P5329	7 AAU/455 (UB – B)	25. 3.42	Control lost when prop came off; crashed at Watchfield, Somerset	F/Lt A H Metcalfe DFC; PO J A Sayer; F Sgt Wilson; Sgt Donohoe. All killed	Transit flight	
P5330	7 AAU/106/420 (PT – L)	24/25. 4.42	Shot down by night fighter NW of Ameland	Sgt J Potter*; Sgt J M Hicks RCAF*; F/Sgt J Smith RCAF*; Sgt F Adams (P). *b Assens, Denmark	Bombing Rostock	
P5331	7 AAU/144/Cv TB/5 OTU	14. 7.43	Stalled on turn after take – off, Long Kesh	F/Lt G W H Allerton, FO H Grimshaw and one other killed	Training	
P5332	7 AAU/44/420 (PT – T)	2/ 3. 7.42	Shot down by night fighter, Ysselmeer	F/Sgt C G Wilde RCAF*; Sgt A D Bond+; Sgt T Crothers RCAF*; F/Sgt J E Gibbs RCAF+. *b.Amersfoort; +b.Vollenhove	Bombing Bremen	
P5333	7 AAU/16 OTU/ 5 OTU/1402 Flt	19.11.43	SOC in UK			

Hampdens of 489 Squadron at dispersal in the early summer of 1943. (Photo: S.B.L.Beattie)

Serial	Units	Date	Cause	Crew	Duty	See
P5334	7 AAU/408	26/27. 7.42	Attacked by Ju 88	Sgt L F Kelly (rear gunner) (K)	Bombing Hamburg	2F20
P5334	408/1406 Flt/519	25. 8.43	Missing on met flight over North Sea after SOS; engine failure	PO R Rowe RNZAF; PO E D Bonin RCAF; PO F Finlater RAAF; Sgt D McNichols RCAF. All missing	Met flight	5D6
P5335	489	13. 6.42	Attacked by Fw 190s, Norway	F/Sgt Dawney and Sgt McGill-Brown RNZAF wounded	Torpedo bomber (day)	2J1
P5335	7 AAU/50/144/Cv TB/489	17. 2.44	SOC in UK			
P5336	Western Cmd RCAF/32 OTU	2. 8.44	SOC in Canada			
P5337	124 (Ferry) Sqn RCAF	24.11.42	Crashed near Dugald, Manitoba	No details	Ferry flight	
P5338	7 AAU/AAEE/TDU/RAE	20. 6.46	SOC in UK			
P5339	7 AAU/5 BGS/1 AAS	17. 2.44	SOC in UK			
P5340	7 AAU/1 AAS	9. 6.44	SOC in UK			
P5341	7 AAU/CGS/Cv TB/1 TTU	16. 8.43	Engine cut; crashed off Dunure, Firth of Clyde	Sgt J R Maxwell; Sgt R Warfield; one crew killed, another injured	Training	6D19 5K4
P5342	7 AAU	24. 3.42	SOC in UK			
P5343	7 AAU/CGS/Cv TB/415	20. 2.44	SOC in UK			
P5344	14 OTU	31.7/1.8.42	Navigator baled out over sea	F/Sgt W E W Holmes RCAF (M)	Bombing Dusseldorf	5A51
P5344	7 AAU/14 OTU/1404 Flt/517	9. 1.44	SOC in UK			
P5345	7 AAU/16 OTU	13.10.41	Landed on Anson R9668, Upper Heyford	Sgt D F Allen (I) FO Paxton injured in Anson.	Training	5C10 6C20
P5346	7 AAU/16 OTU	3. 4.42	Hit ground during low-level bombing practice, Otmoor ranges, Oxon.	PO J E Piggott RCAF; PO J F Anderson RCAF; F/Sgt R St J Gregory RCAF; Sgt W A Skinner All killed	Night training	6E16
P5386	7 AAU/16 OTU	6.11.41	Dived into ground 2m E of Chalgrove, Oxon	Sgt R E Hatherill; Sgt S W Green; Sgt R Goff; Sgt W Dott. All killed	Night bombing exercise	
P5387	7 AAU/16 OTU/14 OTU/Cv TB/489	22. 3.45	SOC in UK			
P5388	7 AAU/16 OTU	20. 4.42	Swung on take-off and hit L4202, Upper Heyford	Sgt W J Wilkinson, Sgt D S Jackson, PO Edmunds, Sgt H A Vallis RCAF all unhurt	Training	6B2
P5389	7 AAU/Cv TB/5 OTU	16.12.43	SOC in UK			
P5390	7 AAU/16 OTU/Cv TB/489/455	19.12.43	SOC in UK			
P5391	7 AAU/16 OTU	3.10.41	Overshot Towyn	Sgt T F Freeman and crew unhurt	Training	6C6
P5392	7 AAU/408 (EQ-W)	14/15.12.41	Dived into ground, New-farm, Longdown, Hants.	Sgt R W Sterling; Sgt J C Tomlin RCAF; Sgt W R Williams; Sgt C C Gibson. All K	Bombing Cherbourg	3E1 2K19
P5393	7 AAU/83 (OL-T)	14/15.12.41	Collided on ground with AE374, Scampton	PO The Hon W R C B Parker (K); Sgt H Standen(I); Sgt G Love (K); one other (I)	Minelaying Brest	5C3
P5394	7 AAU/16 OTU/Cv TB/415 (GX-O)	27/28.10.42	Ditched after engine failure, Padstow Harbour, Cornwall	F/Lt G H Lawrence (I); PO K R Maffre, PO L S Sharp, F/Sgt R E Vokey all unhurt	Night torpedo bombing, Bay of Biscay	
P5395	7 AAU/1406 Flt/519	13. 4.44	SOC in UK			

Serial	Units	Date	Cause	Crew	Duty	See
P5396	7 AAU/3 FPP	7.11.41	Dived into ground on overshoot, Burtonwood	First Officer I J Parades ATA (USA) (K)	Ferry	5K5 5J3
P5397	7 AAU/16 OTU/ 14 OTU	30. 7.42	Collided with AE192 1m NW of Market Deeping	Sgt A R Gruber RCAF and crew baled out safely	Night training	5A38
P5398	7 AAU/14 OTU	7. 4.42	Crashed at Whittle Farm, Brockhampton, Glos. Cause not known	Sgt E J Biddulph (Rhod) and three crew all killed	Night training	
P5399	RCAF/32 OTU	16. 9.43	Undercarriage collapsed on take-off, Patricia Bay, BC	PO E J Stainer unhurt	Training	
P5400	RCAF/32 OTU	5. 8.42	Swung on take-off and broke back, Patricia Bay	Sgt J M McCarrison RNZAF unhurt	Training	
P5421	RCAF/32 OTU	27. 7.43	Undershot and hit obstruction, Patricia Bay	PO J R Ramshaw unhurt	Night training	
P5422	RCAF/32 OTU	2. 8.44	SOC in Canada			
P5423	RCAF/32 OTU	2. 8.44	SOC in Canada			
P5424	RCAF/32 OTU	10.11.43	Overshot landing at Patricia Bay	PO J E L Gilmore RAAF, Sgt J F Bishop, Sgt W T Elliott RAAF, Sgt R J Chartrand all unhurt	Training	
P5425	RCAF/32 OTU	2. 8.44	SOC in Canada			
P5426	RCAF/124 RCAF	29. 5.42	Dived into ground 3m NE of Rockcliffe, Ottawa, on approach	FO E S Parsons (K) on first Hampden solo	Training	
P5427	RCAF/32 OTU	31. 7.43	Abandoned after control lost near Patricia Bay	PO P A Hartman DFC baled out safely	Air test	5F13 5A29
P5428	RCAF/32 OTU	24.10.42	Stalled on landing, Patricia Bay	Sgt R A Perry unhurt	Training	
P5429	RCAF/32 OTU	5. 9.43	Damaged when taxied into by P5428, Patricia Bay		On ground	
P5430	RCAF/32 OTU	26. 5.44	SOC in Canada			
P5431	RCAF/32 OTU	24. 2.43	Forcelanded after engine failure near Patricia Bay	Sgt P O H Herbert (I)	Training	
P5432	RCAF/32 OTU	2.12.42	Swung on landing, Patricia Bay; DBR	Sgt Woodroofe unhurt	Training	
P5433	RCAF/32 OTU	14. 3.43	Crashed in sea, Saanich Inlet, Vancouver Island	PO A T Hunt RAAF (K); PO R K Manitan RAAF (M); PO G L Hall RCAF (M); Sgt H S Piercy RCAF (M)		Training
P5434	RCAF/32 OTU	26. 4.43	Damaged on take-off and crashlanded, Patricia Bay	Sgt J T Hynes RAAF unhurt	Training	
P5435	RCAF/32 OTU	16. 3.44	To GI airframe A386; SOC 27.4.54!			
P5436	RCAF/32 OTU	15.11.42	Developed stabilised yaw after torpedo drop; ditched in Saanich Inlet, Vancouver Island	Sgt L H Robinson, Sgt D W Smith, Sgt K E Blood RAAF all unhurt; Sgt J Fink RCAF injured	Torpedo training	5B29 6E21 5K15

(Note: The wreck of P5436 was raised from a depth of 600 fe et in the summer of 1986; it will eventually be displayed at the Museum of Flight and Transportation, Vancouver)

* * * * * * * * * * * * * * * * *

150 Hampden Is built by English Electric, Preston, and delivered between July and December 1940

Serial	Units	Date	Cause	Crew	Duty	See
X2893	61/AAEE	9. 6.43	Undercarriage collapsed, Boscombe Down; SOC 6.7.44	F/Lt J C Mann, F/Lt Mackie unhurt	Training	
X2894	61	5/ 6. 9.40	Ran out of fuel and crashed near Stradsett, Norfolk	F/Lt How, Sgt R A Brooker, Sgt Dickenson all unhurt. Sgt Knight (K)	Bombing Stettin	
X2895	83	25/26. 8.40	Ran out of fuel and crashed, West Bouldon, Co.Durham	PO Bunker DFC (I); Sgt Vivian (unhurt); Sgt Thomas (I); Sgt Haste (unhurt)	Bombing Berlin	
X2896	50	2/ 3.10.40	Returned on one engine and crashed, Woodhall, East Lothian	PO E M Thwaites, PO Hook, Sgt Day all unhurt; Sgt E S Smith (I)	Bombing Hamburg	5E9
X2897	50	28/29. 8.40	Ran out of fuel and ditched off Skegness, Lincs.	F/Lt Pitcairn-Hill DSO,DFC, FO Watson, Sgt Byrne, FO Stannion all picked up unhurt	Bombing Berlin	5B5
X2898	83/44/16 OTU/ Cv TB/415	19/20. 8.43	Shot down by flak off Egmond, Netherlands	F/Sgt W A Robson; F/Sgt H G Laverick; F/Sgt H Gnam RCAF; Sgt H St Jean RCAF All POW	Torpedo bombing (night)	3D1
X2899	83	6/ 7. 4.41	Missing, presumed ditched	PO A E Jackson DFM; PO A F Phillips; Sgt W A Foster; Sgt M A Hollingshead. All Msg	Minelaying Brest	
X2900	49 (EA-S)	16/17.10.40	Recalled due to weather; overshot landing at Abingdon	Sgt Bates, Sgt Bowden, Sgt Goldie, Sgt Fairbrother all unhurt	Minelaying Bordeaux	
X2901	83 (OL-B)	15/16.10.40	Ran out of fuel and forcelanded on beach, Southwold, Suffolk	F/Lt Barker, FO Anderson, Sgt Evans, PO Fleming all unhurt	Bombing Magdeburg	
X2902	50	29/30. 9.40	Missing, presumed ditched	S/Ldr J C Taylor; Sgt K H Watchous; Sgt A G Hobson; Sgt P A Merrick. All missing	Bombing Stuttgart	
X2903	144/Cv TB	10. 4.44	SOC			
X2904	83/Cv TB/ 455/1 TTU	16.12.43	SOC			

Torpedo – bombers of No.489 Squadron in formation off North Coates. Coastal Command practice was to omit squadron codes on Hampdens. (CH8653)

Serial	Units	Date	Cause	Crew	Duty	See
X2905	83/Cv TB/489	28. 7.43	Engine cut on take – off; bellylanded, Tain	FO B E Stourton unhurt	**Training**	**6D18**
X2906	61/25 OTU/ 16 OTU	13. 3.42	Crashed in flat spin 3m E of Croughton	PO R A S Parker RNZAF (K); PO G F Keele RCAF (K)	**Training**	
X2907	50	5/ 6.11.40	Crashed, Reutum, Neth. Cause not known	PO J C R Walker; Sgt K S Gowland; Sgt D O Cole; Sgt K Emm. All b.Ootmarsum, Neth.	**Bombing Magdeburg**	
X2908	50 (VN – Y)	15/16.11.40	Crashed after engine failure/fuel shortage Dunhill Road, Goole	Sgt Richardson (K); Sgt Brookes (K); Sgt J Anderson (I); Sgt Parker (K)	**Bombing Hamburg**	
X2909	4 FPP	27. 8.40	Overshot forced landing after engine failure, Upwell, Norfolk	F/Lt F C Gibbs unhurt	**Ferry**	
X2910	44	14/15.10.40	Shot down by night fighter near Rowitz, NE of Wolfsburg	Sgt L J Burt*; Sgt D Windle DFM (P); Sgt J Baldwin*; PO R M Carroll* *b.Berlin	**Bombing Berlin**	
X2911	61	24/25. 9.40	Hit P4397 on take – off, Hemswell	Sgt J E Hills (K); Sgt Mills (I); Sgt Keet (I); Sgt Fenwick (I); AC2 G E Palmer injured on ground	**Bombing Calais**	
X2912	61/49/Cv TB/ 455 (UB – W)	3. 3.44	SOC			
X2913	44	11/12. 9.40	Shot down by flak attacking liners in docks	PO D E Stewart*; Sgt D A Joyce (M); Sgt W Goodwill*; Sgt G J Hobson*. *b.Becklingen	**Bombing Bremerhaven**	**2F3**
X2914	106	26/27. 9.40	Ran out of fuel and abandoned, Chilton Polden, Somerset	Sgt W Huggins unhurt; Sgt E H W Shelfer (K); Sgt J Bogue (K); Sgt B Harvey (K)	**Minelaying**	
X2915	144	1/ 2.11.40	Missing. No trace of aircraft or crew	PO T Dawson, Sgt E A Barker, Sgt R L Hayes, Sgt I Cutler. All missing	**Bombing Berlin**	
X2916	44/25 OTU	11. 8.41	Crashed 1½m SW of Misson bombing range; cause obscure	PO A W Genth; PO T G Boucher; Sgt R J Moss; Sgt T W Clarke. All killed	**Night training**	
X2917	44 (KM – R)	6/ 7. 8.41	Broke up in air in bad weather near Marham	Sgt C S Bradbury; Sgt W C McQuadie RCAF; F/Sgt S D Yeomans; Sgt D H Howe. All (K)	**Bombing Calais**	**2K18 5H7**

Serial	Units	Date	Cause	Crew	Duty	See
X2918	44	10/11. 3.41	Shot down by night fighter near Venlo, Netherlands	PO H R Stockings DFC; PO E G M Bond; Sgt E W Dunkling; Sgt R J N Hanslip. All POW	Bombing Cologne	
X2919	50	2/ 3. 9.41	Ran out of fuel and crashlanded 2m S of Wittering	Sgt Mudd; Sgt Glenn; Sgt Dalgleish; Sgt Pullin. Reported as "three injured"	Bombing Berlin	
X2920	61	5/ 6.10.40	Crashed on Yorkshire Moors near Leeming on return	PO Wright; Sgt W A Cannon; Sgt W B Rayment; Sgt A Algar. All killed	Minelaying Elbe Est	2K11
X2921	106/44	7/ 8. 9.41	Crashed after take-off, Waddington	Sgt A A Watt (Rhod)(K); Sgt J R Newcombe (K); Sgt A D Wimbush (K); Sgt E S Cox (K)	Minelaying Kiel	
X2922	61	13/14. 9.40	Shot down by flak, probably in sea	F/Lt D J How*; Sgt A A Brooker(b.Wimereux) Sgt R C Dickenson*; PO T H Lane (M) *b.Boulogne	Bombing Boulogne	2F3
X2959	49/25 OTU	16.8.41	Hit trees on approach and crashed into house, Balderon	Sgt S D Baldachin (K); Sgt T C Wood (K); Sgt D Macdonald unhurt. Six children in house killed	Night training	6C21
X2960	106	18. 9.40	Forcelanded after engine failure, Cantley neat Finningley	PO I H Grant unhurt	Training	
X2961	144/CGS/Cv TB/ 415 (GX-S)	14. 6.43	Missing over Bay of Biscay, pres. ditched	S/Ldr J G Stronach, WO W A Trask, FO G K Crummy, PO A B Clegg, all RCAF. All Miss.	Anti-sub patrol	2J27
X2962	49	20/21.10.40	Crashed out of fuel and lost near Veryan, Truro, Cornwall	PO J H Green (unhurt); PO E Bingham (I); Sgt R W Coutts (I); Sgt W Hirst (unhurt)	Bombing Berlin	
X2963	144	5/ 6.10.40	Undershot in gale and bad weather, Hemswell	S/Ldr Lerwill, Sgt Gibson, Sgt Black, Sgt Duckworth all unhurt	Minelaying Elbe Estuary	
X2964	83 (OL-X)	9/10.11.40	Failed to become airborne and hit nissen hut, Scampton	Sgt A J Garwell, Sgt Kirke, Sgt R J Flux, Sgt White all unhurt. Three killed and one injured in hut.	Bombing Lorient	
X2965	44	1/ 2.10.40	Missing, presumed ditched	Sgt H Day*; Sgt J Tomlinson (M); Sgt G Devlin*; Sgt E R Sillett (M).*b.in UK	Bombing Cologne	
X2966	44	28/29.11.40	Ran out of fuel and forcelanded on beach, Sutton-on-Sea, Lincs.	Sgt Hubbard, Sgt Drakesford, Sgt Boswell, Sgt Williams all unhurt	Bombing Le Havre	6B18
X2967	61	14/16.11.40	Shot down by flak; location not known	PO D A Young; Sgt R I Hollidge; Sgt J I Boyle; Sgt F Johnson. All POW	Bombing Berlin	
X2968	50/16 OTU	20. 2.42	Crashed at Wootton Downs after three attempts at landing, Upper Heyford	Sgt D G Allen RCAF (K); Sgt G C Robertson (K)	Night training	6C19
X2969	83/144	24/25. 2.42	Shot down by night fighter off Terschelling	F/Sgt B T Dundas (Lord Dundas of Orkney); Sgt G A Barnett (Rhod); Sgt W A Croft; Sgt S B Wilson. All missing	Minelaying Heligoland	
X2970	106/25 OTU	12. 8.41	Stalled on take-off, Balderton	Sgt Groves and two crew injured	Training	
X2971	61	25/26.10.40	Ditched in North Sea on return flight	Sgt H C Loadsman; Sgt K R Humphrey; Sgt D I Pallett; Sgt F D Walter. All M	Bombing Kiel	5B31
X2972	83/14 OTU/ 1407 Flt	16.12.43	SOC			
X2973	144	16/17.10.40	Ditched in North Sea after SOS 20m N of Terschelling	Sgt J H Brown (b.Denmark); Sgt R G Young (M); Sgt J A Ferguson (M); Sgt E A G Rands (M)	Bombing Merseburg	
X2974	83/16 OTU/ 14 OTU	13. 7.42	Collided with Wellington and crashed 4m SW of Newark, Notts.	PO Lees (K); Sgt Pickel (K)	Night training	
X2975	61	8/ 9.12.40	Shot down by flak, Grevenbroich	Sgt Williams (P); Sgt R A Attfield*; Sgt J A Waters*; Sgt O Deeham. *b.Reichswald	Bombing Dusseldorf	
X2976	144/16 OTU/ Cv TB/144	12.10.42	Handed over to Russian Navy	Flown out by Sgt Hamar, Sgt Kidby, Sgt Yale, Sgt Dobson, LAC Ellery	Left in Russia	2J15
X2977	83	5/ 6.10.40	Returned early due to bad weather and crashed, Hemswell	Sgt Hawkes (I); Sgt C A Sherwood (K); Sgt Gilmour (I); Sgt Phillips (I)	Bombing Gelsenkirchen	
X2978	83 (OL-K)	3.11.40	Crashed in bad visibility 8m E of Scampton	Sgt B E Redgrove (K); Sgt J H Green (K); One other killed, name not recorded	Air test	
X2979	61	16/17.10.40	Crashed attempting to land while short of fuel, Swaffham, Norfolk	PO W H Clemerson; PO P Gardiner; Sgt W H Hewitt; Sgt D Flanagan. All killed	Bombing Merseburg	2K15
X2980	61/16 OTU 14 OTU/Cv TB	16.12.43	SOC			
X2981	61	20.12.40	Exploded while testing 'Imp'-type mine, Theddlethorpe ranges	F/Lt E H Cooper; PO E N Ward; Sgt S Evans. All killed	Trials	2D1 3F1
X2982	44	12/13. 5.41	Crashlanded out of fuel near Catterick, Yorks.	Sgt E D Tyler, Sgt Greig, Sgt Betts, Sgt Campbell all unhurt	Bombing Mannheim	6B18
X2983	50	14/15. 2.41	Shot down by flak, Pte de la Chambrette, Garonne	PO P G Tunstall (K); Sgt Barclay (K); Sgt F Bailey (P); Sgt Dixon (P)	Minelaying Garonne Estuary	2D8
X2984	50	1/ 2. 3.41	Flew into high ground while lost 9m NW of Bridlington, Yorks.	PO H L Pexton; Sgt Richards; Sgt Wheeler; Sgt Tittley. Reported as 'one killed, two injured'	Bombing Cologne	2K16
X2985	49 (EA-W)	10/11.11.40	Missing, last heard of 30m from Norfolk coast	FO A Bulmer; PO D W Wakeford; Sgt D Herdman; Sgt D W Watt. All missing	Bombing Danzig	5B14

Canadian-built Hampden aboard a merchant ship as its convoy assembles at Halifax, Nova Scotia, in April 1941, for the crossing to Britain.
(Photo: Public Archives of Canada DND/PA-105314)

Serial	Units	Date	Cause	Crew	Duty	See
X2986	106 (ZN-F)	20/21. 4.41	Crashed near Cologne; cause not known	PO R M A Lakin*; PO J Cutmore*; Sgt H E Preston DFM*; Sgt W E Burrell* *b.Rheinberg	Bombing Cologne	
X2987	61	22/23.11.40	Undershot in bad weather and hit bomb dump, Hemswell	Sgt R S C Ivatt (K); Sgt R Pelham (K); Sgt R W Smith (I); Sgt F Bissell (I)	Bombing Duisburg	2F9
X2988	144	16/17.10.40	Abandoned low on fuel in bad weather near West Raynham	PO T Dawson, Sgt E A Barker, Sgt R L Hayes all unhurt. Sgt J J Jackson (K)	Bombing Merseburg	5A3
X2989	61/408 14 OTU	10. 2.42	Control lost in circuit, Saltby	PO W C Anderson RCAF unhurt	Training	2F8
X2990	83 (OL-Z)	26/27.10.40	Missing, presumed ditched	Sgt J G Loveluck; Sgt K E Young; Sgt J M Dall; Sgt G L Middleton. All missing	Minelaying Garonne	
X2991	50	27/28. 8.41	Crashed on take-off, Swinderby	PO Ferrie, Sgt Redwood, Sgt Hawkins, Sgt Holland all unhurt	Bombing Mannheim	
X2992	50/25 OTU/ 14 OTU	13.12.41	Stalled and dived into ground on single-engine flying practice, Blue Point Farm, Cottesmore	PO A P Webb (K); Sgt K F Thornton (K)	Training	
X2993	50	14/15.10.40	Shot down by night fighter, Kalbe	PO A H Davies*; Sgt E J Andrews*; Sgt Hurrell (P); Sgt Lee (P). *b.Berlin	Bombing Berlin	
X2994	50	7/ 8.11.40	Abandoned out of fuel near Bircham Newton	FO E B Liddell, Sgt Leech, Sgt A S Atkinson, Sgt E McConnell all safe	Bombing Essen	5A3 5A14
X2995	44	13/14.11.40	Control lost on take-off, Waddington	PO Cuthbert (I); PO K L Millar (K); Sgt D Pewsey (K); Sgt E Egar (K)	Bombing Hamburg	
X2996	44	14/15.11.40	Missing, possibly blown up by own "Imp" mine	FO R D Perkins*; Sgt J G W Maybury*; Sgt J D Green*; Sgt T W Hawxsby*.*b.Berlin	Bombing Berlin	3F1 2D2
X2997	44	16/17.10.40	Undershot and hit old hangar in fog, Waddington	S/Ldr H P Broad (I); Sgt Hammond (I); Sgt Logan and AC1 (Sgt?) Egar both unhurt	Bombing Merseburg	
X2998	144	25/26.10.40	Throttle levers jammed; crashed on return, Upton, near Hemswell	Sgt R J Curtiss (I); Sgt A Goulder (I); Sgt E H Woods (I); Sgt W C Anderson (I)	Bombing Kiel	2K23

Serial	Units	Date	Cause	Crew	Duty	See
X2999	44	17/18. 4.41	Missing, presumed ditched in North Sea	F/Sgt J E Sneeston*; Sgt E N Brundish (M); Sgt J N Taylor (M); Sgt J W Mulford (M) *b. in UK	Bombing Berlin	
X3000	50	29/30.10.40	Abandoned low on fuel in icing conditions near Linton-on-Ouse	PO C J R Walker, Sgt Gowland, Sgt Cole, Sgt Emm all baled out safely	Bombing Berlin	5A3 5A21
X3001	49 (EA-H)	10/11. 2.41	Shot down by night fighter 10 km N of Alkmaar, Netherlands	PO Green (P); Sgt D A Cruickshank*; Sgt A L Bryceson (P); Sgt H E Fisher (P); *b. Alkmaar	Bombing Hannover	5A43
X3002	106	3/ 4. 3.41	Crashed near Ardres; cause not known	Sgt K Good; Sgt A J W Ward; Sgt D A Crouch; Sgt V T Essex. All B.Schoonselhof, Belgium	Bombing Cologne	
X3003	50/16 OTU	11. 7.41	Engine cut on take-off, Upper Heyford	Sgt C A Snelling and Wireless Operator unhurt	Night training	6C12
X3004	50	7/ 8.12.40	Missing, presumed ditched	FO R T Mulligan; PO M J Preece; Sgt C J Smith; Sgt F A Buckle. All missing	Bombing Dusseldorf	2K25
X3005	61	23/24. 3.41	Ran out of fuel and abandoned near Driffield	PO P H H Pritchard; PO F A Caunter-Jackson; Sgt P J Breene; Sgt O W J Pearce. All unhurt	Bombing Kiel	2F8 5A3 5A10
X3006	61	13/14.11.40	Hit hut on take-off, Hemswell	F/Lt A Moncrieff (K); Sgt E F Edwin (K); Sgt Eborall (I); Sgt Donnelly (I)	Bombing Hamburg	
X3007	144	11/12. 2.41	Abandoned near Kirton Lindsey, Lincs.; crashed 70m away near Bircham Newton	Sgt E Dainty. Sgt L A Wilmot, Sgt T Grosvenor, Sgt G Bottomley all baled out safely	Bombing Hannover	5A3 5A17
X3008	44	16/17.11.40	Overshot landing in bad weather, Bircham Newton	PO Stockings, Sgt Watts, Sgt Moon, PO Bond all unhurt	Bombing Hamburg	
X3021	49/106	29/30.10.41	Damaged by flak and ditched off Lincolnshire coast	Sgt T L Panting (K); Sgt F R Lamin (M); Sgt G E Bradley (M); Sgt D C Cranston (M)	Bombing Schiphol	
X3022	50/Cv TB/	12.10.42	Handed over to Russian Navy	Flown out by Sgt King, Sgt Honan, Sgt Hamlin, Sgt Allen, Cpl Evans	Left in Russia	2J15
X3023	44	19/20.11.40	Crashed in bad weather, South Repps, Norfolk	Sgt J L F Ottaway (K); PO A R Kerr (K) Sgt S F Elliot (K); Sgt Hird (I)	Bombing Lutzkendorf	
X3024	49 (EA-H)	19/20.11.40	Missing, presumed ditched in North Sea	Sgt B I M Fulton; Sgt L Ball; Sgt D G B Steell; Sgt T H Mellon. All missing	Bombing Lutzkendorf	
X3025	44	3. 9.41	Hit by AE157 at dispersal, Waddington	Unoccupied	On ground	5C24 5H7
X3026	44/Cv TB/ 1 TTU	7. 8.43	Hit high ground in bad visibility, Ayrshire	FO J H W Johnstone RAAF and three crew all killed	Training	
X3027	49 (EA-A)	27/28.10.40	Attacked by intruder near Scampton; crashed in sea off Skegness	PO J R Bufton (K); PO K Ballas-Anderson (M); Sgt R F Robertson (K); Sgt F J W Bichard (K)	Bombing Hamburg	5H1
X3028	49 (EA-S)	6/ 7.12.40	Crashed on return near Dunholme Lodge	Sgt Barrier (I); FO K W Mitchie (K); Sgt G V Davenport-Jones (K); Sgt T Goldie (K)	Bombing NW France	
X3029	49 (EA-D)	3/ 4.11.40	Swung on take-off in rainstorm and crashed 5m W of Scampton	Sgt Richman, Sgt Davies, LAC (Sgt?) Upton, Sgt Reid all unhurt	Bombing Kiel	
X3030	144	20/21. 9.41	Undershot landing at North Luffenham and hit HT cables	Sgt E Parker; Sgt H Weaver RCAF; Sgt R F Aberneathy; Sgt J Tautard All killed	Bombing Frankfurt	
X3047	144	1/ 2.11.40	Missing, presumed ditched in North Sea	PO G Morgan; Sgt D A Jenner; Sgt P D Adams DFM; Sgt E J Bushell. All Miss.	Bombing Berlin	
X3048	144	15/16. 2.41	Landed with undercarriage retracted, Hemswell	S/Ldr Lerwill and crew unhurt	Bombing Homberg	
X3049	44	10/11.12.40	Ran out of fuel and crashlanded in France	S/Ldr N H J Tindall; Sgt Martin; Sgt J McEwan; Sgt P R Tebbutt. All POW	Minelaying Gironde	3D7
X3050	49 (EA-N)	6/ 7.12.40	Shot down by flak off Ostend	Sgt S G Greeves; Sgt L Young; Sgt E F Parish; Sgt M H Davies. All missing	Bombing NW France	
X3051	144/408 (EQ-U)	21/22. 1.42	Missing; either X3051 or AE352 shot down by night fighter	PO H W Clarke RCAF; Sgt P B Chapple; Sgt R G Eves; Sgt R Brown. All missing	Bombing Bremen	
X3052	49 (EA-C)	24/25.11.40	Shot down by night fighter, St Maartens-vlotbrug	Sgt F Phillips (P); Sgt D C Dunphy (P); Sgt E Terpey (P); AC2 S R Hoskins (P)	Bombing Hamburg	3D60
X3053	83/Cv TB/ 144	12.10.42	Handed over to Russian Navy	Flown out by PO Bowden, PO Mennill; Sgt Elsmore; Sgt Vosper, Cpl Willcox	Left in Russia	2J15
X3054	49	21/22. 3.41	Flew into hill in cloud, Hamel Down Tor, Dartmoor	PO R D Wilson; Sgt R Ellis; Sgt R Brames; Sgt C Lyon. All killed	Bombing Lorient	2K13
X3055	144 (PL-O)Cv TB	9. 6.44	SOC			
X3056	144	19/20.11.40	Short of fuel in bad weather; abandoned near Withern, Lincs.	PO Hartop and crew* baled out safely. *Possibly Sgts Kirby, Burns, Daniels	Bombing Lutzkendorf	5A3 2F12
X3057	44/49/420	21.12.43	SOC			
X3058	61/106	28/29. 1.42	Ditched, probably out of fuel, off Norfolk	Sgt Brown picked up unhurt. Sgt A G Patrick RCAF (K); Sgt R Biddlecombe (K); Sgt F Baker (K)	Bombing Munster	5B30 5H11
X3059	83	3. 7.41	Crashed after engine failure 2m S of Scampton	Sgt Adams (I); Sgt Foster (K)	Training	
X3060	49	26/27. 6.41	Overshot on return, Scampton	PO Falconer on first op.; Sgt Stuart; Sgt Knowling, Sgt Drinkell all unhurt	Minelaying Frisians	

Bombs arrive at dispersal in the snow. "VN – Z" of 50 Squadron awaits its quota at Swunderby in early 1942. The large bombs on the trolleys may be 1,800 – lb HC (High Capacity or Blast) bombs. (Photo via Cee Wadsworth)

Serial	Units	Date	Cause	Crew	Duty	See
X3061	83/44/420/ 14 OTU/Cv TB/415	22. 6.43	Hit by N9096 while parked, St.Eval; DBF		On ground	5C20
X3062	83 (OL – Z)	8/ 9. 5.41	Shot down off Mablethorpe by intruder on outward flight	PO F K Gill; PO G D G Hudson; Sgt R C Blatchford; Sgt H W Francis. All missing	Bombing Hamburg	5H4
X3063	49	16/18.12.40	Believed crashed off Isle of Wight on return	F/Lt A T N Scoltock; Sgt J M Tiedman; Sgt A W Wood; Sgt E G Rimes. All missing	Bombing Mannheim	5H10 5B15
X3064	61	25/26.11.40	Dived into ground, Spital – in – the – Street, Lincs. on return	Sgt J Hopkins; Sgt G F Williamson; Sgt T P Fenwick; Sgt Clarke. All killed	Bombing Kiel	
X3065	144	6/ 7.12.40	Engine cut on fighter patrol over Bristol; abandoned, Marcham, Berks	PO N J Kerr; PO R N Harrison; Sgt A E Moffat; Sgt J Howell; Sgt G Bottomley. All baled out safely	Fighter patrol	5A3 5A8
X3066	144	10/11. 4.41	Shot down by night fighter 2 km SW of Neeroeteren	PO N J Kerr; PO R J Hamilton; Sgt J Howell; Sgt A R Wickens. All b.Adegem, Belgium	Bombing Dusseldorf	5C23
X3115	AAEE/Cv TB/ 415	22. 6.43	Damaged by exploding bombs when N9096 crashed, St.Eval		On ground	5K2 5C20
X3116	83/25 OTU/14 OTU/Cv TB/455	21. 6.44	SOC			
X3117	50	10/11.12.40	Shot down by flak near Tres, Mosel	FO E B Liddell; PO R G Ash; Sgt E McConnel; Sgt A S Atkinson. All b.Rheinberg	Bombing Mannheim	
X3118	83 (OL – J)	8/ 9. 8.41	Crashed in sea off Amsterdam; cause not known	F/Sgt F J Wood*; PO J E Cunning (M); Sgt E C Matthewman*; Sgt W Mackay*. *b. North Holland	Bombing Essen	5H8
X3119	83 (OL – R)	20/21. 4.41	Crashed at Houthulst, Belgium. Cause not known	F/Sgt R B Hanmer; Sgt G W Whitehead; Sgt J R Ronnie; Sgt E S Phillips. All b. Houthulst	Bombing Cologne	
X3120	61	2/ 3. 6.41	Believed ditched out of fuel	Sgt W Asson; Sgt N M Campbell; Sgt H G Johnson; Sgt C E Hawkes. All POW	Bombing Dusseldorf	
X3121	83 (OL – F)	24/25. 8.41	Collided with AD967 on return, Scampton	F/Lt A G Mills DFC; Sgt J A Somerville; Sgt B V Mason; F/Sgt D I Sharpe. All (K)	Bombing Dusseldorf	5C1
X3122	83/420/ 519 (Z9 – G)	28. 8.43	Missing, pres. ditched NE of Scotland	Sgt B S Cox; Sgt A M Finfer; F/Sgt A Fairall RCAF; F/Sgt D C McKenzie RCAF. All (M)	Met Flight North Sea	

Serial	Units	Date	Cause	Crew	Duty	See
X3123	83/5 AOS/1 AAS	24. 1.44	SOC			5B13
X3124	83 (OL – H)	26/27. 2.41	Flew into high ground, Derrington Cross near Stafford	W/Cdr W W Stanthorpe AFC; PO A McD Melville; Sgt L Emmerson; Sgt D V Weaving. All killed	Bombing Cologne	2K12
X3125	50	25/26.11.40	Ran out of fuel and crashed near Mablethorpe	Sgt H V Crum, Sgt Adams, Sgt McAlwane, Sgt Barley all unhurt	Bombing Kiel	
X3126	61	1/ 2. 1.41	Both engines cut on approach, Hemswell. Possibly out of fuel	F/Lt Powdrell (I); Sgt Horn (I); Sgt Durtnall, Sgt Clelland unhurt	Bombing Bremen	2K28
X3127	61	8/ 9. 8.41	Missing, cause not known	PO J W W Graham (P); Sgt L R Biddlecombe (P); Sgt W Hughes (P); Sgt A Nuttall RNZAF b.Kiel	Bombing Kiel	
X3128	61	16/17.12.40	Missing, presumed ditched in North Sea	Sgt G E Cowan DFM; Sgt H R Richardson; Sgt A J Casey; PO E Reeve. All missing	Bombing Mannheim	2F8
X3129	144	1. 4.41	Shot down by flak or fighters near Brest	PO W C Hartop (P); Sgt F H Sykes*; F/Sgt R G Lambourne*; Sgt J Riley*. *b.12 miles N of Brest	Bombing Brest (day)	2G1 2F12
X3130	144/Cv TB/3 FPP	18. 3.42	Crashed in transit, Kirkbride – Thorney Island	First Officer T C D Bray ATA (Aus), Second Officer N A Berry ATA both killed	Ferry	5J5
X3131	83/106/Cv TB/ 455 (UB – D)	12.10.42	Handed over to Russian Navy	Flown out by Sgt Lord, Sgt Weekes, Sgt Bremner, Sgt Castles, LAC Crawford	Left in Russia	2J2 2J15
X3132	83 (OL – L)	20/21. 3.41	Missing, cause not known	Sgt O B James (P); Sgt W A Miller b.Bayeux Sgt Maccallum evaded capture; Sgt N Weir b.Bayeux	Minelaying Brest	3D31
X3133	50	29/30. 6.41	Shot down by night fighter 3 km SE of Oldenburg	Sgt W D Hughes DFM; Sgt C T Chipperfield; Sgt A P Drane; Sgt C R Wells. All b. Hamburg	Bombing Kiel	
X3134	49 (EA – C)	30.6/1.7.41	Ditched on return; in dinghy for eight days	Sgt B Woolston, Sgt P McKay; Sgt E B Chandler, Sgt G Wood all picked up	Bombing Dusseldorf	5B10
X3135	49	11.11.41	Dived in on steep turn 2m NE of Scampton	Sgt K Bryant and two others killed Names not recorded	Training	
X3136	49 (EA – K)	2/ 3. 9.41	Ran out of fuel and crashed, Thornham Ling Common, Norfolk	Sgt Samuel, Sgt West, Sgt Watchorn, Sgt Tilley all unhurt	Bombing Berlin	
X3137	44/16 OTU/ Cv TB/32 OTU	30. 4.44	SOC in Canada			5K8
X3138	61/5 AOS	4. 9.42	Engine caught fire; crashed, Whitchurch, Salop.	FO Zielinski unhurt	Ferry	6D14
X3139	83 (OL – V)	8/ 9. 7.41	Shot down by night fighter N of Groningen	F/Sgt R C Brutey(M); Sgt M T Matthews(M); Sgt A Macpherson b.Winsum, Neth.; Sgt V W Kent (P)	Bombing Hamm	
X3140	61/408/ Cv TB/455 (UB – T)	25. 1.43	Ditched in North Sea returning to Leuchars	PO E J Gleeson; PO K J Murphy; Sgt H Witte; Sgt R J Booth, all RAAF. All (M)	Torpedo bombing (day)	2J18
X3141	50	28/29.12.40	Shot down by flak	PO D B Ainsworth (P); Sgt E H Battle*; Sgt M Hemmingway*; Sgt G G Smith* *b. near Lorient	Bombing Lorient	
X3142	44/16 OTU/ 14 OTU/Cv TB	4. 6.44	SOC			
X3143	50	1/ 2. 1.41	Overshot in bad visibility on return, Lindholme	PO Burrough, PO Hodgson, Sgt Hobson, Sgt Brooks all unhurt	Bombing Bremen	
X3144	83 (OL – N)	2/ 3. 9.41	Presumed flew into sea off Frisians	Sgt C W Allen (M); Sgt A W Rowan b.Texel; Sgt H T Smith (M); Sgt H Kay b.Harlingen	Minelaying Frisian Is	
X3145	50/Cv TB/ 415 (GX – F)	16/17. 7.43	Missing after SOS heard at North Coates	F/Sgt R Walter; FO V C Steer; F/Sgt H Huntingdon RCAF; F/Sgt J Janisse. All (M)	Anti – ship patrol, Biscay	
X3146	50	13/14. 3.41	Missing, presumed ditched in North Sea	Sgt R L Grainger; Sgt R E Wunderlich; Sgt J Cameron; Sgt R D Watson. All missing	Bombing Hamburg	2K16
X3147	61	1/ 2. 3.41	Crashed short of fuel, Syderstone, Norfolk	Sgt R W Clarke; Sgt Cox; Sgt P Thomas; Sgt Loates. All killed	Bombing Cologne	
X3148	106 (ZN – E)	10/11. 4.41	Shot down by night fighter near Ittersvoort	Sgt R J West (P); Sgt J Spencer*; Sgt J T Wright*; Sgt E J Butler*. *b.Jonkerbos	Bombing Dusseldorf	5C23
X3149	44/420/ Cv TB/32 OTU	30. 4.44	SOC in Canada			5K8 5C16
X3150	44/Cv TB/ 455 (UB – U)	16. 7.43	Spun into ground 2½m S of Leuchars	Sgt R B M Gemmell (K); Sgt A N Owen (K); FO F Bradshaw RNZAF (K)	Training	2J5
X3151	49 (EA – T)	25/26. 7.41	Engine failure due to fuel shortage; crashed 2m S of Scampton	PO Cooke, PO King, Sgt Evans, Sgt Newall all unhurt	Bombing Hannover	
X3152	106/5 BGS	11. 9.41	SOC. Cause not known			
X3153	106	10/11. 4.41	Shot down by night fighter, Kessel – Helden, Netherlands	Sgt W Osborne (P); Sgt J F R Boyall*; Sgt D Skene*; Sgt S Davies (P) *b. Jonkerbos	Bombing Dusseldorf	5C23
X3154	106 (ZN – A)	21.12.40	Flew into hill on night navex, Chapel – en – le Frith, Derbyshire	PO Hubbard; Sgt Perkins; Sgt D J Davey; Sgt Smith. All killed	Night training	

* * * * * * * * * * * * * * * * *

P1272 was delivered to the Torpedo Development Unit on 2 December 1939 and was there for six months during which time it was used for training anti – aircraft gunners (ours!) in aircraft recognition. The undersides appear to be painted yellow, the standard colour to identify training aircraft. (Photo via Dennis Thompson)

425 Hampden Is built by English Electric, Preston, and delivered between October 1940 and October 1941

Serial	Units	Date	Cause	Crew	Duty	See
AD719	49	10/11. 2.41	Shot down by intruder on return, Scampton	Sgt G M Bates (I)*; Sgt Butterworth (K); Sgt Blower (I)*; Sgt Caldwell (K). * Baled out	Bombing Hannover	5H2
AD720	144	10/11. 3.41	Crashed soon after take – off from Hemswell, at Bishops Norton	Sgt T D Leitch; Sgt T W Little; Sgt D Lane; Sgt G F Asbrey. All killed	Bombing Cologne	
AD721	50	12/13. 3.41	Shot down by night fighter near Nordhorn	F/Lt D T Johnston; PO L E Stannard; Sgt H Howell; Sgt W Williams. All b.Berlin	Bombing Berlin	
AD722	83 (OL – X)	11/12. 2.41	Hit building on approach in fog, Finningley	F/Lt D W F Barker DFC (K); Sgt W H Lowson (K); FO Wilson (K); Sgt Targett (I)	Bombing Bremen	
AD723	61	1/ 2. 3.41	Crashed in bad weather on return near Caister	Sgt K G Cooper; Sgt H Windle; Sgt M Gough; Sgt Hall. All killed	Bombing Cologne	2K16
AD724	144/Cv TB/CCDU/ 1406 Flt/1404 Flt/517	11. 9.43	Swung on take – off, St.Eval	Sgt Miller, Sgt Neill, Sgt Nicholson, Sgt Wall all unhurt	Met flight	
AD725	61	1. 2.41	Overshot in snowstorm Hemswell	Sgt N M Lloyd (I); Sgt R Guest (K)	Night training	
AD726	44	31/8.1.9.41	Ditched 25m E of Harwich, probably engine failure	Sgt S Harvey (RHOD)(M); Sgt H Taylor RCAF (K); F/Sgt J Phillips (M); Sgt R Hayes b.Belgium	Bombing Cologne	
AD727	44/61	11/12. 6.41	Shot down by flak off Buelk	FO P Pritchard b.Kappel; PO P Caunter – Jackson (M); F/Sgt P Breene (M); Sgt J Bestwick b.Kiel	Minelaying Kiel	5A10 3E5
AD728	50	28/29. 4.41	Missing; may have hit water in sea – fog	S/Ldr D C F Goode DFC (M); PO Willis*; Sgt Evans*; Sgt Page*. *b La Rochelle area	Minelaying La Rochelle	
AD729	49 (EA – N)	27/28. 5.41	Missing, presumed ditched	S/Ldr D A Smith (M); Sgt J R Butler (M); F/Sgt J G Leslie (M); Sgt G Hadaway (M)	Minelaying Brest	
AD730	50	17/18. 4.41	Flew into hill while lost, Blessington, Co.Wicklow, Eire	PO J K Hill, Sgt J T Lamb, Sgt S Wright, Sgt F H Erdwin all b.Blessington	Bombing Berlin	2K14 5D5
AD731	83 (OL – M)	16/17. 1.41	Ditched 80m ENE of Great Yarmouth	PO P J Strong; Sgt D V House; Sgt D Brand; Sgt A G S White. All missing	Bombing Wilhelmshaven	
AD732	61	18. 4.41	Shot down by flak in attack on airfield near Cherbourg	F/Lt R S E Aldridge DFC*; PO F W Holden*; Sgt A Drury*; Sgt K Downing (P). *b. Cherbourg	Bombing Cherbourg (day)	5A4 3D16 2G2
AD733	49 (EA – B)	28/29/ 9/41	Missing, cause not known	F/Lt J D Mundy DFC; Sgt P C Darwin; Sgt A Winton; F/Sgt F Hibbert DFM. All POW	Bombing Frankfurt	

Serial	Units	Date	Cause	Crew	Duty	See
AD734	83 (OL – K)	11/12. 2.41	Abandoned after hitting balloon cable over Birmingham	F/Lt R Anderson, FO J Badcock, Sgt D O Weaving, Sgt T P Byrne all baled out safely	Bombing Bremen	5E7 5A52
AD735	106 (ZN – R)	7/ 8. 7.41	Shot down by night fighter NE of s'Hertogenbosch	Sgt R Wotherspoon (P); F/Sgt P Hudson (K); Sgt H Tait (K); F/Sgt K Botsford (K) No.97 Squadron crew	Bombing M – Gladbach	2F6
AD736	106/16 OTU/ 14 OTU/1402 Flt/8 FPP	18.12.43	Control lost on approach; hit building, Hawarden	First Officer J D Hurley ATA (K)	Ferry	5J6
AD737	144	1/ 2. 3.41	Returned with engine trouble; undershot landing at Hemswell	PO McVie, Sgt Dyke, Sgt Scouller, Sgt Carter all unhurt	Bombing Hannover	
AD738	106	4/ 5. 4.41	Shot down by flak over Brest	W/Cdr P J Polglaise; PO W Brown; F/ Sgt G Allanson; Sgt E Holman. All buried 7 miles NW of Brest	Bombing Brest	2F4
AD739	49 (EA – A)	6/ 7. 7.41	Missing, believed shot down by flak in sea	PO A J Henderson; F/Sgt I M Fisher; Sgt H Aldridge; F/Sgt K Coney. All missing	Bombing Brest	
AD740	83/25 OTU/ 14 OTU	15. 9.42	Stalled and dived into ground on overshoot, Cottesmore	F/Sgt A Cooper RCAF (K); Sgt Neuls RCAF (K); F/Sgt St Pierre RCAF (K); Sgt Rushford (I)	Night training	
AD741	BDU/14 OTU/ 1406 Flt/519	9. 1.44	SOC			
AD742	50	20/21. 3.41	Undershot flarepath at Lindholme on return	PO Burrough; PO Hodgson; Sgt Russell; Sgt A B Brooks (I)	Minelaying Lorient	
AD743	106/25 OTU/ Cv TB/455 (UB – F)	12.10.42	Handed over to Russian Navy	Flown out by PO Clarke, PO Wilson, F/Sgt Parkhouse, Sgt Richards, LAC Barassi	Left in Russia	2J15
AD744	83/49	6/ 7. 9.41	Crashed out of fuel near Banff	Sgt Robinson, Sgt Sanderson, Sgt Price, Sgt Mossop all injured	Minelaying Oslofjord	
AD745	144	20/21. 3.41	Abandoned in bad visibility near Chelveston, Northants	PO R N Harrison, Sgt E W Smith, Sgt H E Turner, Sgt A Simmons. All baled out safely	Minelaying Brest	5A3 5A9
AD746	106	20/21.10.41	Returned with engine failure; stalled on approach, Coningsby	Sgt Cooke, Sgt Fry. Sgt Goodwin, Sgt Gowan all unhurt	Bombing Bremen	
AD747	44	19. 6.41	Engine cut on take – off; crashed, Ternhill	PO Lauderdale (I); Sgt Drakeford (I); LAC Wall (I)	Ferry	6B18
AD748	83 (OL – M)	3/ 4. 4.41	Flew into Hangingstone Hill, Dartmoor in bad weather	F/Lt Thompson, PO L R Evans, Sgt L R Eden, Sgt Murray all killed	Minelaying La Rochelle	5D4 2K13
AD749	106/14 OTU	17. 3.42	Overshot in bad weather, Newton	Sgt H H Davis RCAF unhurt	Night training	
AD750	106	4/ 5. 2.41	Missing, cause not known	F/Lt W K Burr – Thomas, PO G F H Innis, Sgt J L Franco, Sgt F Colson. All b.Nantes	Minelaying St.Nazaire	
AD751	BDU/14 OTU/ 32 OTU	30. 4.44	SOC in Canada			5K8
AD752	61/144	9. 1.42	Crashed near Buddington, Rutland	Sgt D Bryan (K); Sgt J C Bradshaw (K)	Night training	
AD753	50	5. 4.41	Dived out of control into sea on way to Brest	PO H M Macrossan; Sgt R F Cheesman; Sgt F Worthington; Sgt J Matthews; All missing	Bombing Brest (day)	2G1 5F8 3E3
AD754	408	21/22. 1.42	Attacked by night fighter over Holland	Sgt J P Farrow; Sgt T I May (K); Sgt Millward (I); Sgt V R Baker (K)	Bombing Bremen	2F19
AD754	144/61/408/ Cv TB/32 OTU	30. 4.44	SOC in Canada			5K8
AD755	44	30/31. 7.41	Returned with engine failure; crashed 5m W of Navenby, Lincs.	Sgt Armstrong, Sgt Morley, Sgt Flint, Sgt Edwards all unhurt	Bombing Cologne	
AD756	106	16/17. 8.41	Shot down by night fighter near Ophoven	PO G G Watts*; Sgt L Acres*; Sgt G Lynn (P); Sgt Cook (P); PO P Grant*. *b. Gembloux	Bombing Dusseldorf	2F5
AD757	14 OTU/1404 Flt	16. 7.43	DBR by blast when Whitley BD276 hit P1204, St.Eval		On Ground	
AD758	106/44/ 408/14 OTU	14.12.41	Crashed on overshoot, Jurby, and blew up	PO J T Chrystal and crew unhurt	Training	2F7
AD759	49	25.11.41	Spun into ground near Manby	Sgt R Hough (K); Sgt G E Smith (K)	Training	6E2
AD760	106	7/ 8.11.41	Shot down by flak in Oslofjord	F/Lt J Henderson DFC(P); Sgt D Ingram(M); Sgt J W Steven (P); FO C B Randall b.Oslo	Minelaying Oslofjord	
AD761	144	16/17. 4.41	Shot down by intruder soon after take – off from Hemswell	Sgt H Kirby (K); Sgt W A Tyler (K); Sgt V G Elliott (K); Sgt G Walsh (I)	Minelaying Frisian Is	5H3
AD762	144/Cv TB/ 415 (GX – J)	30/31. 7.42	Returned early with engine failure; over – shot North Coates	WO B D McComb; F/Sgt M Bloomfield; Sgt R M Ennis; Sgt J Labelle. All RCAF. All killed	Torpedo bombing Dutch coast	2J29
AD763	106	1. 3.41	Undercarriage collapsed on landing, Coningsby	PO W A Brown, Sgt Stevens both unhurt	Ferry	
AD764	50/Cv TB/5 OTU	16.12.43	SOC			
AD765	106/144	26. 1.42	Control lost on take – off, North Luffenham	Sgt J H Carr (K)	Night training	
AD766	50/144/14 OTU	5. 7.42	Stalled on take – off, Cottesmore	Sgt C M Ross RAAF unhurt	Training	

Hampden torpedo – bomber AN127 "XA – Y" of No.489 Squadron in July 1942, while undergoing trials at Boscombe Down. The landing light, in the port wing leading edge, appears to have an anti – balloon cable protection strip across its face. (Photo via Dennis Thompson)

Serial	Units	Date	Cause	Crew	Duty	See
AD767	144/Cv TB/415/ 32 OTU	13. 1.44	Dived vertically into ground, Dinner Point, Mayne Is., BC	Sgt R Hayes (K) after only five hours solo on Hampden	Training 5K9	5K8
AD768	106	1/ 2.10.41	Abandoned out of fuel, Glendowan Hill, Donegal	Sgt Newby, Sgt Tisdall, Sgt Wakelin, Sgt Reid all baled out safely and interned	Bombing Karlsruhe	2A6 3D4
AD782	14 OTU/408/ (EQ – A)	25/26. 1.42	Stalled soon after take – off from Balderton and crashed	Sgt V C Dadson RCAF (K); Sgt D Marsden (K); Sgt A B Wright (K); Sgt H Davis (K)	Bombing Brest	
AD783	144/455/ (UB – U)	12/13. 2.42	Returned early with engine failure and crashed near Wigsley	Sgt Kennedy (K); Sgt V J Hopwood (K); Sgt Kelly (K); Sgt Metcalf (K)	Minelaying Frisian Is	2H7 2K22
AD784	144	30/31. 7.41	Crashed in sea off Dutch coast	PO D Brook RNZAF (M); Sgt P T Giblett (M); PO C Hewson RCAF (M); PO F Pratt b. Bergen – op – Zoom, Netherlands	Bombing Cologne	2L4
AD785	106	26/27.10.41	Blew up near Whitby, Yorks.	Sgt E B Smith, PO D R Bowden; Sgt Halward; F/Sgt D B Stables DFM. All killed	Bombing Hamburg	2K17
AD786	14 OTU/455/ 420 (PT – L)	23/24. 6.42	Crashed at Boothby Pagnall 20 minutes after take – off from Waddington	F/Sgt H S Hilet DFM (I); F/Sgt H Germain RCAF (K); Sgt G Johnson (K); Sgt K Little (K)	Minelaying Lorient	2F14
AD787	14 OTU	16. 6.42	Swung on landing and undercarriage collapsed, Saltby	Sgt Featherstone and crew unhurt	Training	6C11
AD788	49 (EA – V)	25/26. 6.41	Shot down by Kiel by flak or night fighter	Sgt S N Hind (P); Sgt C L Ray*; Sgt P E Whiting*; Sgt C W Hancock* *b.Kiel	Bombing Kiel	3D14 5H9
AD789	50	10/11. 4.41	Shot down by night fighter near Venlo	PO C J Cornish (P); Sgt J Ratcliffe*; F/Sgt R Royal*; Sgt S Cassey*. *b. Jonkerbos	Bombing Dusseldorf	5C23
AD790	106	25. 2.41	Engine cut on take – off, Coningsby	Crew names not known. No injuries		
AD791	144	7/ 8.12.41	Missing, cause not known	Sgt F K Calcutt; Sgt A H Lawrence; Sgt R E Hirons; Sgt G Barnes. All b.Heverlee	Bombing Aachen	
AD792	49/14 OTU/ Cv TB/455 (UB – P)	11. 1.43	Flew into Hill of Warren, Angus, on return	PO P J Hill (K); PO Rees (K); Sgt Smithers RAAF (K); Sgt Spohn RAAF unhurt	Torpedo bombing (night)2J18 Norway	5D6
AD793	144/Cv TB/ 5 OTU/1 TTU	20. 1.44	SOC			
AD794	83/144/Cv TB/ 415	17. 1.44	SOC			
AD795	83/50/Cv TB/ 144/489 (XA – K)	14/15. 9.42	Ditched out of fuel 60m E of Rattray Head	FO T Murray (M); Sgt C Brandon (M); Sgt D Newman*; Sgt A Brown*. *b. Kiel	Anti – sub patrol	
AD796	83	26/27. 4.41	Crashed at Eddelak; cause not known	PO R H Crush; Sgt G T Hall; Sgt E G Norman; Sgt W K Thompson. All b. Kiel	Bombing Hamburg	

Serial	Units	Date	Cause	Crew	Duty	See
AD797	50	2/ 3. 6.41	Missing, cause not known	PO P Hodgson; PO P Brown; F/Sgt J Donnelly; Sgt G Cheetham. All b.Leopoldsburg, Belg.	Bombing Dusseldorf	
AD798	14 OTU/1407 Flt	11. 2.44	SOC			5F14 5A30
AD799	106/49/144/ Cv TB/455/415	9. 9.43	Crashed near Thorney Island after developing stabilised yaw	PO R S Arnold; PO C Coons; F/Sgt D K Stroud. All RCAF. All killed	Air test	
AD800	83 (OL – X)	29/30. 3.41	Crashed in sea 36m SSE of Start Point, Devon	PO E W Reynolds; Sgt C B Harrison; Sgt J T Owen; Sgt H R Northwood. All missing	Minelaying Brest	
AD801	144/Cv TB/489	30. 4.44	SOC			
AD802	106/14 OTU	30. 6.42	Lost sight of Saltby flarepath and crashed, Gunthorpe Bridge, Notts.	Sgt R T Pierson and three crew all killed	Night training	6C16
AD803	106/455/106/ 408 (EQ – L)	15/16. 5.42	Shot down by naval flak off Endelave Island	Sgt H Copeman RCAF (P); F/Sgt A Smith RCAF (M); Sgt N W Smith*; F/Sgt W F Millerd RCAF*. *b. Denmark		2D7
AD804	61/144	28/29.12.41	Crashlanded, Doetincham. Possibly damaged by flak	Sgt T E Barnett (P); Sgt T S Jones (P); Sgt W R Cheesman (K); Sgt W Gibley (P)	Bombing Huls	
AD805	49 (EA – R)	8/ 9. 9.41	Lost and ran out of fuel, Carlton, Lincs	PO Jones, Sgt Thorndyke, Sgt Taylor, Sgt Hardisty all unhurt	Bombing Kassel	
AD806	61	5/ 6. 7.41	Missing, presumed ditched	Sgt T A Holden; F/Sgt A A Horn; Sgt D A Johnson; F/Sgt H Mainey. All missing	Bombing Osnabruck	2K29
AD824	50/49/144	7. 2.42	Shot down by fighter off Frisians Five crew aboard	Sgt R F Thompson (K); F/Sgt R N Thompson (K); Sgt D Duce (M); Sgt L Bow (M); Sgt R Rowell (M)	Minelaying Frisians (day)	2H1
AD825	61	18. 4.41	Abandoned near Swindon after flak damage	Sgt L W Metcalfe (I); Sgt F A Armstrong, Sgt D Pearce, Sgt W Butler baled out OK	Bombing Brest (day)	5A13,5A3 2G2,3A3
AD826	61/455/16 OTU	6. 5.42	Dived into ground, Wootton, Beds	Sgt A G Maxwell; Sgt J P Hancock; Sgt D O'Brien; Sgt T Brown all RCAF. All killed	Night navex	
AD827	61	8/ 9. 4.41	Shot down by flak over Kiel	FO D G Glennie; PO G F Chipperfield; Sgt J G Donnelly; Sgt C P Thomas. All b.Kiel	Bombing Kiel	
AD828	50	10/11. 4.41	Shot down by night fighter near Venlo	PO D Powell b.Jonkerbos; Sgt J A Davis (P); Sgt R F Drake (P); Sgt F C Snook (P)	Bombing Dusseldorf	5C23
AD829	155/83/44/ 455/408 (EQ – E)	28/29. 8.42	Probably shot down by night fighter	Sgt P Kemp (P); Sgt S Clark*; Sgt H Gregson*; Sgt E Stansfield*. *b.Sauville, Fr.	Bombing Saarbrucken	2A1
AD830	50	10. 4.41	Crashed into house, Evington, Leics.	Sgt J J Campbell, three crew and one civilian in house killed.	Training	6E2
AD831	16 OTU	19. 6.41	Engine failure soon after take – off, Upper Heyford	Sgt J Goldman (Rhod); Sgt C Davis RCAF; FO W H Corbett; Sgt P Lally. All killed	Training	6B8
AD832	144	17. 2.42	Crashed after engine failure 8m E of Harrington, Northants	PO R P Worthy and crew. One injured	Training	
AD833	16 OTU	1. 9.41	Crashed on approach to Shoreham in fog, Lancing College	PO B S C Fynn (K); PO A S Playfair (K); Sgt C E Butt (K); Sgt G Morffew (I)	Night training	6C5
AD834	50	28/29. 4.41	Engine cut; abandoned over Brittany	FO J A Whitecross evaded capture; Sgt Martin (P); F/Sgt D Ross*; Sgt J O'Hare* *b. 60m ESE of Brest	Minelaying La Rochelle	5A26 3D31
AD835	83 (OL – G)	25/26. 7.41	Shot down by night fighter near Schiermonnikoog, Neth.	Sgt P H Draper (P); Sgt J Tate*; Sgt E Marsden*; Sgt F Ireson*. *b.S'monnikoog	Bombing Hannover	
AD836	50/Cv TB/455	22.12.42	Port engine cut on take – off, Leuchars	WO P H Baker, F/Sgt G S Dun RAAF, Sgt L A Jonas RAAF, Sgt N Bayliss RAAF unhurt	T/bombing Norway	2J17
AD837	83 (OL – E)	17/18. 8.41	Ran out of fuel and ditched 10m E of Holy Island, Northumberland	Sgt Linacre, Sgt Harrison, Sgt Hunt, Sgt Calder all picked up by patrol boat	Bombing Bremen	
AD838	144/16 OTU/14 OTU/Cv TB/489	6.11.44	SOC			
AD839	50	29/30. 8.41	Missing, cause not known	F/Sgt L C Turner; Sgt J G Proctor; Sgt T H Willis; Sgt R Urpeth all b. Abbeville	Bombing Frankfurt	
AD840	44	8/ 9. 7.41	Missing, cause not known	Sgt A W Wilson; Sgt L A Soutar; F/Sgt S J Lytle; Sgt H D Mackenzie all b.Reichswald	Bombing Hamm	
AD841	144 (PL – Q)	15/16. 5.41	Shot down by flak N of Amsterdam	PO W A McVie*; PO S L Bailey*; Sgt J Scouller (P); Sgt C Carter (P) *b. Amsterdam	Bombing Hannover	
AD842	49/408	8/ 9. 3.42	Stalled on take – off, North Luffenham	Sgt D C Hunter RCAF (I); Sgt R Ball (K); Sgt W D Morris (K); Sgt F F MacKinnon RCAF (K)	Minelaying Frisians	
AD843	50	23/24. 7.41	Crashed in North Sea	Sgt H T Holme (M); Sgt L M Megginson (M) Sgt R Evans*; Sgt W M Watt*. * Bodies picked up 8m E of Cromer, Norfolk	Bombing Frankfurt	5B26
AD844	50	16/17. 7.41	Missing, presumed ditched	PO H M Owens; PO R G Mitchell RCAF; F/Sgt D E Russell; Sgt R W Owen. All missing	Bombing Hamburg	
AD845	49/14 OTU	13/14. 9.42	Ran out of fuel and crashed near Cottesmore	Sgt H D Bearnes, Sgt D McKay, Sgt M B Anderson, Sgt W R Jordan all unhurt	Bombing Bremen	2A1
AD846	144	5/ 6.11.41	Shot down by flak in attack on convoy off Terschelling	PO D A Baker (P); FO E L Shea (M); Sgt J S Ansley (P); Sgt R L Frank (M)	Bombing shipping	
AD847	44	26/27. 4.41	Crashed while returning on one engine near West Raynham	Sgt Lauderdale, Sgt Mallen, Sgt Hartley, Sgt R Oliver all unhurt	Bombing Hamburg	6B18

L4098 of No.44 Squadron being refuelled, July/August 1939. It bears the pre-war code "JW". The slotted leading edge of the wing shows up very well. The NCO in the foreground is Sgt Percy Nixon who was lost on ops in L4087. (Photo: Harry Moyle)

Serial	Units	Date	Cause	Crew	Duty	See
AD848	16 OTU/106/ 14 OTU	30. 6.42	Dived into ground, Moorend near Doncaster	Sgt T A Krome and three crew all killed	Night navex	
AD849	16 OTU	11. 7.41	Crashed soon after take-off, Upper Heyford	PO D G L Buckley; PO P L Nash RCAF; Sgt J T Travis; Sgt L A Webster. All killed	Training	6B13
AD850	16 OTU/83 (OL-L)	5/ 6.11.41	Shot down by flak SE of Mano Island	PO S Hartley; F/Sgt A E Ward; Sgt N Cott-erell; F/Sgt R B Turner. All b. Esjberg	Minelaying Kiel Bay	2D6
AD851	16 OTU/14 OTU/ 1407 Flt	11. 2.44	SOC			
AD852	50/Cv TB/ 489/455	8.12.43	SOC			2F23
AD853	50/420/408/ 1 AAS/Cv TB	29. 5.44	SOC			
AD854	50	8/ 9. 9.41	Ran out of fuel and crashed near Bedford	Sgt Rowney (K); Sgt Morgan*; Sgt Barker (K); Sgt Mitchell*. *Baled out safely	Bombing Kassel	
AD855	106/44/420/ Cv TB/489	16.12.43	SOC			
AD856	49 (EA-P)	6/ 7. 7.41	Shot down by intruder off Cromer	Sgt J Flint (I); Sgt A S Beningfield (K) Sgt J D Fitch (K); Sgt E C Atkinson (I)	Bombing Brest	5H12
AD857	106/408/ 7 FPP	18. 5.43	Crashed after engine failure, Sherburn-in-Elmet	Flt Capt J E Martens ATA unhurt	Ferry	
AD858	7 BAT Flt/ 25 OTU	14/15. 8.41	Missing, believed ditched in North Sea	Sgt W K Willcox; Sgt J Smith; Sgt A J G Bull; AC1 T Robinson. All missing	Night navex	
AD859	6 BAT Flt/ 83 (OL-O)	31.8/1.9.41	Crashed Munsterbilsen, Belgium. Cause not known	Sgt F Dacey; Sgt G G Bensley; Sgt J A Clark; Sgt H G Tonks. All b.Heverlee	Bombing Cologne	
AD860	7 BAT Flt/ 25 OTU/14 OTU	4. 3.42	Burnt out at dispersal, Cottesmore; possibly due to aircraft heater		On ground	
AD861	106	6/ 7. 7.41	Hit balloon cable over Plymouth and crashed in sea	Sgt J Field (I); Sgt C Bedell (I); F/Sgt McCormack; Sgt J Wright (K) No.97 Squadron crew	Bombing Brest	5E8
AD862	106	2/ 3. 7.41	Shot down by night fighter 8 km SW of Maastricht	PO R A Walker (P); PO Campbell (P); Sgt J Diggory b.Heverlee; Sgt Henderson (P)	Bombing Duisburg	
AD863	106	15/16. 6.41	Missing, presumed ditched in North Sea	PO D G Dickie; Sgt A L Glaves; Sgt P Somers; Sgt E R Hall. All missing	Bombing Cologne	
AD864	44	2/ 3. 5.41	Missing, cause not known	PO H U H Tripp*; PO J E P Jeff (M); Sgt R J Leaper*; Sgt E Egar*. *b.Hamburg	Bombing Hamburg	
AD865	83/49/Cv TB/ 455/415	20.12.43	SOC			
AD866	144	3/ 4. 7.41	Missing, believed ditched in North Sea	F/Sgt G C Marsh b.Ameland; Sgt D Napier (M); Sgt W Cross (M); Sgt E R Foster (M)	Bombing Bremen	
AD867	50	30. 5.41	Dived into ground on full load test, Lindholme	PO J C Cunningham and one crew both killed	Air test	6E10
AD868	61/44	17/18.12.41	Missing, believed hit by flak and ditched	PO T L Kaschula (Rhod) b.9m SW of Brest; Sgt J Henderson (M); Sgt E Wade (M); Sgt K Hall (M)	Bombing Brest	

Serial	Units	Date	Cause	Crew	Duty	See
AD869	44/420 (PT – L)	20. 4.42	Control lost on take – off from Waddington; crashed, Wispington	Sgt H Davis RCAF; Sgt J Pritchard RCAF; Sgt G Laronde RCAF; Sgt G Player RCAF. All killed	Night navex	
AD870	83/408 (EQ – X)/Cv TB	17. 2.44	SOC			
AD871	144	6/ 7. 7.41	Missing, presumed ditched	Sgt T Porteous; Sgt N V Edwards; Sgt L N Crook; Sgt J T Leitch. All missing	Bombing Brest	
AD872	144	20/21. 9.41	Crashed on return in fog 4m NNE of Coningsby	FO W P Carroll; Sgt C Kellington; Sgt L Smith; Sgt P A Ginett. All killed	Bombing Frankfurt	
AD873	106	2/ 3. 7.41	Crashed near Grosskonigsdorf	F/Lt J F Sharp, Sgt R A Reid RCAF, Sgt D C Dunne, Sgt W F McKay. All b.Rheinberg	Bombing Duisburg	
AD895	106	29/30. 6.41	Shot down by night fighter S of Uetersen	FO M R F Baker, Sgt J H Bevans, Sgt M Ford, Sgt G E Smith. All b. Sage	Bombing Bremen	
AD896	49 (EA – M)	6. 1.42	Engine cut on approach, Scampton	Sgt G L West (K); Sgt G T Atkins (K)	Night training	
AD897	50	19. 7.41	Rolled over on back and dived in S of Lindholme	Sgt E R Bousfield and three crew killed	Formation practice	5F9
AD898	83	21. 5.41	Stalled on take – off, Scampton	Sgt O Stromberg (K); AC2 M Cross (K); one other, name not known (K)	Training	
AD899	44	8/ 9. 4.41	Missing, cause not known	PO J G Curley; PO E T Laverack; Sgt Ashurst; Sgt K S Campbell. All b.Pihen – les – Guines, France	Minelaying Calais	
AD900	144	11/12. 5.41	Shot down by night fighter, Hoogkarspel, Netherlands	S/Ldr C G C Rawlins (P); PO R Featherstone (P); PO R Vaughan*; Sgt S Taylor*; *b. Bergen – op – Zoom, Neth.	Bombing Bremen	3D11
AD901	144 (PL – M)	8/ 9. 5.41	Returned on one engine and crashed near Bircham Newton	S/Ldr Fleming, Sgt Wathey, Sgt Crombie, FO Morgan all unhurt	Bombing Hamburg	5A11
AD902	50	28/29. 7.41	Undershot Swinderby on return	PO Christopher, Sgt Glenn, Sgt Thomas, Sgt Pullin all unhurt	Minelaying Kiel Bay	
AD903	144 (PL – A)	6/ 7. 8.41	Missing, presumed ditched	Sgt P A Hammond; Sgt R T Clark; Sgt J H H Cutmore; Sgt E F Reeve. All missing	Bombing Calais	2L4
AD904	44	19. 6.41	Crashed while circling crash of AD747 at Ternhill	Sgt C F Greig; Sgt J O Walshe; Sgt T K Boardman. All killed	Ferry	6B18
AD905	144	1/ 2. 9.41	Missing, cause not known	FO M T H Adams (P); F/Sgt A J Box*; Sgt G J Hartlett*; Sgt G Morgan*.*b.Rheinberg	Bombing Cologne	
AD906	14 OTU/Cv TB/ 415/RAE/CCDU	27. 4.44	SOC			
AD907	83 (OL – K)	19/20. 8.41	Crashed in sea 6m E of Bridlington. Cause not known	PO R J Potts, Sgt I Ilsley, Sgt G D Clark, Sgt G T Shirley. All killed and bodies recovered	Bombing Kiel	
AD908	50/144/Cv TB/ 455 (UB – V)	12.10.42	Handed over to Russian Navy	Flown out by Sgt Gleeson, PO Murphy, Sgt Witte, Sgt Booth, Sgt Orr	Left in Russia	2J15
AD909	49 (EA – H)	9/10. 1.42	Missing, presumed crashed in sea	PO T F Jacobs; PO P T Close; Sgt W B Robertson; Sgt J Warren. All missing	Minelaying Brest	
AD910	49 (EA – Y)	12/13. 7.41	Ran out of fuel in bad weather and crashed 5m NE of Pocklington	Sgt Batchelor, Sgt Williams, Sgt Calvert, Sgt Betts all unhurt	Bombing Bremen	
AD911	83 (OL – M)	10/11.10.41	Shot down off Harlingen by fighter or flak	Sgt D C Hedley (P); Sgt G Mitchell (P); Sgt W Brooks b.Texel; Sgt E W Gough (P)	Bombing Essen	
AD912	83 (OL – Y)	31.8/1.9.41	Shot down by night fighter near Meyel, Netherland	S/Ldr E Willcox DFC b.Jonkerbos; PO D Organ (P); Sgt H Gabbitas (P); F/Sgt J Lawrence (P)		3D10
AD913	44 (KM – K)	2/ 3. 9.41	Returned on one engine; abandoned near Dorking	Sgt Knight*; Sgt Spanner*; Sgt J Stephens (K); Sgt Churchill*. *baled out safely	Bombing Frankfurt	5A3
AD914	106	4/ 5. 7.41	Shot down by night fighter 3 km NNE of Heesch, Netherlands	F/Sgt N E Bowering; Sgt R S Bagnall; Sgt W H Lapsley; Sgt I L Reis. All b.Eindhoven	Bombing Ruhr	
AD915	44/420 (PT – F)	18/19. 2.42	Shot down by flak near Schiermonnikoog	PO R Kee (P); Sgt W Rutledge (P); Sgt H Baker*; Sgt J Adams*. *b.Schiermonnikoog	Minelaying Frisians	2D10 3D29
AD916	83 (OL – Z)	30.6/1.7.41	Hit by flak; crashed near Duren	S/Ldr N H Svendsen DFC (P); PO D Irving b.Rheinberg; PO S Carter (P); Sgt L Coldwell (P)	Bombing Dusseldorf	3D12
AD917	44 (KM – P)	27/28. 8.41	Ran out of fuel and crashed on decoy site near Waddington	Sgt B W Johnson (I); Sgt S Mallen (K); F/Sgt W H Saville (I); Sgt E A Windsor (K)	Minelaying Frisians	
AD918	144	25/26. 8.41	Missing; may have been over enemy territory before dark	Sgt D A Whiting (K); PO A Hayward (P); Sgt G G Smith*; Sgt E A Windsor (K) *b.Evère, Brussels	Bombing Mannheim	2L7
AD919	106	18/19. 8.41	Engine cut on take – off, Coningsby; crashed at Chapel Hill, Lincs.	Sgt Field (I); PO Finlayson (I); Sgt Norman (I); Sgt Wright (I)	Bombing Cologne	
AD920	44/Cv TB/ 144/455	19. 5.43	Overshot landing at Inverness	Sgt J A Macdonald unhurt	Training	
AD921	144	13.12.41	Missing, presumed crashed in sea	PO D McLaren; Sgt R Pulham; Sgt J Pearce; Sgt A Taylor; Sgt R Atkinson. All missing	Minelaying Brest (day)	2F13,2D5 5A16 2D5
AD922	144	20/21. 9.41	Abandoned out of fuel, Foulsham, Norfolk	Sgt E C W Turner (K); Sgt R J Pulham, Sgt R Atkinson, Sgt T Pearce baled out safely	Bombing Frankfurt	5A3 5A15
AD923	144	20/21. 9.41	Ran out of fuel and crashed, Hutton Moor, Yorks.	FO Kingston; Sgt De Courcy; Sgt Tod; Sgt Tobin. All unhurt	Bombing Frankfurt	2H1

AE196 of No.408 Squadron at Syerston in July 1941. This aircraft was subsequently lost on a daylight minelaying operation off Brest in December 1941 while with 44 Squadron. (Photo: National Archives of Canada PA – 144774)

Serial	Units	Date	Cause	Crew	Duty	See
AD924	144	9/10. 7.41	Missing, cause not known	PO B Rennie evaded capture; Sgt G F Bottomley (P); Sgt T H Marquisse*; Sgt E R Berkey RCAF*. *b.Dilsen, Belgium	Bombing Aachen	3D31
AD925	106/AAEE	22. 7.42	Live 20 – lb fell out when bomb doors opened at dispersal and exploded, Boscombe Down	WO H Norris (I); F/Lt H T W Evans (I)	On ground	
AD926	144	17/18. 6.41	Overshot Hemswell on return	Sgt Caesar – Gordon, PO Pringle, Sgt Horton, Sgt E W Thompson all unhurt	Bombing Cologne	
AD927	50/Cv TB/489	17. 2.44	SOC			
AD928	50/144/Cv TB/ 489	24. 1.44	SOC			
AD929	106/50/144/ Cv TB/489	4. 6.44	SOC			
AD930	44 (KM – L)	15/16. 9.41	Ran out of fuel and crashed between Waddington and Harmsworth	Sgt C E Musgrave (I); Sgt J Endersby (I); Sgt R L Cox (K); F/Sgt J Barry (K)	Bombing Hamburg	
AD931	49 (EA – X)	14/15. 4.42	Missing, possibly shot down by night fighter	PO H G Williams (P); Sgt L W Webbe*; Sgt H S Brown*; Sgt W Ralph (P).*b.Reichswald	Bombing Dortmund	
AD932	106	7/ 8.11.41	Ditched 48m E of Wick; possibly damaged by flak	F/Lt T R Herd DFC (M); FO B G McIver (M); Sgt H Sell DFM (M); Sgt S Jones b.Kirkwall	Minelaying Oslofjord	5B18
AD933	44	27/28.11.41	Ran out of fuel in bad visibility; crashed at Honeybourne	Sgt B D Moss, Sgt Ross, Sgt Seagoe, Sgt Hackett all unhurt	Bombing Dusseldorf	
AD934	83 (OL – T)	22/23.10.41	Windscreen iced up on outward flight; undershot, Swanton Morley	Sgt Mowatt, Sgt Brown, Sgt Riddell, Sgt Rex all unhurt	Bombing Mannheim	
AD935	83 (OL – U)	14. 8.41	Hit tree low flying, Badsley, Broadway, Worcs	PO E R Davies (K); Sgt Wells (I); Sgt G A Newbold (K)	Training	6E2
AD936	144	7/ 8. 9.41	Hit by flak and crashed SE of Amsterdam	PO P Stevens (P); Sgt A W Payne (P); Sgt H Thompson (P); Sgt I R Frazer (M)	Bombing Berlin	
AD937	61	7/ 8. 7.41	Shot down by night fighter near Meerssen	PO J G N Braithwaite (P); Sgt G Owens*; F/Sgt R Wordsworth*; Sgt J Walton* *b. Jonkerbos	Bombing M – Gladbach	
AD938	16 OTU/14 OTU/ 5 OTU	10. 6.43	Developed stabilised yaw and crashed 4m E of Maghaberry	PO Brockett RCAF unhurt	Training	
AD939	44	31. 8.41	Collided with Spitfire of 412 Sqn and crashed near Waddington	PO P R Owen (K); Sgt D G Forbes (K); AC2 F B Prest (K)	Training	5C12

Serial	Units	Date	Cause	Crew	Duty	See
AD959	16 OTU/144	1/ 2. 4.42	Missing, cause not known	F/Sgt L Smith (P); Sgt H Robertson RCAF*; Sgt J Davies*; Sgt A Cassam*.3b.Becklingen	Bombing NW Germany	
AD960	49/420/408/ 1402 Flt/ 1407 Flt	11. 2.44	SOC			
AD961	RAE/32 OTU	2. 8.44	SOC in Canada			5K8
AD962	44	24. 7.41	Shot down in sea by Bf 110 off Brest	FO J F Clayton DFM (P); PO J Grant RCAF (M); Sgt A J Clark (P); Sgt W O'Brien (P). Three baled out of aircraft before crash	Bombing Brest (day)	2G3 3D17
AD963	61/408/Cv TB/ 489/415	24. 1.44	SOC			
AD964	83/49/144/ Cv TB/5 OTU	4. 3.44	SOC			
AD965	144	12/13.10.41	Shot down by night fighter in Ysselmeer	Sgt H Chamberlain b.Wonseradeel; Sgt E Hawkins (M); Sgt C Webster (M); Sgt C Morris (M)	Bombing Huls	
AD966	44 (KM – R)	1. 8.41	Dived into ground, South Park near Lincoln. Cause obscure	Sgt G LeBlanc Smith (Rhod); Sgt G Dodds RCAF; Sgt A Forsythe; AC1 T Jeffcote; AC1 A Clark. All killed	Training	6E7
AD967	49 (EA – H)	24/25. 8.41	Collided with X3121 near Hackthorne, Lincs	Sgt O McMahon; Sgt G Upton; Sgt J McKinnon; Sgt E Welbourne. All killed	Bombing Dusseldorf	5C1
AD968	44/420/49/408	3. 2.44	SOC			
AD969	83 (OL – X)	22/23. 6.41	Pilot dazzled by search-light while flying low; crashed near Bevern	PO R J Heavens; Sgt W G Price; F/Sgt E W Sponder; F/Sgt N E Byrnes All b.Becklingen	Bombing Bremen	2F24
AD970	106	30. 7.41	Crashed in sea ½m S of Skegness Pier, Lincs.	Sgt Lockyer; Sgt G E Thirston (K); AC1 Sissons. Two picked up unhurt	Air test	6E6
AD971	49 (EA – O)	28/29. 8.41	Shot down by night fighter near Ameland	PO B M Fournier; Sgt D H Barrett; Sgt E Palmer; Sgt D Watson. All b.Ameland	Bombing Duisburg	
AD972	408	7/ 8.11.41	Observer baled out	Sgt B W Palastanga b.Hindelope, Neth.	Bombing	5A50
AD972	408/5 Gp TF/ 1 AAS	8.12.43	SOC			
AD973	49/144	12. 6.42	Engine failure; crashed near of Crail	Sgt W H Hood RCAF and crew unhurt	Training	
AD974	61/49/455/ 5 OTU	1.10.42	Misjudged height and crashed in sea 2m NW of Girvan, Ayrshire	Sgt J N Reigate RCAF (I); two crew (I)	Training	6E12
AD975	44	13/14.10.41	Shot down by night fighter, Waasmunster, Belgium	Sgt E Owen (Rhod); F/Sgt G Edmondson RCAF; Sgt G Shearer; Sgt G Ramsay All b.Waasmunster	Bombing Cologne	
AD976	49/Cv TB/455	24. 1.44	SOC			
AD977	50/Cv TB/ 455 (UB – T)	12.10.42	Handed over to Russian Navy	Flown out by W/Cdr Lindeman, F/Lt Oliver, Sgt Bastian, Sgt Rawle, LAC Lacey	See Russia	2J15
AD978	83 (OL – C)	2/ 3. 9.41	Ran out of fuel in fog, Stone Pit Hills, Norfolk	W/Cdr Satterley; Sgt Rogers; Sgt Cross; Sgt Yardley. Minor injuries	Bombing Berlin	
AD979	49	12.12.41	Navigator hit by flak	Sgt Black RNZAF (K)	Bombing Cuxhaven (day)	
AD979	49/144/Cv TB/ 455	13. 3.43	Hit by 235 Sqn Beaufighter while parked, Leuchars		On ground	5C18
AD980	408 (EQ – Y)	6. 6.42	Crashed NE of Balderton	PO H J Randall (I); PO Wishart (I); Sgt H E Elliott (I)	Air test	6E4
AD981	44 (KM – A)	11/12. 9.41	Ran out of fuel and ditched off Cromer	Sgt Dedman, Sgt Birkett, Sgt Norman, Sgt Hugh all picked up unhurt by HMS *Garth*	Bombing Rostock	
AD982	44	2/ 3. 9.41	Attacked by fighter on outward flight	Sgt Shipton (I); Sgt Durnan (I)		
AD982	44/408/ Cv TB/AAEE/RAE	18. 6.44	SOC			
AD983	44	21/22. 7.41	Crashed into Lindum Hall School, Lincoln	Sgt D M Bruce (K); Sgt W R Relyea RCAF (K); Sgt J A Connolly (K); Sgt P J Lynch (K); Miss Fowler killed on ground	Minelaying Frisian Is	2K20
AD984	106	20/21.10.41	Crashed near Borby, Germany; cause not known	Sgt J L Lockwood RCAF; Sgt G F Williams RNZAF; Sgt W Alton; Sgt T G Jones All b.Kiel	Bombing Bremen	
AD985	25 OTU/16 OTU/ 14 OTU/1402 Flt	7.12.43	SOC			
AD986	106	4/ 5. 7.41	Crashed near Brackwede 25m E of Hamm	Sgt A E Brownbill; PO L S Christman; Sgt S Linley; Sgt S C Morse. All b.Reichswald	Bombing Ruhr	
AD987	25 OTU/408/Cv TB/144/455/1 TTU	12.12.43	SOC			
AD988	106/14 OTU	16. 3.42	Dived in on approach, Cottesmore	S/Ldr N S Royle; Sgt D O L Dold (Rhod); Sgt F Hiesler RAAF; Sgt J Stevenson. All killed	Training	6C16
AE115	50/420/Cv TB/ 455/5 OTU/415	4. 6.44	SOC			
AE116	50/Cv TB/489	9. 6.44	SOC			
AE117	144/455	19. 3.44	SOC			5K7
AE118	144	11/12. 9.41	Ran out of fuel while lost and crashed, Barn Farm, Billesden, Leics	PO Booker (I); Sgt A J Pomes RCAF (I); Sgt S J Binns and Sgt Crossley unhurt	Bombing Rostock	
AE119	144	11/12. 7.41	Hit ground in cloud 4m N of Withernsea, Yorks.	Sgt E W Smith (I); Sgt J Nicoll (K); Sgt S J Easton (I); Sgt R Saunders	Bombing Wilhelmshaven	

Hampden UB – C of No.455 Squadron, RAAF. (Photo: Grant Lindeman)

Serial	Units	Date	Cause	Crew	Duty	See
AE120	106	5/ 6. 8.41	Shot down by night fighter; hit house in Gendringen	Sgt Knowles, Sgt McIlwraith; F/Sgt Price; Sgt Rhodes all b.Gendringen 2 killed, 2 injured in house	Bombing Mannheim	
AE121	144	11/12. 9.41	Ran out of fuel and crashed, Empingham, Rutland	Sgt N H Page (K); Sgt J L Longmore (K); Sgt Bailey?; Sgt Bowskill?. One injured	Bombing Rostock	
AE122	144/Cv TB/ 455 (UB – P)	11. 2.44	SOC			
AE123	49/106	21/22. 1.42	Engine failure; crashed at Zwolle, Netherlands	Sgt A W Deere b.Zwollerkerspel; Sgt V Cooper, Sgt T Berry, Sgt J Palmer all POW	Bombing Bremen	
AE124	50	8/ 9. 8.41	Port engine cut on take – off, Swinderby	PO Milner; PO Temperley; Sgt Graver; Sgt Bigby. No serious injuries	Minelaying Frisians	2L6
AE125	144	6/ 7. 8.41	Shot up by Ju 88	Three wounded	Bombing Karlsruhe	
AE125	144/Cv TB/455	12. 2.44	SOC			
AE126	49 (EA – N)	28/29. 8.41	Presumed shot down by night fighter into North Sea	PO T Pratt b.Texel; PO H Tonge, Sgt C Hodkinson both b.Vlieland; Sgt A Willis b.Harlingen	Bombing Duisburg	
AE127	44	12/13. 6.41	Missing, cause not known	F/Lt P Shaughnessy DFC (P); Sgt J Wardrop (M); Sgt C W Townend (P); Sgt J E Hugff (P)	Bombing Soest	
AE128	144/44/455/ 5 OTU	27.10.42	Swung on approach and hit Beaufort X8937, Turnberry	Sgt C M Tapper unhurt	Training	
AE129	44	14/15. 6.41	Returned early and crashed on approach, Waddington	Sgt S P C Saunders (K); Sgt J R Miller (K) Sgt R Greenaway (K); Sgt T J Dickson (K)	Minelaying Frisians	
AE130	44/Cv TB/ 455/144	12.10.42	Handed over to Russian Navy	Flown out by FO Adams, F/Lt Logan, Sgt Wardle, Sgt Waller, Cpl Craddock	Left in Russia	2J3 2J15
AE131	83 (OL – W)	14/15. 8.41	Hit by flak and crashed Pattensen, 5m S of Hannover	PO P H Slater – Eiggert; F/Sgt L S Thorrow – good; PO W C Rasbary; Sgt F Stephens. All b.Hannover	Bombing Brunswick	
AE132	49 (EA – U)	12. 2.42	Shot down by flak or fighters attacking battlecruisers	F/Sgt D C Pollitt (M); Sgt E I S Green – street (M); Sgt J R Cridge (M); Sgt W Smith (M)	Bombing warships (day)	3B1 2H5
AE133	83 (OL – X)	10/11. 1.42	Crashed at Kampen; cause not known	Sgt M Fletcher and Sgt L A Fox both b. Sage; Sgt R Sekine (P); Sgt H Holmes (P)	Bombing Wilhelmshaven	
AE134	106	16/17. 8.41	Shot down by night fighter near Meyel, 20 km W of Venlo	PO Robinson; Sgt R G Woodwards; Sgt J Harker; Sgt H Birtwhistle. All POW	Bombing Dusseldorf	

Serial	Units	Date	Cause	Crew	Duty	See
AE135	61/455/Cv TB 5 OTU/1 TTU	20. 1.44	SOC			
AE136	106	26/27.10.41	Crashed near Selsingen, 40 km WSW of Hamburg	PO Wood (P); Sgt D J Carmichael*; Sgt E Haste*; Sgt R A Gladen*. *b.Becklingen	Bombing Hamburg	
AE137	50	5/ 6. 8.41	Missing, cause not known	Sgt D G Fothergill, Sgt S O Hirschfield, Sgt H Smith, Sgt A Davies all b.Reichswald	Bombing Karlsruhe	
AE138	16 OTU	20. 8.41	Hit by AE147 on take-off Croughton	PO T J Schofield (I); PO J L Roberson (I); Sgt E J Thomas (K); Sgt F Williams (K)	Night training	5C8
AE139	25 OTU/408	25/26. 3.42	Crashed on return at Wootton, Berks.	PO D H Vipond RCAF (K); Sgt H Phalempin RCAF (K); Sgt E V Conwell (I); Sgt G A Medd (K)	Leaflets, Rennes	
AE140	144	6/ 7. 8.41	Presumed crashed in sea probably due to icing	PO S W S Beedie RNZAF; Sgt H E Fear; Sgt J R Cook; Sgt F T Prest. All missing	Bombing Karlsruhe	2L4
AE141	144	12. 2.42	Damaged by flak and crashed near Norwich	Sgt A Nightingale (K); three crew unhurt	Bombing warships(day)	2H5
AE142	144/5 OTU	12. 4.43	Overshot landing on wrong runway, Maghaberry	Sgt D J Smith, Sgt Bacon, Sgt Dobbie, Sgt Evett. All unhurt	Training	
AE143	144	29/30. 9.41	Ran out of fuel on approach, Driffield	Sgt McDermott, Sgt Boucher, Sgt H Smith, Sgt Elliott all unhurt	Bombing Hamburg	
AE144	106	10/11.10.41	Crashed near Lippramdorf, Ruhr. Cause not known	PO G A O Gordon; Sgt P R Martin; Sgt F Morris; Sgt W H A Everson. All b. Reichswald	Bombing Essen	
AE145	49/Cv TB/455	3. 3.44	SOC			
AE146	16 OTU/Cv TB/ 144/455	16. 7.43	Missing from Rover off Norwegian coast	F/Sgt J A R McDonald, FO J D Arthurell, Sgt W E Ramsay RCAF, Sgt L C Stewart RCAF. All POW	Torpedo bombing (day)	
AE147	16 OTU	20. 8.41	Collided with AE138 on take-off, Croughton	Sgt T C Roux, Sgt H Maple, Sgt Urquhart, Sgt I G Stewart all injured	Night training	5C8
AE148	16 OTU/408/ (EQ-B)	29/30. 9.41	Crashed at Sanderum, 5 km SW of Odense, Funen, Denmark	PO W F Hull RCAF, F/Sgt D L Todd RNZAF, Sgt S Jamieson, Sgt L Harding. All b. Odense	Minelaying Kiel Bay	
AE149	25 OTU	22. 7.41	Engine failure; crashed near Finningley	PO W H Andrews (I)	Night training	
AE150	25 OTU/408/ (EQ-M)	12/13. 8.42	Missing, presumed ditched in North Sea Cause not known	FO J S Coulter RCAF; FO R C Hodgson; F/Sgt W Gittings RCAF; F/Sgt W Gartside RCAF. All missing	Bombing Mainz	
AE151	106	21.12.41	Crashed at Schoonebeek on day cloud cover "Sneaker" operation	Sgt R S Hartgroves (P); PO A Carter b. Schoonebeek; Sgt R Yearsley (P); Sgt Martin (P)	Bombing (day) NW Germany	2G4
AE152	44 (KM-J)	2/ 3. 9.41	Missing presumed ditched in North Sea	PO E Thompson b.Ameland; PO H Cook RCAF (M); Sgt E Dyer (M); Sgt W McBeth (M)	Bombing Berlin	
AE153	44	8/ 9. 7.41	Missing, presumed ditched in North Sea	F/Sgt E D Tyler (M); Sgt M Livis RCAF (M); F/Sgt K G Betts (M); Sgt F W Black (M)	Bombing Hamm	5H8
AE154	83 (OL-H)	2/ 3. 8.41	Ran out of fuel and ditched 40m E of Flamborough Head	S/Ldr F Newall, PO Monkhouse, PO Bishop, Sgt Dickerson. All picked up after 11 hours in dinghy	Bombing Kiel	5B11
AE155	83	19/20. 8.41	Navigator baled out over Kiel	PO J D Gould b.Kiel	Bombing Kiel	2L8 5A48
AE155	83/455/420/ 14 OTU	11. 8.42	Crashed on low-level bombing practice, Edenham, Lincs.	Sgt Davis (K); F/Sgt J H O'Connor RCAF (K); Sgt Barton (K)	Night training	
AE156	83/Cv TB/489/ 455 (UB-N)	12.10.42	Handed over to Russian Navy	Flown out by PO Broughton, FO Jeffries, F/Sgt Carter, Sgt Willis, LAC McIver	Left in Russia	2J15
AE157	50	2/ 3. 9.41	Hit X3025 on ground in fog, Waddington	PO Ferrie, PO Bartley, Sgt Holland, Sgt Hawkins all unhurt	Minelaying Kiel Bay	5C24
AE158	50/144	12.10.42	Handed over to Russian Navy	Flown out by PO Brown, PO Vicary, Sgt Hewlett, Sgt Dixon, Cpl Earl	See Russia	2J13 2J15
AE159	50	28/29. 7.41	Missing, presumed ditched	PO W B Burrows (M); PO J M Graham (P); F/Sgt C J Jolin (M); Sgt C D Rose (M)	Minelaying Langeland	
AE184	50	25.10.41	Overshot landing in bad weather, Ratcliffe	Sgt A J Weber unhurt	Training	
AE185	50	17/18. 8.41	Shot down by night fighter 2m S of Groningen	PO Maskell (P); PO H Law*; Sgt P F B Orwin*; Sgt G A Cowell*. *b. Haren, Netherlands	Bombing Bremen	
AE186	207/61/44/ 420/106/408/ 14 OTU	5. 5.42	Swung on take-off and undercarriage collapsed, Cottesmore	Sgt J Page unhurt	Training	
AE187	83 (OL-L)	31.8/1.9.41	Shot down by night fighter, Deurne, Neth.	F/Sgt R Plaistowe, Sgt Somerville-Woodiwis, Sgt J Hughes, Sgt R Scott. All b.Eindhoven	Bombing Cologne	
AE188	83 (OL-D)	8/ 9. 9.41	Ran out of fuel and crashed near Boscombe Down	Sgt Linacre (I); Sgt Speakman, Sgt Galloway-Wilson, Sgt Calder all unhurt	Bombing Kassel	
AE189	61	24/25. 7.41	Shot down by night fighter off Den Helder	FO M Parry (M); PO G Wise DFM (M); F/Sgt A F Hill b.Terschelling; Sgt I Waugh RNZAF b.Kiel	Bombing Kiel	
AE190	25 OTU/408/ 14 OTU/Cv TB	4. 6.44	SOC			
AE191	83 (OL-Z)	7/ 8.12.41	Ditched in North Sea	PO A L Parsons (P); Sgt B A Basevi*; PO D Jacobs*; Sgt G Wiscombe (P) *died in dinghy and buried at sea	Bombing Aachen	

408 Squadron Hampden "EQ – C" being run up at dispersal early in 1942, probably at Balderton. (CH4738)

Serial	Units	Date	Cause	Crew	Duty	See
AE192	207/44/408/ 14 OTU/Cv TB/ 415	14/15. 9.43	Undercarriage collapsed on landing, Docking SOC 17.2.44	PO J A Enns RCAF and crew unhurt	T/bombing (night) Dutch coast	2A38
AE193	106 (ZN – A)	28/29. 8.41	Ditched 60m NW of Texel	Sgt E R H Lyon; Sgt R S K Woodroofe; Sgt G B Stanton; Sgt G Luke. All came ashore in dinghy and POW	Bombing Wesel	3D23
AE194	49/Cv TB/ 455 (UB – P)	12.10.42	Handed over to Russian Navy	Flown out by PO Gunton, PO Edgoose, Sgt Thomas, Sgt Crossley, LAC Stafford	See Russia	2J15
AE195	144/16 OTU Cv TB/455/ (UB – M)	26. 8.43	Crashed in sea on navex after engine failure	F/Sgt R Woodroff (M); Sgt P Taylor (M); Sgt B Martin RCAF rescued; Sgt A McKenzie RCAF (M)	Night training	6D17
AE196	408/44	13.12.41	Shot down in sea off Brest in daylight	W/Cdr S T Misselbrook DSO; Sgt A Gumbley RNZAF; Sgt L Leggett DFM; FO H Jeffcoat All missing	Minelaying Brest (day)	2F10 2G5
AE197	408	28/29. 8.42	Shot down by night fighter NW of Phillipville, Belgium	W/Cdr J Twigg RCAF (K); F/Lt G Fisher RCAF*; F/Lt I Maitland DFC (K); FO R van den Bok DFC* *Evaded and retd to UK	Bombing Saarbrucken	3D31 2E18 2A1
AE198	455/Cv TB	22. 3.45	SOC			
AE199	25 OTU/16 OTU/ Cv TB/5 OTU	4. 6.44	SOC			
AE200	61/144	26/27. 3.42	Shot down by night fighter off Vlieland	Sgt D Towler, Sgt E F Penlington; Sgt J Burns, Sgt W L L Woodford. All missing	Minelaying Frisians	5A44
AE201	44/Cv TB/ 455/415	26. 2.43	Destroyed when P2065 blew up, Thorney Island		On ground	2J33
AE202	61/44/420 (PT – K)	26/27. 7.42	Missing, cause not known	PO R N Payne (P); F/Sgt J H Timmis*; Sgt N F Axford*; Sgt J R Elliott*. *b.Kiel	Bombing Hamburg	2F10
AE203	49 (EA – F)	2/ 3. 9.41	Ran out of fuel and crashed, Stradbroke, Suffolk	Sgt Flint, Sgt Combes, Sgt Jardine, F/Sgt Gadsby all unhurt	Bombing Berlin	
AE218	44/50	27/28. 2.42	Ran out of fuel and crashed, Skellingthorpe	F/Sgt Johnston, Sgt Booker, F/Sgt D L Barrie, Sgt Denman all unhurt	Bombing Kiel	2K26
AE219	207/61/ 408 (EQ – R)	27/28. 3.42	Shot down by night fighter into sea	PO W M Fraser RCAF; WO A Jenkins RCAF; Sgt A F Trawford; Sgt H A B Bird. All (M)	Minelaying Weser Est.	
AE220	106	22/23. 8.41	Shot down by flak near Plankstadt	PO A F McGruer RNZAF; PO J P Early RCAF; Sgt A Rawlinson; Sgt R Goddard. All b.Durnbach	Bombing Mannheim	
AE221	25 OTU	24. 8.41	Hit flock of birds on take – off, Balderton	Sgt A L Smith unhurt	Training	

Serial	Units	Date	Cause	Crew	Duty	See
AE222	16 OTU	1.10.41	Engine cut on take-off, Croughton	Sgt E Phillips unhurt; Sgt D W Markall (I); Sgt R A Buckley (I)	Night training	
AE223	83 (OL-V)	24/25. 8.41	Bomb exploded at dispersal on return, Scampton; burnt out	PO V C C Maybury, PO T G M McIntyre, Sgt E J Clarke, Sgt E Scholles and four ground crew killed	Bombing Wesel	5C2
AE224	49 (EA-Z)	1/ 2.11.41	Missing off Frisians, pres. shot down by flak	S/Ldr D B Drakes; PO V D Beaney; F/Sgt W A Watson; PO W H Cheetham. All missing	Bombing shipping	
AE225	144	24. 7.41	Shot down by fighter off Brest	F/Lt R Barr (P); PO P Anderson*; Sgt J Wiggall (P); Sgt A Cooper*; Sgt D Parkin* *b.13m N of Brest	Bombing Brest (day)	2G3
AE226	50	12/13. 7.41	Shot down by night fighter near Veendam	PO E D Vivian; Sgt K Lord; F/Sgt J Guest BEM; Sgt I Jackson. All b.Veendam, Neth.	Bombing Bremen	
AE227	49/408 (EQ-D)	28/29. 8.42	Probably shot down by night fighter	F/Sgt W Gould RNZAF; F/Sgt L Garner RCAF; Sgt C Sutherby RCAF; Sgt T Murlis. All b.Rheinberg	Bombing Saarbrucken	2A1
AE228	50/455 (UB-P)	11. 7.43	Engine on fire; crashed 1m N of Leuchars	FO R C Barton, Sgt K McLaren, Sgt E M Birkland all unhurt	Training	6D17
AE229	50	29/30. 8.41	Landed on decoy flare-path, Bassingham, Lincs	F/Sgt Mapp; Sgt Furley; Sgt Welford; Sgt James. Two injured	Bombing Frankfurt	6D13 5E10
AE230	50	12/13. 7.41	Shot down by night fighter 2 km E of Weltrup	Sgt J Austin; Sgt D Onions; F/Sgt W Crichton; Sgt P A Mitchell All b.Reichswald	Bombing Bremen	
AE231	50/Cv TB/ 144	12.10.42	Handed over to Russian Navy	Flown out by S/Ldr Foster, FO Barrett, F/Sgt Freeman, Sgt Moderate, F/Sgt Thompson	Left in Russia	2J14 2J1
AE232	106	15/16. 9.41	Shot down by night fighter, Wilhelmsburg near Hamburg	Sgt Richardson (P); F/Sgt J McIntyre RCAF (M); Sgt H D Clapham (P); Sgt Kirk (P)	Bombing Hamburg	
AE233	RAE	19. 9.41	Undershot runway, Farnborough	S/Ldr G J Wright unhurt	Training	
AE234	50	25/26. 7.41	Crashed after take-off, Swinderby	Sgt Montgomery (K); Sgt Rampton (K); Sgt Ellesley (K); Sgt Madden (K)	Bombing Hannover	
AE235	61/144/Cv TB/144/1 TTU	12.12.43	SOC			
AE236	49	7/ 8. 9.41	Missing	PO J W T Bromham; Sgt E V Durham; Sgt P J Duffy; Sgt A Kettlewell. All b.Kiel	Bombing Kiel	
AE237	49/83/408 (EQ-X)	22/23. 4.42	Shot down by night fighter off Ameland	Sgt A McGregor RNZAF; F/Sgt R Wilson RCAF; Sgt E A Corderoy; F/Sgt J Daughney RCAF All missing	Minelaying Heligoland	
AE238	144	7/ 8.11.41	Damaged by flak; ran out of fuel 4m from Lichfield, Staffs	Sgt Woodhall, Sgt Pomroy, Sgt Tod, Sgt Lynch all unhurt	Bombing Cologne	
AE239	44	16/17. 8.41	Missing, presumed ditched in North Sea	Sgt J G Armstrong (M); Sgt C R Morley (M); Sgt J Flint (M); Sgt S C Edwards b.Kiel	Bombing Dusseldorf	
AE240	49 (EA-P)	12. 2.42	Shot down by flak or fighter off Dutch coast	Sgt M H Holt RAAF (M); Sgt C L Lee (M); PO S W A Way (M); Sgt E G Green (M)	Bombing warships (day)	2H4 3C4 3B1
AE241	49/Cv TB/5 OTU	21.12.43	SOC			
AE242	44/455/ Cv TB/415	21.12.43	SOC			
AE243	455 (UB-B)	7/ 8.11.41	Missing, cause not known	Sgt M Jenkins, Sgt D Rawlings, F/Sgt K Morris, Sgt G A McGarvey. b.Heverlee	Bombing Cologne	
AE244	408 (EQ-P)	31.7/1.8.42	Shot down by night fighter near Zewerdint	Sgt O Nelson RCAF (P); Sgt H Warr RAAF*; Sgt C Lloyd RCAF*; Sgt H Price*. *b. Turnhout, Belgium	Bombing Dusseldorf	2F17
AE245	408/Cv TB/ 5 OTU	16. 7.43	Swung on take-off, Long Kesh	Sgt Chester and crew unhurt	Training	
AE246	106/420 (PT-V)	28/29. 3.42	Shot down by night fighter, Heligoland Bight	S/Ldr G Tench; F/Sgt K Hyde RCAF; F/Sgt H Thorne; Sgt F Durnan. All picked up from dinghy and POW	Bombing Lubeck	
AE247	207/61	29/30. 8.41	Missing, presumed ditched in North Sea	Sgt H Richmond (M); Sgt J Spencer (M); Sgt J Archer (M); Sgt A G Clarke (M)	Bombing Frankfurt	
AE248	50/144/420 (PT-A)	2/ 3. 7.42	Shot down by night fighter near Koudum	F/Lt K Brown RCAF (P); Sgt R Williams*; Sgt J Waddington*' Sgt Whytock RCAF* *b. Koudum, Neth.	Bombing Bremen	2F17
AE249	207/455	15/16. 9.41	Shot down by night fighter, Winkeldorf	PO A J Hibell (P); Sgt W N Pratt (M); Sgt H Hobbs (M); Sgt Claydon (P)	Bombing Hamburg	
AE250	50	10/11. 1.42	Ran out of fuel while lost and crashed near Brampton, Cumberland	Sgt Williams (K); Sgt Earnshaw (K); Sgt Bernard (K); Sgt Caunce (I)	Bombing Wilhelmshaven	
AE251	50	13/14.10.41	Crashed after being hit by flak	F/Lt Abbott (P); PO Chase (P); Sgt Wake (P); Sgt Barrett (P)	Bombing Cologne	3D28
AE252	144	30/31. 7.41	Missing, presumed due to icing up	Sgt G Harvey; Sgt J Rogers RCAF; Sgt R Smale; Sgt R Chrisp. All b.Cambrai	Bombing Cologne	2L4
AE253	144	5/ 6.11.41	Probably shot down by flak off Frisians	PO F Kidd*; Sgt P McReed (P); Sgt E Crowe*; Sgt J Blackwood. *b.Sage	Bombing convoy	
AE254	44 (KM-W)	2/ 3. 9.41	Missing, cause not known	Sgt L A Robertson, Sgt H N Unwin, F/Sgt H B Turner, Sgt T P Riddell. All b.Berlin	Bombing Berlin	
AE255	106/5 OTU	18. 8.42	Crashed after take-off, Turnberry	Sgt Burns (K); one crew injured	Night training	
AE256	61/455/50	23/24.10.41	Probably shot down by flak over Kiel	PO K T Cooper; Sgt R C Blamey; Sgt W C Wright; Sgt S Barton. All b.Kiel	Bombing Kiel	

"EA – F", P1333 of 49 Squadron, awaits its load of bombs. There are four bomb trailers hitched to the tractor nearest to the left of the picture. The two rearmost are loaded with two 500 – lb GP (General Purpose) bombs whilst the two trailers immediately behind the tractor each have three 250 – lb GP bombs secured in wooden carrying cradles. (CH272)

Serial	Units	Date	Cause	Crew	Duty	See
AE257	44	21/22.10.41	Missing, presumed ditched	PO W Bidd b.Sage; PO D Schafheitlin RCAF (M); Sgt W E Austin (M); Sgt M Hughes b.Becklingen	Bombing Bremen	
AE258	44/420/408/ Cv TB/1 AAS/ 32 OTU	30. 4.44	SOC in Canada			5K8
AE259	61	8/ 9. 8.41	Hit by AA from convoy off Spurn Head and crashed, Fosdyke, Lincs	PO Metcalfe; PO McCulloch; Sgt Ashurst; Sgt McKeown all unhurt	Bombing Kiel	5E4
AE260	44/420 (PT – O)	2/ 3. 6.42	Missing, cause not known	Sgt E Harrison*; Sgt J Gething*; Sgt L Nall (P); Sgt G Laing*. *b.5m NW of Lorient	Minelaying Lorient	5A45 2L9
AE261	106/49/Cv TB/ 489/5 OTU	20.12.43	SOC			
AE262	49	14/15. 8.41	Overshot in fog on return, Scampton	Sgt Owen, Sgt Donald, Sgt Kettlewell, Sgt Drinkwell all unhurt	Bombing Brunswick	
AE263	61	8/ 9. 8.41	Missing, presumed ditched	Sgt T Craven (M); Sgt V E Baker (M); Sgt G Lodington (M); F/Sgt E Robertson (M)	Bombing Kiel	
AE264	207/455/408/ 1404 Flt	8. 5.43	AT225 blown into aircraft by gust, St.Eval SOC 21.12.43		On ground	
AE265	144	25/26. 8.41	Made emergency landing, Ypenburg airfield, Neth.	PO I G St C Pringle; Sgt J R Blake; Sgt J Erskine; Sgt S C Brown. All POW	Bombing Mannheim	
AE266	61	30/31. 7.41	Undershot landing at Upwood on return	PO Adshead (K); F/Sgt Scott (I); F/Sgt Durtnell (K); Sgt Fillmore (K)	Bombing Cologne	2K30
AE267	408/420	26/27. 7.42	Missing, presumed ditched	F/Sgt A Johnstone RCAF; WO R Law RCAF; PO G E Tilling; Sgt J Price. All missing	Bombing Hamburg	
AE286	61/408 (EQ – T)	10/11. 1.42	Missing, presumed ditched	F/Lt T F Priest RCAF; F/Sgt D Henderson RCAF; Sgt G Glanville; Sgt L Churcher. All missing	Bombing Wilhelmshaven	
AE287	408/Cv TB/455 (UB – B)	17. 2.44	SOC			2F18
AE288	61/408 (EQ – H)	8/ 9. 5.42	Crashed on Mando Island, Denmark. Cause unknown	Sgt J S Norton RCAF; Sgt M S Sinclair RCAF; Sgt G Jefferies; Sgt H Stroud. All b.Esbjerg	Bombing Warnemunde	
AE289	61/408/Cv TB/ 415	29. 4.44	SOC			
AE290	61/44	23/24.10.41	Missing, presumed ditched	Sgt P H Bell (Rhod); Sgt K A Holmes; Sgt R H Atkinson; Sgt A T Laing (Rhod). All M	Bombing Le Havre	
AE291	50/455	10/11. 4.42	Missing, cause unknown	FO S T Manners RAAF; FO C A Harland RAAF; Sgt J Larkin; Sgt C Ormston. All b.Rheinberg	Bombing Essen	

Serial	Units	Date	Cause	Crew	Duty	See
AE292	106	14. 1.42	Spun into ground 4m E of Boston, Lincs	Sgt D Lawson (K); Sgt C W Wight (K)	Training	6E9
AE293	97/106/408/ Cv TB/455	23. 2.44	SOC			
AE294	A&AEE	22. 6.43	Undercarriage collapse, landing, Boscombe Down	G/Capt H Purvis; S/Ldr H Hastings; Mr.Dearnley	Training	
AE295	207/5 Gp TF/408/ 14 OTU/32 OTU	30. 4.44	SOC in Canada			5K8
AE296	207/455/Cv TB/5 OTU	21.12.43	SOC			
AE297	207/408 (EQ – F)	8/ 9. 5.42	Missing, cause not known	F/Sgt J Markle*; F/Sgt C Finkbeiner*; F/Sgt J Dafoe*; Sgt E Killham* Three RCAF All b. Berlin	Bombing Warnemunde	
AE298	44/420 (PT – D)	26/27. 3.42	Shot down by night fighter off Terschelling	F/Sgt W Groff RCAF; F/Sgt L Stalker RCAF; Sgt R Morgan; F/Sgt A Williams RCAF. All missing	Minelaying Frisians	
AE299	207/106	7/ 8. 9.41	Crashed near Schwartenek; cause unknown	Sgt M C Curties (P); F/Sgt F J Matkins*; Sgt J Plant*; Sgt H Mann*. *b.Kiel	Bombing Berlin	
AE300	97/106	11/12. 9.41	Engine cut; abandoned near Hostrup, Jutland	Sgt J A Bannister; Sgt Dunn; Sgt J A S Philson; Sgt F W Davies. All POW	Bombing Rostock	3D18
AE301	97/106	26/27. 8.41	Shot down by flakship 25 km E of Aalborg	FO M Harwood DFC*; F/Sgt N Lusher*; F/Sgt N Powell (P); F/Sgt W Oastler*. *b.Frederikshaven	Minelaying West Baltic	2D5
AE302	97/106	26/27. 8.41	Missing, presumed ditched	Sgt H V Wilkinson; Sgt K F Runcorn; Sgt A P J Grundy; Sgt A Mitchell. All missing	Bombing Cologne	
AE303	97	7/ 8. 8.41	Missing, presumed ditched	PO B Rodwell (M); Sgt G Imison (M); F/Sgt M Harvey (M); F/Sgt R Bell b.Terschelling	Bombing Essen	5H8
AE304	144	7/ 8. 9.41	Engine failure after take – off, North Luffenham	PO R J Roake (K); Sgt B Hemmings (K); Sgt E Horton (K); Sgt S E Taft (K)	Bombing Berlin	
AE305	50	2/ 3. 9.41	Ran out of fuel in fog and crashed near Bircham Newton	F/Sgt Titcombe, Sgt Gifford, Sgt Raynes, Sgt Christy all unhurt	Bombing Berlin	
AE306	50	7. 2.42	Shot down by fighter into sea W of Terschelling	F/Sgt R N Smith (Rhod); F/Sgt C Carroll RCAF; F/Sgt J Stables; Sgt A Bocock. All missing	Minelaying Frisians (day)	5B33 2H1
AE307	106/Cv TB/455 (UB – M)	12.10.42	Handed over to Russian Navy	Flown out by Sgt Baker, Sgt Dun, Sgt Jones; Sgt Bayliss, LAC Williams	See Russia	2J4 2J15
AE308	61/455 (UB – L)	6. 2.42	Shot down by fighter into sea 70 km W of Terschelling	Sgt B E Brown; Sgt R A Billington; Sgt D M Pickersgill; Sgt E Bland All missing	Minelaying Frisians (day)	2H1
AE309	144/Cv TB/ 5 OTU/45 Gp	14.10.43	Engine cut; ditched off Greenland en route to Canada	Capt/ R E Coffman; FO N Greenaway RCAF; Radio Officer R E Snow. All rescued	Ferry flight	6D20 5K10 2F13
AE310	144/Cv TB/144	4/ 5. 9.42	Ran out of fuel in fog and crashed near Afrikanda	Sgt Nelson, Sgt McGregor, Sgt Woodgate, Sgt Kane, Cpl North all unhurt	See Russia	2J6 2J15
AE311	144	9/10.11.41	Hit control caravan on landing, North Luffenham	Sgt Nightingale; Sgt Sparks; Sgt Gailey; Sgt Ward. All unhurt. AC2 F G Malin and AC1 G Roberts killed. One injured	Bombing Hamburg	2H5
AE312	83/455/14 OTU	28. 8.42	Crashed in flames 1m E of Cottesmore.	Sgt Clare (K); F/Sgt G A Gray RCAF (K); Sgt Brownlees (K); Sgt Patny (K)	Night training	
AE313	44 (KM – C)	2/ 3. 9.41	Missing, cause not known	Sgt D N DeBrath (Rhod)*; Sgt R L Cole*; Sgt A Goldston (P); Sgt J Rawson (P) *b. Rheinberg	Bombing Frankfurt	
AE314	83/420/Cv TB/ 455 (UB – L)	21.12.43	SOC			
AE315	83 (OL – X)	2/ 3. 9.41	Hit sandbank and crashed in sea off Borkum	Sgt R B Olliver; Sgt J F Payne; Sgt J D Walshe; F/Sgt H Logan. All POW	Minelaying Frisians	
AE316	50/144	24/25.10.41	Missing, possibly due to icing – up	F/Lt I Kingwell (P); Sgt F Stephenson RCAF (M); Sgt A Smith b.Dunkirk; Sgt L A Parker (P)	Bombing Frankfurt	
AE317	106	26/27.11.41	Crashed in sea off Spickerooge	Sgt W J Moss (M); Sgt R H Breckell (M); Sgt W L Ball (M); Sgt J F Tarran b.Sage	Bombing Emden	5F18
AE318	50	8/ 9. 9.41	Crashed near Oldenburg, Holstein; cause unknown	Sgt D Good; Sgt D F Endsor; Sgt R Williamson; Sgt I A McDonald. All POW	Bombing Kiel	
AE319	83 (OL – J)	6/ 7. 9.41	Shot down by flak in Oslofjord near Sandvika	PO J B Leetham (P); Sgt N B Morrison*; Sgt L W Sadezky (P); Sgt H Yardley* *b. Oslo	Minelaying Oslofjord	
AE320	50	25/26. 8.41	Missing, cause not known	Sgt R J Fisher; PO R F B Temperley; Sgt C Sciville; Sgt K Fyvie. All b.Rheinberg	Bombing Mannheim	
AE352	61/455 (UB – D)	21/22. 1.42	Either AE352 or X3051 shot down by night fighter NE of Lingen	Sgt E Thompson RAAF; PO J Stanfield RAAF; Sgt W Giles; Sgt W E Mabbett. All missing	Bombing Emden	
AE353	144	11.12.41	Probably damaged over Brest; crashed in sea off Sussex coast	PO C B Payne (K); PO A L Smiley (K*); FO A J B Monk RCAF (M); Sgt C E Gailey (K); PO J H Dunn RCAF (K*).*b.Littlehampton	Brest (day) Bombing	2G5 2F13
AE354	49 (EA – S)	18.12.41	Overshot Scampton in poor visibility	Sgt J Bow unhurt	Training	
AE355	420 (PT – A)	31.7/1.8.42	Missing, cause not known	WO W Kaufman RCAF; WO R Stewart RCAF; PO W D Frost RCAF; Sgt A J Greenaway. All(M)	Bombing Dusseldorf	
AE356	83/Cv TB/ 144	4/ 5. 9.42	Shot down by Russian fighters in sea near Murmansk	Sgt Hood, Sgt Turnbull, Sgt O'Neil, Sgt W T Tabor (K), Cpl Barker. Four unhurt	In transit to Russia	2J7 2J15

Hampdens of 14 OTU in line astern near Cottesmore. (CH710)

Serial	Units	Date	Cause	Crew	Duty	See
AE357	49/Cv TB	7. 2.44	SOC			
AE358	83 (OL – U)	1/ 2.11.41	Missing, presumed ditched	PO A E Lloyd; Sgt W Powell; Sgt D F Spindler; Sgt J W Butler. All missing	Bombing Kiel	
AE359	83/144	31.1/1.2.42	Crashed in sea, cause not known	Sgt P L E LeQueux b.Creances Fr 1204; Sgt J K Laing (Rhod)(P); Sgt C Smart (P); Sgt H Saunders (P)	Bombing Brest	3E4 3D25
AE360	408/Cv TB/ 415 (GX – H)	4.10.42	Missing, presumed ditched	Sgt B Boyton RAAF; PO J H Barton; Sgt J G Stuckless; Sgt J F Maher. All missing	Night navex	
AE361	408/Cv TB/TDU	25. 5.44	SOC			
AE362	83	15/16. 9.41	Crashed landing in fog, Scampton, on return	PO Bromiley (I); PO Rowe, Sgt Isherwood, PO Goodman all unhurt	Bombing Hamburg	
AE363	83/Cv TB/144/ 5 OTU/32 OTU	30. 4.44	SOC in Canada			5K8
AE364	83/Cv TB/144/ 5 OTU/415	1. 5.44	SOC			
AE365	83	15/16. 9.41	Crashed landing in fog, Scampton, on return	Crew unhurt; names not known	Bombing Le Havre	
AE366	83/420/408	5/ 6. 8.42	Attacked by night fighter	(see account in text)	Duisburg	2F21
AE366	5 OTU	4. 2.43	Swung on landing, Maghaberry; SOC 7.2.44	PO J W Maxwell unhurt	Training	
AE367	50	12/13.10.41	Missing, cause not known	PO J M Waddell; PO W E Hinchcliffe; Sgt R Curtis; Sgt N Lawson. All b.Reichswald	Bombing Huls	
AE368	49/144/Cv TB/ 415/5 OTU	20. 5.44	SOC			
AE369	50	27.12.41	Shot down by flak near Maaloy Island, Norway	Sgt L Redfern b.Trondheim; Sgt D Roberts (M); Sgt L R Brockett (M); Sgt A V Sherwin (M)	Bombing Maaloy (day)	2G6
AE370	50/Cv TB/415	30.12.42	Collided on ground with Fortress FK188, Thorney Island	Sgt Donald unhurt	On ground	5C15
AE371	83/Cv TB/144/ 489/455(UB – T)	3. 2.44	SOC			
AE372	49/408/ Cv TB/415	17. 5.44	SOC			
AE373	50/408/Cv TB/ NAFDU Wittering	9. 6.44	SOC			2L6
AE374	83 (OL – R)	14/15.12.41	Collided with P5393 taxying at Scampton	Sgt Thompson and crew apparently unhurt	To bomb Brest	5C3
AE375	50/408/Cv TB/415	12. 4.44	SOC			3C4
AE376	49 (EA – E)	28/29. 9.41	Crashed 11 mins after take – off, Burton Estate near Lincoln	Sgt A Walker (K); Sgt T Smith (K); Sgt E R Greenhalgh (K); Sgt A Raine (K)	Bombing Frankfurt	
AE377	44	8/ 9.11.41	Missing, presumed ditched	PO W J Frost (M); PO F L Smith (M); Sgt T Copsey (M); PO J Lindsay (M)	Bombing Dunkirk	
AE378	106/420/408/ Cv TB/455	14/15. 8.43	Missing from shipping strike, presumed ditched	PO V B Finn RAAF (M); FO R T Latham (M); WO C B Campbell RAAF (M); F/Sgt K Marriott (M)	T/bombing Norway (day)	

Serial	Units	Date	Cause	Crew	Duty	See
AE379	44/420/455/ 5 OTU	10.11.42	Dived into ground, Miltonburn, Ayr; cause obscure	Sgt A C Smart RNZAF (K); PO E Greaves (K); Sgt D C Barnie RCAF (K); Sgt J Munro RCAF (K)	Night navex	
AE380	50	15/16.12.41	Missing, presumed ditched	Sgt W J Young b.Ostend; Sgt D A Ferguson (M); Sgt F A White (M); Sgt E W Armer (M)	Bombing Ostend	
AE381	50	21. 1.42	Flew into hill in bad weather, Kinder Scout, Derbyshire	Sgt Heron (K); Sgt Williams (K); Sgt Peters (K); Sgt Tromans (K)	Night navex	
AE382	44 (KM – A)	10/11.10.41	Missing, cause not known	Sgt J P Bonnett (Rhod); Sgt G W Niblett Sgt F E Wootton; Sgt J A Anderson. All(P)	Bombing Dunkirk	
AE383	50	20/21.10.41	Missing, cause not known	PO W Laidlaw; Sgt K S Furley; Sgt W Saunders; Sgt E R James. All b.Becklingen	Bombing Bremen	
AE384	44/420/Cv TB/ 455 (UB – F)	17. 3.45	SOC Shipped to Australia as memorial aircraft but sold for scrap in 1954			2J36
AE385	44/420/408	12. 9.42	Stalled and dived into ground, Balderton	Sgt G Frame RCAF (K); F/Sgt W Keogh RCAF (K)	Night training	
AE386	50/14 OTU/420/ Cv TB/455 (UB – Y)	3. 1.44	SOC			
AE387	50	7. 3.42	Swung on take – off and caught fire, Skellingthorpe	PO P W Knock (I); LAC H Lees? (K); Two others injured	Training	
AE388	83/50/Cv TB/ 144/5 OTU	4. 6.44	SOC			
AE389	83/420 (PT – D)	7/ 8. 5.42	Missing, probably shot down by flak, Kiel	PO M F Carson (P); Sgt G C Williams*; Sgt A S Urquhart*; Sgt W A McDonnell* *b. Kiel	Minelaying Kiel Bay	
AE390	44/420 (PT – Z)	12/13. 7.42	Missing, presumed shot down by flak in sea	Sgt T Hannah RCAF (M); Sgt W Chapman*; F/Sgt P Bull DFM (M); F/Sgt J Thompson* *b. Lorient	Minelaying Lorient	
AE391	106	12.12.41	Crashed near Oberhausen – Osterfeld on day raid	F/Lt H Webber; Sgt S Crowhurst RNZAF; Sgt F Norman; Sgt T Robson – Scott. All b. Reichswald	Bombing Gelsenkirchen (day)	2G4
AE392	144	7. 2.42	Shot down into sea off North German coast	F/Lt W Kingson (M); Sgt R DeCourcy (M); Sgt J Tobin (M); Sgt A Gibson b.Sage; Sgt A Fulton (M)	Minelaying Frisians (day)	2H1
AE393	44/420/408 (EQ – G)	15/16. 1.42	Ran out of fuel in bad weather and hit high ground, West Burton, Yks	S/Ldr W J Burnett DFC (I); WO J Bunting (K); Sgt J R Appleby (K); Sgt M Jones (K)	Bombing Hamburg	2K11 5D3
AE394	50	21/22. 2.42	Ran out of fuel in bad weather and abandoned, York	WO I P Mapp baled out safely; Sgt Thompson, Sgt Law and Sgt Ballard killed	Bombing Koblenz	5A3 5A20 6D13
AE395	144/Cv TB/ 415 (GX – S)	23/24. 3.43	Shot down by flak in attack on convoy off Dutch coast	FO J McLeod RCAF (P); PO C Chambers RCAF (M); PO D Wraith RCAF (P); Sgt W Haddock RCAF (M)	T/bombing Dutch coast (night)	2J26
AE396	49 (EA – W)	12. 2.42	Shot down by fighters or flak	Sgt E W Phillips (M); Sgt K W Heard (M); F/Sgt A Jackson (M); Sgt L C Toghill (M)	Bombing warships (day)	3B1 2H4
AE397	49 (EA – G)	14/15. 2.42	Engine cut; ditched off Isle of Wight	PO Allsebrook, Sgt Stanbridge, Sgt Wilkinson, Sgt Woolgar.All picked up OK	Bombing Mannheim	5B27
AE398	44	26/27.10.41	Crashed in circuit in bad weather after engine trouble, Coltishall	Sgt M Gruber and crew. One seriously injured	Bombing Hamburg	
AE399	44/420 (PT – P)	30/31. 5.42	Swung on landing and hit Lancaster R5842, Waddington	F/Sgt W McDermid RCAF (I); F/Sgt L Gardiner RCAF (I); Sgt Waddington, Sgt Wilkinson unhurt	Bombing Cologne	5C17 2F16
AE400	50	7/ 8. 3.42	Missing, presumed ditched near Brest	PO A G Davidson (M); Sgt I Follows*; Sgt W M A Eardley (M); Sgt G Shackleton (M) *b. near Brest	Minelaying Lorient	
AE401	50/420	19/20. 6.42	Hit by flak off Dutch coast	Sgt R Davidson RCAF, lower gunner (K)	Bombing Osnabruck	
AE401	Cv TB/415	4. 6.44	SOC			2F15
AE418	408/Cv TB/ 415 (GX – B)	19/20. 3.43	Dived into sea on approach, Thorney Island	Sgt W Hurl, PO B A Johnson, Sgt W C Foerster, Sgt R H O'Brian, all RCAF. All b. Thorney Island	Torpedo bombing (night)	
AE419	49 (EA – T)	28/29.12.41	Missing, presumed ditched	Sgt A C Watt; Sgt E G Sawdy; F/Sgt E C Atkinson; Sgt H W Wisdom. All missing	Bombing Huls	5H13
AE420	83/50	14/15. 1.42	Missing, presumed ditched	Sgt R A Baddeley, Sgt J M Horley, Sgt D W Cockburn, Sgt L A Angel. All missing	Bombing Hamburg	
AE421	83/49 (EA – P)	10/11. 4.42	Shot down by night fighter 28 km NE of Den Helder	FO R Worthy b.Amersfoort; Sgt A Considine (K); Sgt J Wilkinson b.Stavanger; Sgt T McGrenery (M)	Bombing Essen	
AE422	50/420/Cv TB/ 415/489 (XA – A)	24. 1.44	SOC			2G6
AE423	50/Cv TB/ 5 OTU	27. 4.43	Undercarriage collapsed on landing, Maghaberry	PO T Sandberg on first Hampden solo unhurt	Training	
AE424	144	5/ 6.11.41	Presumed ditched after attack on convoy off Dutch coast	PO L N Evans (M); PO H Walter (M); Sgt T G Miskin (M); Sgt G H S Bennett (M)	Bombing convoy	
AE425	106	11/12. 2.42	Port engine cut on take – off, Woodhall Spa	PO R Ayton and crew unhurt	Bombing Mannheim	2H7
AE426	106/408	28/29. 4.42	Missing, cause not known	Sgt R L Gordon RCAF; F/Sgt J Romas RCAF; Sgt W Alderdice; Sgt T Hindle. All b.Hamburg	Bombing Kiel	

AE184 VN–Z at Lindholme in the summer of 1941. It was written off in a crash at Radcliffe on 25 October 1941. (Photo: F/Lt Dent, 50 Squadron)

Serial	Units	Date	Cause	Crew	Duty	See
AE427	50	6/ 7.11.41	Shot down by flak near Oslo	Sgt J Howett; PO P F Boreham; Sgt M Bloomfield; Sgt H W Fewkes. All b.Oslo	Minelaying Oslofjord	
AE428	44/50	27.12.41	Shot down by flak near Maaloy, Norway	Sgt R N Smith (Rhod) picked up; Sgt J Williams (M); PO R Watson (M); Sgt D Bell (M)	Bombing Maaloy (day)	2H1 5B32 2G6
AE429	50	24/25. 3.42	Crashed near Brest; cause not known	PO D R Fair; PO I O Davies; Sgt J Hudson; Sgt J H Pearson. All b.Brest	Minelaying Lorient	
AE430	44/455	30.11/1.12.41	Missing, cause not known	Sgt C J H Blunt; Sgt L G Manning; Sgt A Gee; Sgt E Waller. All b.Kiel	Bombing Hamburg	
AE431	50	14/15. 1.42	Missing, presumed ditched	PO E V Hore; PO L M Jacob; Sgt G A Gale; Sgt P C Pettersson. All missing	Bombing Hamburg	
AE432	408 (EQ–S)	15/16. 8.42	Missing, presumed ditched	F/Lt J Cardell RCAF; PO L G Staniar; Sgt E Mitchell; Sgt H L Jewell. All missing	Bombing Dusseldorf	
AE433	408 (EQ–S)	8/ 9.11.41	Shot down by night fighter, Maasbree	PO E Houghton RNZAF; PO J Monkhouse RCAF; Sgt A Gallan; Sgt J Woodward. All POW	Bombing Essen	
AE434	455 (UB–G)	4/ 5.11.41	Crashed after engine failure on return, Bassingham, Lincs.	Sgt Jones (I); Sgt Kirk (I); Sgt Baynes (I); Sgt Shimmons (I)	Minelaying Frisians	
AE435	50/Cv TB/455/ 415 (GX–U)	18/19. 2.43	Crashed on take–off, Docking	Sgt B Campbell (K); FO K R Mafre (K); F/Sgt R Vokey (K); Sgt Z M Niblock (K)	T/bombing (night)	
AE436	Cv TB/144	4/ 5. 9.42	Flew into Tsatsa Mtn., Kvikkjokk, Sweden	PO D Evans (I); PO W H Bowler RCAF*; F/Sgt J Jewitt RCAF*; F/Sgt J Campbell RCAF*; Cpl B J Sowerby. *b.Gothenburg, Sweden	In transit to Russia	2J12 2J15 3F1
AE437	408 (EQ–U)	27/28.11.41	Misunderstood W/T bearings; ditched 370m W of Brest	FO J A Caldwell (M); Sgt H Marshall RCAF (M); Sgt E H Harry (M); Sgt R Crawley (M)	Bombing Dusseldorf	2F17 5B9 3F3
AE438	408 (EQ–N)	9/10.11.41	Missing, cause not known	PO J C Wilson RCAF; Sgt E B Robertson RNZAF; Sgt D Norton; Sgt D Markall All b.Westkerke, Belgium 102	Bombing Ostend	2F18
AE439	408/AAEE	2.10.44	Engine caught fire; crashed, Boscombe Down	F/Lt S W Bannister, LAC J J Roman unhurt	Training	6D14
AE440	144	27/28.11.41	Flew into high ground, Birdlip, Cotswolds	PO S T Farrington, Sgt Robinson, Sgt Martin unhurt; Sgt G Cornish (K)	Bombing Dusseldorf	
AE441	144	14/15. 1.42	Hit chimney on approach after engine failure, Langham, Norfolk	F/Lt W J W Kingston, Sgt DeCourcy, Sgt Tobin, Sgt Gibson all unhurt	Bombing Hamburg	
AE442	455/14 OTU/ Cv TB	22. 3.45	SOC			

12 Hampden Is delivered in March 1942 by Canadian Associated Aircraft, St.Hubert PQ

Serial	Units	Date	Cause	Crew	Duty	See
AJ988	RCAF/32 OTU	25.10.42	Hit sea on low flying practice and crashed near Sidney Island	Sgt J Orrell (K); Sgt H J Thorn (K)	Training	6E20 6E23
AJ989	RCAF/32 OTU	10. 7.42	Engine caught fire; crashed into sea off Sidney Island	Sgt E Williamson (I); Sgt A F Vaughan (I); Sgt A McKinnon RNZAF (I); Sgt C Johnson RNZAF (I)	Training	6D23 6E20 5G2
AJ990	RCAF/32 OTU	10. 8.43	Collided with trailer, Patricia Bay	Sgt O Williams unhurt	Training	
AJ991	RCAF/32 OTU	27. 5.43	Engine cut; overshot landing, Tofino	Sgt G A Webster (I); Sgt K Rawlin, Sgt H E Hollitt RAAF, Sgt L R Coghlan RAAF unhurt	Night training	
AJ992	RCAF/32 OTU	14. 6.43	Crashed in sea in bad weather 20m W of Cape Flattery	FO G G Douglas-Home (K); PO C Sugden (K); Sgt A Lynch RAAF (K); Sgt E Ritchie RAAF (K)	Night training	6E22
AJ993	RCAF/32 OTU	19. 9.43	Dived into sea in poor visibility off Port Renfrew	PO H H Frost RCAF (M); Sgt A S D Dobie (M); F/Sgt A France RCAF (M); F/Sgt J Bateman (M)	Training navex	6E21
AJ994	RCAF/32 OTU	28. 9.43	Spun into ground on single-engine flying, Galiano Island	Sgt H D McLeod (K)	Training	
AJ995	RCAF/32 OTU	12. 8.43	Swung on landing and undercarriage collapsed, Patricia Bay	PO A Shippen, Sgt J H Oblein, Sgt W Park RAAF, Sgt D J Oliver RAAF unhurt	Training	
AJ996	RCAF/32 OTU	23. 7.43	Undercarriage collapsed in heavy landing, Patricia Bay	PO J Thomson RAAF, PO A Lockwell, Sgt R Bruce RAAF, Sgt N R Chaffey RAAF all unhurt	Night training	
AJ997	RCAF/32 OTU	2. 8.44	SOC in Canada			
AJ998	RCAF/32 OTU	4. 9.43	Undershot landing at Patricia Bay	Sgt J R Baillie (I) on first Hampden solo	Training	
AJ999	RCAF/32 OTU	2. 8.44	SOC in Canada			

* * * * * * * * * * * * * * * * *

68 Hampden Is delivered between March and May 1942 by Canadian Associated Aircraft, St.Hubert, PQ

Serial	Units	Date	Cause	Crew	Duty	See
AN100	RCAF/32 OTU	4. 6.43	Missing from navex cause not known	Sgt W F Barker; Sgt J Ralph; Sgt D Maddrell RAAF; Sgt G Senger RAAF. All(M)	Training	
AN101	RCAF/32 OTU	9.10.43	Crashed in sea on torpedo drop NE of Stuart Island	PO A L Warner; PO K Shaw; Sgt N Weekes RAAF; Sgt F Maiden. All missing	Training	6E17
AN102	RCAF/32 OTU	23. 7.43	Undercarriage collapsed after landing, Patricia Bay All unhurt	PO C A McPherson RAAF; PO E J Lipscombe; Sgt D Davies RAAF; Sgt W Christie	Training	
AN103	RCAF/32 OTU	2. 8.44	SOC in Canada			
AN104	RCAF/32 OTU	2. 8.44	SOC in Canada			
AN105	RCAF/32 OTU	4.10.42	Crashed on Mount Tuam, Saltspring Island	PO N Thomas (K)	Night training	5D7
AN106 to AN108	RCAF/32 OTU	2. 8.44	SOC in Canada			
AN109	RCAF/32 OTU	3. 7.42	Swung on take-off and undercarriage collapsed, Patricia Bay	PO Davies and three crew unhurt	Training	
AN110	RCAF/32 OTU	24. 1.43	Crashed landing on one engine, Patricia Bay	Crew unhurt	Training	5F19
AN111 and AN112	RCAF/32 OTU	2. 8.44	SOC in Canada			
AN113	RCAF/32 OTU	18. 5.43	Stbd engine cut on take-off, Patricia Bay	Sgt P H Griffiths (I)	Training	
AN114	RCAF/32 OTU	6. 5.43	Damaged in heavy landing, Patricia Bay; to GI airframe A385; SOC 21.2.45	PO W Marsden unhurt	Training	
AN115	RCAF	6. 1.43	Crashed at Regina and reduced to spares, 22.9.43		Ferry flight	
AN116 to AN119	RCAF/32 OTU	2. 8.44	SOC in Canada			
AN120	RCAF/32 OTU	19. 1.44	To GI airframe A366; SOC 27.4.54			
AN121	RCAF/32 OTU	13. 4.43	Dived into sea inverted off Cowichan Head	PO J S Peterkin (K); Sgt R T Barrow RAAF (K); Sgt S MacGillivray (M); Sgt K E Thompson (I)	Training	5F12 6E18
AN122	RCAF/32 OTU	21. 1.44	Swung on take-off and undercarriage collapsed, Patricia Bay	Sgt J S T Budgen unhurt	Training	

Demonstration day at Radlett as three Hampdens are put through their paces for the benefit of inspecting Government officials. (Photo: RAF Museum 5836 – 9)

Serial	Units	Date	Cause	Crew	Duty	See
AN123	7 AAU/Cv TB/489 (XA – D)	23. 1.43	Missing, presumed ditched	PO W C Salmond RAAF; Sgt J H Sheppard; F/Sgt J Maguire; F/Sgt A G Brown RCAF. All missing	T/bombing, Norway (day)	
AN124	7 AAU/Cv TB/ 415 (GX – M)	10/11. 8.42	Engine failure; crashed in sea off Bell Rock, off Firth of Tay	W/Cdr R Dennis, PO N Altstedter, PO P Houlsworth, Sgt D A Stallard all picked up unhurt	T/bombing Norway (night)	
AN125	7 AAU/Cv TB/ 5 OTU	7.10.42	Stalled into sea off Girvan, Ayrshire	Four crew picked up by safety boat; one dead	Training	6E12
AN126	7 AAU/Cv TB/ 144	12.10.42	Handed over to Russian Navy	Flown out by S/Ldr Hird, Sgt Hepplewhite, Sgt Burrow, Sgt Coles, AC1 Tungate	Left in Russia	2J15
AN127	7 AAU/Cv TB/489	1. 4.44	SOC			
AN128	RCAF/32 OTU	26.10.42	Undercarriage collapsed on landing, Patricia Bay	Sgt Jones and three crew unhurt	Training	
AN129 and AN130	RCAF/32 OTU	2. 8.44	SOC in Canada			
AN131	RCAF/32 OTU	30.11.43	Hit hill in fog 6m SW of Ladysmith, Vancouver	Sgt J F Fearnley (K); Sgt L Pettit (K)	Instrument training	5D7
AN132	RCAF/32 OTU	16.10.43	Hit hill NW of Maggie Lake, Ucluelet, BC; found 6.7.45	Sgt R Allcorn; Sgt P Hornbrook; Sgt R Porter RAAF; Sgt M Smith RAAF All killed	Night navex	3F1 5D7
AN133	RCAF/32 OTU	15. 4.43	Swung on landing and undercarriage collapsed, Patricia Bay	Crew unhurt	Training	
AN134	RCAF/32 OTU	2. 8.44	SOC in Canada			
AN135	RCAF/32 OTU	18. 5.43	Stbd engine cut; crashed in sea near Gulf Island	Sgt J Birch (I)	Training	6E19
AN136	RCAF/32 OTU	14. 1.44	Flew into Mount Tuam, Saltspring Island	Sgt H C Birch (K) on solo flight	Night training	5D7
AN137	7 AAU/Cv TB/ 455/489	14. 5.44	SOC			
AN138 and AN139	RCAF/32 OTU	2. 8.44	SOC in Canada			
AN140	RCAF/32 OTU	30.11.42	Collided with crash tender, Patricia Bay	Sgt Clasper and crew unhurt	Night training	6C7

Serial	Units	Date	Cause	Crew	Duty	See
AN141	RCAF/32 OTU	18. 8.43	Stbd engine cut; ditched off James Island	PO R K Westhorp, PO K H Webster, F/Sgt J M Booth RCAF, F/Sgt J A Buckham, all picked up	Night navex	6D21
AN142	RCAF/32 OTU	23. 5.43	Engine cut; crashed on landing, Patricia Bay	PO C Davies RAAF; Sgt A Marlow RAAF; Sgt R Hughes RAAF; Sgt Bishop. All killed	Navex	6C19 6D22
AN143	RCAF/32 OTU	10. 6.42	Swung on take – off, Patricia Bay; to GI airframe A183	PO J M Ireland unhurt	Training	5F19
AN144	RCAF/32 OTU	25.11.42	Swung on take – off and undercarriage collapsed, Patricia Bay	Sgt F Guss unhurt on solo flight	Training	
AN145	RCAF/32 OTU	16. 3.43	Swung on take – off and undercarriage collapsed, Patricia Bay; caught fire	Crew unhurt	Training	
AN146	RCAF/132 OTU	17. 9.43	Spun into ground on approach, Patricia Bay	Sgt A W V Fear (K)	Training	6C19
AN147	RCAF/32 OTU	4.10.43	To GI airframe A320 24.5.44 SOC in Canada			
AN148	7 AAU/Cv TB/ 455 (UB – K)	16. 5.43	Spun into sea after engine failure off Shetlands	F/Sgt G Courtney RAAF; FO A C Smith; Sgt J A King; F/Sgt C H Smith RAAF All missing	Night training	6D17
AN149	7 AAU/Cv TB/ 415/455 (UB – X)	23.12.43	SOC			
AN150	RCAF	20. 7.42	Forcelanded near Kirkland Lake, Ont. To GI airframe A367	Name(s) not known	Ferry flight	
AN151	7 AAU/Cv TB/ 5 OTU/1 TTU	12.10.42	Crashed in sea on torpedo drop, Firth of Clyde	Sgt G Howell and three crew all killed	Night training	6E12
AN152	7 AAU/Cv TB/ 144/489	20. 5.44	SOC			
AN153	7 AAU/Cv TB/ 5 OTU/415 (GX – B)	17/18. 5.43	Shot down by flak from convoy off Dutch coast	F/Sgt L McGee b.Hamburg; FO L Burgess (M) WO E Ramey RCAF (M); Sgt Jackson RCAF b.Kiel	T/bombing Dutch Coast (night)	2J26
AN154	7 AAU/Cv TB/489	20. 2.44	SOC			
AN155	7 AAU/Cv TB/ 489/5 OTU	4/ 5. 4.43	Missing, presumed ditched off Ireland	Sgt Hobbs and three crew; all missing	Night navex	
AN156	7 AAU/Cv TB/ 489/(XA – K)	28/29. 3.43	Spun into sea after jettisoning torpedo N of Rundo	F/Sgt A Grant b.Trondheim; Sgt W Nicholson (P); Sgt F Brotherton RAAF (P); Sgt R Holloway (P)	T/bombing Norway (night)	
AN157	7 AAU/Cv TB/ 455 (UB – N)	14. 9.43	Shot down by ships' flak off Lister	FO W Austin RAAF; WO J Mellor; WO R S Esmay RCAF; Sgt E W Smith. All missing	T/bombing Norway (day)	
AN158	7 AAU/Cv TB/ 5 OTU	8.12.43	SOC			
AN159	Cv TB/5 OTU	22. 6.43 4. 5.44	Damaged SOC			
AN160	7 AAU/Cv TB/ 144/455	20.12.43	SOC			6C7
AN161	7 AAU/Cv TB/ 5 OTU/455	4. 6.44	SOC			
AN162	7 AAU/Cv TB/	25. 5.44	SOC			
AN163	7 AAU/Cv TB/ 455/(UB – K)	5.11.43	Overshot landing at night, Fraserburgh; flarepath unlit (night)	PO S J Cliff, FO J C McGhee, F/Sgt D McLean RCAF, Sgt J McFeely all unhurt	T/bombing Norway	
AN164	7 AAU/Cv TB/ 455/489	6. 3.44	SOC			
AN165	7 AAU/5 OTU	23. 3.45	SOC			
AN166	7 AAU/Cv TB/ 455 (UB – R)	11. 1.43	Engine lost power after take – off; crashlanded near Leuchars	Sgt S A Vincent RAAF (I); F/Sgt A Watt RCAF (I); Sgt K Applekampf RAAF (I); Sgt D Martin RAAF (K)	T/bombing Norway (day)	
AN167	7 AAU/Cv TB/	4. 6.44	SOC			

* * * * * * * * * * * * * * * * *

120 Hampden Is delivered between October 1941 and March 1942 by English Electric, Preston

Serial	Units	Date	Cause	Crew	Duty	See
AT109	50/Cv TB/ 455 (UB – C)	4/ 5. 9.42	Hit by flak and crashed near Vardo	S/Ldr Catanach; FO Anderson; F/Sgt Cameron; Sgt Hayes; F/Sgt Davidson All POW	In transit to Russia	2J11 3D27 2J15
AT110	83/144	14/15. 4.42	Crashed near Bognor Regis en route outwards	Sgt S W C Culmer (K); Sgt R E Nation (K); Sgt J A Hutchinson (K); Sgt D Manning (K)	Bombing Dortmund	
AT111	49/Cv TB/ 489	10. 7.43	Overshot landing at Inverness; SOC 3.2.44	F/Lt G E Willis unhurt	Training	
AT112	83/49	14/15. 2.42	Ran out of fuel and crashed, Glossmore Farm near Upwood	Sgt R N Hamer, Sgt Haddock, Sgt Minchin, Sgt Watchorn all unhurt	Bombing Mannheim	
AT113	455/408 (EQ – A)	29. 7.42	Stalled and dived into ground SE of Balderton	S/Ldr L B Price RCAF (K); Sgt I Hughes (K) ATC Cadets G Hughes, K Cousin – Wood both K	Air test	6E5

Two 50 Squadron Hampdens at Waddington before the war, the code "QX" being changed to "VN" in September 1939. QX – D was L4076. (Photo via C.G.Jefford)

Serial	Units	Date	Cause	Crew	Duty	See
AT114	455/Cv TB/ 415 (GX – D)	11. 3.43	Missing returning from reconnaissance	WO J N Reigate RCAF; Sgt D T Hodgson; F/Sgt J Kerr RCAF; F/Sgt J St Laurent RCAF. All missing	T/bombing Norway	2J26
AT115	106	30.11/1.12.41	Missing, presumed ditched	Sgt F E Rolfe (M); Sgt A E Hill (M); Sgt L Goodbrand (M); Sgt H Griffiths (M)	Bombing Hamburg	
AT116	144 (PL – F)	22/23. 3.42	Hit Stirling N3674 taxying at Exeter	Sgt D J Towler, Sgt E F Penlington, Sgt J Burns, Sgt W Woodford all unhurt	Minelaying Lorient	
AT117	144/Cv TB/ 5 OTU/1 TTU	3. 5.43	Dived into sea on approach, Turnberry	Sgt L F Barrett and one other killed	Night training	
AT118	49/50	26. 3.42	Engine cut; forcelanded 5m N of Penzance	S/Ldr F R Jeffs and crew unhurt	ASR search	
AT119	455 (UB – Y)	21/22. 1.42	Shot down by night fighter, Schlocteren	Sgt G F Poulton (P); Sgt I M Ince RAAF* Sgt S Williams*; Sgt Wilkinson* *b.Holten, Netherlands	Bombing Emden	
AT120	408	12/13. 4.42	Engine cut; crashed in North Sea 40m off Dutch coast	Sgt W Meach*; Sgt J Davenport*; Sgt G Lantze (M); F/Sgt J Wilkie (M) All RCAF. *b. in UK	Bombing Essen	
AT121	106	28/29. 1.42	Engine cut over Ysselmeer; crashed off Den Helder	FO H J Bareham (P); PO T Whiting (P); Sgt E J Snelling b.Kiel; Sgt S Dungey b.Bergen – op – Zoom, Neth.	Bombing Munster	
AT122	106	28/29. 1.42	Crashed at Lievelde, Netherlands Cause not known	PO R V Selfe, F/Sgt J Jacobson RCAF; Sgt D Hodgkinson, Sgt S Harding. All b. Lichtenvoorde	Bombing Munster	
AT123	106	3/ 4. 1.42	Ditched in North Sea, pres. engine failure	PO D A Howard DFM; PO A J Fraser; F/Sgt L McKenzie DFM; Sgt A Gardner. All (M)	Minelaying Frisians	5B19
AT124	49 (EA – C)	16/17. 2.42	Missing, presumed ditched	FO G R Jenkins; PO J B Clough; Sgt J A Woodroffe; Sgt J A Steele. All missing	Minelaying Frisians	
AT125	50/Cv TB/144/ 5 OTU/1 TTU	18. 4.43	Collided with LB238 and crashed 3½m ENE of Ailsa Craig, F.of Clyde	Sgt R E Cordingley; Sgt E Vevers; Sgt J B Reid; all killed Wellington LB238 landed safely	Training	5C21
AT126	49 (EA – Z)	6/ 7. 4.42	Missing, presumed ditched in North Sea	F/Sgt L Davis; Sgt D P Walmesley; Sgt M Griffiths; Sgt C Hartley RAAF. All missing	Bombing Essen	2L4
AT127	83 (OL – L)	8/ 9. 1.42	Crashedlanded at Sedge – brook, Lincs on return	Sgt Price, PO Rayne, Sgt Osborn, Sgt Lown all unhurt	Bombing Brest	
AT128	44/420 (PT – G)	14/15. 4.42	Missing, cause not known	F/Sgt B W Johnson; Sgt H S Vosper; Sgt J F Shipton; Sgt J C D McHardy. All b.Rheinberg	Bombing Dortmund	
AT129	83/49 (EA – O)	25/26. 1.42	Crashed on take – off in ice and snow, Scampton	Sgt C D S Stewart; Sgt K Northrop; Sgt A Hibbitt; Sgt L Jardine. All killed	Bombing Brest	2L1
AT130	44/420 PT – S)	21/22. 1.42	Hit by flak and crashed 6 km NNE of Harlingen	S/Ldr T V L Wood; Sgt D D Grealy; Sgt D Semple; Sgt R L Bott. All POW	Bombing Emden	

Serial	Units	Date	Cause	Crew	Duty	See
AT131	106/455/Cv TB/ 489 (XA – V)	12. 7.42	Hit sea on practice torpedo drop, Firth of Clyde	FO D J Nilsson RNZAF, PO H W Cowan, Sgt Scott, Sgt Lander all picked up by range safety boat	Training	
AT132	44/420/ 455 (UB – K)	24. 5.43	Collided with Beaufighter JL502 5m off Fraserburgh	F/Sgt J Freeth RAAF (M); F/Sgt A Wheatcroft (K); FO I Whitson RAAF (M); F/Sgt H Downing RCAF (K)	Training	2J38
AT133	408 (UB – K)	14/15.12.41	Missing, presumed ditched	Sgt J Vaughan RCAF; F/Sgt R Thompson RCAF Sgt B Howells; Sgt W Oates. All missing	Bombing Cherbourg	
AT134	44/420 (PT – K)	12. 2.42	Shot down by fighters or flak attacking warships	S/Ldr G Harris b.Amsterdam; FO H Miller RCAF (M); F/Sgt A Brunt*; Sgt P Rothery* *b.The Hague	Bombing warships (day)	2H3
AT135	44/420/Cv TB/ 455/5 OTU/ 415 (GX – S)	18. 8.43	Port engine fell out; crashed off Scillies	FO R Armstrong, FO H R Cuddon, F/Sgt A T J Maker RCAF; Sgt J Bileski RCAF all unhurt	Anti – sub patrol	5G1 2J32
AT136	44/420 (PT – N)	8/ 9. 6.42	Shot down by night fighter, presumed in North Sea (M)	F/Sgt I Reid RCAF (M); Sgt A Grabham RAAF (M); Sgt H R Copeland (M); Sgt C J Bunn	Bombing Essen	
AT137	455	8. 6.42	Destroyed by fire Leuchars		On Ground	
AT138	408/Cv TB/ 144	4/ 5. 9.42	Shot down by two Bf 109s over North Finland	Sgt J Bray (P); Sgt J Smith (M); Sgt G Kirby (M); Sgt R S Otters (M); AC2 L Mallinson (M)	In transit to Russia	2J8 2J15
AT139	50/408 (EQ – A)	21/22. 7.42	Crashed into Waddensee Cause not known	F/Sgt R Wishart*; Sgt R Doherty (M); PO T R Ashton b.Bergen – op – Zoom; F/Sgt J Pleasance*. *both RCAF b.Harlingen	Bombing Duisburg	
AT140	50/Cv TB/ 144/489	24. 1.44	SOC			
AT141	106/408	14/15. 4.42	Missing, possibly shot down by night fighter	Sgt A McClinton RCAF; Sgt F Hagel RCAF; F/Sgt A Boisvert RCAF; Sgt J Johnstone All b.Reichswald	Bombing Dortmund	
AT142	50	22/23. 1.42	Crashed on landing, Cottesmore on return	FO D B Peace; Sgt Ady; Sgt Brooke; Sgt Jory. All killed	Bombing Munster	
AT143	144/408	28/29. 8.42	Overshot landing at Booker; SOC 3.2.44	F/Sgt J T Craig, F/Sgt H Treherne, Sgt J Galloway, Sgt K Godfrey all unhurt	Bombing Saarbrucken	
AT144	44/420 (PT – A)	8/ 9. 5.42	Hit by flak over Rostock and crashed	S/Ldr G Campbell DFC (P); F/Sgt O Petersen RCAF*; Sgt R Parry*; Sgt G Soper RCAF (P) *b.Berlin	Bombing Warnemunde	
AT145	144/Cv TB/ 144	12.10.42	Handed over to Russian Navy	Flown out by W/Cdr LcLaughlin, PO Wilson, Sgt Matthews, Sgt Denzey, F/Sgt Church	Left in Russia	2J15
AT146	50/106	22/23. 1.42	Missing, presumed ditched	Sgt H S E Ellis (M); F/Sgt S Morrison RCAF (M);Sgt A S Chantrell (M); Sgt W Holland (M)	Minelaying Frisians	
AT147	50/Cv TB/ 489/32 OTU	30. 4.44	SOC in Canada			5K8
AT148	49 (EA – S)	20/21. 1.42	Shot down by night fighter, Roode Haan	FO A M Harvey, Sgt W Gray, Sgt J Hallam, Sgt R Knapman all b.Noorddijk, Netherlands	Bombing Emden	
AT149	144	31.1/1.2.42	Missing, probably shot down by flak	WO A N Mowlam, Sgt G J Linegar, Sgt W Black, Sgt C J Stringer all b.Brest	Bombing Brest	
AT150	49/Cv TB/ 455	29. 5.43	Overshot aborted take – off, Leuchars SOC 20.2.44	FO R C Barton unhurt	Training	
AT151	50	25/26. 3.42	Shot down by night fighter off Vlieland	PO H P Roblin; Sgt E D G Lumsden; Sgt J D Beattie; Sgt R D Howarth. All missing	Bombing Essen	
AT152	50/Cv TB/ 415/455 (UB – F)	4. 4.43	Hit by flak off Norway and crashed on return, Leuchars	FO M S Humphrey RAAF (I); FO D R Hawkes (I); F/Sgt T Perry RAAF, PO Ackroyd RAAF both unhurt	T/bombing Norway	
AT153	50/Cv TB/ 5 OTU/1 TTU	8.12.43	SOC			
AT154	408 (EQ – B)	2/ 3. 6.42	Shot down in Ysselmeer by night fighter	PO K G Taylor RCAF (P); PO P L Edwards (M); Sgt E Ronson b.Hindelopen; Sgt McKenzie b.Amsterdam	Bombing Essen	
AT155	144	12. 4.42	Stalled and dived into ground, Ridlington, Rutland	F/Sgt D L C Liddell RNZAF and one other killed	Air test	
AT156	49 (EA – C)	5/ 6. 4.42	Shot down by night fighter, Philippeville, Belgium	PO D Kay DFM (P); Sgt R Brown*; Sgt S Ainger (P); Sgt J Waddell*. *b.Charleroi, Belgium	Bombing Cologne	
AT157	144	14/15. 4.42	Missing, cause not known	FO S Booker (P); F/Sgt W Browne*; Sgt J Gledhill*; Sgt D F Stephen (P). *b.Rheinberg	Bombing Dortmund	
AT158	50	25/26. 3.42	Missing, cause not known	PO J D Crombie; Sgt H Deckman; Sgt C R Perry; Sgt F Manley. All b.near Lorient	Minelaying Lorient	
AT172	144/Cv TB/415	3. 2.44	SOC			
AT173	50	10/11. 3.42	Ran out of fuel in fog and abandoned, Norfolk	PO Stone, Sgt Hudson, Sgt Mills, Sgt O'Connor all baled out safely	Bombing Essen	5A3 2L6
AT174	49 (EA – E)	10/11. 3.42	Missing, cause not known	PO W Andrews; PO J Dunne; Sgt J Gauld; Sgt S Drew. All b.Reichswald	Bombing Essen	
AT175	144	12. 2.42	Shot down by fighters or flak	W/Cdr G Simonds (M); FO H Wathey DFM (M); Sgt J Baness (M); Sgt W Walters (M); Sgt J Canavan (M)	Bombing warships (day)	5A11 2H5
AT176	408 (EQ – A)	27/28. 3.42	Shot down by night fighter N of Terschelling	F/Lt W B Brown RCAF (M); Sgt C D Dunn RCAF (M); Sgt R F Howell (M). Sgt C Brown (M)	Minelaying Weser Est.	

Serial	Units	Date	Cause	Crew	Duty	See
AT177	50	12. 2.42	Shot down by fighters or flak	FO D G Carter (M); Sgt D M Symes (M); Sgt F G Hancock (M); Sgt D W Meeking (M)	Bombing warships (day)	2H6
AT178	106/49/408/ 1404 Flt	24. 1.44	SOC			
AT179	49/408/Cv TB/455	21.12.43	SOC			
AT180	49/408 (EQ – B)	4. 6.42	Crashed 5m E of Balderton; cause not known	PO H J Randall, PO R P Wishart, Cpl K E Godfrey, AC1 J H Chalmers all unhurt	Air test	6E4
	Cv TB	3. 2.44	SOC			
AT181	455	21/22. 2.42	Missing from nuisance raid on SW Germany	Sgt J R Moody (M); PO K H Daly (M); Sgt J Davey RAAF (M); Sgt C Scott RAAF (M)	Bombing SW Germany	
AT182	455/144/455/ 408/1404 Flt	2. 8.43	Missing, presumed ditched	PO W C Liebermann; PO A A Pottinger; Sgt J Cunningham; Sgt W M Crosbie. All missing	Met Flt	
AT183	RAE	6. 3.44	SOC			
AT184	Cv TB/455 (UB – U)	27. 3.43	Crashlanded at Low Eldring, Wigtown	Sgt J O'Connor RNZAF, Sgt D Simes, Sgt R Wilson all unhurt	Ferry flight St.Athan to Leuchars	6D16 2J19
AT185	49/420 (PT – A)	20/21. 6.42	Shot down by night fighter N of Ameland	F/Sgt G Ellis RCAF b.Ameland; PO H Waddell RAAF b.Sage; Sgt B Nidelman RCAF (M); Sgt L Still (M)	Bombing Emden	
AT186	408	11. 4.42	Engine cut; crashed, Norton Pastures, Notts.	Sgt N Marment RNZAF and crew unhurt	Training	6D12
AT187	144	10/11. 4.42	Engine cut; crashed near Sutton Bridge	Sgt H D Smith (I); Sgt W Allen RAAF (K); Sgt G Johnstone (K); Sgt F Woodfield (K)	Bombing Essen	
AT188	144	4/ 5. 5.42	Missing, presumed ditched	F/Sgt J Long RCAF (M); Sgt R Lewin (M); Sgt R F Birchnall (M); Sgt H Webb (M)	Bombing Lister	
AT189	455/408 (EQ – G)	18/19. 6.42	Crashed in field, Servel, N.France	PO S J Baranek RCAF, Sgt E C Squire, Sgt E V Connel, Sgt D Haddock all b.Bayeux	Minelaying Lorient	
AT190	106/49 (EA – A)	10/11. 4.42	Missing, cause not known	F/Lt C Marshall (P); Sgt S Woodhead*; Sgt F Webster (P); Sgt W Collins* *b.Reichswald	Bombing Essen	
AT191	106/49/408 (EQ – A)	1/ 2. 6.42	Shot down in Ysselmeer by night fighter	PO W Charlton b.Harderwyk; PO C I Sandland*; F/Sgt Womar DFM*; Sgt Marland* *b.Amersfoort, Netherlands	Bombing Essen	
AT192	455 (UB – R)	10/11. 3.42	Missing, possibly iced up	PO O Gandy; PO L A Wright RNZAF; Sgt E M Jennings; Sgt G Skinner. All b.Reichswald	Bombing Essen	
AT193	Cv TB/144/ 415 (GX – R)	1.11.42	Hit lorry when landing in fog at St.Eval	FO J H Godfrey (K); PO R G Frederick (I); F/Sgt D M Coates (I);Sgt R A Clarke (I)	Anti – ship B of Biscay	2J30
AT194	144	24/25. 2.42	Shot down by night fighter N of Terschelling	Sgt K Eklund RAAF (M); Sgt G Anderson RAAF*; Sgt J Dobson (M); Sgt P Thomas (M). *b.Lemvig	Minelaying Heligoland	
AT195	16 OTU/14 OTU/ Cv TB/489 (XA – W)	3. 4.43	Developed uncontrollable yaw and crashed off Kecis Pier, Caithness	F/Sgt Price and crew. One injured	Training	
AT196	49	12/13. 4.42	Both engines cut; abandoned, Fitling, Yorks.	Sgt James baled out safely; Sgt Marshall, Sgt Taylor, Sgt J T Smith killed	Bombing Essen	5A3 5A23
AT216	50	5/ 6. 4.42	Crashed after take – off, Thorpe – on – the – Hill, Lincs.	FO Seely – King (K); Sgt Quincy (K); Sgt Skelton (K); Sgt J Hardy (K)	Bombing Cologne	
AT217	49 (EA – S)	19/20. 4.42	Crashed in sea off North coast of Netherlands	Sgt F Sligo b.Sage; Sgt I McLaren (M); Sgt R Webley (M); F/Sgt E Jackson (K)	Minelaying Terschelling	
AT218	144	10/11. 4.42	Missing, cause not known	Sgt H Weeks RAAF*; PO D Johnson*; Sgt S Rickard (P); F/Sgt M Kennedy RNZAF* *b.Hannover	Bombing Essen	3D5
AT219	106/420 (PT – C)	14/15. 4.42	Engine cut after take – off; crashed, Hykeham, Lincs.	PO W J Murray RCAF (K); PO W F McCarthy (K); Sgt K A Johnson; Sgt K A Birch (K)	Bombing Dortmund	
AT220	408 (EQ – G)	3. 6.42	Burnt out during re – fuelling, Balderton		On ground	
AT221	455 (UB – D)	10/11. 4.42	Missing, cause not known	PO R C Roberts RAAF; F/Lt F Keck; Sgt J Wright; Sgt F H Canning RAAF. b.Rheinberg	Bombing Essen	
AT222	144/16 OTU/ 14 OTU/Cv TB/ 5 OTU	6. 5.44	SOC			
AT223	14 OTU	18. 6.42	Port engine cut; under – shot landing, Cottesmore	PO R H C Boosey and three crew injured	Night training	
AT224	408 (EQ – A)	15/16. 5.42	Shot down by warships off East Jutland	F/Sgt J Dillon RCAF*; Sgt R Dreyer (M); PO C Cresswell (M); Sgt W Palmer* *b.Denmark	Minelaying Gt.Belt	2F22 6E1 2D7
AT225	420/408/ 1404 Flt	3. 8.43	Undercarriage collapsed on landing, St.Eval	F/Sgt Bartrum unhurt	Training	
AT226	144	5/ 6. 4.42	Ran out of fuel while lost; crashedlanded Grendon Underwood, Bucks	Sgt Weeks, PO Johnson, Sgt Rickard, Sgt Gowland. Crew set fire to aircraft	Bombing Cologne	3D5
AT227	49/408 (EQ – L)	19. 7.42	Collided with German fighter 50 km NW of Vlieland	PO E Swatton b.Denmark; Sgt R Burley (M); PO H Pearce (M); Sgt E Haville (M)	Anti – ship Dutch coast (day)	5C24
AT228	49/420/ 408 (EQ – U)	28/29. 8.42	Missing; cause not known	PO L Lyon RAAF; PO K Pellett RCAF; Sgt M S Brighouse RAAF; Sgt C Chalken RCAF All b. Liesse, France 759	Bombing Saarbrucken	2A1

Serial	Units	Date	Cause	Crew	Duty	See
AT229	415 (GX – V)	10. 8.42	Port engine failed; overshot landing on early return, Leuchars	PO C D Cross RCAF, PO W H Main RCAF, Sgt G K Summers, Sgt R Zumar all unhurt	T/bombing Norway (night)	
AT230	415 (GX – L)	23. 5.42	Spun into hill Hambledon, Hants.	F/Sgt G Jay; Sgt P G O'Brien; Sgt M Ramsay; Sgt R Oliver. All RCAF and (K)	Training	
AT231	14 FPP	7. 2.42	Crashed at Weston Park, Staffs.	First Officer E L Renicker (USA) ATA (K)	Ferry flight	5J4 5F16
AT232	Cv TB/415 (GX – A)	23. 3.43	Spun in on beach near Thorney Island	W/Cdr G H D Evans and one other (I)	Air test	
AT233	Cv TB/415 (GX – B)	24. 9.42	Lost control and crashed near Leuchars	Sgt K Coates (K); Sgt D R Bird (K); Sgt A H Gibbons (K); Sgt A Higson (K)	Formation training	5F10
AT234	Cv TB/415	24. 1.44	SOC			
AT235	415 (GX – P)	1/ 2. 7.42	Crashed in sea, cause not known; probably flak	FO G James RCAF (M); F/Sgt A Farnie RCAF b.Sage; F/Sgt G Pearce RCAF (M); Sgt T Gibbons (P)	T/bombing Dutch coast (night)	2J24
AT236	415 (GX – R)	29/30. 5.42	Crashed in sea off Vlieland; cause not known; probably flak	FO D Sargent RCAF (M); PO M Edwards RCAF (M); F/Sgt A Conway RCAF b.Sage; F/Sgt J McWilliams RCAF (M)	T/bombing Frisians (night)	2J1 2J23
AT237	415	8. 6.42	Stalled on take – off, Abbotsinch	F/Sgt D B McGregor RCAF; F/Sgt J C Black; both killed	Training	
AT238	Cv TB/415/ 489 (XA – O)	25/26.11.42	Returned early with engine failure; crashed, Skitten	F/Sgt Dubbury, F/Sgt J Lavin and two others; one injured	T/bombing Norway (night)	
AT239	415 (GX – T)	1/ 2. 7.42	Engine cut; hydraulics failed; crashlanded at Donna Nook	F/Sgt B D McComb, F/Sgt M Bloomfield, Sgt R Ennis, Sgt J Labelle all unhurt. All RCAF	T/bombing Dutch coast (night)	2J29
AT240	415 (GX – D)	4/ 5. 6.42	Crashed in sea on return and crew took to dinghy	PO F H Mahn (rescued); F/Sgt J Stirling (M); F/Sgt W Peebles (M); F/Sgt E Thomas (M). All RCAF	T/bombing Frisians (night)	5B11 2J31
AT241	Cv TB/415 (GX – F)	8.11.42	Last seen heading for French coast with engine on fire	PO R F McBride, PO A Robertson, PO P Ramage, Sgt G W Clubb, all RCAF. All POW	T/bombing Biscay (day)	2J25
AT242	Cv TB/415 (GX – F)	24/25. 6.42	Shot down in sea between Rottum and Borkum by ship's flak	WO J A Ridley (M); F/Sgt R Veit (M); F/Sgt G Klarner (M); Sgt H A Clarson (M) All RCAF	T/bombing Frisians (night)	2J24
AT243	Cv TB/415/ (GX – N)	11. 6.43	Collided with AT250 near Tain	Sgt F U Batchelor; PO D Viatkin; Sgt E Johnstone RCAF; LAC E Dockendorff RCAF All killed	Training	5C14
AT244	Cv TB/415 (GX – M)	24. 4.43	Collided with Spitfire BM354 near Tangmere	F/Sgt J C Donald RCAF (K); Sgt C A Brown (I); Sgt A Glass (I); two ATC cadets (I)	Training	5C13
AT245	Cv TB.415 (GX – U)	27/28. 6.42	Shot down by ships' flak off Schiermonnikoog	WO I Garfin b.Sage; F/Sgt V Whelpley*; F/Sgt D Pearce*; F/Sgt B Dakin*. All RCAF *b.Schiermonnikoog	T/bombing Friesians (night)	2J24
AT246	489/415	22.10.43	SOC			
AT247	415 (GX – W)	12. 3.42	Crashed in sea off Isle of Wight; cause unknown	PO Hebert RCAF (K); FO W L Mackay RCAF (M) Sgt Murtha RCAF (K); Sgt H R Hatfield RCAF (M)		Training
AT248	Cv TB/415/5 OTU	4. 6.44	SOC			
AT249	Cv TB/415/489	9. 6.44	SOC			
AT250	Cv TB/415 (GX – K)	11. 6.43	Damaged in collision with AT243 near Tain	Crew unhurt		5C14
AT250		24. 1.44	SOC			
AT251	Cv TB/489 (XA – O)	28/29.10.42	Missing from shipping reconnaissance off Karmoy Island	PO S C Carlson (M); Sgt S Vincent (M); PO J Lawrence (M); F/Sgt H Fisher RCAF (M)	T/bombing Norway (night)	
AT252	Cv TB/489	23. 2.44	SOC			
AT253	Cv TB/489	8.12.43	SOC			
AT254	Cv TB/489	24. 1.44	SOC			
AT255	Cv TB/489	9. 6.44	SOC	Was last Hampden to return from an operation, 0941 on 26 October 1943		
AT256	Cv TB/489	8.12.43	SOC			
AT257	Cv TB/489 (XA – B)	18. 1.43	Shot down by ship's flak off S Norway	FO J Richardson DFC; F/Sgt B McKenzie; F/Sgt A Hyde; Sgt G Gaskell. All RNZAF All POW	T/bombing S Norway (Day)	3D24
AT258	Cv TB/489 (XA – T)	9. 4.43	Dived into sea and blew up after attacking ships; presumably flak	FO S Latta RNZAF (M); FO S H Walpole (M); Sgt J McAllister (M); F/Sgt W Douglas RNZAF (M)	T/bombing Norway (day)	2J22
AT259	Cv TB/489 (XA – W)	4. 4.43	Hit by ship's flak and dived into sea	FO C J Freshney RNZAF (M); F/Sgt W Lanigan (M); Sgt R Ward*; F/Sgt F Bacon RNZAF* *b. Trondheim	T/bombing Norway (day)	2J21
AT260	Cv TB/489	24. 1.44	SOC			

* * * * * * * * * * * * * * * * *

"Gawd, these Pommie bombs ain't 'arf cold, Mum!". Four lads from Oz secure a 250 – lb bomb on the wing bomb – rack of a 455 Squadron Hampden at Swinderby early in 1942. The one at the rear right is literally wearing two hats, with a woollie hat beneath his bush hat.
(Photo: Cee Wadsworth)

A bush – hatted Australian armourer prepares a 1000 – lb GP bomb for loading into the aircraft whose starboard wing bomb rack, at top right, has already been loaded with a 250 – lb GP bomb. The "UB" painted on the bomb rack signifies that it belongs to 455 RAAF Squadron.
(Photo: Cee Wadsworth)

Armourers of No.50 Squadron at Swinderby in early 1942 prepare to load what may be a 1,800 – lb High Capacity bomb. A 250 – lb wing bomb would be hung on a bomb – rack beneath each wing. (Photo: Cee Wadsworth)

A 1000 – lb GP bomb being made ready for loading. Half – a – dozen parachute flares are on the ground at the right of the group. The headgear is of interest; on the left, a Warrant Officer RAF wears a glengarrie, behind him an officer wears a RAF peaked cap, in the centre is an erk with a beret and on the right we have an Aussie bush hat and a RAF sergeant in a glengarrie. (Photo: Cee Wadsworth)

Cecil Mannix and a fellow erk polish the perspex on "D – Dopey" at Waddington. (Photo: Ray Leach)

Jim Taylor, of 44 Squadron, mans the Wireless Operator/Air Gunner's twin Vickers "K" guns, which are without ammunition pans, in summer 1940. (Photo: Jim Taylor)

Pre – war Handley Page photograph of a navigator at his chart table. (Photo via Dennis Thompson)

The wireless transmitter (right) and receiver (left). The receiver had to be installed on its side due to lack of space. Until about the end of 1940, the Transmitter/Receiver 1082 – 3 was in use before being superseded by the Marconi T/R 1154 – 5. (Photo: Nairn Robertson)

Sgt Buchanan, of 455 Squadron, installs his twin Vickers "K" guns into the lower gun position. (Photo: Cee Wadsworth)

PART FIVE: INCIDENTS

5A ESCAPES BY PARACHUTE

Posed photo of a Hampden crew. The pilot, standing on the ladder, has a "Pilot" type parachute slung over his shoulder. The parachute pack became his seat cushion; harness and parachute were an entity. The three men on the wing are wearing their parachute harnesses and each carries an "Observer" type parachute in his left hand. The fastenings at the back of the 'chute that clipped on to the harness hooks can be seen on the parachute of the man on the right; on the other two, the D – shaped handle of the rip – cord which operated the 'chute shows up well, as do the clip – hooks on the front of their harnesses into which the parachute was clipped. The circular face of the quick release box at the front of the harness is also apparent. After landing, it was essential to get rid of parachute and harness as quickly as possible. This was achieved by turning the front of the release and then hitting it with the fist which unfastened the four ends of the harness so that it became free. *The occupants of the upper and lower rear positions would normally enter and leave the aircraft via the door to the lower position. Most navigators clambered into their position through the escape hatch in the floor of the nose. It was usually only the pilot who climbed onto the wing to reach his cockpit. (Photo: Ray Leach)*

It can be seen from a letter that "Bomber" Harris, when Air Officer Commanding the Hampdens of No.5 Group, wrote to Air Commodore Cochran, AOC No.7 (OTU) Group, that he was concerned about the fact that far too many invaluable crews were being lost when the pilots should have ordered them to bale out. Harris took the view that the crews were far more valuable than the aircraft, and that the risk of an abandoned aircraft killing civilians when it crashed, was so remote as to be acceptable. He did not feel that a pilot should attempt a forced landing unless there was a 75% chance of getting away with it, and should not keep his crew aboard unless there was a 99% certainty of success.

The evidence for the AOC's concern may be seen elsewhere in this book, particularly in the "crashed on return" section (2K). It must be pointed out, however, that these sad incidents did not reflect lack of concern, on the part of the pilot, for his crew. The option of changing the comparative security of the inside of the fuselage of a Hampden, with the reassuring voices of crewmates on the intercom, for a lone tumble into the unknown was not an attractive one, and the crew understandably remained in the aircraft in the hope that salvation, in the form of a safe landing, would materialise. From the reports of incidents in which Hampden aircrew saved their lives by the use of their parachutes, it can be seen that there were basically two sets of circumstances in which aircraft were abandoned.

In the first of these, a "dire catastrophe" would arrive with terrifying suddenness in the form of a well – aimed shell from an AA gun, a burst of cannon fire from a fighter or a collision with a balloon cable. The luckless Hampden with, perhaps, its tailplane or half a wing shot away and often on fire soon plunged with ever – increasing speed to its doom. The absence of any response to the rudder pedals or the control column forced the pilot to realise that he had no control over the aircraft. The only course left to him would be to tell his crew to bale out and make his own escape.

The second set of circumstances allowed a more orderly departure, being the culmination of a gradual train of events, often involving a combination of such factors as fuel shortage, engine failure, being lost, adverse weather conditions or battle damage. In the "dire catastrophe" losses, the pilot was often the only survivor and available reports provide only the barest details of the escape.

There is no indication of the appalling swiftness with which a vertically – diving Hampden came to earth. It is worth bearing in mind that a speed of 240 mph is 352 feet per second, and when a diving aircraft reached that speed it was descending at a rate of 1,000 feet every three seconds. Thus, a pilot who baled out at 1,000 feet in such a case, missed being killed by three seconds. The violence of the change from normal flight to diving down out of control, would be such that the navigator, wireless operator and the lower gunner would be drastically affected by the force of gravity as the Hampden dived at an ever – increasing speed. They would be rendered incapable of even reaching for a parachute, let alone clipping it on and making good an escape.

Occasionally, a more experienced (or nervous) crew member would have anticipated trouble and put his parachute on; some, lower gunners in particular, wore their parachutes in any case, in the hope that it might afford extra protection from any bullets that came their way. One other factor that increased the chance of survival for a pilot was the armour plating at the back of his seat; it is self – evident that any bullets that reached him had very likely put an end to the two men in the rear positions.

Before describing some of the incidents in which escapes were made by parachute, it is relevant to describe briefly the differences in the types of parachutes worn by the crew of a Hampden. The pilot had a normal "Pilot" type parachute and harness combined, this harness being buckled on when he took his seat in the cockpit and the parachute pack becoming his seat cushion. A "Sutton" safety harness strapped the pilot into his seat; this fastened over the parachute harness and was quite separate from it, being part of the cockpit seat fittings and capable of being quickly released. All the other crew members used "Observer" type parachutes. The harness of this type of parachute was usually donned in the crew room, the separate parachute being carried to the aircraft and placed in the appropriate stowage position. When the order "Prepare to abandon aircraft" was received, the parachute was clipped on to two hooks on the front of the harness and rested on the chest of the wearer. Thus it will be appreciated that, under normal flying conditions, if operational flying over Germany can be so described, only the pilot would be wearing his complete parachute and harness.

There was a standard drill for the cases in which a relatively organised departure from the aircraft was possible. The first order from the pilot was "Prepare to abandon aircraft"; on receipt of this order, the crew clipped their parachutes on and made preparations to leave the aircraft such as checking that the escape exit was clear. The order to bale out was given as "Jump! Jump!". Each crew member called up the pilot to say that he was leaving the aircraft, then removed his helmet (with its intercom and oxygen leads) and departed. The lower gunner and the wireless operator went out through the door to the lower gun position which was jettisoned by pulling an emergency release cable that withdrew the hinge pins at the top of the door; the navigator opened the hatch in the floor of his position, and dived out. In straight and level flight, neither of these exits presented any difficulty in leaving the aircraft.

The pilot, when satisfied that his crew had left the aircraft, then made his own escape, having first, if flying over friendly territory, set the automatic pilot on a course that would take the aircraft away from populated areas, preferably towards the sea. It was then a matter of sliding his cockpit hood back, removing his helmet, releasing his Sutton harness and departing. One major problem that faced the pilot as he left his cockpit, was that of avoiding the tailplane as he slid down the wing; doing this feet first seemed to be the safest way, although Bob Pearman, who baled out of a Hampden on two occasions, favoured diving out of the cockpit towards the port engine (see P2079 (17); P4359 (18). At least one pilot fell to his death with his parachute streaming out above him, probably with the canopy torn, maybe by shell splinters or with rigging lines snarled up (P4375 (1)). There was also a danger that if a pilot pulled his ripcord too soon, he would become entangled with the tailplane with fatal consequences, as happened to Pilot Officer Snook in P4392 (2).

Abandoned on Return

Twenty-seven Hampdens (3) were abandoned by their crews on return from operations, mainly due to running out of fuel. Fifteen complete crews abandoned successfully from L4104, P2079, P4359, P4402, P4407, X2994, X3000, X3005, X3007, X3056, X3065, AD734, AD745, AD825 and AT173. In five instances, P1340, P4394, P4399, AE394 and AT196, the pilot was the only one to escape and in P1326, P4391 and AD922, he was the only one to be lost. L4054 was the only case of a complete crew known to have perished after baling out, but others may have suffered the same fate. In AD913, the wireless operator was the only casualty, while the lower gunner of X2988 failed to survive.

The pilot of L4067 (12) flew his damaged aircraft back to England and baled out safely over Essex after ordering his crew to abandon the aircraft over Germany. Of the twenty-seven Hampdens abandoned on return, two were from 44 Squadron, four from 50, three from 61, five from 83 and twelve from 144 Squadron, which was thus top of the "abandon league". All of 144's bale-outs occurred while it was a bomber squadron and it is interesting to see that there was no instance of a torpedo-bomber Hampden being involved in a bale-out accident of any kind. There were also no bale-outs from 408, 420 or 455 Squadron Hampdens in accidents or on return, although some of the crew members who survived as prisoners of war no doubt accomplished this by the use of their parachutes.

An official file includes a summary of the results of abandonment of aircraft for the period January to September 1941; thirteen Hampdens were involved from which 35 crewmen baled out, one being killed on landing and seven injured; in seven instances difficulty was experienced in getting out of the aircraft. It is obvious that several baled out because the pilot believed that a crash-landing was inevitable, and in the event the aircraft either landed safely or was repaired afterwards and so was not written off.

Bale-out Incidents

The following accounts of bale-out incidents should enable the reader to appreciate more fully the problems involved, and the varying circumstances in which they arose. In his book "Into the Silk", Ian Mackersey describes three dramatic Hampden incidents (4) in which Hampden aircrew were involved, Pilot Officer Davis's escape from the blazing P4343 from a height of fifty feet over Wilhelmshaven (see 2E3), Sergeant Ken Downing from AD732 (see 3D16) and Pilot Officer Vollmer from P2087 (see 3D15).

L4054 (5) was the first Hampden to be abandoned on return from operations. It had taken off from its base at Scampton at 1915 hrs on 6 April 1940 for a "Security Patrol" of German seaplane bases; nine hours later the crew abandoned their aircraft because of fuel shortage near Whitley Bay on the Northumbrian coast. A matter of two or three minutes more flying time and they would have floated down to earth; instead they splashed down in the cold waters of the North Sea. In the darkness they would first of all have had to release their parachute harnesses before struggling free from the encumbrance of parachute rigging lines and canopies, then inflate their "Mae West" life-jackets which in those early days had to be orally inflated. The final opponent in their fight for survival would have been the deadly enemy - hypothermia. At that time of year, the sea temperature would not be above 40 degrees F and even assuming that no injury had been sustained, that the weather was good and the sea state smooth, their life expectancy was, at most, two hours. The crew of L4054 lost their fight for survival; three bodies were picked up next day; that of the wireless operator was never found.

AD768 (6) was another aircraft in which the destiny of its crew was decided by a few minutes flying time. During the course of their flight to Karlsruhe they had encountered static electricity, which may have had an effect on the aircraft's compasses; but whatever the reason, they became lost on the flight back and failed to find any sign of England in the darkness beneath them. With fuel gauges flickering on the "Empty" mark the pilot was beginning to wonder how much longer they would be able to stay aloft when the navigator reported that he believed he could see hills below them.

There seemed little likelihood that a successful forced landing could be made on such unfavourable-looking terrain so the pilot decided that the sensible course was to abandon the aircraft while they were at least sure they were over land and before shortage of fuel forced the aircraft down. The navigator "guesstimated" their position as being either in the north of England or in Scotland. The crew all made their escape safely and landed on the heather-clad hills. In the ensuing years of internment they had plenty of time to speculate on their ill-luck in having reached the ground on the north-west corner of Ireland at Glenveigh in County Donegal, in neutral Eire, just a few miles over the border from Ulster. The thought might also have entered their heads that they had been lucky to see those hills as a few more minutes flying time and they might have ended up in the Atlantic.

International Law in regard to a crew coming down in a neutral country was interpreted to the letter in the case of AD768 and its crew was interned. Another incident involving neutral Eire, although nothing to do with an escape by parachute, is worth mentioning. It concerned L4045 (7) of 49 Squadron which, on 16 May 1940, overflew England on return from operations and landed on the Curragh race-course near Dublin. The aircraft was undamaged and on this occasion "was given 200 gallons of fuel and told to clear off before they were interned". This offer was accepted with alacrity and the Hampden was soon airborne for Aldergrove, on the other side of the border in Northern Ireland. The crews of AD768 and L4045 fared much better than their comrades in AD730 who were all killed when their aircraft flew into high ground near Blessington, Co.Wicklow, in Eire and were later buried with full Military Honours in the local churchyard. It should also be put on record that many young men from Eire enlisted in the RAF as aircrew and that many gave their lives in the fight against Hitler's regime.

X3065 (8) could lay claim to two unique distinctions on 7 December 1940, the first of which was to be the only Hampden to be lost on operations over Britain as it patrolled the Bristol area in the forlorn hope that it might be able to intercept the Luftwaffe night bombers. Problems, in the form of engine failure, materialised and the pilot, realising that X3065 was failing to maintain height on one engine, reluctantly decided that the best course of action would be to abandon the aircraft while they were still at a safe height, rather than risk making a forced landing in darkness with one engine out of action. X3065 carried an extra air gunner to operate the beam gun position and when the order "Prepare to abandon aircraft" was given, he joined the wireless operator and lower gunner with the intention of leaving by the lower door. This, however, stubbornly refused to open, probably because the hinges were frozen up, despite being thumped, kicked and in desperation finally attacked with the emergency axe The only other way out from the rear of the aircraft was over the side from the upper rear position, which was a risky business because of the danger of hitting the tailplane but the three sergeants managed it successfully and landed safely as did the pilot and navigator. X3065 crashed at Marcham in Berkshire. The second distinction had been that five men had baled out safely from a Hampden and incorporated in that feat was the fact that three of them got out from the upper rear position.

Two of those who baled out from X3065 had pulled their ripcords again within the next three months. In AD745 (9), it was the navigator from X3065; Pilot Officer R N Harrison, who had graduated to the pilot's seat, was beset by troubles which began when one engine began running badly on the return flight from Brest. Bad weather added to his worries and when, as fuel began to run low, the crew saw barrage balloons all around them, the pilot decided that enough was enough and he and his crew baled out, then landed safely; AD745 crashed near Chelveston, Northants. Sgt Harry Edward Turner, the wireless operator on AD745, had previously baled out of P2079 and P4359 so it was not surprising to find that he did remember to release the pigeons before he jumped. What was surprising was that he was using a parachute harness that was several sizes too big for him so that when his chute opened he received a nasty jolt to his reproduction facilities. The harness had "pulled up and covered his eyes so that he was unable to see as he descended". Maybe the jolt had brought tears to his eyes anyway.

Harry Turner, like so many of his Hampden colleagues, graduated on to Avro Manchesters and was one of the crew of a 61 Squadron Manchester, L7396, when it went missing on a Brest operation. Harry had become a Flight Sergeant by this time, his name appearing on the Runnymede Memorial as having no known grave. His tour of operations on Hampdens, with its three bale-outs, does not appear to have made a sufficient impression on his superiors for him to have been awarded a Distinguished Flying Medal which, to an impartial observer, would seem to have been well deserved.

The pilot who was flying as navigator of AD745 was Sgt W E Smith. On that occasion he was lucky to get away with it when his parachute tangled with a balloon cable on the way down but his good fortune deserted him later when, flying as pilot of AE119 (9), he was seriously injured when he crashed, on return from a night bombing raid against

Wilhelmshaven, at Monkwith, four miles north of Withernsea.

The crew of X3005 (10) had a harrowing flight back across the North Sea after their port engine seized up solid when the aircraft was hit by flak over Kiel. After depositing a trail of ammunition pans, guns, the bomb sight and everything else that could be thrown overboard to lighten the aircraft, they eventually reached England and abandoned their aircraft near Driffield in Yorkshire. All landed safely. Three of X3005's crew continued to fly together and failed to return in AD727 ten weeks later. The fourth man, wireless operator Sgt O W J Pearce, had joined another crew and less than a month after his experience in X3005 had baled out again, this time from AD825 near Swindon because his pilot had been wounded and did not feel that he would be able to land the aircraft.

In May 1940, P1326 (11) and L4067 (12), both of 144 Squadron, were lost after being badly damaged by flak. In each case, the pilot, after ordering his crew to bale out, attempted to fly his crippled aircraft back to base.

At 0030 hrs on 12 May 1940, Wing Commander A N Luxmore's Hampden P1326 was hit by severe anti-aircraft fire as he approached his target, Moenchen-Gladbach, at a height of 6,000 ft. The rudder control was shot away, the starboard engine hit and put out of action, the port engine began to lose power; it was impossible to steer an accurate course but the Wing Commander managed to coax his crippled aircraft towards the west. After flying nearly an hour on the ailing port engine, P1326 had lost height until it was down to 1,500 feet and the pilot decided it was time to order his crew to abandon the aircraft while it was still safe to do so. The Germans had at this time not yet occupied France and the three crew landed safely on the French side of the Maginot Line, returning to Hemswell within two days. The pilot's efforts to keep P1326 in the air must have ended tragically in a crash in Belgium as he is buried in Finnevaux Cemetery, 15 km south of Dinant.

Corporal Jolly, the wireless operator of P1326, was lost a month later in P4345 with Wing Commander J J Watts who had taken over as CO of 144 Squadron when Luxmore failed to return. The lower gunner of P1326 was Sgt H Wathey, an air observer, who later survived crash landings when flying as navigator in P2117 and AD901. His luck ran out on 12 February 1942 when, as Flying Officer Herbert Wathey DFM, he failed to return in AT175 from daylight operations against the German battlecruisers in the English Channel. His name, together with the other four members of the crew, appears in the Runnymede Memorial Register.

Just after P1326 was lost, Pilot Officer E Coton took off from Hemswell in L4067 to bomb railway targets between Krefeld and Aachen. Flak hit the Hampden, shooting the rudders and ailerons away and the pilot, no doubt fearing imminent disaster, ordered his crew of three to bale out; two of them landed safely and became prisoners of war but the third man, Sgt C J Shewry, was killed due to problems with his parachute. The pilot, discovering that L4067 would still just about remain aloft, headed in the general direction of England. He eventually reached the English coast where "friendly" searchlights probed the sky and soon held the crippled bomber in the apex of their beams. Realising that there would be little chance of making a safe landing, even if he found an airfield with its flare path lit and also no doubt influenced by the thought that more lethal "friendly" defences in the form of AA fire or RAF night fighter might decide to finish the job that the Germans had started, Coton baled out and landed safely. L4067 crashed at Raynham in Essex.

AD825 (13) also suffered from the attentions of German anti-aircraft guns and was badly damaged by flak during a daylight attack on Brest with its pilot, Sergeant Metcalfe, being wounded in the arms and leg. With the assistance of his navigator, he managed to fly the aircraft back to England, but decided that it would be physically impossible for him to land the aircraft because of his wounds. The crew abandoned the aircraft and landed safely; AD825 crashed near Swindon at 1800 hrs.

Winter 1940/41 and 1941/42

The problems that faced crews on their return to England in harsh winters of 1940/41 and 1941/42 are evident from the reports of crews that baled out. Flying Officer E B Liddell of 50 Squadron took off from Lindholme in X2994 (14) for an attack on Essen; on the return flight, the Hampden was damaged by a heavy AA shell as a result of which the aircraft could not be flown straight and level. To add to his problems, the radio ceased to function. The pilot must have breathed a sigh of relief when he made landfall on the Suffolk coast and later found the aerodrome at Bircham Newton, where to their dismay, an Aldis lamp flashed a red signal to indicate that they were not being given permission to land there. Undaunted, the optimistic pilot headed for West Raynham some ten miles away, where, receiving no reply whatsoever to his signals, he nevertheless continued to circle around hopefully. When, thirty minutes later, the starboard engine failed through lack of fuel, the pilot flew back to Bircham Newton where he was again refused permission to land. Realising that the port engine would not keep going indefinitely, the somewhat despondent pilot ordered his crew to bale out while they were still at a height of 3,000 feet and then continued flying around Bircham Newton in the hope that the port engine would keep going until it became light enough for him to land without a flarepath. However, after a further 25 minutes, the port engine gave up the ghost so there was nothing left for the pilot to do but bale out, which he did at 0700 hrs. None of the crew was any the worse for their experience, but three of them were less lucky on a Mannheim raid a month later when, in X3117 (14) they again suffered the attentions of flak, and were shot down and killed.

The problem of fuel shortage arose, for Sgt E C W Turner of 144 Squadron in AD922 (15), as he reached Norfolk on the return flight from Frankfurt. After coughing and spluttering a little in protest at being deprived of petrol, the starboard engine packed up and when the Hampden refused to maintain height on one engine, the pilot ordered the crew to bale out. They did so and landed safely near Swanton Morley, but Sgt Turner remained at the controls and was killed when the aircraft crashed at nearby Foulsham. The three survivors from AD922 remained together and were lost in AD921 in a daylight minelaying operation off Brest on 13 December 1941.

Sergeant Bob Pearman was one of the many RAF Volunteer Reserve pilots who were called up at the outbreak of war. After a period of training, Bob arrived, in August 1940, on 144 Squadron at Hemswell and, after flying his first six operations as navigator, graduated to pilot. Bob recalled one of his early trips as pilot of P2079 (17) on 8/9 November 1940. The target had been Munich "to disrupt some celebration by the Fuhrer. Extra fuel tanks had been fitted but after a flight of almost ten hours and after experiencing engine trouble all the way home, we ran out of fuel and had to bale out just before we reached Hemswell". A month later Bob overshot when landing at St.Eval with a load of two 2,000-lb bombs on board in P2094 which was later repaired.

P4359 (18) was another Hampden in which problems arose for Bob on 8/9 February 1941. The target had been Mannheim and engine trouble developed on the way back so in order to shorten the flight across the North Sea, they hugged the German and Dutch coasts before crossing the sea. They were not at all certain where they were when their engine problem was compounded by fuel shortage and Bob and his crew were forced to bale out. It was an immense relief for them to see land appear beneath their feet as they floated down. P4359 crashed and burst into flames in the grounds of a large country house near Tavernham in Norfolk, hitting the ground between two Nissen huts occupied by the Army; luckily nobody was injured although the lads in the huts had a rude awakening. Bob actually had to tug at his rigging lines to avoid descending into the blazing wreckage of P4359 and, as happened on other occasions to aircrew who parachuted to safety over their homeland, had to satisfy an irate farmer, armed with a shot-gun, that he was "one of ours".

X3007 (19) piloted by another VR pilot, Sgt Dainty, had problems three nights later when, in the early hours of 12 February, he returned from bombing Hannover. The radio was out of action and the weather decidedly murky, with visibility so poor that, when he did see the lights of a flarepath below, he lost sight of them as he descended into the fog. Climbing to a safe height, he ordered his crew to bale out at 0245 then flew around in the hope that the weather would improve. After stooging around for another half-hour, with his fuel gauges reading "Empty" and with no sign of a landing becoming possible, he reluctantly undid his safety harness, set the autopilot on a course that he hoped would take the aircraft out to sea and took his leave of X3007. He returned safely to terra firma, as had his crewmates, while X3007 flew on into the night for another half-hour before crashing near Bircham Newton, seventy miles away.

The escapes of two of the crew of AD719 (18) and one from AD761 (19) are described in Part 5H which deals with intruders (5H2 and 5H37).

AE394 (20) piloted by Warrant Officer I P Mapp of 50 Squadron had problems with the weather throughout the flight to Koblenz. Severe icing and general bad weather had been encountered, which was not entirely surprising as the whole of Europe was experiencing Arctic conditions with heavy snow over the land and ice floes drifting in the North Sea. During the flight, the navigator lost consciousness but had recovered by the time they arrived back over England where they were unable to locate their own airfield, or any other. Their problems increased when an engine cut

Caterpillar Club Membership Card

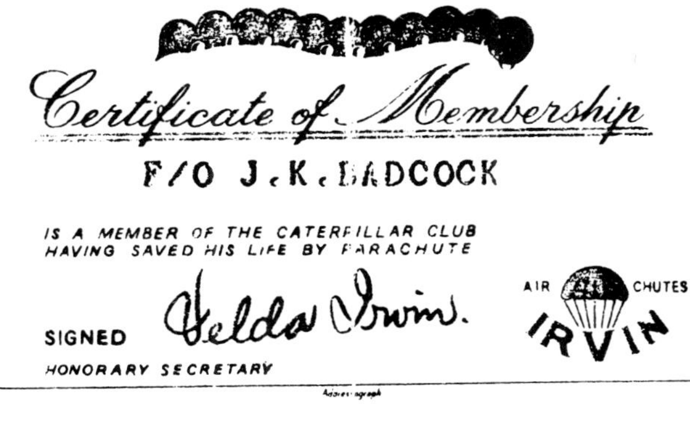

out as they flew around seeking a landing place and when the other engine failed, the pilot could only order his crew to abandon the aircraft and make good his own escape. As the other three crew members were still in the aircraft when it crashed outside the Rowntree Theatre in York, it is believed that the rear door refused to jettison. This could well have been due to the hinges being frozen and the navigator's hatch may have been similarly affected. AE394 appears to have been WO Mapp's fourth Hampden write-off in fourteen months (see also L4120, P4314 and AE229) unless by a remarkable coincidence there were two pilots of that name flying Hampdens at the same time.

Two other cases of severe weather being at least a contributory factor in losses on return were X3000 and P4399. X3000 (21), flown by Pilot Officer C J R Walker of 50 Squadron had experienced severe snow and icing and had turned back for home when within 50 miles of the target, Berlin. After almost ten hours in the air he was back over England and suffering intermittent failing of both engines. Icing up of the carburettor intakes could have been one of the causes of the problem but with fuel running out, the crew successfully abandoned their aircraft at 0430 near Linton-on-Ouse.

P4399 (22): Sergeant Baker of No.61 Squadron was not so fortunate on his return from Cologne when he ran into a severe thunder-storm over England and his Hampden went down out of control. When he realised that he was not going to be able to pull the aircraft out of its dive, he gave the order to bale out but none of the other crew members got clear and were still in the Hampden when it crashed at Deast Hill, Dartford. After this tragic incident, Sgt Baker may have been sent on his next trip as navigator since a Sgt V E Baker, another of the VR pilots, is missing in AE263 a week later.

AT196 (23) appears to be the last of the Hampdens to be abandoned on return. Sgt James of 49 Squadron had brought AT196 back from Essen and was over Yorkshire when both engines cut, probably from loss of fuel. James landed safely by parachute after giving the order to bale out but the other three members of the crew were killed when the Hampden crashed at Fitling, Yorkshire.

P4394 (24): Pilot Officer D H Skinner of 144 Squadron was flying towards the Lincolnshire coast at 2215 hrs on 1 March 1941, returning with engine trouble that had developed on the flight across the North Sea. Thunderstorms had been encountered on the outward flight and these had to be endured again on the way back. An extract from the pilot's report on his horrific experience reads: "My instruments ceased to function on climbing into cu-nim cloud", cumulo-nimbus, or thunderstorm clouds. "The aircraft dived out of control. I tried vainly to regain control, at the same time telling the crew to jump. At the last moment I stood up in the cockpit. The airspeed was around 350 mph and the aircraft in a vertical dive. I got out at approximately 800 feet". P4394 crashed at Wainfleet St.Marys on the Lincolnshire coast, three of the crew being killed in the crash. PO Skinner had been lucky to get out of the aircraft, and even luckier to parachute safely on to land and not end up in the sea a couple of miles away where his chance of survival would have been almost nil.

The official reports of pilots provide only the bare details of their escapes as can be seen in the report of Sgt E Spencer of 44 Squadron in P1340 (25): "I struck a balloon cable (in the Harwich area) The aircraft swung violently to port. The starboard wing dropped. The aircraft dived steeply and I was unable to regain control of the aircraft which was going down in a very steep dive at 260 mph with speed increasing. I baled out at 1,000 feet and the parachute opened just before I hit the sea". There is nothing in the report to describe how he got rid of his parachute and harness and had then found that it was impossible to swim against the strong tide in the River Orwell and that he had had to hang on to a buoy until his cries for help were heard by a Royal Navy Petty Officer, Ben Driver, who dived into the water and came to his rescue.

The "Parachute Emergency Report" written by Flg Offr J A Whitecross DFC of 50 Squadron illustrates how AD834 (26) came to grief over enemy territory without suffering the attention of flak or fighter. He took off from Lindholme at 2020 hrs for a mine-laying sortie to La Rochelle; the last wireless message received from the aircraft reached base at midnight saying that the port engine had failed. The pilot's report explained: "At about 0015 the port engine seized up and caught fire. I extinguished the fire by operating the engine fire extinguisher. The starboard engine began heating up. When it failed, I told my crew to jump. My action prior to jumping was to open the cockpit hood and disconnect my seat harness and oxygen and intercom leads. I held the stick back to reduce speed. The aircraft was at a height of 1,000 to 1,500 feet in a slight spiral and airspeed of 100 to 120 mph when I abandoned by going over the starboard side. I slid down the wing feet first sitting on my parachute so that I avoided the tailplane. After I was clear of the aircraft, I pulled the ripcord. Before the canopy opened it felt as though I was floating on a soft mattress. I was slightly shaken when I landed but unhurt".

The last question on the report asking about "damage to parachute" is answered "Parachute missing". The report was written by PO Whitecross when he returned to England after evading capture and reaching Spain; it is to be hoped that the fact that he had not carried his parachute back across the Pyrenees did not cause him any problems! As for the other three members of AD834's crew, the two wireless operators in the rear of the aircraft are buried in Brittany at Loudeac, some 23 miles from St.Brieuc. They are the only servicemen buried in this cemetery and as the name of the navigator Sgt Martin does not appear on the lists of those killed or missing, it has to be assumed that he became a prisoner of war. James Whitecross,a Canadian from Winnipeg, returned to flying duties with 50 Squadron and only a few weeks later, he and his crew were lost in P4408 when, as a result of engine failure on a search for the crew of a Wellington, he crashed into the sea.

Escapes by parachute were not the sole prerogative of the operational squadrons as may be seen in the following accounts of instances in which Hampden aircrew became members of "The Caterpillar Club" whilst at Operational Training Units.

Flight Lieutenant B P Jones of 16 OTU seems to have been the first man to bale out of a Hampden and survive. He did so on 22 April 1940 from L4148 (27) after getting into difficulties immediately after taking off from Upper Heyford when, although he applied correction, the aircraft did not respond so he jumped from about 800 feet as the aircraft went down out of control, suffering a fractured spine when he hit the ground. Two airmen who were in the Hampden did not get out in time and were killed in the crash.

The problem of an uncontrollable Hampden features in a number of reports and there is a sinister similarity in the details. After Pilot Officer G E Weston took L4169 (28) on an air test, the controls became locked and when the aircraft rolled over on to its back and continued to lose height, the pilot, who fortunately was flying solo, baled out successfully. Three years later almost to the day and 5,000 miles away at Patricia Bay near Vancouver in Canada, Flg Officer P A Hartman DFC, a staff pilot at 32 OTU, was test flying P5427 (29) when the elevators failed to respond. As the Hampden continued its uncontrollable descent, the experienced pilot decided, at 1,000 feet, it was time to go and he parachuted to safety.

Both the previous incidents appear to have been due to the "stabilised yaw" phenomenon which is dealt with in Part 5F. Another such incident occurred to AD798 (30) of 1407 Met Flight when Flying Officer Park struggled to regain control of his Hampden as it wallowed earthwards in a typical stabilised yaw. He ordered his crew to abandon the aircraft but, soon after they departed, managed to get the aircraft back under control and landed safely at Reykjavik where 1407 Flight was then based. FO Ludlow and PO Johnson landed safely but LAC Dixon broke a leg as he hit the ground; they all had good reason to be thankful that it had happened in midsummer as Iceland is not many miles from the Arctic Circle.

Pilot Officer C D Magyer of 16 OTU in P1148 (31) lost sight of the flarepath when carrying out night circuits and landings when visibility deteriorated. His wireless operator did not exactly ease the tension when he reported that he could see tracer being fired in the distance and that he had loaded his guns in case it turned out to be a German intruder on the prowl. After searching in vain for the flarepath, the two occupants of the aircraft baled out and landed safely; there must have been two very red faces when they discovered that the distant "tracer" had in fact been the signal rockets that were being fired to guide them back to base.

Sergeant Butterworth, another 16 OTU pilot, flying P1185 (32), also had a problem with fog blotting out the land beneath him and with R/T and W/T not working properly, he ordered his crew, Sergeants Enderby, Collinson and Hollingsworth, to bale out which they did successfully. Butterworth decided to remain at the controls until fuel ran out in the hope of finding a landing place; his perseverance was rewarded when the airfield at Fairwood Common came into view and he was able to land safely. The good fortune of three of the crew of P1185 came to an end a week later, Sergeants Butterworth, Enderby and Hollingsworth being killed when P5303 came down in a near-vertical dive near Leamington after being recalled to base while on a cross-country flight.

Hereford L6016 (33) of 14 OTU developed engine trouble and the pilot, Sergeant D J Fielder, ordered his three pupils, LACs D Sharp, J L Brooks and A L Wright to bale out; when doing so LAC Sharp suffered a fractured spine when he hit the ground. The pilot crash-landed L6016 at Tugby, twelve miles east of Leicester. On the same day, another 14 OTU pupil, LAC J Gerry, made a precautionary bale out from P1273.

Sergeant Breckell of 16 OTU had a nightmare few minutes when he lost control of P4319 (34) at 800 feet while carrying out practice bombing at the Otmoor range. He ordered his crew to bale out. Sergeant G P McLean from Canada went out at 400 feet and lived to tell the tale but the Bombing Instructor, Pilot Offr A T Greenway, apparently had difficulty in jettisoning the rear door and was found dead through breaking his neck on hitting the ground after baling out at 200 feet. The other member of the crew, the wireless operator, had remained in the aircraft, apparently not having heard the order to bale out for which he had good reason to be thankful as the pilot managed to regain control and landed safely.

The starboard engine of P2142 (35) of 16 OTU failed as Sgt E A Robinson awaited permission to land at night at Upper Heyford but all aboard baled out safely. It was a different matter just over a fortnight later when P5319 (36), another 16 OTU Hampden developed engine trouble and went down out of control on a night cross-country flight. On this occasion, the pilot, Pilot Officer J Maura, was the only one to escape before the Hampden crashed just after midnight near Bourton-on-the-Water in Gloucestershire. Six months later, Wellington R1075 of 16 OTU collided with a balloon cable in the Birmingham barrage; the pilot was PO

Maura. He and two of the crew baled out; three others were killed as the aircraft crashed on the bank of a canal at Castle Bromwich on the outskirts of the city. The last of 16 OTU's fatal Hampden accidents occurred on 16 May 1942 to P2120 (37) when Pilot Officer W H Parr lost control at 10,000 feet while on a night cross-country flight and P2120 spun down and crashed near Saxilby in Lincolnshire at midnight. The lower gunner, whose name is not on record, managed to bale out and came down safely but his three crew mates died in the crash.

Collisions

In the busy skies above the United Kingdom during the war years, it is probably not so surprising to find records of a number of mid-air collisions involving Hampdens. On three such occasions, at least one of the crew saved his life by the use of his parachute.

Two Hampdens of 14 OTU engaged in separate training exercises collided near Market Deeping at 0315 hrs when flying at a height of 4,000 feet. "One aircraft, caught in searchlights, saw the other approaching but was unable to avoid hitting the top of it. The pilot of AE192 was commended for landing his aircraft under difficulty as was the pilot of P5397 (38), Sgt A R Gruber RCAF, for ordering his crew to bale out and for remaining with the damaged Hampden until they had baled out before he made his own escape", noted the OTU's Recording Officer.

Squadron Leader Falconer of 14 OTU had almost reached home in P5321 (39) after taking part in the first "Thousand-Bomber" attack on Cologne. Disaster, in the shape of the black mass of a Halifax returning from the same raid, descended from the cloud layer above the Hampden and literally sliced it in half just behind the pilot's position. On recovering from this terrifying experience and finding that, although he was still strapped in his seat he was no longer in the aircraft, he instinctively released his seat safety harness and pulled the ripcord, landing safely a few seconds later in a field at March, near Ely, in Cambridgeshire. His three crew had no chance of escaping and were found dead in the wrecked Hampden.

One of the other two Hampdens lost by 14 OTU on the same raid, P2116 (40) had been almost shot to pieces by a night fighter over Holland and the pilot, PO T E P Ramsay, had been the only one to survive the onslaught; he baled out of his wrecked and blazing Hampden, and became a prisoner of war.

Hampden P4354 (41) of the Aeroplane and Armament Experimental Establishment collided in mid-air with Buffalo AS430. The pilots of both aircraft, Flg Off W J Carr in the Hampden and Flt Lt S Wroath in the Buffalo, parachuted to safety before the two aircraft crashed at Bower Chalke, 12 miles south-west of Salisbury. The two civilians from the A&AEE, Mr J W Parsons and Mr E L Oxley lost their lives, another instance of the disadvantages of not actually wearing a parachute when disaster struck unexpectedly.

Another rear gunner who mistook the order was Sgt Wilkinson of 420 Squadron RCAF who baled out of AE260 (45) over England when the pilot was having problems through heavy icing. The Hampden went down out of control for several thousand feet and in the confusion, the lower gunner baled out. He landed safely and, as he was flying with the same crew a few days later, must have been forgiven by his crew mates who, no doubt on every subsequent flight asked the question "Well, are you with us for the whole trip this time?".

Several Hampdens landed after operations with one less crew member than when the aircraft had taken off. L4051 (46) of 83 Squadron returned without its navigator, Sgt Newton, who baled out over Mannheim when he believed the order to jump had been given. He became a prisoner of war. Another 83 Squadron crew member who failed to return with his aircraft, P4392 (47), was Pilot Officer J C Bowman. According to the Operations Record Book "While over Kiel Harbour on the way to the target, PO Bowman either jumped or fell out of the aircraft. He has been posted as missing". PO Bowman did in fact become a POW. A third member of 83 Squadron went missing from AE155 (48). He was Pilot Officer J D Gould and he disappeared when the aircraft was thrown about violently in cumulo-nimbus cloud. Sqn Ldr Freddie Newall, the pilot, regained control to find his navigator had gone, his premature evacuation having cost him his life as he lies in Kiel War Cemetery.

Three other men who went missing lost their lives.Air observer Sgt Bill Bissett DFM, a veteran of 61 Squadron, was found to be missing from P2082 (49) over Calais after the Hampden, piloted by Flt Lt Webster, had been turned upside down when attempting to avoid intense flak over the target. His body is at rest in Pihen-les-Guines War Cemetery, ten kilometres south west of Calais.

AD972 (50) from 408 Squadron RCAF came under attack from a Bf 110 over Holland. The pilot, Sgt Dadson, gave the order to stand by to abandon the aircraft and when the attack was over and after violent evasive action, the observer, Sgt B W Palastanga, was found to have baled out. He is buried at Hindeloope on the Dutch coast.

The circumstances under which Flt Lt W E W Holmes RCAF, the navigator of P5344 (51), left the aircraft on the flight across the North Sea after bombing Dusseldorf are obscure. The aircraft was from 14 OTU and piloted by Sgt R L Featherstone who made a forced landing at Mildenhall on return. The body of the unfortunate RCAF navigator was never found and his name is on the Runnymede Memorial Register although, as seems to be the practice, his unit is not given as it was a training unit, even though he was lost on operations.

The average reader may find it difficult to understand how these crew members came to leave their aircraft but anyone who flew on operations will know how easy it was for an intercom plug to pull out of its socket when an aircraft was being thrown about in an intentional evasive action manoeuvre, or due to loss of control from some involuntary cause. If a crew member whose intercom had become disconnected called up the pilot and received no answer, he might come to the conclusion that the pilot and other crew members had already baled out and would hasten to make good his own escape. On one occasion when a Hampden was being thrown about the sky as its pilot attempted to escape some particularly accurate flak, one of the crew is reported to have heard the word "parachute" spoken on the intercom and baled out. It must have been quite mortifying for him to discover, after he had spent the rest of the war as a POW, that the word "parachute" had been part of a request from the pilot to the rear gunner for a parachute flare to be dropped.

Precautionary Bale-outs

Another situation in which a pilot might order his crew to bale out could arise if a pilot had reason to believe that a crash on landing seemed possible because, for example, the warning lights indicated that the undercarriage was not locked down properly. Sergeant G B Dakin of 83 Squadron had this problem and ordered his crew to bale out of L4070 (42) and then made a landing with the undercarriage retracted. The aircraft was subsequently repaired and the crew, Sgts R Bucknell and J H Green and LAC A W Ingham, came down safely so both aircraft and crew lived to fight another day.

Pilot Officer Green, of 49 Squadron, also believed that he had undercarriage problems with X3001 (43) when he returned from operations and decided it would be a wise move to send his crew down in advance. Two of them, Sgts Hirst and McIver, reached the ground none the worse for the experience but the third, Sgt W D F Huggett, broke an ankle on landing. The Hampden landed quite normally. Both PO J H Green and X3001 came to grief over Holland on 10/11 February 1941 after an encounter with a night fighter. It is quite likely that Sgt J H Green who had been with Sgt Dakin in L4070 and the PO J H Green of X3001 were one and the same person.

Mistaken Orders

There are several incidents on record in which one or more member of a Hampden crew mistook the instruction "prepare to abandon aircraft" for the order "Jump! Jump!".

One such incident occurred as AE200 (44) of 144 Squadron returned from a raid on Hamburg. The pilot, Pilot Officer Raw, realised that as they were unsure of their position with weather conditions none too good and the fuel tanks almost empty, it might be necessary to abandon aircraft in the not too distant future. He accordingly alerted the crew to the situation and told them to "prepare to abandon aircraft" so that they would at least be prepared when, as seemed inevitable, the engines failed. Sergeant Carrie, the lower gunner, misunderstood the order and baled out, whereupon the wireless operator followed suit. A few minutes after the rear half of the crew had departed, the front half breathed a sigh of relief as the flarepath of Linton-on-Ouse, in Yorkshire, came into view and AE200 had soon touched down normally. The wireless operator had landed safely and no doubt had some explaining to do but Sergeant Carrie's parachute harness failed in some way and he was killed when he hit the ground.

To conclude this chapter, the following account by Julian Badcock is a good example of how, even when comparatively leisurely preparations for baling out had been possible, an orderly exit by every crew members of the crew was not assured. Julian was flying as navigator of AD734 (52) of 83 Squadron. After completing their mission, a bombing attack on Bremen, they had returned to find thick fog covering their homeland and, with the wireless operator having difficulty with his set and being unable to provide the vital "QDMs" (the course to steer for base), Julian and Flt Lt "Andy" Anderson decided that it might be better to descend to a lower height in order to make it easier to identify a feature such as a river or railway which would help to establish their position. As they did this, there was a violent shudder and the Hampden lurched to starboard, then seemed to leap forward before slewing back on to an even keel. Andy eased the stick back and climbed to a safer height, mystified about what had caused the erratic behaviour of AD734 a few moments earlier and equally anxious about fuel running low. There was no point in waiting until the engines actually failed and Andy gave the order "Prepare to abandon aircraft" then when each of the three crew members had reported that he was ready, Andy spoke the magic words "Jump! Jump!"

The lower gunner pulled the emergency release of his door and rolled out into the blackness of the night, closely followed by the wireless operator. Julian, however, in the nose of the Hampden, discovered that he had a problem. It was impossible to open the escape hatch in the floor of the navigator's position. Andy, meanwhile, was setting the autopilot on to a steady westerly course preparatory to making his own escape when an unexpected thump on the shoulder alerted him to the fact that his navigator was still on board. Julian hastily outlined his problem and indicated that he intended to make his way aft to escape from the lower gun position. After struggling through the obstacle course of the narrow fuselage, Julian arrived at the wireless operator's position where he sat on the floor and slid his feet down to the lower position, pausing for a moment to check that his parachute was properly clipped on to the harness. Satisfied that all was in order, he turned to signal to Andy that he was about to leave the aircraft but the eerie darkness in the forward part of the fuselage prevented him from seeing whether Andy was still at the controls. In the faint light that emanated from the wireless set above him, Julian could see the two carrier pigeons shuffling about in their basket; the thought that the invisible "George", the autopilot, might already be in control of the aircraft added urgency to his movements and leaningdown towards the space where once the lower door had been he tumbled into the cavern of darkness beyond it. He waited a few seconds before pulling the ripcord. The canopy and rigging lines streamed out with bewildering precision, giving Julian good reason to offer a silent "Thank you" to whoever had packed his parachute. The chute took shape and arrested his fall with such a jolt that his badly – adjusted harness jerked taut in the region of his private parts, made his eyes water and caused him to wonder whether his masculine virility had been affected! As he floated peacefully down, Julian had time to spare a thought for those pigeons and was annoyed with himself for not having released them so that they too could have shot a good line when they returned to their loft. Instead they would remain in the Hampden until the fuel ran out and it plunged to its fate after its brief existence as an aerial *Mary Celeste*.

There is no record of seven tons of Hampden and its two carrier pigeons having hurtled to earth in any part of the British Isles so AD734's last resting place is almost certainly at the bottom of the Irish Sea. Julian and his three companions all landed safely in the outskirts of Birmingham, Julian himself landing in Northfield. They discovered that the cause of the aircraft's erratic behaviour had been a collision with a balloon cable. Fortunately the cable cutter in the Hampden's wing had functioned properly and cut the cable, setting the balloon free to drift away across the Midlands. No doubt it ultimately became the target for one of our fighters which would on this occasion be justified in shooting something down that had no Swastika painted on it.

5B IN THE SEA

Members of 408 Squadron survey an inflated dinghy. The dinghy is still secured by its lanyard to the housing in the port engine nacelle; the cover of the housing has been replaced. The spaniel no doubt has its own canine notion of the use for which the side of the dinghy will prove suitable. Left to right: Wg Cdr N.W.Timmerman DFC of Kingston, Ont.; PO T.W.Dench of Kingston, Ont.; Cpl Bangs RAF; FO Houston RAF; Flt Lt P.A.C.Clayton DFC of Vancouver, BC; Gp Capt R.T.Taaffe; Sqn Ldr H.Fowler, Medical Officer, of Toronto, Ont. (Photo: Public Archives of Canada PL – 4545)

Walrus W3026 "AQ – N" of 276 Squadron taxying to pick up a man in a dinghy. The Walrus aircraft of the Air – Sea Rescue Units, affectionately named "Shagbats", were based at coastal airfields and performed valiant service in increasing numbers as the war progressed. The crew of Hampden P1157 were rescued by a Walrus of 277 Squadron; an account is given in this section which also describes the most fortuitous rescue of P5436 by a Stranraer flying boat. The work of other ASR aircraft, including the Lysanders used for dinghy dropping, must not go unmentioned. Nor must the fact that it was at the Lindholme base of 50 Squadron's Hampdens that the "Lindholme" rescue apparatus materialised. This was basically a spare dinghy and a container of supplies that were loaded into a Hampden's bomb – bay and dropped close to a ditched crew. (Photo: RAF Museum P1175)

At least 271 Hampdens ended up at the bottom of the sea, no crew member surviving from 203 of these losses which include 125 instances in which no trace was ever found of the aircraft or crew while in the remaining 78, the body of at least one of the crew was picked up at sea or washed ashore. Remaining are 68 aircraft that went down in the sea from which at least one of the crew survived. Crew members from 28 of these became POWs; the incidents involving P4350, AE193, AE359 and AT257 are mentioned in the "Kriegie" chapter (see Section 3D). Four members of separate crews from L4121, L4126, L4132 and L4134 became the first Hampden POWs on 29 September 1939 when five of 144 Squadron's Hampdens were shot down by German fighters (see 2C2/5).

L4089 and L4090 of 44 Squadron had also come to rest on the sea bed before 1939 had ended, being shot down by Spitfires off North Berwick on 21 December 1939 (see 2C8/9). Seven of the eight crew members were picked up by fishing boats after being kept afloat for half an hour by their "Mae Wests". These last two Hampdens are included in the forty which remain after the 203 "no survivor" and 28 "Kriegie" incidents have been subtracted from the overall total of 271 "in the sea" losses; all forty came down close enough to a friendly shore for one or more of the crew to be rescued and brought back safely home.

Reports from seven of the pilots who ditched Hampdens between 5 August and 5 September 1940 provide a good outline of how the operation was carried out. It should be borne in mind that these successful ditchings all took place in daylight, in summer and in generally good weather, with the added advantage that the fuel tanks were completely empty and that the pilots knew that they were going to have to put the aircraft down in the sea and so the crew were at ditching stations.

Sgt J Unsworth took off from Scampton at 2055 hrs in P4351 (1) to bomb the battlecruiser *Gneisenau* at Kiel. At 0640 hrs next morning, the Hampden force – landed in the sea when its fuel ran out. The pilot put it down a mile from the steamship *Sheraton* ten miles east of Skegness on the Lincolnshire coast, making a wheels – up, flaps up landing at 90 mph.

"The Hampden slid along the surface of the sea and both engines fell out. The aircraft floated for fifteen minutes which allowed plenty of time for the dinghy to be pushed out through the astro – hatch and inflated on the wing. The dinghy was launched over the trailing edge and three of the crew stepped into it. The pilot jumped into the sea and swam, pushing the dinghy away from the sinking aircraft, but was then unable to climb into the dinghy without the help of those already in it because of the weight of his water – logged flying clothing".

The last sentence is a grim reminder of the fact that if a dinghy was capsized by a wave and all its occupants ended up in the water, there would be no helping hand to haul them back on board the dinghy. The crew of P4351 were all picked up by *Sheraton* after a brief stay in their dinghy.

P1324 (2), P2124 (3) and P4380 (4) had taken part in the first raid on Berlin on 25/26 August 1940; all went down in the sea through shortage of fuel on return. Flt Lt N H Svendsen in P1354 had been airborne for 10 hours, 40 minutes and went down in The Wash. "The engines began to splutter ... I told the crew to prepare for a landing on the sea ... we gradually lost height. ... the wireless operator continued to keep in touch with base. At a height of 1,000 feet, I sighted two ships close together and decided to land as close to them as possible. We were challenged by one of the ships which failed to see my reply and, as we were about to land on the water, a four – inch shell was fired at us but luckily it just missed.

The landing was made by holding off just above the water with wheels and flaps up until at a speed of about 60 mph the aircraft stalled. The tail struck the water first and as the machine fell into the sea, the engines were wrenched off. There was quite a swell running at the time but very little water was shipped into the aircraft ... it was difficult to lift the dinghy out of the fuselage due to its bulk and weight but it was eventually thrown off the leading edge and inflated. The four of us stepped into the dinghy without getting wet and were picked up by a boat from a Naval vessel after a few minutes in the dinghy. The Hampden floated for about ten minutes". They were taken aboard the naval ship that had been protecting a Trinity House vessel that was laying buoys and after a day at sea arrived at Harwich naval base where the adventure ended in an extremely boozy night.

On his return to Scampton, Neil Svendsen was questioned by "Butch" Harris, then AOC 5 Group, about whether Neil thought Berlin was too distant a target for his Hampdens. His reply was "No Sir, so long as the wind didn't change to our disadvantage while we were in the air".

It must have been an immense relief to Pilot Officer Pott's crew in P2124 when they at last sighted the coast of England after struggling back from the Dutch coast with an engine out of action after their trip to Berlin. "All ancillary equipment was thrown overboard. A landfall was made at Scarborough and, after flying a few miles inland and finding no suitable landing place, the pilot returned to Scarborough Bay and alighted on the water close to a fishing trawler at 0750 hrs, a quarter of a mile from the shore. The trawler promptly picked up the crew and attempted to tow P2124 to the shore. The Hampden floated for twenty minutes but sank 200 yards from the shore".

The last part of the report has its humorous aspect, although it is not made clear whether the trawler skipper's interest in taking the Hampden in tow was patriotic or pecuniary. Whatever the motive, the action involved a nifty act of seamanship.

Pilot Officer A J G Mills' Hampden P4380 went down in the sea in much the same manner as did Neil Svendsen's P1354. "After it became apparent that a forced landing on the sea was unavoidable, the crew was told to jettison all loose equipment such as the guns and ammunition. By rocking the aircraft on a rolling plane, I was able to keep the aircraft in the air for fifteen minutes after the engines had started to splutter". (Was this the origin of "Rock and Roll", one wonders?). "We ditched with tail down, flaps and undercarriage up. The aircraft came to rest in a very short distance. The waves were about four feet high. The navigator and rear gunner came out through the astro-hatch on to the port wing and the pilot climbed out of his cockpit on to the port wing. The dinghy was inflated and pushed off the trailing edge while the wireless operator, who was in a semi-conscious condition, was helped into it. The Hampden sank in three minutes. We were in the water for seven hours, baling out the dinghy with a flying boot and using the hand pump to keep it inflated. Hand flares were fired to attract the attention of searching aircraft without success. One of the flares failed to ignite. An RAF rescue launch eventually arrived and picked us up."

X2897, 28/29 August 1940 (5): Flt Lt J A Pitcairn-Hill DSO, DFC, who had been awarded the DSO for leading the attack on the Dortmund-Ems Canal on 12/13 August, took off at 2110 hrs from Scampton for Berlin. At 0620 he ran out of fuel and landed on the sea off Skegness close to a trawler which picked up all the Hampden's crew safely after they had taken to their dinghy. Less than three weeks later, this gallant officer's luck ran out when he was shot down in a low-level night attack on Le Havre.

P2123, 31 August/1 September 1940 (6): Pilot Officer Dave Romans, DFC, took off from Waddington at 2005 hrs for Berlin and at 0620 next morning ran out of fuel within a quarter of a mile of Cromer. After a copy-book ditching, the crew took to their dinghy and paddled ashore where they nonchalantly ambled through a minefield in the sand dunes to safety.

P4290, 5/6 September 1940 (7): Four days after the above effort, Dave Romans took off for Stettin at 2035 hrs. At 0620, almost ten hours later, he had to put P4290 down in the sea when land was only a quarter of a mile away. On this occasion they were off Lowestoft and a Navy patrol vessel soon picked them up none the worse for their experience.

Dave Romans was one of a number of Canadian pilots who were flying on Hampdens during the first year of the war; Australians and New Zealanders also featured in the crews. After completing his tour on Hampdens Dave graduated to Avro Manchesters, the Rolls-Royce Vulture engines of which had already earned the aircraft an evil reputation. He must have believed his lucky star was again ascending when he learned that he had been selected as one of the pilots who were to train for operations on the B-17, the new daylight high-altitude bomber that was being delivered from the USA. This "wonder bomber" had been dubbed

A rapid expansion of Air-Sea Rescue (ASR) services took place in 1940 to cope with the increasing number of crews who were going down in the sea, many of whom were being lost because of the shortage of suitable rescue craft. The barrage balloon in the background of the photograph and the airman manning the front gun turret underline the necessity for vigilance when, as was the case with HSL (High Speed Launch) 122 based at Dover, the enemy coast was only twenty miles away. The two gun turrets, similar to those used on the Anson, were manually operated. Each housed a single 0.303 machine gun, a very inadequate defence against the cannon and machine guns of German fighters.

HSL 122 was one of five Dover-based boats that were involved in the Dieppe Raid rescue operations on 19 August 1942, when it was sunk by cannon fire from Fw 190s, as was HSL 123 which came to its aid. HSL 177 went to the rescue of the crews of both craft and picked up fourteen survivors, seven of whom were wounded. Six of the crew of HSL 122 had been killed in the action and six from HSL 147, which had also been sunk, were killed or missing. Five of the crew of HSL 123 became prisoners of war, as did Corporal Bert Noy and Corporal H.Greenhalgh from HSL 122. Bert Noy described how he and his fellow corporal had taken to a Carley float after their craft had been set on fire; they were captured when it drifted ashore two days later. One little-known anomaly of POW existence mentioned by Bert was that whereas aircrew promotion continued while they were in captivity, RAF groundcrew and ASR boat crew remained in the same ranks they had attained when taken prisoner. Bert also wryly remarked that duty on ASR boats had not provided the soft service option he had anticipated.

HSL 122 was built by British Power Boats of Hythe. Powered by three Napier Sea Lion engines, her maximum speed was over 40 knots. The design was known as the "Whaleback" and the craft was 63 feet long. The normal crew was Skipper, Coxswain, three or four deckhands, two wireless operators, a nursing orderly and two fitters for the engine room

(Photo: RAF Museum P2177)

"The Flying Fortress" and began operations with the RAF in the middle of 1941 as the Fortress I. On 8 September 1941, Dave, now a Flight Lieutenant with 90 Squadron, was flying in AN525 when it was intercepted by a Bf 109 at 30,000 feet over Oslo and shot down. None of the crew escaped and all are buried in Oslo at Bygland. AN525 was the second of 90's Fortresses to be shot down over Norway, AN533 having suffered the same fate two days earlier. They had the dubious privilege of being the first of many thousands of American – built four – engined bombers that were to tumble out of the daytime skies over Europe before the war ended, most of these daylight losses being sustained by the B – 17 and B – 24 Groups of the US Eighth and Fifteenth Air Forces in the later years of the war.

P4392, 28/29 August 1940 (8): On the night that Pitcairn – Hill put X2897 in the sea off Skegness, one of his squadron mates, Pilot Officer Clayton, came back from the same Berlin raid and, after being in the air from 2110 to 0750 hrs – a total of ten hours and forty minutes – he just managed to make it to the coast of Norfolk where, like a weary homing pigeon, P4392 landed on the beach on its last drop of fuel. The aircraft was later refuelled and flown back to Scampton. Lessons were learned from the experiences of the crews who survived these early ditchings. One particular improvement was the housing of the dinghy in the nacelle of the port engine instead of being stowed inside the fuselage; an automatic release was also incorporated.

In February 1941, the various services involved in the rescue of crews that were believed to be down in the sea were co – ordinated under the control of a Director of Air Sea Rescue Services at Headquarters, Coastal Command. The manner in which this service operated is evident in a Distress Report concerning Hampden "S" of 144 Squadron. On 24 April 1941, a look – out at Sidmouth on the South Devon coast reported that he believed he had seen a Hampden go down in the sea. The DASRS at Coastal Command at once ordered a lifeboat, a seaplane tender and a drifter to search the area and sent a Blenheim from Thorney Island and a Lysander from Warmwell to assist. The search was later called off when it was proved to have been a false alarm as "S" had landed back at Hemswell oblivious of the fact that it had been the centre of so much activity.

There are also reports of our submarines on patrol off the enemy coast being alerted to keep a look – out for a crew believed to have gone down in the sea in their area as when AE437, 27/28 November 1941 (9), apparently because directional signals from Heston were misinterpreted, went down in a position over 300 miles to the west of Brest. The actual whereabouts of the ditching could be pin – pointed with no greater accuracy than that it was within a forty – mile radius of a position obtained by bearings from a shore station. It was an impossibly large area to search at such a distance from land and the chances of the crew surviving the ditching would not in any case have been rated very high. It is no surprise to discover that no trace was ever found of aircraft or crew.

X3134, 30 June/1 July 1941 (10): Sergeant Woolston reluctantly abandoned his flight to Dusseldorf when engine trouble developed. The Hampden gradually lost height until at 0230 hrs the pilot had no option but to land in the sea. This he achieved successfully and the four men were able to step into the dinghy which released automatically and inflated before the aircraft sank.

Although the wireless operator did not believe that his SOS call had been received at base, they were all quite optimistic about their chances of being rescued within a few hours, but nevertheless decided to ration their small supplies of a few ounces of chocolate, some boiled sweets, thirty – six Horlicks tablets and one – and – a – half pints of water. Four days later they were still in the dinghy and the "food" ration ran out; the water bottle dried up on the seventh day. Their efforts at catching fish with makeshift fishing lines proved to no avail so that once the rations had run out, they were literally living on hope. A number of RAF aircraft were seen going about their business; on the fourth day, three passed about 200 yards from them but failed to see them. Neither of the flares in the dinghy operated and the only means of signalling that remained to them were two small hand mirrors for which sunshine was a prime necessity.

Their luck seemed to be in when three MTBs appeared on the horizon and headed towards them, only to turn away when they were about two miles distant. The reason for their change of course came to light a little later when they noticed a large barnacle – encrusted sea mine just below the surface. No wonder shipping was non – existent! The discovery of the mine spurred their efforts at paddling the dinghy westwards where they confidently expected land to appear at any time. Two of the sergeants used the paddles for a couple of hours, after which they handed over to the other two who had been resting. The dinghy had to be kept pumped up with a hand pump and when for a time the sea became rough enough for the water to wash over the side, it had to be baled out so that even during the night two men kept watch while two tried to sleep. All were affected by cramp through lying uncomfortably on the wet floor of the dinghy, the only protection from the elements being from the wireless operator's Sidcot suit which had been ripped up to serve as a cover.

On the eighth evening, a Hampden escorted by two Hurricanes flew over them at a height of 2,000 feet but failed to see the frantically – waving men in the dinghy. The four sergeants, without drinking water or food, were by now becoming exhausted and their tongues were beginning to swell and crack so that the situation was really quite desperate. At 0820 on the morning of the ninth day, a Hampden passed about a quarter – of – a – mile away at about 2,000 feet and the four men waved and flashed the two mirrors at it. The aircraft turned towards them and flew over before making a slow turn to drop a dinghy in the water nearby. They paddled towards it and eagerly drank from the water bottle after they had hitched the two dinghies together. After signalling "Help coming" on its Aldis lamp, the Hampden that had been their saviour continued to circle the dinghies until two Blenheims and two fighters arrived on the scene and a short time later the ASR launch hove in sight to take them on board. All four at first had difficulty in standing up, which was no great surprise after being huddled in the dinghy for eight – and – a – half days but, after drinking a lot of water and hot tea, they recovered sufficiently to go on deck and take an interest in what was happening. The happy outcome of this particular incident came about in part because the crew in the dinghy acted with good sense allied to the fact that the crew in the Hampden from Waddington that spotted them had been keeping a good look – out. The major factor had to be that it happened in midsummer when the general conditions for survival were good.

The crew of X3134 could not claim the record for having spent more time in a dinghy than any other Hampden survivor as Pilot Officer Mahn of 415 Squadron was adrift for fourteen days a year later, after ditching AT240 (11) (See 2J31).

AE154, 2/3 August 1941 (12) Squadron Leader Freddie Newall was a flight commander on 83 Squadron when he put AE154 down on the sea at 1710 hrs on 3 August. His wireless operator had sent out a SOS when one engine cut out and had maintained contact with Scampton until the other engine packed up. A perfect landing was made on a calm sea.

"The dinghy, housed in the port engine nacelle, failed to release automatically when the Hampden hit the water and had to be released by breaking the perspex cover and manually operating the release. We all got into the dinghy without getting wet and at about midday a Wellington flew close enough to see the flare that we fired and remained with us for a short time. A Beaufort arrived later in the afternoon and stayed with us until the ASR launch from Grimsby arrived and picked us up at 1740 hrs. Our ten – and – a – half hours in the dinghy might have been even less had it not been for the fact that one of the launches was unserviceable". A fine example of an efficient ditching and subsequent rescue, the malfunction of the automatic release being the only slight problem. The rescue was not entirely incident – free, however, as X3123 (13), which had been engaged in the search for the dinghy, crashed at 1030 hrs after hitting a barrage balloon cable at Patrington in Yorkshire. Fortunately only one of the crew suffered injury and the aircraft was later repaired.

The accounts so far of the crews who were picked up from their dinghies show the bright side of the picture, the tragic dark side of which was that most of the crews who went down in the sea in those early days were lost. Distress signals were the last that was heard of many of the Hampdens and a few examples illustrate the nature of the messages that were picked up at base.

X2985, 10/11 November 1940 (14) returning from Danzig was last heard of when it obtained a fix 30 miles from the Norfolk coast; X3063, 16/17 December 1940 (15) was believed to have gone down in the sea off the Isle of Wight; P4322, 5/6 January 1941 (16) was believed to have ditched 10 miles south of Sidmouth; L4185, 4/5 July 1941 (17) sent out an SOS

Ralf Allsebrook, third from left, with his crew of 49 Squadron Lancaster ED587 in 1942. He was one of many ex – Hampden aircrew who graduated to Lancasters, some by way of its predecessor, the ill – starred Manchester. By the time Ralf was killed in action with 617 Squadron on 16 September 1943, he was a Flight Lieutenant with a DSO and DFC. In his Hampden days, he had a lucky escape from P1248 of 25 OTU which went down out of control after hitting a cable in the Sheffield balloon barrage. On 14/15 February 1942, he and his crew were rescued from the sea after ditching in AE397 of 49 Squadron (See 5B 27). (Photo: Julia Allsebrook, sister of Ralf)

but crashed in the Thames Estuary; AD932, 7/8 November 1941 (18) sent at 0845 hrs "SOS; landing in sea 48 miles east of Wick"; AT123, 3/4 January 1942 (19) sent a signal "Landing in sea"; L4180, 29/30 October 1940 (20) came down in sea about 20 miles from Spurn Head at 0700 hrs; P1329, 26/27 June 1940 (21) signalled "Going down 15 miles from the Dutch coast at 0359 hrs". These were but a few of the many Hampdens that went down in the sea with the loss of the complete crew.

L4064, 12 April 1940 (22): The action in which this aircraft was severely damaged is described in Section 2C13. After limping back across the North Sea, shepherded by Sqn Ldr D C F Good, until the last of the fuel had drained from its punctured tanks, it was forced to alight on the sea.

The crew were seen to have taken safely to their dinghy by Sqn Ldr Good who then circled around while his wireless operator transmitted so that the dinghy's position could be accurately fixed. He stayed there a few minutes longer before his own fuel gauges reminded him that it was time to head for home. Next morning, a Hudson of 220 Squadron located the dinghy at 0725 hrs in a position 100 miles east of Flamborough Head and remained over it until 1240 hrs when it flew back to Thornaby and landed with almost empty tanks. The search was resumed next day with two Ansons from Waddington assisting in the search; the one from 50 Squadron ran out of fuel through staying in the search area too long and crashed near Grimsby, all of its crew being injured. An empty dinghy was seen in the search area a few days later. The bodies of three of the missing crew were never found but that of the lower gunner lies in Kiel War Cemetery. Whether he was picked up by a German patrol craft or was washed ashore on the German coast is uncertain.

No.50 Squadron's sad story that began with the loss of four out of five Hampdens on 12 April, followed by the injuries to the Anson crew on 13 April, continued when L4065 failed to return from the first minelaying flight to the Baltic.

L4065. 13/14 April 1940 (23): An aircraft that was believed to be L4065 was plotted by radar as it flew back towards the Lincolnshire coast until it disappeared off Mablethorpe; no trace of the aircraft or crew was ever found. The pilot, Flt Lt Robert Cosgrove, was the son of the Premier of Tasmania. By a sad coincidence, Pilot Officer J B Bull, who had been the pilot of L4064, originated from South Australia while Sqn Ldr Good, who survived the debacle of 12 April was also an Australian. His luck ran out just over a year later when he came to grief in AD728; the name of all three Australians appear on the Runnymede Memorial register.

P4352, 3/4 July 1940 (24): P4352 took off from Waddington at 2122 hrs to plant a mine to the south of Langeland Island in the Baltic. At 0130 hrs, its W/Op reported the Hampden as flying on one engine. He obtained a number of fixes, the last of which was 45 miles north – west of Terschelling after which no more was heard and the aircraft was presumed to have gone down in the sea. At 0550 hrs, two Coastal Command Hudsons from 206 Squadron took off to search for P4352; they reported later that they had found no trace of the Hampden and were flying to the north of Texel. That was the last that was heard of Hudsons N7386, piloted by PO Henderson and P5162 by PO Lester; both almost certainly fell victim to German fighters and the names of their crews and three of the crew of the Hampden are on the Runnymede Memorial. The body of PO W S Bull who was the lower gunner of P4352 was washed ashore at Klegod on the west coast of Denmark on 27 July and was buried at Lyngvig Cemetery on 4 August.

P4383, 31 July/1 August 1940 (25): The wireless operator of this Hampden had signalled that they were returning on one engine at 0410 hrs and maintained contact with base until the aircraft went down. A search aircraft located the dinghy with the four crew aboard at 1900 hrs on the same day in a position 30 miles east of Flamborough Head.

Twenty – one search flights were made on the next day but no trace of the dinghy or its occupants could be found. Further flights were made on the following day but eventually after forty – six rescue flights had been flown, the crew of P4383 was given up as lost. The body of the pilot, PO S S Bell, was washed ashore six weeks later at Petten on the Dutch coast, over 150 miles from where the dinghy was first reported and he is buried at Bergen – op – Zoom; no trace was ever found of the other three in the crew.

AD843, 23/24 July 1941 (26): A routine wireless message from AD843 had fixed the aircraft's position as being off the Norfolk coast as it returned from Frankfurt; nothing further was heard and the Hampden failed to return to base. The bodies of the wireless operator and the lower gunner were picked up eight miles east of Cromer but no trace was found of the pilot or navigator.

The exact number of Hampden aircrew who went into the sea and took to the dinghy but perished before help arrived will never be known but one particularly tragic incident occurred during the attack on German warships in the "Channel Dash" on 12 February 1942. A Hampden was seen to crash off the Dutch coast after being damaged in the attack and at least two of the crew were seen to have taken to the dinghy before the aircraft sank. The survivors were quite close to the shore but the layer of broken slushy ice that covered the sea prevented a boat from setting out to pick them up. It also made it impossible for the dinghy to be paddled ashore to safety and yet the ice was not thick enough for the men to walk across. In the Arctic conditions that prevailed, the men died of exposure. Two days later, the crew of AE397 were lucky not to suffer the same fate.

AE397, 14/15 February 1942 (27): An engine had failed on the return flight from Mannheim but the Hampden struggled on until the English coast was almost in sight before the other engine failed and it went down in the sea west of the Needles on the Isle of Wight. The crew successfully took to their dinghy but were soon suffering from the bitter coldness of that Arctic February night and the effects of sea and spray. They owed their lives to the crew of a Beaufighter returning from patrol who "noticed lights on the water" and alerted the rescue organisation. A few hours later, "a Coastguard boat" arrived and picked the crew up, three of them suffering from frost – bite. The account of the rescue in the report of services of the Royal National Lifeboat Institution offers a slightly different account of the rescue, as follows: "Ventnor, Isle of Wight. At 2.35 in the afternoon of the 15th February 1942, a member of the coast life – saving corps reported what appeared to be a rubber dinghy four miles SSE of Woody Point. Three men put out at once in a pulling boat. The coastguards were told and so too were the military in case the boat was not back by sunset. An air/sea rescue launch also put out and a Walrus amphibian. The launch rescued four airmen whose aeroplane had come down in the sea at 2.30 that morning while returning from a raid on Mannheim. The launch towed the pulling boat back to Ventnor. Rewards to the three men – 2.5s.0d"

The truly amazing story of how the three man crew of AE309 survived for eleven days after ditching off Greenland is related in Part 5K14.

P1157 (28) took off from Docking for a night attack on a convoy, in the course of which it was badly damaged by flak and subsequently went down in the sea on the return flight; the crew took to their dinghy at 2257 hours on 18 February. The record book of RAF Martlesham Heath and of 277 Squadron report the rescue, on 20 February, with justifiable satisfaction, thus:

"Two Walrus aircraft and a Defiant conducted a search for the crew of a bomber reported down in the sea. Walrus X9526, crew Flying Officer L J Brown, Flying Officer D Sheppard and Sergeant D Lunn, took off from Martlesham Heath at 1530 hours and at 1745 hours was on the point of abandoning the search when the dinghy was spotted. The Walrus alighted and picked up the four men who were in surprisingly good shape considering the fact that they had been in the water for 38 hours. The Walrus had great difficulty in taking off with its total of seven men on board but it did so and landed back at base at 1845 hours".

The report concluded proudly "A RED LETTER DAY for the "They also serve who only stand and wait" lads", a sentiment that was, no doubt, heartily endorsed by Brown and his crew, who appear to have been the only Hampden crew to have been picked up by a flying boat after going down in the sea whilst flying from the United Kingdom.

P5436 (29): the records of No.3 Operational Training Unit at RCAF Station, Patricia Bay, reveal that the crew of P5436 from 32 OTU, also based at Patricia Bay, was rescued by one of 3 OTU's Stranraer flying boats; the Unit Daily Diary for 15 November 1943 reads:

"At 1135 hours, a Hampden aircraft of No.32 OTU crashed into the water whilst on a training flight. Squadron Leader Cowan, who was airborne in Stranraer No.947, witnessed the crash and immediately landed on the sea and taxied to the scene of the crash and, with the help of his crew, rescued the four Hampden men".

The Hampden crew may have been even luckier than they thought as the same 3 OTU diary entry refers to flying training being interfered with by high winds and rough water; it is obvious that, had conditions been just that bit worse, the Stranraer would not have been on hand to assist.

X3058 (30) is a grim example of how other Hampden crews were not so fortunate as some of those previously mentioned who survived after taking to their dinghy in the depth of winter. The aircraft went down in the sea off the Norfolk coast in the early hours of 29 January 1942, probably through running out of fuel as it returned from an attack on Munster. A rescue boat found the dinghy sixteen hours later; only the pilot was alive, his three companions having already succumbed to the intense cold.

X2971 (31): The crew of this aircraft were completely out of luck; according to radar reports it had probably gone down in the sea off the Lincolnshire coast. No sign of the crew or aircraft wreckage was found next morning when three of 61 Squadron's Hampdens carried out a search of the area, but one of the searching aircraft encountered a Heinkel He 115 floatplane which it shot down "after an exciting twenty – minute chase". The victorious Hampden pilot, PO Massey, watched the Heinkel crash onto the sea and break up. The crew of the floatplane sat disconsolately on the wreckage with no sign of a dinghy, so PO Massey dropped his own dinghy near his erstwhile foe. The Germans used the dinghy to good advantage and paddled ashore at Hemsley in Norfolk, 22 hours after they had been shot down.

Sergeant Horace C Loadsman, pilot of X2971, whose name together with those of his crew appear on the Runnymede Memorial, was a 23 – year – old RAF Volunteer Reserve pilot from Brighton; his name had already appeared in the records when he was carrying out his flying training on Hampdens with 76 Squadron at Upper Heyford on 6 March 1940. His natural exuberance obviously over-rode his better judgement thus:

"Whilst carrying out unauthorised low flying in P1179, Sergeant Loadsman flew into an 11,000 volt electricity supply line near the Oxford to Witney by-pass at Eynsham. The aircraft was damaged. Sergeant Loadsman was arrested and charged".

At a subsequent court-martial, he was reduced to the rank of Aircraftman Second Class. An entry in the 16 OTU record books shows Horace having resumed flying training on 20 June 1940 as a sergeant and, after successfully completing his training with no further electrifying experiences, he was posted to 106 Squadron, which was at that time the Aircrew Pool Squadron, on 15 July 1940 and thence to 61 Squadron.

The rapidity with which the human body succumbed to hypothermia, and the unrecognised gallantry of a sergeant pilot, are equally evident in an account from one of his 50 Squadron contemporaries, wireless operator/air gunner Nairn Robertson. Nairn describes how AE428 had been hit by flak as it dropped smoke-bombs to cover the dawn landing of troops from landing craft at Maaloy in northern Norway.

AE428 (32) crash-landed on the water and sank so quickly that only three of the crew were able to escape; the dinghy failed to release and went down with the aircraft. The three sergeants, kept afloat by their "Mae West" lifejackets, clung together in the icy water. The pilot, Sgt R N Smith, described by Nairn as "a tall super-fit Rhodesian" was the only one of the trio who could swim and, after telling the other two to hang on to him, he struck out valiantly towards the shore. The water was intensely cold, with a strong cross current making progress difficult, but the sergeant, with his two buddies in tow, swam doggedly on until one of the craft that had been involved in the landing operation hove into view. The three men waved and shouted frantically and to their immense relief the boat turned towards them and soon had them safely aboard. Sergeant Smith made a speedy recovery but his two chums had been so affected by the cold that they died soon afterwards, even though they had only been in the water for a matter of fifteen minutes or so.

Nairn Robertson met up with Smithy when they were both back at Skellingthorpe a week or so later and heard the sad story of how the other three men in the crew had come to grief; he particularly remembered Smithy telling him how the Naval chap who had first heard their cries for help had suggested that aircrew should have a whistle attached to their "Mae Wests" as it would be a much better means of attracting attention than shouting. Shortly afterwards all aircrew were issued with whistles and Nairn believed there may have been a connection with the AE428 incident. The Rhodesian pilot's gallant and unselfish action went unrecognised, as may be seen in the details in the Runnymede Memorial Register which read:

"SMITH, Flt Sgt Reginald Newbald, 778353. R.A.F.(V.R.). 50 Sqdn 7th February 1942. Age 23. Son of Charles N Smith and Beatrice Eleanor Smith of Wankie, Southern Rhodesia."

The operation on 7 February 1942 on which Flight Sergeant Smith and his crew were missing in AE306 (33) was a daylight minelaying operation off the Friesian Islands, for which they took off from Skellingthorpe ar 1211 hours. This operation, in which AD824 and AE392 of 144 Squadron were also lost, is described in Section 2H.

The incidents recorded in the book of services provided by the Royal National Lifeboat Institution during the war years give proof in profusion of the aid that the lifeboatmen, sometimes using private motor boats, gave to crews of aircraft that had gone down in the sea; AE397 (27) has already been mentioned. Two incidents concerning Hampdens P1321 and L4085 appear on the same page of the records for 1940, although of course the identity of the aircraft is not given.

P1321 (34) at Haisborough (i.e.Happisburgh), Norfolk: "About 3.30 in the morning of 26 July 1940, an aeroplane crashed in the sea off Haisborough Gap. The Cromer lifeboat and a motor boat manned by two brothers and an R.A.F. officer put out to the rescue. They found nothing but wreckage and then, as the tide ebbed, the dead airmen were found among the wreckage." Two paragraphs later another aeroplane incident turns out to be that of L4085.

L4085 (35) at Aberystwyth, Cardiganshire: "At 9.30 in the morning of the 1st August 1940, the coastguard asked the owner of the motor boat *Emerald Star* to go out to a spot where aircraft were searching for survivors from an aircraft that was down in the sea. The weather was fine and the sea smooth. Three men put out in the motor boat. They picked up one man alive, then another unconscious who could not be revived, and then a third man who was very exhausted and whom, with much difficulty, they rescued from a rubber boat. A motor boat from New Quay also went out and the Aberystwyth lifeboat picked up the body of the fourth crew member."

The report from Aberystwyth on August 1st adds more details: "At 6.30 a.m. the coastguard reported Very lights several miles to the S.W.by W. Police helped in getting the life-boat crew assembled and with the help of soldiers the motor life-boat *Frederick Angus* was launched at 7.26 a.m. The life-boat picked up one body".

It is reported, in one of the other R.N.L.I. service reports, that "under wartime conditions the firing of maroons to summon the lifeboat crew was forbidden". This aspect was mentioned by Don Seager, the observer/lower gunner of L4085, who had been the man "rescued with much difficulty from a rubber boat" by the *Emerald Star*. Don had been informed that a delay in launching the lifeboat had been caused by the maroons not being fired "because there is a war on". A man on a bicycle had been sent to round up the lifeboat crew!

The circumstances of L4085's ditching were that, on returning to where England ought to have been, dense fog enshrouded the landscape beneath it until a coastline appeared just ahead which was identified as Southern Ireland. The prospect of internment did not appeal to the crew, so the pilot turned on to a reciprocal course and headed eastwards towards Wales. The engines began to falter just as the mountains of Wales appeared on the horizon, then cut out altogether. Sergeant Farmer had no option but to put L4085 down on the sea, which was calm with just a slight swell.

There was sufficient buoyancy in the empty fuel tanks to keep the Hampden afloat for several minutes; the pilot took advantage of this to fire off Very cartridges whilst the wireless operator fired the rear upper guns as an extra means of attracting attention. The Very lights were seen by the coastguard, but that was the only good news for the crew. The bad news came when the dinghy burst as it inflated; Don Seager hung on to what was left of it until help arrived. The pilot and wireless operator tried to swim for the shore and were drowned; the navigator was picked up alive by the *Emerald Star*.

After a spell in hospital at Aberystwyth, Don recovered sufficiently to be sent home to South Norwood for a spell of sick leave but the story does not end on that happy note. The Gods of Destiny, apparently resentful that Don had survived, arranged a more bitter ending to the affair which began as he stood on the platform at King's Cross station in London, awaiting the train that would return him to Lincoln and 44 Squadron at Waddington. The air raid sirens wailed the prelude to the last act while Don quite rightly assumed that "Jerry was after the docks again" and remained on the platform quite happy to be a spectator of bursting anti-aircraft shells and not, as on his adventures over the Fatherland, the target for their angry explosions.

The end to the Aberystwyth story and to Don's service in the RAF arrived on the platform close to where he stood, in the form of an anti-aircraft shell that had failed to explode in the air against the enemy. This malevolent missile chose instead to detonate on the platform and it caused such severe injuries to his legs that one had to be amputated above the knee so that Don was subsequently invalided out of the RAF.

One slight consolation for Don was that he became one of the founder members of the "Goldfish Club", the membership requirement for which was that the applicant's life had been saved by the use of a dinghy or "Mae West".

The Lindholme Rescue Dinghy

It is appropriate to end this section with a mention of the apparatus that became known as the Lindholme Rescue Dinghy; it was developed in late 1940 when Group Captain E F Waring was Station Commander at Lindholme, where 50 Squadron was based. The device consisted of a large packed dinghy to which, on either side, had been secured by a 30-yard length of buoyant nylon cord, a float chamber containing survival rations and warm clothing. A further cylindrical float was spaced between the dinghy and the float chambers.

The packed dinghy and its connected floats and float chambers were loaded on to the bomb racks in the Hampden's bomb bay so that they could be dropped on a timed sequence in the order: 1: flotation chamber, 2: float, 3: dinghy, 4: float, 5: flotation chamber. Before releasing its load the Hampden dropped a smoke-float to ascertain the wind direction to ensure that, when the Lindholme Dinghy apparatus was released, it would enter the water downwind of the crew in the dinghy and at right angles to its drift path. Once the dinghy pack entered the water a soluble plug dissolved rapidly and brought a gas cylinder into operation to inflate the dinghy and, if all went according to plan, the distressed crew drifted down towards the apparatus. Once the nylon line had been picked up they were able to transfer to the larger dinghy and avail themselves of the food and clothing in the float chambers which also contained flares and other signalling equipment.

Instructions regarding the Lindholme Rescue Dinghy were circulated to all RAF Commands and "Gus" Walker (later Air Chief Marshal Sir Augustus Walker), who was flying from Lindholme in 1941, recalled a successful drop of a dinghy in June of that year. He was most dismayed when the crew of the dinghy that he was trying to assist made frantic efforts to paddle away from the apparatus as it drifted towards them. The Air-Sea Rescue launch for Grimsby eventually arrived and picked up both Lindholme dinghy and the non-co-operative dinghy occupants who turned out to be the crew of a German Heinkel He 111 bomber. The Germans no doubt mistrusted this fiendish contraption that the wicked British had dropped to ensnare them.

5C COLLISIONS

Collisions in the air were the cause of a number of Hampden casualties. They frequently occurred in the vicinity of their own aerodrome, involving aircraft that were taking off, landing or preparing to land. Although any loss of life through a collision in the air was tragic it seemed particularly so in the case of two Scampton Hampdens.

X3121 (1) of 83 Squadron and AD967 (1) of 49 Squadron had returned from operations against Dusseldorf on 24/25 August 1941 and were awaiting permission to land at Scampton when they hit one another at 0250 hours and crashed to the ground at Whale Jaws Farm near Hackthorne with the loss of both crews. The intruders of I/NJG2 were regular visitors to the Hampden bases at this time and it was common practice for our own aircraft to fly with their navigation lights switched off if the Ju 88s were thought to be prowling around the area. It is not certain whether one or both of the Hampdens did have their lights switched off but it is obviously a factor that has to be borne in mind.

A cruel twist of fate followed the loss of the previous two aircraft just over an hour later on that sad August night, when AE223 (2) landed at 0400 hours after carrying out an intruder operation in the Wesel area. The Hampden had taxied back to its dispersal when, as the crew prepared to leave the aircraft, it blew up, killing four of 83 Squadron's ground crew in addition to the four aircrew of the Hampden.

The explanation for the accident was thought to have been that one of the wing bombs had failed to drop over the target and had remained "hung up". The bomb release switches had been left "on" after the attack; they remained inoperative whilst the bomb doors were closed but when the bomb doors were opened at dispersal the live bomb dropped off the wing and exploded with grievous consequences.

Death came to the men in the previous incidents at Scampton at the very end of an operational flight. To two of the crew of P5393 (3) a few months later it arrived before they had even reached the take-off point. The incident highlights the problem of taxying around the airfield from dispersal in the days when there were not even perimeter tracks, let alone perimeter track lighting. P5393 collided with another 83 Squadron Hampden AE374 (3) as both aircraft taxied towards the flare path on 14/15 December 1941. The mine carried by P5393 exploded before the crew was clear of the aircraft, killing the pilot and the wireless operator and injuring the navigator and the lower gunner. The crew of AE374 escaped uninjured.

The miraculous escape of Squadron Leader Falconer from P5321 after a collision with a Halifax over Cambridgeshire is described in "Escapes by Parachute" (5A 39), as also are the incidents involving the collision between AE192 and P5397 (5A 38) and that of P4354 of the Aeroplane and Armament Experimental Establishment and a Buffalo fighter. (see 5A 41)

The busy Hampden Operational Training Units had their full share of collisions, most of which involved aircraft that were taking off or landing, although the first of the OTU collisions on 26 May 1940 involved L4156 and L4158 from 14 OTU which collided over the Bristol Channel as they carried out an Air Gunnery exercise, with the loss of the seven men in the two crews (4).

The problem of controlling take-offs and landings from the grass aerodromes of the training units is evident in the reports of some of the accidents. L4138 of 14 OTU and P4339 of 61 Squadron took off from Cottesmore on converging courses and collided when they were about twenty feet from the ground. Of the seven men involved, two in P4339 were killed, two in each aircraft were severely burned and one in L4138, probably the wireless operator, escaped with cuts and bruises (5).

P1302 of 16 OTU was coming in to land at Upper Heyford when it collided with P1149 also of 16 OTU which was taking off on a converging path. The two Hampdens locked together, burst into flames and crashed onto the aerodrome. Wing Commander W C Sheen and Pilot Officer J C Cunningham, who had witnessed the accident from the ground, rushed across to the blazing wreck of P1302 and dragged the badly injured pilot from his cockpit. Their heroism proved to be of no avail as the unfortunate pilot, Flying Officer S Moxham, died later. The only survivor from the four men involved was the injured pilot of P1149 (6).

There was another bad collision at 14 OTU when P1301 and P4303 (7) hit one another at Saltby, the Cottesmore satellite airfield. Not one of the eight men in the crews survived the crash. The pilot of P4303 was the holder of a DFM and probably "resting" after a tour of operations and two of the crew had come all the way from New Zealand to meet such a sad end.

The collision between AE138 and AE147 (8) both of 16 OTU on 20 August 1941 is grim proof of how a simple misunderstanding could result in a routine night flying training programme being marred by an awesome accident. It is recorded that "AE138 taxied onto the flare path at Croughton, the Upper Heyford satellite, and was struck by AE147 which was taking off. Both pilots apparently believed that they had been cleared for take-off." Two men died, four were seriously injured and two slightly injured as a result of this mis-interpretation of a green light signal from an Aldis lamp. A terrible price to pay.

Other accidents at the OTUs involved Ansons being assailed by their heavier brethren, the Hampdens or Herefords. Sergeant Forster for example of 16 OTU in Hereford L6027 (9) on 13 January 1941 must have been rather pleased that he didn't have to pay for the damage that an engine failure on take-off caused when he "collided with Hampden P2141 then ploughed through a hedge and hit Anson N5267. All three aircraft were severely damaged." No one was hurt. P5345 (10) of 16 OTU on 13 October 1941 "landed on top of Anson R9668 which was stationary on the airfield. The Hampden pilot was injured and the pilot of the Anson suffered slight head injuries".

14 OTU was not to be outdone in the "Anson bashing" business as P1289 proved on 15 July 1941 when it "landed without permission and crashed into Anson R3386 which was taking off, causing injuries to the Anson crew" (11). The Hampden's association with the Spitfire on the other hand, seemed to be even more unfriendly in that it was shot down by them, shot at by them and, of course, in collision with them.

AD939 (12) of 44 Squadron took off for a routine air test on 31 August 1941 and was involved in a collision with a 412 RCAF Squadron Spitfire from Digby, piloted by The Hon P Hughes. The Hampden crashed near Waddington and three occupants lost their lives; one of them, AC2 F B Prest, had more than likely only gone along for the ride.

P1204 (19) of 1404 Met Flight was minding its own business at dispersal when a Whitley, BD276 of 10 OTU, crashed into it at St Eval. St Eval was also the scene of a spectacular accident that involved four Hampdens of 415 Squadron when N9096 came in to land after completing an anti-submarine patrol at 1735 hours on 22 June 1943. It swung off the runway and crashed into X3061 which was burnt out; X3115 was also damaged beyond repair and L4144 damaged. Not surprisingly N9096 did not survive this debacle but what is surprising is that no one is reported to have been killed in the incident (20).

The circumstances in which AE157 (25) of 50 Squadron was written off on 3 September were shrouded in mystery until photographs taken by Eric McCabe, who was Engineer Officer at Waddington in 1941, were handed to Squadron Leader Ray Leach, the Waddington and 50 Squadron historian. The crash card for AE157 confirms that it was Cat EFB, i.e. written off in an operational crash, on 3 September 1941, yet according to the 50 Squadron record book it was not one of the fourteen Hampdens operating from Swinderby that night of which five crashed on return due to fog. The record book does relate that AD852, piloted by Pilot Officer Ferrie crashed into a dispersed aircraft at Waddington at 0320 hours owing to fog. The answer to the mystery would appear to be that the number of PO Ferrie's aircraft is incorrect in the ORB. Possibly AE157 was the reserve aircraft that night and PO Ferrie had changed to it after finding a problem with AD852. The identity of the 44 Squadron aircraft KM-? is not apparent from a study of the record books, but the only real possibility for the serial number of 44 Squadron's aircraft is that it is X3025, the movement card write-off date for which is 8 September 1941. X3025 had been regularly flying on operations in July and August 1941 and last appeared in the 44 Squadron ORB on 31 August. The photographs suggest that AE157 was still moving at a fair speed when its starboard wing hit X3025's port wing at right angles to its fuselage and then demolished both engines and the nose of the aircraft, in the process of which its own starboard wing was ripped off, the fuselage broken aft of the lower gun position and the tail badly damaged. The evidence that X3025 had been peacefully picketed on the dispersal can be seen in the form of the securing strop under the port wing.

After the Hampden ceased operating with Bomber Command it was used by Coastal Command as a torpedo-bomber from mid-1942 until the end of 1943. One of the units involved in training the crews in this technique was 1 Torpedo Training Unit at Turnberry and on 18 April 1943 one of its Hampdens, AT125 (21), collided with Wellington LB237 and crashed into the sea with the loss of its crew; the Wellington landed with no injuries to the crew. The flying life of P5316 (22) was probably the briefest of any Hampden. It had arrived at 7 Aircraft Assembly Unit at Hooton Park after crossing the Atlantic by ship and, after being assembled and fitted with new Pegasus engines, it was taken up for a test flight on 24 July 1941 with Captain Bailey at the controls and two civilians on board. The Hampden collided with Tiger Moth N5447 of 19 EFTS near Hooton Park and both aircraft crashed to the ground with the loss of the three men in Hampden and also the two in the Tiger Moth.

There were a number of accidents involving Hampdens in collisions on the ground in which the Hampden concerned suffered neither casualties nor irreparable damage and listing them would have been repetitive and unnecessary as far as this book is concerned.

Collisions with German Aircraft

Two incidents that must be mentioned in concluding this section are ones in which a Hampden is known to have collided with a German fighter; it is possible that there were other such instances. On the night of 10/11 April 1941 the German night fighters took full advantage of the brilliant moonlight and shot down five Hampdens X3066, X3148, X3153, AD789 and AD828 all of which were operating against Dusseldorf (23). In the course of one attack Bf 110C-4 serial number 3300 collided with a Hampden and both aircraft crashed near each other at Neer in Holland. The German pilot, Lieutenant H Reese of I/NJG1, although wounded, managed to bale out but his wireless operator Unteroffizier W Roitzaak was killed. It is difficult to identify the particular Hampden involved as all five Hampdens lost that night crashed in Holland, although AD789 does seem to be the candidate for this dubious honour.

The second such incident involved AT227 (24) of 408 RCAF Squadron which took off from Balderton at 0840 hours on 19 July 1942 for a "Mole" anti-shipping bombing operation off the Dutch coast. The code name "Mole" was quite apt, as the aircraft made use of cloud cover to remain hidden and dropped below cloud base from time to time in the hope of spotting a suitable target. All the other aircraft detailed returned because of lack of cloud cover. AT227 was 50 km north west of Vlieland when it was intercepted by Unteroffizier Brejczek of II/NJG2 who was probably flying a Bf 110 and operating from Leeuwarden; the two aircraft collided and fell into the sea with the loss of both of the crews. The fact that, although no one lived to describe what happened, the incident was attributed to a collision may seem surprising. The explanation could be that the operators in the German coastal radar stations, after guiding Uffz Brejczek towards the incoming "blip" of the Hampden, had watched the two traces on their screens gradually converge until they met, at which point both dots of light dramatically switched off. The lives of the young men in both crews faded out at that same moment.

AT244 (13) of 415 RCAF Squadron was crashed into by 485 Squadron Spitfire BM354 at 1050 hours on 24 April 1943 near Tangmere. Although the Hampden landed afterwards it had to be written off; one of the crew, Flight Sergeant J C Donald, who may have been the aircraft's wireless operator, was killed and the other two sergeants and the two Air Training Corps cadets were injured. No.415 Squadron also suffered the loss of AT243 and its four-man crew when it collided with AT250 near Tain while on a daytime training flight; AT250 landed safely with the crew being uninjured (14).

AE370 (15) of 415 Squadron collided with 59 Squadron Fortress FK188 at Thorney Island on 30 December 1942 but fortunately both aircraft were on the ground when this rather one-sided argument took place and although AE370 was a write-off there were no casualties. Another RCAF squadron, 420 based at Waddington, had P2094 (16) written off and X3149 damaged as they roosted peacefully at their dispersal. A 44 Squadron Lancaster, L7581, that failed to become airborne on take-off, caused the destruction which resulted in the death of AC2 Commins, who was working at the dispersal. Just over a week later 420 Squadron appears to have had revenge when AE399 (17), returning from the Cologne "1000-bomber" raid swung on landing at Waddington and hit Lancaster R5892 of 44 Squadron's Lancaster Conversion Flight. Two of the crew of AE399 were injured in the accident. Other innocent victims of another aircraft's error included 455 RAAF Squadron AD979 (18) which "was struck by a Beaufighter of 235 Squadron when picketed at dispersal and burned out".

5D HIGH GROUND CRASHES

A number of Hampdens were lost through flying into high ground. Such accidents incurred a high proportion of fatal casualties because they usually occurred when the crew were busying themselves with their normal tasks, quite oblivious to the fact that the height at which the Hampden was flying was too low to enable it to clear the high ground that lay in their path. Darkness, low cloud, fog or general bad weather were usually contributory factors and a number of the aircraft lost in this way, particularly on return from operations, were many miles from where they ought to have been.

The hills of the Isle of Man, reaching up from the middle of the Irish Sea to a height of over 2000 feet and often shrouded in cloud, claimed many victims during the war years, two Hampdens amongst them, both on navigational training exercises from training units. P1260 (1) of 7 Squadron was flying in cloud when it hit Snaefell just 150 feet below its 2,030-foot summit. Snaefell is the highest point of the Isle of Man and it was a sad twist of fate that brought about the deaths of three of P1260's crew and caused the fourth man to be badly burned. Just a couple of hundred feet higher or a course that was a degree or two different and they would all have missed this New Year's day appointment with disaster. P4313 (1) flew into a hillside just under a mile east of Laxey on the Isle of Man. Three men died from injuries, a fourth survived.

The fact that even comparatively low hills were high enough to cause accidents can be seen in the loss of two of 16 OTU's Hampdens within a month in 1941. Both had been built in Canada and were lost within a few weeks of their arrival at 16 OTU. P5308 (2) flew into a hill at Corble Farm, near Bicester, Oxon, whilst carrying out bombing practice at the nearby bombing range. P5318 (2), whilst on a night cross country exercise, crashed into a 600-foot-high hillside near Wiggington Church, Bloxham in Oxfordshire. Four men lost their lives in P5308; the crew of P5318 also perished.

A number of Hampdens crashed on the Yorkshire Moors, most being abandoned by their crews as they ran out of fuel on return from operations. AE393 (3) crashed with its crew still on board after being in the air for almost ten hours at Dalefoot in Wensleydale on 15/16 January 1942 and its pilot was the only survivor. P2072 (3) was more fortunate on 5 October 1940 when it crashed into a hill near Richmond in Yorkshire in bad visibility at 2000 hours whilst on a night cross-country flight from 14 OTU. On this occasion two of the crew were injured and two appear to have escaped without being more than shaken up. AD748 (4) crashed into the ominously-named Hangingstone Hill on Dartmoor as it returned from a minelaying sortie to La Rochelle, with fatal consequences for the crew.

A casual visitor to the churchyard at Blessington in Co.Wicklow, Eire, may have paused to consider how four men from the Royal Air Force had found their last resting place there on 18 April 1941. The sad answer is that this 50 Squadron Hampden crew in AD730 (5) had overflown England and Wales on its return flight from a visit to Berlin and crashed into a nearby hillside. The Hampden crews operating from the bases on the east coast of Scotland were well aware of the danger of the rugged hills and mountains that stood ready to claim them if they made a bad landfall at too low a height.

The wreckage of AE436 on a mountain at Tsatsa, Sweden.
(Photo: RAF Museum 696/854)

AD792 (6) of 455 RAAF torpedo-bomber Squadron took off from Leuchars at 1556 hours on 11 January 1943 to attack shipping off the coast of Norway. The crew had probably seen AN166 crash a few minutes earlier when one of its engines failed on take-off and must have been relieved when they themselves were safely airborne. Just over six hours later, at 2213 hours to be precise, the wireless operator asked for a course to steer for base and two minutes later a further message from AD792 ended abruptly. An ominous silence descended and nothing further was heard from the aircraft. The record book for 13 January 1943 provides the answer to AD792's fate. "News has been received that AD792, missing from operations on the night of 11 January, was located in remote country on the north side of the Hill of Warran, north of Edzell. PO Hill and PO Rees are killed. The RAAF lower gunner Sergeant Spohn suffered only slight injuries and walked for forty hours in search of assistance. Sergeant Smithers RAAF had a fractured leg". A later entry reveals that the injured leg had to be amputated and that the unfortunate Sergeant Smithers died on 19 January. The brief account gives no hint of the weather conditions which were likely to have been bitterly cold with heavy snow on the mountains. Sergeant Spohn's gallant quest for help would certainly have been no pleasure trip nor would the task of the rescuers have been easy, particularly that of bringing the badly-injured sergeant down from the crash site.

Another Hampden crew that had the misfortune to fly into the side of a mountain in Scotland was that of P2118 of 519 Met Squadron on 25/26 August 1943. P2118 (6) had taken off from Wick to search for any trace of P5334 from their own squadron which had sent out an SOS as it flew back across the North Sea and had then failed to return. The crew of P2118 had found no sign of the dinghy and as they headed back towards base the weather gradually deteriorated until severe thunderstorms developed with heavy rain and lightning. Their wireless link with base

AE436 of 144 Squadron was one of the thirty-two Hampdens which took off from Sumburgh in the Shetlands to fly to Afrikanda, 111 miles south of Murmansk, in Russia. The aircraft were airborne between 2000 and 2100 hrs on 4 September 1942 but the pilot of AE436 experienced engine problems due to icing and descended. Unfortunately for the crew. the aircraft crashed into a mountainside at Tsatsa in Sweden at 5,600 feet. The pilot, PO Evans, and a groundcrew wireless operator survived the crash with only minor injuries but their three crew-mates died in the crash. The two survivors walked for two days before reaching the Swedish village of Kvikkjokk and convinced the Swedish authorities that they had crashed in Norway, which meant that the Swedes interned them and later permitted them to return to England before the end of September. During August 1976, the wreckage of the Hampden was discovered by two Swedish mountain walkers. The bodies of the three dead men still lay in the wreckage and were buried at Kviberg (Gothenburg) with full military honours. The remains of AE436 were brought down from the mountainside and transported to Malmslatt (Linkoping). A considerably amount of wreckage has since been moved to the RAF Museum and recently has become the basis for the Hampden fuselage reconstruction that Brian Nicholls is undertaking at East Kirkby airfield. (RAF Museum photos 613/12 and 613/13)

became increasingly more distorted by electrical interference from the storms but P2118 flew on until, just before midnight, its progress was halted by the hard rocky slope of Ben Loyal in Sutherland some fifty miles to the west of Wick. The violence of the impact rent the Hampden asunder transforming three of its crew from vibrant manhood into mangled corpses in that brief moment of extinction. Only the wireless operator Flying Officer C A Faulkes, survived the crash, suffering serious injuries to his legs and face and yet by some quirk of fortune being thrown clear of the wreck. To the rescuers who struggled up the mountainside and found the remains of P2118 and its crew it seemed most unlikely that the lone survivor would be alive for long enough for him to even reach hospital, but he refused to succumb to the injuries and once he had been delivered into the care of the hospital medical team he began a slow and painful progress to a recovery that later allowed him to resume flying duties.

The grim examples of what happened when an irresistible force in the form of a Hampden met an immovable object such as a hillside in the home countries is matched by similar incidents five thousand miles away at 32 OTU at Patricia Bay near Vancouver in Canada. The aerodrome there has the lower reaches of the Rocky Mountains almost in its circuit on the eastern side. The accidents to AN105, AN103, AN132 and AN136 from which there were no survivors give further proof of the calamitous results of such encounters.

AN105 (7) on 4 October 1942: "Pilot on solo night flying training made a wide circuit and came into the vicinity of Mt Tuam in a position and height that he should have known were not high enough to avoid hitting the mountain. The aircraft hit the mountain and travelled 200 yards through the trees, exploded and burnt out. Quite an expedition was involved by the Mountain Troops to determine the fate of the pilot and recover his body."

AN132 (8) on 16 October 1943: Took off from Patricia Bay for a night navigation exercise with a crew of four. Last radio contact was at 1949 hours after which no trace of the aircraft was discovered until July 1945 when its remains were found on a hilltop NW of Maggie Lake near Ucluelet BC. The crash site indicates that the Hampden was flying up the west coast of Vancouver Island, too far to the right of its intended track, and had struck the first available piece of high ground north of the entrance to Barkley Sound.

AN131 (9) on 30 November 1943: "Crashed 6 miles SW of Ladysmith. Both crew killed and aircraft totally destroyed by fire. Pilot had descended below 3,000 feet in heavy fog and struck tree-tops at 2,800 feet. The aircraft crashed into the ground at a steep angle".

AN136 (10) on 14 January 1944: "Crashed into Mt Tuam, Saltspring Island at 2010 hours. Pilot killed instantly, aircraft exploded and burned. Pilot was carrying out solo night flying circuits and landings. He swung wide of circuit on crosswind leg, placing himself in line with Mt Tuam on downwind leg. He failed to see the mountain and flew into the side of it at 1500 feet cutting a 200-foot swathe through the trees while flying due west in level attitude."

5E MISTAKEN IDENTITY

The Hampden, with its twin rudders and slim fuselage, was on a number of occasions mistaken for a Do 17 or a Bf 110 by "friendly" defenders of Britain, particularly in the early stages of the war. The defences concerned were the fighters, anti-aircraft guns and balloon barrages whilst "Decoy" sites laid out as flare paths attracted RAF aircraft as well as their intended Luftwaffe victims.

Attacks by Fighters

The shooting down of L4089 and L4090 by Spitfires into the sea off North Berwick and the subsequent incredible misfortune of Tony Reid navigator of L4090, in being shot down in flames by Hurricanes when flying in a Hudson over Gravesend some ten weeks later, is described in Part 2C 8-9. It is small consolation to Tony that the aircraft recognition of some fighter pilots at that time was so bad that a Spitfire even shot down a 12 Squadron Fairey Battle near Mablethorpe on 1 August 1940 causing the loss of its three crew. Insult was added to injury in the case of AE356 when it was shot down over North Russia by a Russian Air Force pilot flying a Hurricane (See 2J7).

On 22 November 1939 Flying Officer Shaughnessy of 106 Squadron was entrusted with the task of flying two civilian artists from No.1 Camouflage Unit on an aerial survey to assess the effectiveness of the camouflage of various installations in the Thames Estuary. What the artists thought of the camouflage is not on record but they must certainly have formed a very poor opinion of the manner in which the war was being conducted, because before the flight ended their Hampden had been shot at by anti-aircraft guns and intercepted by Spitfires, one of which opened fire and hit the Hampden in the tail and wing but fortunately caused no serious damage or any casualty.

L4100 (1) 6 October 1940: this 106 Squadron Hampden was on a cross-country training flight when two fighters, "believed to be Spitfires" attacked it near Stradishall just after midday. Sergeant K S Powers, who would seem to have been flying as navigator, was killed but none of the other crew members suffered injury nor was the Hampden badly damaged and it was able to land normally.

On 14 November 1940 Flying Officer Powdrell of 61 Squadron in P4338 (2) came back to his native shore in the grey morning light after a visit to Hamburg which had been made in appalling weather; his reward for the night's work was to be set upon by three Spitfires as he reached the English coast. Fortunately neither the Hampden nor its crew suffered mortal damage in the attack, although when the aircraft subsequently landed at Leconfield the undercarriage gave way "due to the hydraulic system having been shot away".

The Hampdens were still attracting the attention of Spitfires in 1942. In one instance on 29 March a 420 Squadron aircraft that had been engaged on a search for the dinghy of AE426 (2) which was believed to be down in the sea, was shot up by Spitfires when it returned and had to crash-land at Driffield. On 10 July AT252 of 489 Squadron came under attack from two Spitfires as it flew from Abbotsinch in Scotland to Thorney Island. The attacks ceased when the correct recognition signal was fired and there were no casualties in either Hampden.

"Friendly" Anti-Aircraft Gunners

Three days after war broke out two of 61 Squadron's Hampdens were fired at by the anti-aircraft defences as they flew up the River Mersey on their way to Speke airport which was being used as a dispersal 'drome for Hemswell. Jock Galloway from Hemswell's 144 Squadron remembered how the same defences had banged away at his own Hampden at about the same time. Fortunately in neither case was the gunners' aim good. Within a month that memory had been overlaid by a much more vivid one which resulted in Jock being badly injured and becoming one of the first Kriegies after suffering the attention of enemy fighters on 29 September 1939. (See 2C2)

All RAF aircrew were constantly warned of the danger of flying too close to convoys or in fact ships of any kind around the coast of beleaguered Britain. The golden rule observed by the ships was "Shoot first and ask questions afterwards". Even the Coastal Command Ansons arriving on the scene for escort duties were likely to be greeted with a barrage of shells, so it is not surprising to find that Hampdens had their share of such incidents. AE259 (4) returned from a night attack on Kiel and flew too close to a convoy off Spurn Head. The convoy escorts opened fire and the Hampden was hit in the starboard engine which resulted in the aircraft crash-landing at Fosdyke. AE259 was written off but its crew escaped unhurt.

To conclude this brief account of eager Ack-Ack gunners, an account by a Mr Harold Jones explains how one of the Hampden valiants almost came to grief. "The Hampden, I remember, was piloted by Flight Lieutenant Learoyd VC. He flew over my home at Hythe on the day that he was being given the Freedom of New Romney, roaring low across the countryside. It appears that the local ack-ack gunners had not been notified of his flight and as they seemed unable to differentiate between a Hampden and a Dornier 17, every gun in the district opened up on him. He eventually landed at West Malling after his aircraft had been hit over thirty times and completed his journey by car". A deserving case of fortune favouring the brave! Harold's account has since been verified by Rod Learoyd VC who recalled that, after lowering the undercarriage to land at Hawkinge, he had looked around him as he banked to make his approach; he noticed a lot of white puffs above him but had no idea they were shell bursts. The first intimation that he had a problem came when a bullet shattered his instrument panel, followed immediately by a squawk from the second pilot/navigator, sitting on the wing spar behind him, to the effect that he had been wounded in the leg. This positive evidence that the natives were hostile brought about an abrupt change of plan and caused Rod to get the wheels up and hedge-hop across to Detling where he landed safely.

The Queen chats to Tony Reid after he had received his Distinguished Flying Medal. Tony had, in the space of ten weeks, been shot down by Spitfires when flying in Hampden L4090 and then by a Hurricane in Hudson N7334 of the Photographic Development Unit on 3 March 1940. Tony was the only member of the crew to escape from the blazing aircraft and recovered completely after extensive surgery. He was commissioned and awarded the DFM for his earlier exploits. The two pilots to his left are PO Desmond Sheen from Sydney and Flt Lt Eric Le Mesurier, a PRU Spitfire pilot who was killed later in the war. (Photo: Dennis Knight via Tony Reid)

Balloon Barrage Victims.

The balloon barrages which protected ports, coastal convoys and factories vital to the war effort in the industrial areas were a particular problem to RAF aircraft returning from operations and also on normal training flights over Britain. The general situation in regard to the danger posed by the balloon barrages is aptly summed up in a report dealing with "Collisions with balloon cables by aircraft in the period from 1 October 1940 to 31 December 1940". During that three month period 38 aircraft had collided with balloon cables. Ten of these had been enemy aircraft, 26 were "friendly" and two unidentified. Out of the total of 38 collisions, 13 had resulted in aircraft being brought down of which 4 were Luftwaffe and 9 RAF.

Asked for comments on how to improve the efficiency of our balloon barrages, one realist replied. "At the present rate of destruction it will not be profitable to increase the effectiveness of balloon cables until the RAF is more that twice the size of the Luftwaffe!" It was obviously fortunate that not every case of a collision with a balloon cable resulted in an aircraft being brought down. It is also surprising to discover that pilots did on occasions run into a balloon cable and were quite unaware that they had done so until they landed and discovered a length of cable wrapped around the back of the propeller. This happened to the pilot of L4049 who collided with a cable over Middlesbrough, and another pilot from the same squadron, Guy Gibson, came back from Hamburg with a length of German balloon cable round the prop boss shaft. Guy was later to earn immortal fame and the Victoria Cross for leading the 617 Squadron "Dambusters".

Hampden losses caused by British balloon defences are as follows: –

P1336 (5) 23 May 1940 hit a cable in the Coventry barrage while on a training flight, it crashed with the loss of its three crew.

P1340 (5) 3/4 June 1940 hit cable near Harwich. Three of crew "Missing". See 5A25.

P4345 (6) 12/13 June 1940 hit cable near Harwich and crashed at Felixstowe. Part of the port wing fell into the water of a dock basin, the main part of the Hampden crashing into a flour mill which caught fire. One civilian in the flour mill was killed and another injured and the crew of the Hampden all perished. The Hampden pilot, Wing Commander J J Watts, had a Sergeant R Jolly as his wireless operator. The sergeant had baled out of P1326 when Wing Commander Luxmore, who lost his life on that occasion, had been pilot.

P1248 (7) 19 April 1941 had part of its port wing cut off when it hit a cable in the Sheffield barrage. The Hampden went down in an uncontrollable dive and only the pilot managed to escape by parachute. The other crew member lost his life when the aircraft crashed in Concord Park, Rotherham Road, Sheffield.

AD734 (7) 11/12 February 1941 hit cable in Birmingham barrage and its crew baled out safely (See 5A52).

AD861 (8) 6/7 July 1941. A 106 Squadron aircraft with a 97 Squadron crew ran into Plymouth barrage as it returned from Brest and crashed into the sea. Two of the crew lost their lives, two survived injured.

In an attempt to lessen the danger from balloon cables the slotted flap in the leading edge of the Hampden's wing was dispensed with and cable cutters fitted. The cable cutter was an explosive cartridge device in the wing that operated if a balloon cable lodged against it and exerted pressure so that the cartridge fired and severed the cable. "Squeaker" warning transmitters were also installed in the balloon barrages and these transmitted a warning signal that the aircraft received on its TR9 receiver.

The outer section of the port wing of P4345 of 144 Squadron which hit a balloon cable in the Harwich barrage on 12/13 June 1940. The cable would appear to have run along the leading edge of the slotted wing until it lodged in the gap at the end of the slot and then literally sawed through the wing. The pilot would have no control whatsoever over the aircraft in such circumstances and in this case all the crew perished. P1340 of 44 Squadron came to grief in the same barrage on 3/4 June 1940 and one consequence of these losses was that cable – cutting devices were fitted into the wing leading edge and the slotted leading edge was dispensed with. "Squeaker" transmitters were also installed in the barrages which sent a signal that could be picked up on the aircraft's IFF (Identification Friend or Foe) receiver.
(Photo: RAF Museum)

Decoy Sites

In an attempt to attract the bombs of the German night bombers that were attacking Britain a series of decoy sites had been established; these included the "Q" sites or dummy flare – paths which looked realistic enough from the air. Although the Q – sites included a combination of lights that was intended to warn RAF aircraft of their deceptive purpose, three Hampdens came to grief when trying to land on them.

X2896 (9) took off from Lindholme at 1840 hours to bomb Hamburg and whilst over the target had the misfortune to have its starboard engine fail. The Hampden laboured back across the North Sea and eventually made landfall on the east coast of Scotland and then at 0200 hours the lights of a dummy flare path appeared below them. Fuel was running low and in the hope that he would be able to make some sort of emergency landing the pilot made his approach to the line of lights. As he neared the ground he realised that he was heading for trees but, with the Hampden flying on one engine, he stalled and crashed as he tried to climb to avoid the danger. The crash occurred at Woodhall Innerwich south west of Dunbar in Scotland and fortunately the crew, with the exception of the lower gunner who had a broken leg, escaped injury.

AE229 (10) 29/30 August 1941: this 50 Squadron Hampden returned from operations at 0430 hours and attempted to land on the Q – site at Bassingham, Lincs. In the crash that followed two of the crew were injured and the aircraft written off.

L6011 (11) began life as a Hereford at 14 OTU and after conversion to a Hampden it resumed flying at 5 Air Observers School at Jurby on the Isle of Man. On 25 May 1942 three Hampdens took off from Jurby to fly to Syerston "for operational experience". At just after midnight on 26 May, L6011, whilst making an approach towards a decoy flare – path, hit high tension cables, caught fire and crashed at Malby near Doncaster. The pilot and three other crew members were injured, a fifth man killed. The "operational experience" for which they had been intended was to participate in the first 1000 – bomber raid on Cologne on 30/31 May 1942. The remaining two No.5 AOS Hampdens failed to operate on that historic occasion however, being withdrawn because they were not considered to be in a good enough state of serviceability to undertake such an arduous task.

5F LOSS OF CONTROL - STABILISED YAW

The Hampden was generally considered to be an easy aircraft to fly, as is evident from the pilots' comments in the Air Transport Auxiliary section, but no record of the Hampden would be complete without mentioning the "Stabilised Yaw" problem. In the months preceding the outbreak of war investigations had been carried out, at the Central Flying School at Upavon and at the Aeroplane and Armament Experimental Establishment at Martlesham Heath, into accidents in which Hampdens had crashed after the pilot had apparently lost control of the aircraft.

Test flights were made in L4032 (1) by Squadron Leader Stainforth of CFS and Squadron Leader Cunningham of 144 Squadron on behalf of the squadron pilots. These test flights mainly investigated the procedure to be adopted if an engine failed which was obviously considered to be the major cause of the crashes. The test report confirmed the findings of previous tests i.e. "That it was preferable, if an engine failed, to turn toward the good engine" and concluded "It is considered that there is no particular difficulty in flying the Hampden on one engine".

There were more enquiries after L4128 (2) crashed at Hemswell on 2 June 1939 when attempting to land on one engine and one of the questions put to Squadron Leader Stainforth was "Is the effort needed to fly straight on one engine such that a small or comparatively weakly pilot might be expected to lose control in a very short time?" His answer was "Yes". The report concluded that "CFS and A & AEE were in complete agreement regarding handling characteristics and performance. The primary cause of the trouble appears to be the smallness of the rudder bias tabs which render the pilot's legs fatigued after about ten minutes flying, it is suggested that urgent action be taken to have this surface enlarged".

On 8 September 1939 the Commander-in-Chief of Bomber Command E R Ludlow Hewitt wrote to the Under-Secretary of State about the accidents to the Hampdens which he asserted "Generally take the form of diving into the ground, and no reason for this lack of control has become apparent in subsequent investigations. In view of the similarity of most of the accidents it is felt that there must be a single cause". "Loss of control" was the term largely used to describe these early accidents and words "obscure cause", "unknown reason", "stalled", "flat spin" or "dived into the ground vertically" featured in the records as the explanation of the manner in which so many young men died.

Formation flying seemed to generate the conditions of speed and aircraft attitude that favoured this particularly dangerous condition as the following incidents tend to prove. L4082 (5), returning from Armament Practice Camp at Evanton on 20 March 1939 and flying in formation with other 50 Squadron Hampdens, suddenly broke formation and crashed near the Open Air Swimming Baths at Boultham, near Lincoln. None of the four crew survived. L4046 (6) was in a formation of twelve 44 Squadron Hampdens that had taken off from Waddington on 18 July 1939 to carry out a long distance practice flight over France. The formation climbed into cloud and L4046 came down out of control and dived into the ground near Corby, Lincs. Again four men were killed and the accepted explanation was that the pilot's instrument flying had been at fault.

L4199 (7) of Central Gunnery School was flying in formation with two other CGS Hampdens near the Isle of Man on 23 June 1941 when, at 1413 hours it "broke away at a height of about 300 feet, turned upwards into cloud and was not seen again". Four men "Missing".

No.50 Squadron lost two Hampdens in just over three months in similar circumstances in 1941. On 5 April AD753 (8), on its way to carry out a daylight bombing attack on Brest and in formation, suddenly broke formation and dived into the sea off the Scillies. Four "Missing". On 19 July AD897 (9) was carrying out formation flying practice at 1100 hours between Lindholme and Finningley when "it rolled over, flew on its back for a few seconds and then dived into the ground".

The torpedo-bomber squadrons were not immune to this sort of accident either. On 24 September 1942 AT233 (10) was practising formation flying when it "broke away from the formation and crashed with the loss of four lives". There seems to be more than an element of coincidence in these crashes but as there were no survivors to report on what happened it can only be a matter of conjecture that the loss of control came about because of the stabilised yaw.

Several pilots at 32 OTU, undergoing Torpedo Bombing training suffered accidents that could have been the result of the stabilised yaw. On 13 April 1943 AN121 (12) "had just completed a low level bombing run. It turned away at what appeared to be a slow speed, turned over onto its back and flew into the sea upside down. The range safety boat picked up two members of the crew, one of whom died." Flying Officer Paul Hartman DFC became a well known RCAF test pilot in later years and also as a civilian pilot flew such aircraft as the Avro 504K, the Nieuport 17 and the Sopwith Snipe, Camel and Triplane for the Canadian National Aviation Museum. On 31 July 1943 Paul Hartman was a staff pilot at 32 OTU and had taken P5427 (13) up for a routine air test. "The aircraft was put into a turn to port. On straightening out the pilot got no response from the elevators and the aircraft commenced to dive from 2,000 ft. When still no control had been regained at 1000 ft the pilot abandoned the aircraft which crashed at Sidney, B.C. and burned out. Cause of the accident is obscure." Whether in the subsequent Court of Enquiry the accident was attributed to the stabilised yaw is not known, but one thing is obvious and that is that Paul was quick to realise that "When you've got to go you've got to go" and he went.

Another Hampden yaw incident that happened in distant parts was to AD798 (14) of 1407 Met Flight when it was based at Reykjavik in Iceland on 4 June 1943. On this occasion the pilot, Flying Officer Park, lost control of the aircraft "when it developed a stabilised yaw" and fearing the worst ordered his three crew to abandon the aircraft (see 5A 30).

There are at least two cases of Hampdens which were trying to land with an engine out of action having crashed in an inverted position. On 7 December 1940 P4292 (15) of 16 OTU flew low over Weston-on-the-Green airfield at 1835 hours at which time it was no doubt dark. The Hampden was given permission to land and "whilst turning, the aircraft rolled onto its back and dived in from 300 feet". The five men on board were killed.

On 7 February 1942 First Officer E L Renicker, an ATA pilot from the USA, was killed when AT23l (16) crashed at 1325 hours as he was approaching to land; in this case it is less definite that an engine was out of action. The aforementioned incidents offer evidence that suggests that the stabilised yaw did exist as one of the Hampden's less attractive features, although the term "stabilised yaw" does not appear to have come into use until about mid-1941 and the very existence of this phenomenon is still disputed by some Hampden pilot veterans. Wing Cdr Jimmie Bennett, for instance, logged over 1000 hours flying on Hampdens between January 1939 and April 1942 which included tours of operations with 50 and 144 Squadrons between which he had a "rest" at 16 OTU; he finished his Hampden flying as Commanding Officer of 144 Squadron. Jimmie was quite adamant that pilot inexperience was the cause of the accidents that were attributed to the stabilised yaw.

When Wing Cdr Grant Lindeman was posted from 102 Squadron Whitleys to take command of 455 RAAF Squadron in January 1941 he recalled that he "was delighted with the change, you could almost believe you were flying a fighter; the Hampden was not a docile aircraft however, you had to watch your flying, but it did not deserve to be called a death-trap as some people referred to it ... The stabilised yaw was not so much a vicious streak in the Hampden as a ham-fisted streak in the pilots who experienced it. The aircraft had very little keel surface aft and by coarse use of the rudder, coupled with insufficient bank, it was possible to convert a flat turn into a terrifying sideways skid which rendered the ailerons useless (so you couldn't increase the bank) and locked the rudders hard over (because of the strong sideways blast of air on them)."

The Wing Commander recalled one particular experience on 1 April 1942 when he was shooting up and dropping 250-lb bombs on rail traffic on the Rhine-Munster line in bright moonlight. "We stopped four trains that night, but in the excitement of making steep turns near the ground and getting jolted by the blast from our bombs I neglected my flying and got into a sideways skid from which I could not recover. With both feet on one side of the rudder bar I could not shift it. The control column was floppy, ailerons useless, and the aircraft losing height. In the nick of time I woke up to the dodge of closing the starboard throttle and opening up the port throttle fully to twist the aircraft out of its situation by brute force. Luckily it worked, but it had taken a great effort of will to close a throttle in such desperate circumstances and we could not have been more than a couple of hundred feet above the ground by the time I regained control." On this occasion a pilot's skill saved his own life and that of his crew who all lived to fly and fight on other days instead of ending up in a War Cemetery in Germany.

P1208 (15), another RAAF Hampden, suffered much more damage when its pilot lost control; its navigator, Flying Officer Ray Lassiter, who was sitting in his normal take-off position on the wing spar behind the pilot, described what happened.

"The port undercarriage leg failed to retract properly, and at a height of 800 feet and in a climbing turn to port the airspeed dropped alarmingly and the nose started to drop. The airspeed indicator showed 40 knots with the aircraft's controls completely inoperative. Opening and closing the throttles had no effect until the pilot managed to regain some brief measure of control at a height of about fifty feet and make a belly crash-landing at speed to career across a field demolishing telephone poles and anything else that stood in its way. A sunken roadway ripped open the bomb doors and the load of eight depth charges scattered about the hillside as the Hampden slithered onwards towards a crofter's cottage and only the intervention of a particularly stout telegraph pole which slewed the aircraft round at right angles prevented the destruction of that humble abode." P1208 was damaged beyond repair but the crew suffered no serious injury.

Theories about the cause of the stabilised yaw focus on the twin fins and rudders which had very little area below the tailplane so that in certain conditions of flight the rudders were blanketed from the slipstream and failed to exert any control on the aircraft; a large side – slip would then develop after which the aircraft became nose heavy and either developed a spiral dive or rolled over onto its back.

The outspoken "Bomber" Harris was not enthusiastic about Handley Page in general and the Hampden in particular. His prophesy that the next Handley Page product would have problems was borne out when the Halifax came into production, as R N Roberts relates in "The Halifax File": "For the first two years of Halifax service the problem of rudder overbalance could not be cured ... many aircraft were lost following flat spins or inverted dives, especially at low altitudes. When rectangular fins were tried difficulties no longer arose ... and the problem was finally resolved".

The early investigations in June 1939 into the Hampden loss – of – control crashes suggested that the rudder bias tabs appeared to be too small and should be made larger. With hindsight it appears to be unfortunate that the complete rudder was not redesigned to give more area below the tailplane which may have cured this problem.

One other cause of the "loss of control" crashes had also been identified in the pre – war investigations and that was that pilots with short legs had difficulties in controlling the aircraft in an emergency when it was necessary to exert more pressure than usual on the rudder pedals. A policy decision about the selection of pilots for Hampdens, made in December 1939, concluded: – "Pilots earmarked for Hampden aircraft should not be too large. It should also be noted that pilots should not be too small as they cannot reach the instrument panel". The last sentence was no doubt intended to discourage African pygmies or Irish leprechauns from being let loose at the controls of a Hampden.

Entries in the OTU record books tend to demonstrate that this early policy decision had not been implemented, as the following incidents, which could have been entitled "The Long and the Short and the Tall" clearly show.

On 3 June 1941 Sergeant R H Breckell lost control of P4319 (18) whilst on a practice bombing exercise at Otmoor bombing range and ordered his crew to abandon aircraft. They were flying at a height of only 800 feet at the time and the observer baled out at 400 feet and survived but the bombing instructor was killed as he went out from a height of only 200 feet. Sergeant Breckell regained control of the Hampden and flew back to Upper Heyford and landed safely, where he discovered to his amazement that his wireless operator, who had not heard the order to bale out, was still on board. Four days later the same pilot developed a swing on take – off in P1246 (18) as a result of which the undercarriage collapsed. The Hampden was repairable and the crew unhurt. After this accident it was ordained that "he should be given further practice" but in the case of the unfortunate Sergeant Breckell practice did not appear to make him perfect because on 21 June he had the undercarriage of P2100 (18) collapse as he made a night landing. Again the aircraft was repairable and no one was injured, but it resulted in the pilot being "removed from training pending decision" on 9 July 1941.

Ralph Breckell was posted to 14 OTU on 3 August and thence to 106 Squadron. He failed to return from his first operational flight, to Emden, on 26/27 November 1941 when he flew as navigator in AE317 (18). The aircraft is believed to have gone down in the sea off Spickerooge. None of the crew survived, all are listed in the Runnymede Memorial register. Ralph had been a member of the Auxiliary Air Force, was 23 years old when he died and came from Waterloo area of Liverpool. It would have been interesting to discover whether he was in fact a not very tall man as, in July 1941, when he was having troubles at 16 OTU, an order was issued that "short legged pilots were to have a cushion at their backs to enable them to apply full rudder".

The records of 32 OTU at Patricia Bay near Vancouver in Canada reveal problems connected with crew stature, too. On 6 July 1942 Pilot Officer Ireland wrote off AN143 (19) after developing a swing on take – off. A few weeks later he had the undercarriage collapse on take – off in AN110 (19) after which "owing to his legs being too short for the effective control of the rudder he was taken off the course". Shades of Sergeant Breckell!

The other extreme of the stature saga can also be seen in the 32 OTU records. In April 1943 Sergeant M H Sutcliffe, navigator, ceased training because "He has proved to be too tall for the cramped space in a Hampden. He is recommended for posting to a Flying Boat or Liberator OTU". The Sergeant's height was 6 ft 7 inches. The following month "Pilot Officer R L Griffiths ceased training, he is too tall for Hampdens. Recommended for Flying Boats". He was half an inch taller than Sergeant Sutcliffe! The crews of the wartime flying boats must indeed have been an impressive sight if they included many such lofty Hampden rejects. No doubt they would answer, one and all, to the call of "Shorty".

5G MECHANICAL FAILURES

From the statements in accident reports and in the unit record books it is evident that the Hampden was not immune from mechanical defects. One problem was the engine bearer brackets which were an early source of trouble as the entry in the 83 Squadron records for 20 September 1939 shows. "Only one aircraft serviceable. Fourteen found to have badly cracked engine bearer brackets". Once such a defect had been discovered there would have been an order that the brackets were to be inspected more closely for cracks; the design may also have been modified to try to remedy the problem. That the problem with the engine bearer brackets still persisted to the very end of the Hampden's history is seen in the 32 OTU records. On 31 January 1944, the day on which it was decreed that the Hampden should be withdrawn from service, these words appear. "No.2 course has been held back considerably by reason of the lack of Hampdens due to cracked engine bearers."

Whether the loss of AT135 was due to failure of bearer brackets cannot be said for certain but it certainly gave the crew the best line to shoot about such an incident. AT135 (1) of 415 RCAF Squadron had taken off from St.Eval at 0602 on 18 August 1943 for an anti – shipping patrol and the trouble began at 0625 when "the pilot noticed that the port engine was vibrating. The port engine fell off at 0645 and the pilot ditched the aircraft five miles from the Scilly Isles. The crew were picked up unhurt less than an hour later."

The reduction gear on the propeller is also known to have failed on a number of occasions usually resulting in the loss of the entire propeller. AJ989 (2) was salvaged after it crashed at 32 OTU; the report on the condition of its starboard engine suggested one possible cause of the accident:

"... oil may have escaped from the thrust race bearing in the reduction gear casing, especially as considerable wear was found in the tail shaft bearing. The oil may have been blown on to the exhaust stub pipes and so have caught fire ...". The prop of AJ989 had not fallen off and the engine actually catching fire made it necessary for the pilot to put the Hampden down on the sea.

Major Cordes, the Handley Page Chief Test Pilot, nearly had his head chopped off in the first of the accidents in which a propeller came off in flight. The fact that in his case the fuselage was actually damaged when the prop hit it suggests that it would be possible for a pilot to be mortally injured in such an incident.

Pilot Officer Gordon Beaver had travelled all the way from Auckland, New Zealand, to fly with the RAF. He ended up as a pilot with 408 RCAF Squadron and must have had doubts about whether the effort had been worthwhile when the prop flew off one of the engines half – an – hour after he had taken off from Balderton in L4204 (3) to bomb the Renault works at Paris on the night of 3/4 March 1942. He managed a crash – landing at Grafton Underwood without injury to himself or any of the crew. L4204 was repaired and next saw service with 14 OTU. Gordon Beaver and the same crew in L4140 (3) were not so lucky three weeks later when they were shot down by a night fighter on a Wilhelmshaven minelaying op. They crashed into the sea and no trace was found of the crew who are all listed on the Runnymede Memorial. P5310 (4) with a crew of four was on a daytime cross – country exercise from 16 OTU on 17 October 1941. As they flew over South Wales the prop and reduction gear came adrift from the starboard engine and when its pilot realised that the aircraft was not maintaining height on one engine he was forced to crash – land five miles south east of Aberayron. The pilot and navigator both lost their lives in the crash and the other two crew members were injured.

P1230 (5), at 32 OTU in late 1943, is the last recorded incident of a propeller on a Hampden parting company from its engine. On this occasion the prop flew off in flight causing damage to the pitot head. "Pilot carried out single engine procedure and landed the aircraft safely on one engine." In addition to the happy outcome of this particular incident it is worthy of mention that there was another redeeming feature in that the propeller at least waited until it reached 32 OTU before severing its connection with the engine. Had it decided to drop off as P1230 slogged its way across the 5,000 miles from Prestwick to Patricia Bay between 18 and 22 August 1943 then Don McVicar might not have lived to describe the experience in his book "North Atlantic Cat", an extract from which appears in "The Canadian – Built Hampdens" (See 5K).

Gilbert Haworth, a 50 Squadron air observer, was flying as lower gunner on a Hampden on 7 August 1939. It was a bright summer's day as they headed out across the North Sea on a routine training flight and Gil settled down to make the most of what looked like being a rather boring trip and watched the sea slipping by 4,000 feet beneath him.

The serenity of their progression across the sky was suddenly shattered as the agitated voice of the navigator spoke on the intercom. "The starboard propeller is wobbling like hell" was his message to the pilot, who could hardly have failed to see it for himself. Before Gil or the wireless operator had had time to appreciate the implication of this announcement they felt a slight jolt and heard a thud as the propeller broke away and hit the side of the fuselage near the navigator's position, fortunately without injuring anyone. Gil's apprehension about his immediate survival prospects were not improved when he moved forward to the midships position to check that the dinghy was ready for use. To his dismay he found that it was conspicuous by its absence; all that he could do was to pray that they would make it back to land.

The navigator had given the pilot a course for the nearest aerodrome which was Martlesham Heath and after what seemed like an eternity, during which the temperature of the good engine rose steadily whilst the oil pressure and altitude gradually dropped, they sighted the coast. Gil still remembers the immense feeling of relief that he felt as the Hampden staggered across the beach at a height of barely two hundred feet. The crowds of holidaymakers on the sands below them waved their appreciation of this entertaining display of low flying, blissfully unaware of the peril that faced the aircraft's crew if the port engine were also to fail; the ensuing crash landing on the beach with both engines out of action would have ended a few holidays in a sudden and tragic manner. Their luck stayed good however and Martlesham Heath appeared on the horizon directly ahead of them. The pilot had not time or height left to allow the nicety of a circuit and a conventional approach and he slipped the Hampden in over the boundary hedge to make a safe landing. An examination of the damaged engine revealed that the rear bearing of the reduction gear tailshaft had melted away. Without this support the fast rotating shaft had wobbled about on a gradually increasing orbit, causing the reduction gears attached to it to cut into the reduction gear casing, until nothing remained to hold the propeller and reduction gear in position. At this point the propeller flew off on its own bouncing against the side of the fuselage in farewell before twirling down to splash into the sea.

5H INTRUDERS

One that got away. AD982 of 44 Squadron returned to base after being shot up by a night fighter on the night of 2/3 September 1941. The aircraft was airborne for almost six hours so it seems probable that it was over enemy territory when the attack took place. Sergeants Shipton and Durnan in the rear positions were wounded but the pilot, Sgt Dobbs, and the navigator, Sgt Armstrong, were unhurt. AD254, AE152, AD913 and AE313 failed to return, AE192 and AD767 crashed on return and X3025 was written off on the ground on the same night, all from 44 Squadron.
(Photo: Eric McCabe via Ray Leach)

A brief outline of the formation and organisation of the German night fighter force is given in the Appendix of Hampdens shot down by fighters. The activity of one of these units is worthy of special mention. I/NJG2 was formed in June 1940 and operated until October 1941 with the specific task of harassing RAF aircraft and bases at night. The unit claimed many victories over England where the busy training units in particular provided many targets for the guns of the Ju 88s with which they were mainly equipped. The intruders claimed a number of Hampdens and Herefords amongst their victims, but in fact no Herefords were shot down and only five Hampdens can be definitely credited to I/NJG2 over the English mainland. These were X3027, AD719, AD761, X3062 and P5314.

X3027 (1) was attacked near Scampton by Lt Volker as it returned from operations on the night of 27/28 October 1940, and eventually crashed into the sea off Skegness with the loss of its crew. AD719 (2) arrived back over Scampton at 0630 hours on 11 February 1941 having completed its bombing operation to Hannover; the four sergeants on board were no doubt eagerly looking forward to their egg and bacon breakfast. Oberleutnant Kurt Hermann arrived on the scene and a withering blast of shells from his Ju 88 ripped into AD719, inflicting severe damage and setting the Hampden on fire. The pilot had no option but to order his crew to bale out and he and his wireless operator, Sgt Blower, who had been wounded in the right arm by an explosive bullet, managed to escape from the blazing Hampden as it plunged earthwards and dived into the ground near Scampton. The navigator and lower gunner, both of whom may have been mortally wounded in the attack, were still in AD719 when it crashed and both were killed. Kurt Hermann and his crew had the fortunes of war turned against them a month later when a Hurricane of 79 Squadron shot down their Ju 88 over East Anglia and they became prisoners of war.

AD761 (3) was one of the Hemswell Hampdens taking off for operations at around midnight on 16/17 April 1941; intruders were known to be in the area and all pilots had been warned to take off with their navigation lights switched off. Whether Sergeant Kirby in AD761 had indeed received the warning is not known for certain but, when he roared away down the flarepath, his navigation lights were on. Four minutes later the pilot of a Ju 88 had the Hampden in his sights and opened fire; only wireless operator Sgt G Walsh, who may have been flying as the lower gunner, managed to bale out before the blazing Hampden crashed two miles ENE of Hemswell. Walsh had been badly wounded in the attack but reached the ground without further injury. Two of I/NJG2's pilots had each claimed "a Hereford" (Lt Volker and Fw Hahn) but AD761 was the only similar aircraft to be shot down. X3062 (4) became the last of the operational Hampdens to be shot down by an intruder when Uffz Koster sent it crashing into the sea off Mablethorpe on 8/9 May 1941. P5314 (5) of 16 OTU was the final definite victim of I/NJG2, it was blasted out of the night sky by the cannons of Oblt Semrau's Ju 88 on 20 September 1941.

On 11/12 June 1941 Pilot Officer L B Gunter of 16 OTU, in P1341 (6), had a lucky escape after being attacked by an intruder. An engine caught fire in the attack and the propeller fell off and as he tried to land at Akeman Street landing ground he overshot and ended up in a field two miles past the airfield after hitting a high tension cable and going through a hedge before coming to rest with the undercarriage collapsed. The crew escaped injury and P1341 was later repaired.

No Hampden from Waddington went down to the attacks of the intruders but there were a couple of very near misses there on the night of 15/16 February 1941. Flying Officer Penman had his aircraft X2917 (7) holed in several places in one attack and an hour later Squadron Leader Smales had his port engine put out of action as he prepared to land in X3025 (7). Both pilots, from 44 Squadron, landed their aircraft safely and doubtless reflected on their good fortune in being still alive.

The activity of the German intruders was not confined to attacks on aircraft near their bases, they also patrolled off the English coast and over the North Sea in the hope of intercepting the RAF bombers on their outward or return flights. It is difficult to be certain exactly how many Hampdens fell victims to these attacks over the sea; none involved lived to relate their experience. X3118, AE153, AE159 and AE303 were all likely victims but there were no survivors to confirm how their aircraft had come to grief. (8)

A further complication in verifying the Hampdens that were shot down by the German intruders lies in the problem that they had in positively identifying the vague shape of a twin-engined bomber at night, Hudsons, Wellingtons, Whitleys and Blenheims all being mistaken for Hampdens or Herefords. Another problem arises when, because there was a Hampden lost on a particular night and a German pilot claimed to have shot down a Hampden, it seems to have been automatically assumed that two and two made four. At least two of the Hampdens credited to intruders are open to question, these being AD788 (9) and X3063 (10).

According to one account AD788 was shot down off the Yorkshire coast at 0015 hours on 26/27 June 1941 by the Ju 88 of Oblt Paul Bohn of 4/NJG2. The starboard wing of the Hampden was set on fire but its rear gunners continued firing and, just as the wireless operator of the Ju 88 was sending news of their victory back to base, their bullets wounded the pilot. Uffz Lindner took over the controls and flew the aircraft back towards Holland and, when he judged that they were near their home base, set the Ju 88 on to autopilot. With the help of Gfr Engmann he pushed the wounded pilot out of the aircraft, pulling the pilot's ripcord as

Damage to the tail assembly of AD982 on the leading edge of the starboard side of the tailplane, close to the fuselage.
(Photo: Eric McCabe via Ray Leach)

he did so. The gunner and the wireless operator then made their own escape and came down safely near Charleville in Belgium. The pilot's body was later found in a wood near Eclaibes in France; he had apparently been already dead when he left the aircraft. The crewless Ju 88 continued flying on auto-pilot; with almost unbelievable luck it actually flew across the Alps and finally ran out of fuel and crashed at Milan in northern Italy. An intriguing story as far as the Ju 88 is concerned.

AD788 was the only Hampden that could possibly have been involved as the only other Hampden to be lost that night, also from 49 Squadron, was X3060 which overshot Scampton on landing on return from the same operation. AD788 (9) took off from Scampton at 2140 hours so that by 0015 it would have been in the vicinity of the coast of Denmark. All the indications are that AD788 was actually shot down by a night fighter at about that time and another Hampden pilot reported seeing a Hampden shot down. Sergeant S N Hind, the pilot of AD788, survived as a prisoner of war and in a letter that he wrote home he almost seemed to be apologising for having been badly burned and injured after baling out of the blazing AD788 when the tailplane disintegrated. He was obviously most distressed that his three crew mates had not been able to get out although all may have been killed in the attack. The fact that the three who lost their lives are buried in the War Cemetery at Kiel confirmed the belief that their Hampden came to grief in that area, so if Paul Bohn did shoot AD788 down it was over Denmark, not off Yorkshire.

It is also difficult to reconcile the loss of X3063 (10) with the "Hereford" claimed by Oblt Paul Bohn who also features in the previous account. The "Hereford" was claimed near Scampton, but X3063 which was the only Hampden to be lost that night is believed to have gone down into the sea off the Isle of Wight; the fact that all of the crew of X3063 are listed in the Runnymede Memorial Register tends to confirm this.

There is proof that the intruders did not have things all their own way as the following accounts demonstrate. X3058 (11) took off from Hemswell at 1655 hours on 23 November 1940 to attack a target in the Ruhr and was 90 miles out across the North Sea when the sinister shapes of three Ju 88s appeared on the scene and, no doubt anticipating an easy victory, roared in to the attack. Sergeant Oakley desperately manoeuvred X3058 to avoid the onslaught, losing height from 9,000 to 2,000 feet in the process and, in a last desperate effort to escape, jettisoning his bomb load. Sergeants Durtnall and Clelland manning the rear upper and lower gun positions used their twin Vickers "K" guns to such good effect that one of the Ju 88s with its cockpit ablaze was sent diving down into the North Sea. The other two attackers decided that discretion was the better part of valour and gave up the attack leaving X3058 to return to Hemswell where it landed safely with its battle scars from the cannon shells and bullets as ample evidence of its encounter. None of the crew of X3058 was any the worse for the experience and on this occasion it was the crew of the Ju 88, piloted by Ofw Kurt Schlict, who disappeared beneath the waves and next-of-kin in Germany who mourned the loss of three brave young men.

Sergeant James Flint of 49 Squadron was another Hampden pilot who suffered the attention of one of the intruders. Jim took off from Scampton at 2230 hours on 5/6 July 1941 in AD856 (12). His ordeal began when he was attacked by fighters on the outward flight but although his aircraft was damaged he managed to shake off his assailant and continue to his target. The return flight was comparatively uneventful until they were well on their way back across the North Sea and had the misfortune to be sighted by two German twin-engined fighters which carried out a series of persistent and destructive attacks. Cannon shells ripped into the unfortunate Hampden mortally wounding the wireless operator Sergeant J D Fitch who was manning the upper guns; his comrade Sgt E C Atkinson in the lower position was also wounded as was the navigator Sergeant A S Beningfield. Very cartridges inside the Hampden were ignited in one of the attacks and in the final onslaught the port engine of AD856 was put out of action.

Luckily for the crew of the Hampden the German fighters broke off their attack at this point; no doubt the sight of the English coastline and the likelihood of RAF fighters arriving on the scene discouraged further pursuit. After the Germans turned for home Jim Flint strove valiantly to keep his stricken Hampden in the air in the hope that a forced landing on land might be made. AD856's wounds proved to be too severe however and with the Norfolk coast only half a mile away Jim had to put the Hampden down on the sea off Cromer. The ordeal of Sergeant Flint continued as after he had made his own escape from the aircraft he realised that his navigator was still in the Hampden, lying seriously wounded and unconscious on the floor in the nose of the sinking aircraft. It was only through a superhuman effort at considerable personal risk that Jim was able to drag the inert body of his colleague through the cramped space under the pilot's seat and extricate him from the Hampden only a few seconds before it slid beneath the waves. Jim's heroism had no happy ending for the navigator who subsequently died from his wounds as did the wireless operator.

The lower gunner, Sergeant E C Atkinson, was also a wireless operator and when he had recovered from his wounds he resumed flying with 49 Squadron and was in AE419 (13) on 28/29 December 1941 when it failed to return from an attack on Huls. The crew of AE419 are all listed in the Runnymede Memorial Register and almost certainly crashed into the North Sea.

Although none of the other members of the crew of AD856 received an award for their part in the action, the efforts of Sergeant Flint, who was another of that gallant band of "VR" (RAF Volunteer Reserve) pilots resulted in him becoming one of the few aircrew to receive two decorations in the course of one operational flight; he was awarded the Distinguished Flying Medal for his airmanship and general devotion to duty and the George Medal for his gallantry in rescuing his navigator. On the night of 4/5 February 1941 one of II/NJG2's Dornier 17Z-10s piloted by Oblt Otto-Heinrich Hauser is reported as having attacked a Hampden over the North Sea at 1837 hours; a well-aimed burst from the Hampden's rear guns sent the Dornier down in flames to crash 30 miles from Lowestoft. The identity of the Hampden involved has not been discovered to date.

A natural consequence of the unwelcome visits by the intruders was an increase in the ground defences at RAF aerodrome. On 21 December 1940 the defences at Manby, home of No.1 Air Armament School near Louth, Lincolnshire, shot down a Ju 88 at 0800 hours. It was apparently piloted by the Commander of NJG2. The defence gunners at Scampton were not to be outdone and they bagged a Ju 88 at 0115 on 12 May 1941.

The operations of the special intruder unit I/NJG2 lasted from July 1940 until October 1941, and during that time at least 50 RAF aircraft were destroyed over England and probably another 30 over the North Sea; the enemy losses including aircraft that crashed as a result of these operations amounted to only half that number. As has been said on a number of occasions, the mere presence of these deadly night visitors in the sky above the RAF bases contributed to many accidents and the decision by Hitler to move the intruder unit to the Middle East theatre of operations in October 1941 is regarded by many historians as one of the great tactical blunders of the air war in Europe. The RAF aircrew of that period were truly grateful for this malfunction of the Fuhrer's much publicised intuition.

5J THE AIR TRANSPORT AUXILIARY

The Air Transport Auxiliary, formed in the summer of 1940, included a number of lady pilots amongst its members and the wide variety of aircraft that these ATA pilots flew enabled them to give an unbiased opinion of the flying qualities of the Hampden and Hereford in comparison to other RAF twin-engined aircraft of that time.

Several of the lady pilots were kind enough to offer their opinion on the Hampden. To their credit none of these ATA "women drivers" did any damage to a Hampden and all considered it to be one of the easier "twins" to fly. Mrs Helen Harrison Bristol, the first Canadian lady ferry pilot "thought the Hampden was alright, but preferred Spits and Mossies". Mary Hunter "enjoyed flying them much more that the Wellington or Whitley, they seemed light to handle", whilst Mary Willis remembered them as "a somewhat gentle aircraft compared with other types". The redoubtable Miss Lettice Curtis made a number of Hampden ferry flights and also delivered one Hereford; she wrote "I seem to recall the old Hampden especially for its good all-round view from the cockpit, but I was not too happy with the Hereford with its unreliable Napier Dagger engines".

The ladies were undoubtedly quite confident about flying the Hampden and consequently earned the respect of the service pilots who obviously noticed other qualities in them besides their flying ability. John Enns DFC, who was a flight commander on 415 (RCAF) torpedo bomber Squadron at Thorney Island in 1943 ended his account of his Hampden experiences thus "On or about 15 October 1943 a group of beautiful lady pilots came to Thorney Island and flew our Hampdens away". An unsolicited testimonial there, girls. Diana Barnato Walker flew Hampdens away from Thorney Island on 17 October and again on the 18th, so she was undoubtedly one of the "beautiful ladies" referred to by John Enns. In addition to impressing the pilots with her feminine charms Diana may also have put the fear of God into those who witnessed one of her take-offs which is still a vivid memory for her. "I thought it was curtains for me when one of the throttle levers came back and jammed half open just after I had become airborne. I just missed a hangar and the wind sock on the top of it must have changed directions as I went past". Nice one, Diana.

The majority of the ATA pilots were mere males and a number of them offered their comments on the Hampden. Peter Garrod flew Hampdens from 7 FPP Sherburn-in-Elmet and summarised its characteristics thus:- "The Hampden was generally regarded in the ATA as a pleasant, easy aeroplane. It could be flown without a crew because everything was within easy reach of the pilot except the main fuel cocks by the wing spar, but for ferry flights there was enough fuel without changing tanks. The twin rudders made it easy to keep straight on take-off (unlike the Wellington) and it had no tendency to pull to the left like the Whitley. The cockpit was unique amongst the larger twin-engined bombers, with a superb view and a hood to slide back, just like being in a fighter. But if you did slide the hood back unwarily, your maps could easily disappear in the draught. The layout was good and the Handley Page slotted flaps gave excellent low speed lift, so that the stall was forgiving, and even a high "hold off" generally resulted in just a gentle sink to the ground".

The slotted flaps were later dispensed with as they were blamed for some accidents on take-off, and furthermore the advent of balloon cutters in the leading edge rendered them impracticable. The snags as Peter remembered them were:- "Engine oil temperatures always tended to be high, sometimes near the maximum permissible. The oil coolers did not "hang out in the draught" and were fed by air scoops that became distorted with time and fouled the air flow. The hydraulic system required a "Power On" control bolt to be operated, as well as the normal undercarriage and flap controls; this control bolt tended to be forgotten after take-off and if it was the pump went on delivering full pressure and usually overheated and gave out. One ATA pilot, who had better remain nameless, made a belly landing at Sherburn, having been caught this way and left without hydraulics. The propellers were non-feathering, bracket type, and the single engine performance suffered accordingly. "In spite of its slim profile I found the Hampden to be no faster than the much larger and fatter Wellingtons fitted with the same engines. Both did around 165 mph (indicated) in typical ferry "cruise" conditions. I also flew one Hereford. The 24-cylinder "Dagger" engine had a take-off rpm of 4,200 compared with the 2650 of the Hampden's Pegasus engines. The noise in a Hereford was incredible."

Two ATA pilots crashed in Hampdens equipped with Air-to-Surface-Vessel Radar (ASV). John E Martens, who was in 1984 the Chairman of the ATA Association, described how P2113 met its end. "P2113 (1) had been repaired at Tatenhill after being force-landed there by an RAF crew. I have never in my life seen such an array of very large "Christmas Tree" aerials on wings and fuselage." After carrying out the routine ground checks of the engines etc John took off, needing a considerable distance before becoming airborne. Once he had reached a safe height he tried flying on one engine to see how the aerial array

Diana Barnato Williams, one of the many Air Transport Auxiliary pilots who flew Hampdens, posing on the wing of a Spitfire. (Photo: Diana Williams)

affected the Hampden's performance, he was quickly convinced that it would not maintain height on one engine.

Having satisfied himself that temperatures and pressures were within acceptable limits, John set a north westerly course past Liverpool and thence northwards towards the Solway Firth and his destination, Kirkbride; flying just below the cloud base at 3,000 feet. There was a light drizzle about but visibility was generally good; John continued: "After crossing Morecambe Bay and when about ten miles to the north of it, the port engine stopped and could not be restarted. I increased the revs on the starboard engine to 2400, boost from +2 to +3 and trimmed the aircraft to best advantage. Under these conditions I was losing 200 to 250 feet per minute which meant that, with a bit of luck, I might just reach the airfield at Silloth which was just visible ahead.

At this time I was about 1200 feet above the hilltops and all was going reasonably well, although the drag of the ASV aerials reduced the airspeed to 12O mph. After five minutes or so the starboard engine started faltering, gradually lost power and then died. By this time I was down to about 200 feet above the hills, the sides of which sloped steeply down to the sea. The only thing to do, as speed was falling so rapidly, was to land in a field straight ahead. The field was a small one, surrounded as were all the others, by a stone wall. I turned off the petrol cocks and ignitions and with the airspeed down below 80 mph, staggered across the wall with flaps and wheels up. As P2113 hit the ground the port airscrew broke off and catapulted over the tail. The Hampden slithered across the rough ground for a hundred yards or so and finally came to rest within a few yards of the far boundary wall."

The subsequent Court of Enquiry commended John for his action in general and in particular for the manner in which he had extinguished several small fires that broke out after the crash. John's final comments on the accident, which could easily have ended his life were that "P2113 was the most difficult Hampden I ever flew. It was heavy, being full of petrol and armed with guns, although no bombs or torpedo were carried. The drag of the aerials was considerable. I have never seen such an array, not even on a Wellington. I would love to have known what it was used for."

Ron Pass, another ATA pilot, crash-landed L6019 (2) at Bamburgh in similar circumstances. The findings of the accident committee in his case revealed these brief details. "Aircraft crash-landed in a field, with complete failure of the starboard engine, which was probably due to the failure of a connecting rod. The aircraft failed to maintain height on the remaining engine, probably owing to the fitment of external aerials (Christmas Trees); the pilot is not held responsible for this accident."

In spite of their general opinion that the Hampden was an easy pleasant aeroplane to fly, it nevertheless claimed the lives of five ATA pilots, two of whom were from the United States of America. First Officer Isidro J Parades (USA) of 3 FPP, in P5396 (3), stalled in at Burtonwood "when attempting to go round again" on 7 November 1941. He was, in fact, from the Philippines.

First Officer E L Renicker (USA) of 14 FPP crashed at Weston, Staffs, on 7 February 1942 in AT231 (4) cause unknown. First Officer T C D Bray with Second Officer N A Berry, both 3 FPP, were killed when X3130 (5) crashed on a flight from Kirkbride to Thorney Island on 18 March 1942. The cause and location of the crash are not known.

First Officer J D Hurley of 8 FPP was the last ATA Hampden casualty on 18 December 1943. He was killed when AD736 (6) crashed at Hawarden at 1325 hours. The crash report relates that "The pilot approached rather low and made a turn to the left. The machine rolled over and crashed in an inverted position, caught fire and was destroyed. The evidence was insufficient to establish the cause with certainty, although it indicated that the port engine failed during the final approach." The particular tragedy of this pilot's death was that the Hampden was almost certainly on its way to be scrapped. The words "The machine rolled over and crashed in an inverted position" have a familiar ring about them. They can be seen in several accidents attributed to the "Stabilised Yaw" one of the Hampden's less attractive features (see 5F).

It is apparent from the following account from Count Zichy, one of the ATA pilots from the early days, that out of the fifteen Hampdens that he flew the memories of one in particular are not exactly pleasant.

Zichy arrived at Shawbury on 26 June 1941 to collect P1290 (7) and fly it to Preston. He recalled the saga of that summer day as follows: "I remember well being confronted with this battered monster P1290 at Shawbury dispersal. Some erks were patching up a jagged hole in the tail, whilst a bod in the cockpit was wiping the seat and controls with a damp rag, of what? – oil? – or blood?? During the run-up and pre-flight check nothing seemed to work. The controls moved as if embedded in mud, half the lights wouldn't come on unless you thumped them. The starboard engine at first refused to start then billowed clouds of black smoke and, in coarse pitch, ran rough enough to almost shake the prop off. Was that why an erk had frantically replaced the cowling when I appeared? What was he trying to conceal? Never having been very good at pulling rank I started to protest feebly "Now look here my good man" sort of thing to the head erk who was crouching behind me in the cockpit during the run up and who was obviously determined to get rid of the damned wreck at any price. He bellowed in my ear "Wot're you on abaht, sir? It flew back from Germany and landed here in one piece, didn't it?" Probably I was just too much of a coward to argue with this overpowering and bloody-minded erk, who'd got my number as a complete mechanical moron; besides, he was a pro and had checked the crate out as airworthy, so how could I prove that it wasn't, apart from pranging which seemed to be a high price to pay, or alternativelyrisking a rocket from Up High for refusing to fly it?

Finally, being a weak-minded twit and having visions of those brave RAF lads nursing this wreck home over the North Sea, I squared my shoulders and informed myself severely "If they can do it, I can do it. (Famous last words). Actually the erk was perfectly right. P1290 took off like a dream, albeit perhaps an unpleasant one and flew like a crow – well perhaps a slightly senile one – in a straight line from A to B. Exactly as a crow should do. The only trouble was that unlike a crow, on arriving at B, P1290 would not – repeat not – turn. Nohow. However I only found this out when I got into the circuit at Preston during my last turn on to pre-final with wheels down, flaps down, pitch in fine, mixture rich – and found myself heading in a direct line straight past the airfield on my left.

I kicked, pushed and stamped on the left rudder with both feet, but to no avail. The Hampden skidded, juddered, yawed, pitched nose up and nose down and wallowed, but firmly refused to answer the rudder, which might as well not have been there. I began to wonder whether an erk had nailed it back on the tail with six-inch nails and a batten to stop it from falling off."

Zichy described how his efforts at bringing P1290 under control, left him "drenched in cold sweat, with bone dry lips and a hideous grin glued to my gums that must have made me look like Poor Yorick as we headed away from the airfield like the Flying Dutchman sailing off into the sunset! Six or seven times I opened up and, sobbing with terror and exhaustion, bullied that damn crate around the circuit. By now the lads in Flying Control were hysterically firing green flares at me and each time I opened up I could see the fire engine and blood wagon racing into position. But eventually I got the hang of it. P1290 COULD be manoeuvred – by banking steeply under full power, rudder centralised, and with the stick pulled right back, the kite actually turned. While doing so it skidded like a sausage on a non-stick frying pan and lost height like a hungry swallow swooping on a mayfly. I pointed P1290 in the direction where the runway ought to be and -- Eureka! – it loomed up out of the mist and I touched down with a perfect three point landing. The lads on the 'drome wanted to know what the Hell I had meant by scaring the shit out of them by going round and round for the past hour ... and would I kindly do my effing landing practice elsewhere in the future. My language on Hampdens, and P1290 in particular, ought to have been tape-recorded for posterity. It would go down with "Friends, Romans, Countrymen ..." ("The Hampden is an honourable aeroplane" sort of thing) as a literary masterpiece. But perhaps those with a weak spot in their hearts for the Hampden might not have appreciated it."

Air Transport Auxiliary Pilot's Notes

Hampden I Air Transport Auxiliary Pilot's Notes

Engines:	Two Pegasus XVIII. Two-speed blowers Fuel: 100 Octane.
Propellers:	D.H.constant-speed, counterweight type
Undercarriage Operation:	(Pump on starboard engine)
Hydraulic:	A sliding bolt below starboard window supplies hydraulic power for ALL services when pushed ON (forward). MOVE BACK TO "OFF" IN FLIGHT
Normal:	Common selector for U/C and flaps. Hydraulic ON, select up or down, return to neutral on completion.
Reserve:	Hydraulic power bolt OFF. Use hand-pump
Emergency:	Pull red handle near right knee and hold till green lights show. This lowers U/C by compressed air regardless of the position of selector. Hydraulic power-bolt should be OFF. DO NOT ATTEMPT TO RAISE AGAIN.
Indicators:	One green light for main wheels, one for tail wheel. One red shows locked up. Warning horn.
Flaps:	Same selector (and power bolt). Leave selector DOWN when landing, as flaps will not go right down until just before landing. Two electric indicators on dash.
Gills:	Winding handle at left of seat.
Tanks:	Three each side - INBOARD (130 gals), OUTBOARD REAR (110 gals) and OUTBOARD FRONT (87 gals). Two cocks each side in upper rear gunner's station. Upper cock for outboard tanks SHOULD BE TURNED OFF (to position labelled FRONT AND REAR TANK BALANCE ENGINE OFF); and lower cock ON (INBOARD TANK TO ENGINE). See that inboard tanks are filled and DO NOT USE OUTERS. Pilot-cocks situated behind throttles, incorporate cut-outs for stopping engines. Electric indicators at bottom of dash.
Starting:	12 volt direct. Starter socket on starboard fuselage. Dopers in nacelles. Starter-mag. switches between starter buttons on dash.
Ballast:	Pilot only, no radio, 4 weights under tail. Pilot only, with radio, 2 weights under tail.

FLYING PARTICULARS

Take-off:

Hydraulics:	Blowers	Boost:	R.P.M.:	Mixture:	Gills:	Trim:
ON	M	+6½	2650	Rich	Closed Flaps up	Elevator rudder & aileron neutral

Safety speed 120 mph

Climb (Max):

Boost:	R.P.M.	A.S.I
+2½	2200	130 mph

A.T.A. Cruise:

Hydraulics	Boost	R.P.M.	Mixture	Gills	A.S.I.	Consumption
OFF	0	1900	Weak	Watch temps	170 mph	40 gals/eng.hr

Single Engine:

	Boost	R.P.M.	Mixture	Dead prop	A.S.I
Level:	+2½	2200	Rich	Coarse	125 mph
Climb:	+2½	2200	Rich	Coarse	115 mph

Stall:
Flaps and U/C up: 76 mph Flaps and U/C down: 68 mph

Glide:
Flaps and U/C up: 110 mph Flaps and U/C down: 105 mph

Approach and Land:

Hydraulics	Flaps	Effect	Max speed for flaps	Final approach
ON	Down	Slight	110 mph	90 mph

AJ993 stuck in the mud on 2 February 1943 at 32 OTU at Sidney, BC. (Photo: Public Archives of Canada PA 144777)

5K THE CANADIAN HAMPDENS

In November 1938 Canadian Associated Aircraft Companies signed a contract to build 160 Hampdens in Canada. Assembly plants were built in Ontario (Malton) and Quebec (St.Hubert) each utilised three sub-contractors for fuselage, wing and tail assemblies. The Malton sub-contractors were Fleet Aircraft, National Steel Car and Ottawa Car Manufacturing, whilst those for St.Hubert were Canadian Vickers, Canadian Car and Foundry and Fairchild Aircraft. Four Hampdens, L4208 to L4211 (1), were shipped out to Canada in June 1939 as "Pattern" aircraft for the new production lines. L4210 and L4211 returned to Britain by sea in mid-1941; L4208 and L4209 were struck off charge in Canada without operating on any service unit.

According to the Hampden "Contract Cards", the Quebec group built 40 Hampdens (P5298 to P5337) whilst the remainder were built by the Ontario group (P5338-P5346, P5386-P5400, P5421-P5436, AJ988-AJ999 and AN100-AN167). A Canadian source credits AN128-AN167 to the Quebec group, in which case the contract would have been evenly divided between the two groups. The last 80 of the Canadian Hampdens were originally intended to be fitted with Wright Cyclone engines, which had proved to be satisfactory alternatives to the normal Bristol Pegasus XVIIIs in trials in Britain in October 1939 on L4032 (2) and later on X3115 (2). The Cyclone was not used in service on Hampdens, mainly due to the great demand for them for other aircraft being built in the United States, and all the Canadian Hampdens were Pegasus-powered.

P5298 (3) was the first Canadian-built Hampden to fly; on 8 August 1940, piloted by J H Lymburner, it took off from St.Hubert. Ten Hampdens had been completed by the end of 1940, 94 more during 1941 and the final 56 in 1942. The Canadian Hampdens that were destined for Britain were completely assembled and test flown, after which their engines were removed and retained in Canada. The fuselage and centre section of the wing, complete with the undercarriage in the locked down position, travelled as deck cargo, suitably protected from the elements; the tail unit and outer wing assemblies, in crates, were loaded into the ship's hold. The ships departed from Montreal in the summer months and from Halifax in the winter.

The port of entry in England was Birkenhead and after being unloaded the Hampden components went by road to Hooton Park, a few miles south of Liverpool, where No.7 Aircraft Assembly Unit (7 AAU) had been formed; civilian contractors Messrs Martin Hearn carried out the re-assembly work and initial test flying. One young man who joined the work force at 7 AAU in November 1940 was Peter Johnson. He remembered how the Hampden parts arrived at Hooton Park loaded on to 60 foot-long low-load articulated lorries which were already known throughout Britain as "Queen Marys". A traffic roundabout at the entrance to Hooton Park had to be modified to enable the lorries to enter; Peter also confirms that the Pegasus engines arrived from the Bristol works.

Before the assembly work proper could commence the hard sticky "goo" that covered the fuselage as protection from the effects of sea spray had to be removed and a team of young girls worked, in very bad conditions, washing this protective film away with hot paraffin and removing the tape that had been glued on to protect the many apertures; the combination of the cold, very wintry weather and the penetrating aroma of "Eau de Paraffin" failed to dampen the ardour of these war workers. At this time Liverpool and Manchester were being regularly visited by the Luftwaffe and those engaged on the Hampden assembly work no doubt worked all the harder in the hope that their labours would result in retribution being meted out to the Germans.

Five of the 85 Hampdens that arrived at Hooton Park from Canada never left 7 AAU for acceptance into RAF service. P5316 (4) collided with a Tiger Moth whilst on a test flight with the loss of Captain Bailey, the Martin Hearn test pilot, and two other civilians in the Hampden. The two occupants of the Tiger Moth also lost their lives. P5299, P5306, P5307 and P5342 (4) remained at Hooton Park until they were struck off charge in April 1942 and it seems likely that they suffered damage on the voyage from Canada. The fact that Peter Johnson recalls working in deep snow on the fuselage of a Hampden that was parked on a dispersal and cannibalising it for spare parts tends to confirm this.

P5396 (5) was another Canadian-built Hampden that had a very brief flying career. On 7 November 1941 when being flown away from Hooton Park it crashed when it overshot at Burtonwood with the loss of its American ATA pilot, First Officer I J Parades.

Of the 80 Canadian-built Hampdens that reached service units in the United Kingdom, 22 were lost on operations, 25 were written off in training accidents in the UK and 30 were struck off charge in UK. Two were handed over to the Russians and the last of the 80, P5298, was flown back to Canada. P5298 (6) had been the first Canadian-built Hampden to fly and was the first to arrive at 7 AAU, after which it saw service with 5 BGS at Jurby on the Isle of Man until the end of 1942. It then languished at 44 MU Edzell for several months until it was decided that it was to be one of the 25 Hampdens intended to be flown back to Canada. P5298 arrived at 32 OTU Patricia Bay, Vancouver on 7 November 1943 and after three months flying on training duties it made its last flight to the RCAF station at Sea Island BC on 21 February to be scrapped. According to some accounts 200 Hampdens were flown out to Canada but this is not borne out by contemporary records. The 32 OTU Record Book refers to 25 Hampdens being ferried by air from the United Kingdom and the Hampden "Movement Cards" show that 25 aircraft arrived at 3 OAPU, Llandow, in South Wales to be prepared for the flight to Canada. "OAPU" stood for Overseas Aircraft Preparation Unit and not Old and Ancient Planes Unit, which might have been an appropriate description of most of these battle-weary Hampdens.

Canadian – built Hampdens ready to begin the voyage from Halifax, Nova Scotia to the United Kingdom. The outer wing and tail sections were stowed in the ship's hold. The aircraft had been test flown in Canada, after which the engines were removed and remained in Canada, new engines awaiting the aircraft at No.7 Aircraft Assembly Unit, Hooton Park, Cheshire. The photo was taken in April 1941 and as P5306 – 5308 and AN152 – 154 all arrived at 7 AAU at the end of April, it is likely that the two illustrated came from this shipment. (Photo: Public Archives of Canada PA 105317)

A torpedo is loaded into the bomb – bay of a Hampden at 32 OTU at Sidney, BC, on 2 October 1942. (Photo: Public Archives of Canada PA 144773)

Of the 25 Hampdens that arrived at 3 OAPU, P1254 and P5320 (7) progressed no further. AE117 (7) flew as far as Prestwick, which was the aerodrome from which the Hampdens took off on the first leg of the 5000 – mile flight, but failed to leave UK. Serviceability problems were obviously the reason for these three non – starters which were eventually scrapped at 44 MU Edzell. Twenty – two Hampdens took off from Prestwick on various dates, the first arriving at 32 OTU being P1230 on 22 August 1943 and P5298 was the last on 7 November 1943. The Hampdens to safely make the trip were L4142, L4145, L4157, L6069, X3137, X3149, P1167, P1200, P1230, P1311, P2067, P2133, P5298, AD751, AD754, AD767, AD961, AE258, AE295, AE363 and AT147 (8). P1200 and AD767 (9) were written off in accidents at 32 OTU. One Hampden (AE309 (10)) failed to arrive; the grim ordeal of its crew of three after it went down in the sea off Greenland is given at the end of this section.

The task of flying these Hampdens to Canada fell to No.45 (Atlantic Transport) Group. No.45 Group was made up of Service pilots and very experienced civilian crews. One of their pilots was Jack Fitzgerald. Jack had flown Liberator BZ816 to England and was given P1167 (11) to ferry out to Vancouver. He left Llandow on 18 August 1943, flew the more southerly route across Canada and arrived at 32 OTU on 25 August after a seven leg flight that took 38 hours flying time. Don McVicar, another senior pilot of 45 Group, has written several books about his experiences when flying a wide variety of aircraft on delivery flights. In his book "North Atlantic Cat", published by Airlife in 1983, Don describes two Hampden ferry flights and the publishers have kindly allowed a few extracts from the book to be quoted.

Don begins the Hampden chapter thus:
"I suppose it made good sense to fly these clapped – out Hampdens, no longer useful on ops, to Canada, to keep the RCAF happy. But I was not altogether happy, the aircraft had some good feature but quite a few bad ones ... It had excellent visibility and light controls. Until you got rudder lock if you inadvertently got into a sideslip. Ailerons wouldn't get you out. When the fuel was low the engines would quit if you made a skidding turn. And the fuel selectors were controlled by the radio operator, definitely not good. On the other hand at 10,000 feet it had about a ten – hour range at an indicated 150 mph. It had an auto – pilot and built – in tail wheel shimmy. The Pegasus XVIII engines were said to be reliable. Both big cowl gill installations were controlled by one hand crank, so, if you lost one engine and had to pull extra power out of the good one and it heated up, you had double drag. The props were non – feathering, more drag. It had one generator and one hydraulic pump on the starboard engine, so make sure that it was the port engine that failed. I felt some madman had designed the landing gear and flap controls. They were hidden away in a tin box with a spring loaded lid".

On August 17, after all the snags had been sorted out, we took off in P1230 (12) for Prestwick. Next day we flew to Reykjavik in Iceland, refuelled and headed for BW – 8 on the West side of the Greenland ice – cap. After climbing above the overcast the port engine decided to add to my grey hair. It belched a couple of times and I thought the engine was going to jump out of its mounting. After jerking the controls violently the engine smoothed down and from then on I never had a murmur from either engine."

Don was flying with a radio operator as crew and doing the navigation himself, he continued his story, "We climbed away from BW – 8 and 3 hours 45 minutes later landed on the gravel runway between the hills at the head of Frobisher Bay. In the face of a poor weather forecast we got away after a delay of five hours; it took 2 hrs 25 mins to reach Coral airport. Next morning we took off, the weather forecast was bad, we flew over Hudson Bay and I got a pin – point on Chesterfield Inlet. Then across the featureless tundra, unmapped and with banks of threatening cloud ahead and below – nothing. The magnetic compass was swirling around, the Magnetic Pole was only 500 miles away. I got a sun shot. The clouds closed in overhead, I couldn't get another sun shot so I descended below the clouds, close to the tundra, and fired my Very pistol into the bog to see which way the wind was blowing. The forecast wind was north west, the wind on the ground seemed to be south west so I altered course 13 degrees to the southward and eventually identified the south shore of Lake Athabasca, I was 120 miles north west of track. I turned due south and landed at Edmonton after being in the air for 7 hours 20 minutes. We stayed there overnight and all next day."

"On the morning of 22 August we took off for the final leg. We climbed to 13,000 feet and had got into cloud just west of the Selkirk Mountains. The air quickly became turbulent and with astonishing suddenness we acquired a coating of ice. The propellers threw chunks of ice against our slab – sided fuselage and I felt as if I was being pummelled physically. By now I had the throttles wide open and the supercharger in high blower. We were climbing. (But oh, so slowly!) Finally, at about 17,000 feet I saw welcome blue breaks in the cloud, and we staggered into the clear. I daren't try to engage the auto – pilot because the aircraft felt as if it were balanced on the edge of a very sharp knife. Even with full power, and the gas being gulped so fast that I could see the gauges moving, we were right on the fatal stall and spin. From time to time a gruesome shudder ran through the airframe as the little tailplane tried to stop flying. I shuddered in complete sympathy. The estimated time of arrival for Vancouver came and went, the westerly winds behind the front must have been almost 100 mph. Then, as I came over a vertical bank of cloud I saw Vancouver harbour and began a violent dive into warmer air. The ice began to slide off in large chunks and I began to breathe normally again. I looked back and found Johnson grinning at me with his thumbs up. Good man, I suppose Vancouver being his home town made it doubly welcome." Don would not have been amused if he had known that the 32 OTU Record Book on 22 August 1943 stated "1230 arrived on this unit having been flown from UK. Flying time 30 hrs 10 mins. Pilot S/Ldr J R Johnson."

Don made the return flight to Prestwick with the delivery of a Marauder and on 5 September was again at Llandow where, after L4142 (13) had been thoroughly tested, he flew it on to Prestwick and after a quick refuel carried on to Iceland. On 6 September Don made it to BW – 8 in Greenland where "the airfield guards seemed to think that the Hampden was a Do 17. Next day the flight from BW – 8 to Fort Smith in the North West Territories was made in 11 hrs 50 mins with a refueling stop at Coral airport on Southampton Island at the north end of Hudson Bay. The weather was beautiful, with a big high pressure area centred over Frobisher. The weather on 8 September was again ideal, after a short hop to Fort St John we set out on the final leg to Patricia Bay, Vancouver, and on this flight Mount Robson in Alberta and Mount Baker in Washington could be seem simultaneously. It occurred to me that we were due for something rotten soon. It happened. The wheels refused to come down. I pumped and pumped while we droned around. No joy. Finally I took the last resort and pulled the emergency air bottle. I guess they could hear Johnson's and my own sighs of relief on the ground as the wheels groaned into place. Our total flying time had been 27 hrs 45 mins. I had plenty of time later to realise how lucky we'd been. An obsolete aircraft with no de – icing and no single – engine performance, over a route where rescue would be almost impossible. No one in their right mind would have done it. You can stretch your skill and knowledge just so far. After that plain Luck has to be on your side."

Don's reflections on the need for Luck to be necessary on these long delivery flights is amply borne out by the experience of the crew of AE309 on 14 October 1943. AE309 (14) crewed by Captain R E Coffman, Flying Officer N E Greenaway RCAF and Radio Officer R E Snow, had been two hours out from Iceland and heading for Greenland when the starboard engine failed; they were at a height of 9000 feet and for an hour continued flying towards Greenland on one engine, gradually losing height as they did so. Captain Coffman's efforts to keep the Hampden in the air almost ended in disaster when the aircraft fell in a spin towards the sea and they were down to 1000 feet before he regained control. Twenty minutes later the port engine began to falter and eventually drew in its last gasp of petrol, spluttered, coughed and finally died. The sea beneath them was a mass of icebergs and smaller ice floes but a kindlier sort of Luck materialised in the form of a relatively clear patch of water in the lee of an iceberg. Bob Coffman glided down towards it and made a perfect belly landing; the emergency drill carried out by the crew was faultless, the impact was surprisingly gentle and the dinghy co – operated by automatically releasing from its housing in the port engine nacelle and inflating.

The three men clambered into the dinghy with their emergency kit and survival rations then cut the line that still secured it to the sinking Hampden; a minute or so later AE309 slid gracefully beneath the surface of the Arctic Sea. They took stock of their situation; both the pilot and navigator agreed that they must be about 15 miles from the southern tip of Greenland but Ron Snow expressed doubts about whether his SOS calls had been received as he had had no acknowledgement signal. There was only one course open to them and that was to try and paddle the dinghy through the ice floes to that distant shore. Twenty hours later, after having suffered the intense cold of an Arctic night, they were close to the shore but dangerous tidal eddies threatened to overturn the dinghy as they tried to reach the mainland proper. They spotted a small beach at the foot of a massive outcrop of rock that rose almost vertically out of the sea to a height of 3,000 feet and paddled desperately towards it. Having landed on the beach they looked across the angry turbulent water that lay between them and the mainland and all agreed that it would be courting disaster to try to paddle across it.

The three men dragged the dinghy up the face of the rock on to a ledge, where it formed their only shelter, then awaited the arrival of the search planes and ships which they knew would have been sent out to look for them. Bob Coffman, assuming that it might be seven days before they were rescued, drew up a schedule of rations of six malted milk tablets and a cupful of water per man per day. Their main hope of attracting attention was the Very pistol with its 27 cartridges which when fired should be easily seen by searching aircraft or ships. The weather began to worsen; a storm of sleet and snow blew at them, the wind tore their distress flag away and huge waves crashed against the rock forcing them to seek a safer spot on a narrow ledge over 200 feet above the sea. Days dragged slowly by, until on the ninth day they sighted a small ship in the distance, but to their dismay it failed to see their distress signals.

Another blizzard swept over them; they were now in a state of complete exhaustion and were too weak to stand up. After another night had passed by, an aircraft was heard approaching as dawn was breaking but it had begun to climb to clear the 7,000 ft mountains to the north of them and again their signals were not seen. The men huddled together through the cold cloudy day and awaited the coming of night which each of them secretly thought would be their last. The next day dawned, as had the previous ten, grey and bitterly cold. Ron Snow crawled out from under the shelter of the dinghy and glanced seawards. To his astonishment he saw a small ship a few miles out to sea and apparently not moving. The three men watched the last of their Very cartridges soar skywards, then as the sun began to break through flashed their helio signalling mirror towards the vessel.

A United States Army Major, John T Crowell, saw the last of the Very lights, at first believing it to be a seagull, then when he saw the flash from the helio mirror he decided to investigate. To the desperate trio on the rock it had seemed an eternity before a boat was launched from the ship, the *Polar Bjorn* and headed slowly in their direction and rescued them. It could truly be said that they had been snatched from the very brink of extinction. "Lady Luck" had indeed had a hand in their salvation as the *Polar Bjorn*, a converted Norwegian whaling ship, had only stopped so that repairs could be made to her engine and she was the last ship that was likely to be in that area before the onset of winter.

Canadian Hampden Postscripts: Ken Kenton, an RAF pilot with 1100 hours flying on twin-engined aircraft, arrived at 32 OTU on 25 January 1944, to be greeted with the question "Have you ever flown a Hampden?". On replying in the negative he was enthusiastically advised "Well now is your chance, we have to fly our Hampdens to the scrapyard". After doing a few circuits and bumps to familiarise himself with the aircraft Ken was detailed to join in on the last rites next day. His main memory of the flight across the eighty miles of water was that the chance of at least one of the aircraft not surviving the flight had been appreciated; none of the Hampdens carried anyone but the pilot and a string of flying boats (Stranraers from 3 OTU) had been positioned along the route.

In the event all the Hampdens made it across the water where the pilots were told to taxy and park their aircraft close behind the aircraft that had just landed, ready for the attention of the wrecking gang.

The final words on the Canadian Hampdens must be the welcome news that the Canadian Museum of Flight and Transportation, at Vancouver, recovered most of the remains of P5436 (15) from 600 feet down on the sea-bed of Saanich Inlet, just west of Patricia Bay airfield. The recovery took place during the summer of 1986 and was achieved by using a remotely controlled submarine with the aid of a video monitor as the wreckage lay at a depth that was beyond the reach of divers. The task of restoring P5436 has already commenced. Time alone will tell whether it will one day become the most impressive exhibit in their Museum.

Canadian-built AJ991 of 32 OTU, Patricia Bay, drops a practice torpedo. The flame damper exhaust of the port engine and the monoplane air tail of the torpedo show up well in this photo which was taken on 29 April 1943. AJ991 came to grief a month later when it overshot at Tofino airfield on landing with a failed starboard engine. (Photo: Public Archives of Canada PA 144776)

L4098 of 44 Squadron landed at Waddington with undercarriage not fully locked, 1615 hrs, 2 May 1940. (Photo: Harry Moyle)

PART SIX: NON-OPERATIONAL WRITE-OFFS

6A SUMMARY OF NON-OPERATIONAL WRITE-OFFS

The "Non – operational Write – off" total includes eighteen pre – war accidents involving L4039, L4046, L4047, L4052, L4056, L4061, L4082, L4101, L4107, L4114, L4122, L4128, L4130, L4136, L4155, L4177, L4179, L4192 (1). It should also be borne in mind that from the outbreak of war 7, 76 and 185 Squadrons functioned as training units. Between them they wrote off eleven Hampdens during the war months until April 1940 when 7 and 76 became 16 Operational Training Unit at Upper Heyford and 185 formed the basis of 14 OTU at Cottesmore.

Circumstance Cause	Hampdens Day	Night	Herefords Day	Total
On Take – off (55) (see 6B)				
Engine failure	13	2	5	20
Stalling	9	2	–	11
Failure to become airborne	15	9	–	24
Landing (134) (see 6C)				
Undercarriage collapse	31	10	4	45
Heavy landings/swings	14	3	–	17
Stalling	12	4	1	17
Undershoot	7	8	1	16
Overshoot	18	20	1	39
In flight (90) excluding above (see 6D)				
Due to engine failure	51	51	2	68
Collisions (see 5C)	22	–	–	22
In flight (115) attributable to particular aspects of training (see 6A, 6B, 6E)				
Pilot training	11	26	–	37
Bombing, Gunnery, Torpedo training	17	–	–	17
Cross – country navigation training	26	31	–	57
Single – engine flying practice	4	–	–	4
In flight (53) due to loss of control (see 5F)				
Stabilised Yaw	14	–	–	14
Spun in	13	–	–	13
Stalled and dived in	6	–	–	6
Loss of control in cloud	3	–	–	3
Broke up in air	4	–	–	4
Cause not known	10	3	–	13
In – flight miscellaneous (16)				
Air Test Flights (see 6E)	9	–	–	9
Unauthorised low flying (6E)	5	–	–	5
Enemy action (see Intruder 5H)	–	2	–	2
On the ground (25)				
Miscellaneous 25	–	–	2	5*

Total comprising 458 Hampdens
30 Herefords 488

Casualties arising from Non – Operational losses
Killed: 570 servicemen 10 non – service
Missing: 38 servicemen
Injured: 118 servicemen 2 non – service
Interned: 4 servicemen (in Eire)
29 servicemen baled out successfully
36 servicemen picked up unhurt after ditching.

106 Squadron served as a training squadron for the first twelve months of the war and wrote off seven Hampdens in that period before commencing flying on operations from Finningley in September 1940. The six Hampden squadrons that had been operational at the outbreak of war were 44 and 50 Squadrons from Waddington, 49 and 83 from Scampton and 61 and 144 from Hemswell. Over 300 of the non-operational write-offs occurred on training units, another 100 plus in training accidents on operational squadrons and 24 on Ferry Pilot Pool, Maintenance Unit or Experimental Establishment crashes. In the "Summary of Non-Operational write-offs" the "Cause" given is that which appears to have been the main factor in the accident; pilot inexperience and bad weather often combined to produce a write-off that a more experienced pilot would have been able to avoid.

A few examples from the various categories illustrate the situations that arose and should assist the reader to a better understanding of the problems that faced these young aircrew. Many of the non-operational write-offs appear in other appropriate sections of this book describing training flight incidents which involved baling out, crashing into high ground, running into our own balloon barrages, going down on the sea; bad weather incidents, intruder victims and accidents to Air Transport Auxiliary pilots. The incidents are mainly described under the aircraft serial number only, details of date, unit, crew names having been included in the aircraft tables. Besides the Hampdens that are listed as written off in crashes there were literally hundreds of others that were put back into service after accidents through the efforts of the maintenance staffs on the aerodromes. More seriously damaged aircraft were sent to Maintenance Units or Civilian Repair depots such as the one at Tollerton which specialised in bringing these mangled wrecks back to life, and which for a while had a good supply of airframes from the Herefords that were withdrawn from service.

It is not surprising to discover that over 40% of all "Training" accidents occurred as aircraft were taking off or landing, and pilot inexperience could justifiably be given as the cause of most of them. No RAF Hampdens were fitted with dual control and the pupil pilots could only become familiar with the take-off and landing procedures by sitting on the wing spar behind the instructor pilot and watching what happened. After several such flights the pupil would be cleared to make his first Hampden solo which would be followed by several hours "circuits and bumps".

The pupil pilot's introduction to night flying invariably took place at dusk, being preceded by a couple of landings as the daylight faded then continuing into the early darkness; their first solos were usually carried out without a wireless operator and the within sight of the aerodrome and flare-path. Once the pilot had demonstrated his ability to handle the Hampden he would venture further afield, with a wireless operator. Some of these wireless operators were "resting" after a tour of ops but many would have preferred the risk of operational flying. There were nevertheless a number of crash landings due to training aircraft becoming lost, particularly at night.

6B ACCIDENTS ON TAKE-OFF

Accidents on take-off varied from the "Failed to get airborne" variety, which although often accompanied by such impressive details as:- P2110 (1): "Crashed into paint store and caught fire" and P5388 (2): "Swung on take-off, hit L4202 (2) and burst into flames" were nevertheless often without serious injury to the crew. The accidents on take-off that were most lethal were the ones in which an engine failed, and it is noticeable that there seem to have been many more cases of the port engine failing than the starboard one. In some cases the engine failed before the aircraft had ended its take-off run and in many of these the crew escaped without serious injury as in such examples as P4307 (3): "Port engine failed on take-off causing aircraft to swing." "It struck P4312 (3), ran into a hedge, caught fire and burnt out". Hereford L6027 (4): "Port engine failed on take-off, collided with P2141 (4) which was written off then crashed through a hedge and hit Anson N5267". P1181 (5): "Destroyed at Upper Heyford after engine failure on pilot's first solo on a Hampden". There were in fact two occupants in P1181, the pilot Wg Cdr K Elliott and F/Sgt Couch, an Instructor pilot.

The reports of "No survivors" appear with a tragic regularity against those unfortunate pilots and crews who suffered an engine failure soon after take-off in the two or three minute period between an aircraft leaving the ground and actually becoming safely airborne with undercarriage and flaps up and a few hundred feet on the altimeter. P1278 (6) for instance "Port engine failed and was on fire soon after take-off. The pilot on his first Hampden solo was unable to control the aircraft and it dived into the ground and caught fire. Both crew were killed". Whether the second casualty was an instructor pilot or a wireless operator is difficult to ascertain. L4058 (7): "Port engine failed during take-off. Aircraft went into a spin and crashed one mile west of Upper Heyford". A few weeks earlier AD831 (8) "Crashed shortly after take-off from Upper Heyford after port engine failed". In both of these cases four men were killed. P1251 (9) "Crashed and caught fire after port engine had failed on take-off" There were five men on board P1251 and although there was no fatal injury the words "Seriously injured" appear against two men, "Burned" against another and "Injured" against the other two.

Other crashes soon after take-off are attributed to various causes and there were no survivors from any of the following take-off crashes. P1155 (10): "Dived into the ground shortly after take-off and crashed near Saltby satellite 'drome". Four killed. L6096 (11): "Stalled after taking off". Two killed. Hereford L6067 (12): "Crashed soon after take-off. Four killed". AD849 (13) "Crashed soon after take-off. Cause obscure". Four killed. P4319 (14): "Climbed steeply and went into stabilised yaw on take-off". Four killed. L4035 (15): "Climbed steeply on take-off and dived in almost vertically". Four killed.

Another hazard that faced the learner pilots at Upper Heyford in particular seems to have been the presence of large trees in the vicinity of the aerodrome. P1179 (16): "Turned to the left and hit a tree soon after take-off on first Hampden solo". P1180 (17): "Hit tree soon after take-off for solo night flying and crashed four miles away". Neither pilot survived the crash. One can only imagine the desperate struggle of the pilot of P1180 for those last few minutes as he tried to control his damaged Hampden and make some sort of emergency landing.

Crash on take-off. The pilot of P1208 "UB-C" of 455 Squadron lost control a few minutes after take-off from Wick on 27 October 1943 and made an involuntary return to earth. The aircraft bellylanded at speed, ripping its bomb-bay open and scattering its eight depth charges as it careered across a field until it hit a telegraph pole. This slewed the Hampden round and stopped it from hitting a crofter's cottage towards which it had been heading. Ray Lassiter, the navigator, is in the foreground. The point of impact with the pole can be seen at the junction of the starboard wing and fuselage. The top perspex section of the nose has broken away, which seems to have been typical of such a crash and may well have been the reason why Bob Galloway went clear of L4121 when it crashed in the sea on 29 September 1939 (see 2C 2). (Photo: Ray Lassiter)

AD747, AD904 (18): No.24 MU at Ternhill witnessed a double tragedy involving two 44 Squadron crews who had arrived there to collect a couple of brand new Hampdens on 19 June 1941. After the crews had checked their aircraft thoroughly and signed the necessary paperwork the pilots started their Hampdens up and prepared for take-off. Sergeant Greig, in AD904, was the first to become airborne and he circled the aerodrome awaiting the other Hampden, AD747 piloted by Pilot Officer Lauderdale, as they intended flying back to Waddington together. Sergeant Greig watched AD747 gather speed across the aerodrome but was distressed to see it begin to lose height soon after it had become airborne. An engine had failed and Pilot Officer Lauderdale had no option but to attempt a crash landing in which he suffered a fractured spine and was badly burned about the hands and face, as were his two crew members. The tragic scene that he was witnessing as he circled the wreck of AD747 from which flames and smoke were billowing may have caused Sergeant Greig to lose his concentration on his flying. The true reason will never be known, but whatever it was it resulted in AD904 suddenly stalling and in the crash that followed the three men on board lost their lives. A tragic end indeed to what should have been an interesting day out and a welcome change from the rigours of operational flying, during which three of the aircrew involved in the two crashes had previously featured in crashes on return from operations and survived. PO Lauderdale whilst a sergeant had crashed in AD847, Sergeant Drakeford in X2966 and Sergeant Greig in X2982 (18).

L6090 of 455 Squadron overshot at Tain on 11 August 1943. Both undercarriage legs were ripped off. The aircraft was on a daytime training flight and the crew were uninjured. (Photo via Dennis Thompson)

6C ACCIDENTS ON LANDING

The fact that it was twice as dangerous to return to earth as it was to depart from it is demonstrated by the fact that 50 Hampdens and 5 Herefords were written off in the process of taking off, whereas landings claimed 127 Hampdens and 7 Herefords.

Operational Training Units in particular give brief details of many landing accidents. "Heavy landing, undercarriage collapsed", "Undercarriage collapsed after swing on landing", "Landed with undercarriage not locked", "Wheels up landing" "Damaged tail on landing", "Port (or starboard) undercarriage failed to lock, one wheel landing". Two Herefords (1) managed the "one wheel landing" feat on successive days at 16 OTU, L6068 being written off, L6079 suffering repairable damage. In this sort of landing accident the aircraft was in most cases repairable and serious injury to any of the crew was the exception rather than the rule. There is a difficulty in categorising landing accidents because quite often other factors such as bad visibility or engine trouble were involved. Learner pilots often found difficulty in judging height and airspeed, particularly at night, and many unorthodox three point landings materialised in which a wing tip became one of the three points involved.

It was not always pilot error that caused such landing accidents. "Chiefy" Clayton wrote P4393 (2) off at Waddington when he came in to land after collecting the aircraft after repair. His arrival in the circuit was, to say the least of it, a bit erratic but he made a reasonable approach until the last moment when, instead of levelling off, the nose dipped towards the grass and his three – pointer became the nose of the aircraft and two wheels. The port undercarriage collapsed and three figures emerged from the Hampden and ran away from it just before black smoke and flames blossomed forth. According to some knowledgeable "erks" the elevator controls had been connected up wrongly.

Amongst the landing accident write offs are three Hampdens that were on their "Maiden Voyage" i.e. on their delivery flight to a service unit. P2086 and P2069 (3) were being delivered by RAF pilots of 4 Ferry Pilots Pool to 16 OTU. P2086 had engine trouble and attempted to land at Honeybourne aerodrome but hit a telegraph pole and crashed in an orchard. P2069 did manage to reach its destination but crashed as the starboard engine failed as it came in to land. AD763 (3) had been collected from 24 MU by a 106 Squadron crew and was written off when its undercarriage collapsed as it landed at Coningsby, its destination.

At the end of the Hampden era the obsolete aircraft were flown to MUs to be scrapped. P1217 (4) was amongst the last to go, being flown to Tollerton in January 1944 where it duly crashed on landing. None of these last – mentioned four accidents resulted in physical injury to the crew, although one suspects that there may have been a few red faces.

Attempts at landing on unfamiliar aerodromes when lost or in difficulties resulted in other accidents besides P2086 at Honeybourne. AD833 (5) tried to get down at Shoreham airfield at night when lost and in bad visibility and crashed in the grounds of Lancing College after hitting a tree, with the loss of its four crew. The crew of P5391 (6) had better luck when attempting a forced landing at Towyn aerodrome after having engine trouble when on air firing practice over the nearby ranges. The aircraft overshot and crashed on the beach without injury to the crew.

Swings on landing resulted in varying degrees of damage; P1314 and AN159 broke their backs, others collided with stationary objects and AN140, for example, hit the crash tender (7).

Undershoots and overshoots brought about a number of "interesting" accidents, at least from the onlooker's point of view. In undershoots P1268 hit a fuel bowser, P2132 demolished a stone wall and then struck a house. L4123, P2076 and P1184 merely "hit trees" (8). When overshooting P1298 hit a lorry as the pilot opened up to go round again, L4050 hit a petrol bowser and caught fire whilst P1285 hit a contractor's lorry and burned out (9).

In several crashes prompt action by people on the ground resulted in injured aircrew being dragged from wrecked and sometimes burning aircraft as happened when P1269 (10) overshot at Upper Heyford and caught fire after hitting a hangar. The pilot of P1269 was assisted from the burning wreck by Warrant Officer J B McGinn. When AD787 (11) swung on landing at Saltby and caught fire on the aerodrome the crew escaped injury "but Pilot Officer Higginson DFM was injured when trying to save equipment from the blazing aircraft".

The catalogue of accidents on landing includes many lucky escapes by aircrew from what must have been pretty horrific crashes, as the following incidents demonstrate. The pilot of P1270 (12), carrying out solo night flying practice, stalled on his approach and just managed to flatten out as the aircraft hit the ground and caught fire. He was lucky to survive the crash although he suffered considerable burns. The two crew of X3003 (12) also had a close call when their aircraft "struck the ground violently and caught fire" whilst on night circuits and bumps. P1225 (13) flew straight into the ground on its final approach when landing at Croughton after fog had come in unexpectedly. The undercarriage broke off and the aircraft caught fire N9065 (14) burst into flames after a fast, heavy, landing at Croughton.

The fact that so many crews did survive these crashes is in itself a tribute to the robust construction of the Hampden but not all were so lucky, as some examples of landing accidents that resulted in fatal casualties show only too well. The pilot of P1274 (15) opened up to go

L4042 overshot landing at Stormy Down on 23 April 1940. The aircraft had been transferred to newly-formed 16 OTU on 18 April and still bears the "MG" code of 7 Squadron which, with 76 Squadron, was incorporated into 16 OTU. (Photo: Philip J Davies)

L4039 of 49 Squadron overshot at Scampton on 1 August 1939. The pilot, Flt Lt Lerwill, was unhurt. Note the "XU" pre-war code for 49 Squadron (Photo: Peter Tunstall)

around again after a heavy landing and lost control, "probably due to the pilot's inexperience of instrument flying at night". AD988 (16) dived into the ground from 400 feet when approaching to land, whilst AD802's pilot "would appear to have lost control whilst looking for the flare-path". P1323 (17) "lost control whilst asking permission to land and dived into the ground". P1247 (18) dived into the ground on an overshoot at night. X2968 (19) crashed at night after making three attempts at landing at Upper Heyford. No survivors from any of these nor from two accidents on landing at 32 OTU in Canada when AN142 stalled and dived into the ground as it returned to base on one engine, and AN146 which spun in after turning on to its final approach (19).

P1289 and P5345 (20) were, when landing, involved in collisions with Ansons. P5345 of 16 OTU actually "landed on top of Anson R9668" at Upper Heyford, whilst P1289 of 14 OTU "after landing without permission, hit Anson R3386 which was taking off". None of the aircrew in the Hampdens suffered serious injury, although the crews of the Ansons were less fortunate.

A most tragic accident occurred to X2959 (21) at 0025 hours on 16 August 1941 when this 25 OTU Hampden crashed, apparently through being too low on its landing approach when returning to Balderton. The three sergeants in the crew were the pilot S D Baldachin, wireless operator T G Wood and lower gunner David Macdonald. The horrific consequences of the crash, and the subsequent inquest and funeral were reported at the time in the Newark Advertiser. Harry S Dring kindly provided copies of several relevant articles, extracts from which describe the sequence of events.

L4052 of 83 Squadron overshot when landing at Northolt on 28 February 1939. The pilot, FO A.O.Bridgeman, was unhurt and later became a prisoner of war when L4049 was shot down by flak. L4052 was on detachment with other 83 Squadron Hampdens for duties with the Air Fighting Development Unit at Northolt and was landing after giving a demonstration for the benefit of a visiting Arab delegation, who were no doubt suitably impressed. Most of the airmen in the photograph are wearing the old "choker" collar type of uniform tunic; the lad on the right seems to find the whole business highly amusing. The AFDU detachment ended with another write-off on 11 April 1939 when L4056 crashed as it was taking off to return to Scampton. (Photo: Douglas Garton via Peter Green)

"In the dark and still hours, just after midnight, a British bomber aircraft crashed on to 84 London Road, Balderton, in which Mrs Brompton and her seven children were sleeping. Mr Brompton worked on the railway and was on night duty. Next day a visit to the scene revealed that in the last minute of its flight the plane had first caught the tops of some tall trees, as was evident from a number of broken branches that were scattered about the area. It had then plunged downwards, severing two overhead electricity cables before lopping off the tops of two medium-sized fir trees in the garden of No.84 and then hitting the house. Fire broke out almost at once".

Describing her terrifying experience inside the house, Mrs Brompton said "I heard a bang, and thought it was a bomb. I had my little boy, Roy, with me and I jumped out of bed and ran to the bedroom door to call to my other children in the other rooms. But I found the doorway blocked and could see flames. Smoke poured into the room and I tried to open the window which has always been difficult to open. Eventually I did manage to open it and lowered Roy to someone who was outside. I then got out myself and tried to get round to the back of the house, but was unable to do so". In No.80, the house two doors away, "Mr Joseph Atkin who is 86 years of age had a remarkable escape when a propeller hurtled through the wall of his house and actually fell by his bedside".

David Macdonald, the Hampden air gunner, described his own remarkable escape. "Just immediately before we crashed I felt a bump and heard a crash (hitting the trees?) and this was followed at once by the crash. I did not know what we had struck. The plane burst into flames. The floor of the fuselage had been ripped open and I got through it and dropped to the ground. Sergeant Wood was sitting directly behind me, I heard him groaning and shouted to him to jump, but got no reply. I turned to go back to the 'plane, but owing to the flames I could not do anything. The petrol tanks exploded a few minutes later. I was taken away by ambulance".

Mr W H F Judson who lived in a nearby house heard the crash and rushed out clad only in his pyjamas. He described what followed. "The plane burst into flames as I approached it. I went to the back of the house, snatched a dustbin lid and climbed on to an outhouse then on to the wing of the burning aeroplane. I got as far as the pilot in the cockpit, I tried to release his harness but couldn't, and I had nothing to cut it with. The flames by this time had got so fierce that I had to run back along the wing. Before I got clear of the aircraft the petrol tanks exploded."

"Police Constable Allen, on duty in London Road, heard the crash which was followed almost immediately by fire and an explosion. The rear of the three houses had been demolished and all three were burning furiously. The occupants of the houses, with the exception of the six children, had got safely away. The children were sleeping in the bedrooms into which the aircraft crashed. Newark Fire Brigade were quickly on the scene but it was 0300 before the fire was brought under control. A rescue party arrived at 0415 and the bodies of the six children were later recovered.

The father of the children, Mr Brompton, was on shunting operations on the railway nearby when he saw the flames. He realised that they must be near his house and so was not altogether surprised to find that the disaster was at his home". The account of the funeral gives the ages of the children who died; they were Geoffrey 18 years, Joyce 14, Mary 13, Alan 11, Lucy 8 and Sheila 5. The two RAF sergeants who died were aware of the risk involved as Death and Flying were constant associates in the skies over England at that time, but it was a cruel twist of fate that snuffed out the lives of those six sleeping youngsters. One of the poignant messages of sympathy ended with the words "We wondered why you had to die, without a chance to say "Goodbye"". Anyone who reads this tragic story will surely say "Amen" to that.

6D ENGINE FAILURE IN FLIGHT

L4047 of 49 Squadron was the first Hampden to be written – off. It crashed 10 miles north of Nottingham when attempting a forced landing in bad weather on 17 January 1939. The pilot would seem to have been trying to make a wheels – down landing but the aircraft tipped on to its nose while still moving at a speed that was fast enough for the impact to rip both engines out and shear the undercarriage legs off. Nobody aboard suffered injury; the pilot, PO B.A.Mitchell, later became a prisoner of war.
(Photo: Peter Tunstall)

The incidents described under this heading concern the 56 Hampdens and 12 Herefords that came to grief after having an engine fail when they were safely airborne. In addition to the 68 write – offs the records reveal many other instances in which aircraft either made a successful landing on one engine at the nearest aerodrome or crash – landed at the earliest opportunity without serious injury to the crew and repairable damage to the aircraft.

The sight of the reassuringly vibrant, albeit noisy, bulk of an engine being suddenly transformed into a mass of flame and black smoke must have quickened the heartbeat of even the veteran pilots; to those less experienced it proved to be a harsh introduction to one of the major threats to survival in the air. In addition to providing an awesome spectacle the sudden loss of power from the engine necessitated prompt action by the pilot to maintain control of the aircraft and prevent it rolling over onto its back and going into a spin from which recovery would be almost impossible.

Problems with the cooling system of the Napier "Dagger" engines were never overcome, over half of the Hereford write – offs being directly attributable to engine failure with the words "engine caught fire" regularly appearing in the crash reports. L6016 and L6075 (1) are two examples of how pilots faced with an engine on fire emergency managed to keep their aircraft under control while their crews escaped by parachute. The pilot of L6016 then made a crash landing in a field and escaped without injury, although the Hereford burnt out. After the engine of L6075 had caught fire whilst carrying out a bombing exercise from Jurby its pilot ordered his crew to bale out, which they did; he then escaped himself "when the flames reached the cockpit".

Not all of the Hereford engines in fire incidents ended so happily however as the tragic crash reports of L6067 and N9074 demonstrate.

L6067 (2): "Probably had engine failure in flight, it stalled and hit the ground in a vertically banked condition at Souldern near Bicester with the loss of life of the four men on board".

N9074 (3): "The Port engine caught fire when the aircraft was flying at 6,000 feet on a practice bombing detail from Jurby. The aircraft dived into the sea. Although two of the crew were seen to bale out there were no survivors from the six men on board".

Two Air Transport Auxiliary pilots had the good fortune to walk away from Herefords L6070 and L6100, when they had crash landed after having engine failure. L6070 (4) failed to maintain height on its remaining good engine and made a belly landing in the course of which the Hereford careered across a field and "slid over a canal before coming to rest". L6100 (5) was actually being delivered to Cranwell where it was to end its days as an Instructional Airframe. The aircraft arrived in the circuit and had an engine fail, crashing one mile south of Wellingore. It is most unlikely that its battered remains would have been of any use to the training school at Cranwell!

N9091 (6) crashed through engine failure near Lodminton village when being flown from 5 MU to 20 MU. Its pilot took advantage of a corn field to make a belly landing from which he escaped unhurt. N9091 was hardly interesting enough to merit a mention but the comment of the Officer Commanding, 5 MU, at the subsequent crash investigation summed up the sentiments of pretty well all the pilots who flew Herefords. His words were brief and very much to the point. "Test pilots of this unit have no confidence in this type of engine".

Many of the aircrew who flew on Hampdens on operations carried out part of their training on Herefords. Those who survived a tour of ops on Hampdens must have felt that the wheel of fortune had turned full circle when they were posted to Avro Manchester squadrons for further operational duties. The Rolls – Royce Vulture engines of the Manchester suffered from the same cooling problems as did the Napier Daggers of the Hereford with the attendant evil habit of catching fire in the air. It was a great tragedy that many experienced ex – Hampden aircrew perished in accidents and on operations before the Manchester was withdrawn from service. It is also a matter of history that the failure of the Manchester led to the birth of the Avro Lancaster which was virtually a Manchester with its two Vultures replaced by four Rolls – Royce Merlin engines.

The Hampden's Bristol Pegasus XVIII engines were infinitely more reliable than the Hereford's Napier Daggers but even so there were many Hampden engine failure incidents which created the same basic problems already described in the Hereford paragraphs. No useful purpose would be served by enumerating too many similar incidents but some of the Hampden engine failure write – offs are worthy of mention.

L4179 (7) crashed into a fir tree plantation on Culloden Moor, Inverness, after both engines failed on a flight from Finningley to Evanton for Armament Practice Camp training. Although the pilot suffered a broken leg in the crash, he and the other crew members were clear of their wrecked Hampden before, as the records relate, it caught fire. There is the added comment that "The fire started in a small way under one of the engines and then rapidly spread to the whole aircraft – an automatic fire extinguisher fitted to the engine would probably have prevented this". The comment about an automatic fire extinguisher is fair enough, but the real problem in an aircraft crash was that petrol pipes were likely to be fractured and fuel tanks ruptured after which it only required a spark from a damaged electrical supply circuit to send the whole lot up in smoke.

P1161 (8) had been carrying out practice bombing when an engine failed and the pilot headed back towards Manby on one engine. The aircraft was three miles out to sea and flying at 500 feet when the pilot lowered the undercarriage and the Hampden stalled and crashed into the sea with the loss of its four crew. The record seems to infer that the pilot erred in lowering the undercarriage too soon, but had he delayed doing so for another two or three minutes, when he would have been approaching Manby, it is not all that obvious that the likelihood of the Hampden diving in would have been any less.

P2142 (9) was circling Upper Heyford at night, awaiting permission to land, when its starboard engine failed. The pilot ordered his crew to bale out and then made his own escape. He obviously did not relish the prospect of circling around at night on one engine and then having to make a single engine landing as a finale.

The crew of P5319 (10) was not so lucky when an engine failed on a night cross country flight. The pilot lost control of the aircraft and ordered his crew to bale out before he made his own escape from a height that was only just sufficient to allow his parachute to open. His three crew mates were still in the Hampden when it crashed.

P4292 (11) illustrates the problem that could arise when a pilot attempted an emergency landing at night with a failed engine. "It flew low over the airfield and was given permission to land. While turning, the aircraft rolled onto its back and dived in from 300 feet". None of the five men on board survived the crash.

AT186 (12) had an engine fail whilst carrying out practice low – level bombing. The pilot, after a rapid appraisal of the situation, made a crash landing from which he and his crew escaped unhurt before the Hampden caught fire and burnt out. L4120 and P4314 (13): Although there was nothing spectacular about either crash they merit mention as the same pilot, Sergeant I P Mapp, achieved them both within a matter of three weeks. He put L4120 down in a field half a mile from Finningley when its port engine failed on a daytime air – test, then made a forced landing at night a mile from Finningley when P4314 suffered engine failure. Two other Hampdens appear in the serial number lists as having been written off by a pilot named Mapp, AE229 by Flight Sergeant Mapp and AE394 by Warrant Officer Mapp and it is quite conceivable that all four Hampdens are attributable to the same gentleman (13).

X3138 and AE439 (14) are two examples of Hampdens that had an engine catch fire in flight. X3138 crash – landed at Whitchurch on a ferrying flight whilst AE439 achieved the distinction of being the last Hampden to be written off in a flying accident when it crash – landed after its starboard engine caught fire on a communications flight from Boscombe Down to Hendon on 2 October 1944.

The torpedo – bomber Hampdens had a fair share of engine failures on training flights, but the chances of an aircraft reaching a safe landing place on one engine dropped to zero when some TB Hampdens were fitted with Air – to – Surface Vessel radar (ASV) in 1943. L6019 and P2113 (15), both fitted with ASV, were written off on delivery flights by ATA pilots and are included in the ATA section. (See 5J1)

AT184 (16) had been flown from Leuchars to St Athan for "RDF mods". RDF (Radio Direction Finding) was the term by which radar had been known in its early days and the old abbreviation conveniently cloaked the true nature of the top secret ASV equipment. A 455 Squadron crew flew AT184 back from St Athan but never reached Leuchars as engine failure brought about a crash – landing near Wigtown.

Other TB Hampdens were written off after engine failure during the 1943 ASV era and although there is no proof that they were fitted with ASV it at least seems to be a possibility. AE195 AE228 and AN148 (17), all of 455 Squadron, were all written off in engine failure incidents. AE195 went down in the sea on a night exercise, AE228 crash – landed near Leuchars after an engine caught fire and AN148 spun into the sea after engine failure at night.

X2905 (18) of 489 RNZAF Squadron crash – landed after engine failure. 489 also had ASV Hampdens. P5341 (19) of 1 Torpedo Training Unit ditched in the Firth of Clyde after engine failure and although there is no record that training in the use of ASV was carried out at 1 TTU it is not unreasonable to expect that it was. No.32 Operational Training Unit at Patricia Bay near Vancouver on the west coast of Canada made use of Hampdens, mostly Canadian – built, from May 1942 to the end of January 1944, its main training task in that period being the provision of crews for torpedo – bombers. In the autumn of 1943 twenty – one Hampdens were flown out to 32 OTU from the United Kingdom and arrived safely. AE309 (20) was also intended for 32 OTU but it had engine failure and went down in the sea off Greenland. That story and other information on the Canada Hampdens is dealt with in Part 5K, but it is appropriate to describe some of the engine failure incidents in this section.

Three 32 OTU engine failure accidents show very clearly how the fortune of the crew in similar incidents could vary from amazingly good to tragically bad. AN141 (21) was flying at 4000 feet when an engine failed; the aircraft gradually lost height. The other engine failed and the Hampden went down in the sea off James Island. All four members of the crew were picked up from the water uninjured. Good fortune for the crew of AN141. AN142 (22) After sending distress signals stating that one engine had failed, the aircraft returned to base. It flew across the aerodrome at approximately 400 feet with only the port engine running and yawed to starboard then stalled and crashed to the ground and caught fire. Three of the crew were killed, the fourth was dangerously injured and died later. The luck of AN142's crew could not possibly have been worse.

AJ989 (23) was observed descending with one engine stopped and with smoke and flames issuing from it. There were no eye witnesses to the crash in the sea but the pilot Sergeant Edward Williamson, who was slightly injured left no doubt about what had happened when he made out his statement for the Court of Inquiry. It says:

"At approximately 1115 hours I was flying at 3500 feet and decided to carry out a practice dive bombing attack. I chose a small island and dived down towards it at about 200 mph. I levelled off at about 1500 feet and continued on down in a shallow dive, practising evasive action then levelling off at about 200 feet. I then noticed that the starboard engine oil temperature was 95 deg.

I opened the gills and continued to fly level. At this point the aircraft swung to starboard and I noticed that the starboard engine was on fire. I immediately throttled back both engines slightly to regain control and pressed the starboard fire extinguisher button. I then opened the cockpit hood and prepared to force land on the sea. I was about to land on the sea when I noticed a large flat rock straight ahead in the path of the aircraft. I opened the port throttle to avoid the rock and the aircraft swung violently and the starboard wing, went down and the aircraft slipped to starboard. The starboard wingtip hit the water and the aircraft turned over onto its back as it entered the water and sank immediately. The crew were all inside the aircraft until it was completely submerged, when they managed to struggle to the surface. We then swam to some rocks nearby where we rested until we were picked up by a passing boat."

The report concluded with the possible cause of the engine failure as ascertained from an examination of the engine after it had been salvaged. "... It is possible that oil may have escaped from the thrust race bearing in the reduction gear casing, especially as considerable wear was found in the tail shaft bearing. The oil may have been thrown onto the exhaust stub pipes and so caught fire, but this is thought unlikely owing to the high flash point of the oil." The report concluded that: –

"The evidence given in court, as well as the examination of the engines after salvage, indicates that the fire was due to ignition of fuel from the broken priming line." The crew of AJ989 had a miraculous escape and were extremely fortunate to all have been sufficiently conscious immediately after the crash for them all to have struggled clear of the upturned aircraft.

6E LOW FLYING AND AIR TEST FLIGHTS

VN – Z, AE184 of 50 Squadron, at Lindholme in the summer of 1941. This aircraft was written – off when it overshot at Ratcliffe on a non – operational flight on 25 October 1941. (Photo via C.G.Jefford)

A number of Hampdens crashed when carrying out low flying. Most were engaged on authorised training to acquire the skills necessary for low level and torpedo bombing, and air gunnery. John Enns, a 415 Squadron pilot, comments on Hampden low flying in the Torpedo Bomber section (see 2J37), and Hubert Clark who flew with 44 Squadron mentioned that in the early summer of 1940 he carried out trials at night in which two spot – lights were fixed in the aircraft to enable height to be assessed with some degree of accuracy. The Dambusters later used a similar idea, and it may be just a coincidence that Guy Gibson, who led the attack, had been a Hampden pilot during the first year of the war.

There is no doubt whatsoever about the fact that pilots considered the Hampden to be a marvellous aircraft for low flying and they had plenty of experience of it in the numerous daylight exercises involving large formations of Hampdens that took place in July and August of 1939.

In the course of these exercises the squadrons, flying at low level, vied with one another to see which could fly lowest across the sea or over the flat fen land of Cambridgeshire; the sections within each squadron sought to outdo one another and each aircraft in a section of three tried to fly a few feet lower than his companions. This rivalry inevitably resulted in a few close calls as propeller tips touched the sea or wings came too close to trees, and on one occasion an aircraft survived after having sea water forced into its engine air intakes. The redoubtable "Peggies" after coughing, spluttering and belching smoke in protest resumed their normal rhythm. The wheat in the flat fenland fields thrashed about madly in the blast from over a hundred slipstreams, telephone wires fluttered in protest and lofty electricity supply pylons looked down in disbelief at Hampdens zooming beneath their outstretched cables. Horses and cattle took fright at these roaring monsters and stampeded.

The civilian populace reacted in different ways. Some waved excitedly, others equally excited wrote letters of complaint at this nerve – shattering intrusion into their privacy and one silver fox breeder even sought compensation on the grounds that the experience had upset the love life of his foxes and played havoc with the natural process of procreation. The sense of mutual security engendered by this low level activity proved to be an unfortunate illusion, as the daylight debacles of 29 September 1939 and 12 April 1940 demonstrated. The disadvantages of flying very close to the surface of the sea in formation were threefold. In the first place the lower gunner had no opportunity of using his guns, secondly a formation was unable to take violent evasive action and, lastly, the attacking pilots were able to correct their aim by observing the splashes of their cannon shells or bullets on the sea.

The pre – war pilots were not unique in their appreciation of the exhilaration of low flying; the RAF Volunteer Reserve pilots who began flying Hampdens after war had started soon developed a taste for it. The records reveal that occasionally one had the misfortune to be disciplined for indulging in the forbidden practice of unauthorised low flying. Sergeant

s posted from 106 Squadron to 44 Squadron for duty
ing below power cables, and Sergeant Loadsman, who
training with 76 Squadron, was court martialled and
ks for a similar misdeed. Low flying over enemy
thorised or not, was much less likely to incur the
but Flight Sergeant Dillon's encounter with an
e over north west France proved to be something of
itizen who had enlisted in the RCAF "was flying at
ticed a huge shape in front of the aircraft, which
steel tower cable support". The aircraft hit a cable which damaged the starboard wingtip and aileron, smashed the pilot's windscreen, and left ten feet of cable dangling from the starboard wingtip. The good fortune of this Yank deserted him a fortnight later when he failed to return from a mine-laying flight to Denmark in AT224 (1).

Unauthorised low flying

Incidents actually attributed to unauthorised low flying include AD759, AD830 and AD935.

AD759 (2) spun into the ground with the loss of its two crew.

AD830 (2) struck houses at Evington near Leicester when low flying over the house of the pilot's girl friend and carrying out unauthorised aerobatics. Two of the Hampden's crew were killed and one injured and a civilian in the house killed.

AD935 (3) hit trees while low flying near the house where the pilot's girl friend lived at Badsley, Broadway, Worcester. The pilot and one of his crew were killed, another injured. Other crashes may have been the result of pilot's having this understandable urge to treat their loved ones or friends to an impromptu exhibition of aerobatics.

Air Test Flights

Routine test flights, such as were made after major maintenance had been carried out or before operations or night flying practice, provided a heaven-sent opportunity for a pilot to put an aircraft through its paces. It also afforded a chance for ground crew or visiting Air Training Corps cadets to take to the air. Several Hampdens crashed on such flights and in many cases no one survived to explain what happened, but it is at least as likely that the pilot lost control while demonstrating his flying ability and the manoeuvrability of the aircraft, as it is that anything went wrong with Hampden. Aircraft lost on air test flights including AD980 and AT180 (4). Pilot Officer H J Randall can claim the unique distinction of writing off these two Hampdens on air test flights in the space of three days, both of them within sight of Balderton where he was based. On 4 June 1942 he crashed in AT180 some three miles east of Balderton. He and his navigator PO R F Wishart were uninjured but two ground crew passengers were injured. On 6 June Pilot Officers Randall and Wishart took off for an air test in AD980. They had with them Sergeant H E Elliott who was proceeding on posting after completing his tour of operations. AD980 crashed into a field near Balderton and was a write-off; its three occupants all suffered injury and were taken to Newark hospital. AT113 (5) was another Balderton Hampden that came to grief on an air test. Squadron Leader L B B Price took off for an afternoon test flight on 29 July 1942; he had a wireless operator and two Air Training Corps cadets with him. All lost their lives when the aircraft stalled and dived into the ground. As in so many of these crashes "the aircraft burned and was just a mass of wreckage". Besides the tragic deaths of the two ATC lads it is sad to see that the RCAF pilot had been recommended for the award of the Distinguished Flying Cross a few weeks before he crashed but never lived to receive it.

Other aircraft which crashed on air test included AT970 (6) which stalled and crashed into the sea off Skegness while low flying; fortunately two of the three men on board survived. AD966 (7) dived into the ground at South Park near Lincoln. Two ground crew airmen were amongst the five men who lost their lives. P4389 (8) lost height rapidly after taking off for a test flight from Lindholme, and crashed near Finningley. Five were killed, three of whom were airmen passengers. AE292 (9) stalled in a turn and crashed into the ground at Poynton Hall, Butterwick, Boston, Lincs. Both occupants were killed. AD867 (10) stalled in a turn two miles from Lindholme aerodrome. The two men on board were killed.

Authorised Low Flying

The height at which the Hampden pilots flew their aircraft is evident in an accident involving L4077 on 21 October 1939 at Tollerton, which at the time was in use as a dispersal aerodrome for the Waddington squadrons. L4077 (11) took off from Tollerton to carry out low-flying attacks on a machine gun post that had been erected on the centre of the aerodrome for the purpose of giving the men of 53 Light Anti-Aircraft battery practice in the use of their weapons. Sergeant G M Wild, the Hampden pilot, flew that little bit too low and hit the gun. Army Sergeant Fretwell was killed and Gunner Wardele injured, but the Hampden landed safely, having been only slightly damaged. The Hampden crew were unhurt but obviously appalled by the tragic consequence of the incident. Sergeant Wild did not live very long to reflect on what had happened, as he was pilot of L4073 (11) on 12 April 1940 when it was shot down in flames off Norway with the loss of its crew.

The crews of torpedo-bomber Hampdens that hit the sea while practising their art suffered varying fates as the following incidents demonstrate. Between 1 October and 13 October 1942, three torpedo training Hampdens went into the sea off Girvan, Ayrshire (12). AD974's crew of three were picked up unhurt by a recovery launch and the crew of four of AN125 were all picked up by the Marine Craft section, but one died. When AN151 went into the sea on night torpedo practice, the four men picked up by the launch were all dead or dying.

Three more Hampdens went into the sea on daytime torpedo practice flights off the west coast of Scotland. P2091 (13) dived in three quarters of a mile west of Rillane Castle, Ayr and only one of the four crew survived, injured. When P1222 (14) hit the sea a mile west of Artach lighthouse, Argyll, it bounced and caught fire before plunging into the sea with the loss of its four crew, of whom one was RNZAF, two RCAF and one RAF. The crew of P1198 (15) fared better when they all lived to tell the tale, although injured, after they too had bounced after hitting the sea and then stalled and crashed into the sea off Lady Island in the Firth of Clyde.

It is no surprise to find that the chance of a crew surviving a crash which occurred when they were carrying out low-level bombing practice over land was almost zero. P1216 (16) for instance stalled and dived into the ground near Wigtown. None of the three crew survived, nor did the four men in P5346 which dived into the ground whilst carrying out low-level bombing practice at night on the Otmoor bombing range.

Incidents in Canada

The Hampdens at 32 OTU in Canada also suffered their share of losses on low-level practice flights as in its Hampden era the unit was mainly training crews for torpedo-bomber operations.

AN101 (17), which had been practising torpedo attacks, crashed into the sea off Turn Point light, NE of Stuart Island, when a mile away from the CPR steamship *Princess Alice*, its four crew perishing. The 32 OTU report on the crash of AN101 mentions that in this and other Hampden crashes the aircraft seemed to have had a problem with a climbing turn to port, often after making a practice torpedo drop, there being a tendency for the turn to develop into a stall from which the pilot was unable to recover control, with a resultant crash into the sea.

AN121 (18) crashed into the sea off Cowichan Head after completing a low level bombing run. It pulled up into a steep climbing turn to port, lost speed, turned over onto its back and flew into the water upside down. The range safety boat picked up two members of the crew, one of whom died; two others were killed. Sergeant K E Thompson was the only survivor, slightly injured, and he seems to have been flying in the rear lower gun position. The pilot of AN135 (19), flying solo, was carrying out authorised low flying over the sea when, while practising turns at 100-150 feet, the starboard engine cut out causing the aircraft to yaw to starboard and crash into the sea near Gulf Island. The pilot, suffering from shock, was rescued by a United States Coast Guard cutter. AJ988 and AJ989 (20) both crashed into the sea off Sidney Island near the 32 OTU aerodrome at Patricia Bay. AJ988 was carrying out low flying when its starboard engine caught fire. It crashed into the sea where it exploded on impact, both men on board being killed.

AJ989 also went down on the sea with an engine stopped and exuding flames and smoke, but its crew were picked up by a fishing boat some 20 minutes later. The crash report mentioned that "most Hampdens still did not have Graviner fire extinguisher rings fitted" and also commended Wireless Operator Sergeant C F Johnson for his courage and unselfishness in saving the lives of two of his crew mates. All four crewmen suffered some degree of injury.

P5436 and AJ993 (21) were two more aircraft that went into the sea after making a turn. P5436 stalled after turning to port at 800 feet and crashed on the sea off Saanich Inlet. The crew, none of whom suffered more that minor injuries, got clear of the aircraft and, although the dinghy failed to operate, were all picked up by a Stranraer flying boat that had seen the Hampden ditch, within fifteen minutes of going into the water. The crew of AJ993 were far less fortunate when their Hampden was seen to dive into the sea in the Cape Flattery area from a turn at about 400 feet, as although wreckage from the aircraft was found there was no trace of the four men.

AJ992 (22) had earlier that year gone missing on a night navigation exercise and wreckage had been picked up also in the Cape Flattery area. All the evidence suggested that the pilot, Flying Officer G G Douglas-Home, had lost control of the aircraft in bad weather and that it had then dived vertically into the sea with the loss of its four crew.

The assistance of J E (Jerry) Vernon, Squadron Leader (retired) and in his spare time Research Director of the Canadian Museum of Flight and Transportation at Vancouver, is gratefully acknowledged. He forwarded copies of many Hampden crash cards, amongst which was a report concerning AJ988 (23) accidentally dropping eight 11 lb practice bombs from 4000 feet. One of the bombs narrowly missed Mrs Bramhall as she stood in the kitchen of a house; the bomb apparently hit the kitchen sink before penetrating to the soil below the building. The damage caused was approximately 150 dollars ... but there is no indication that the errant bomb aimer had to foot the bill!

Major Cordes, Senior Test Pilot of Handley Page, handing over a Hampden to the Royal Air Force at Radlett during an inspection of Handley Page factories by Captain Harold Balfour, MP. (Charles E.Brown/RAF Museum 5836 – 14)

APPENDIX I

HAMPDEN AND HEREFORD PRODUCTION BATCHES AND MANUFACTURERS

HAMPDENS Quantity	Manufacturer	Delivery Dates	Serial Numbers (see Note 1)
500	Handley Page, Cricklewood	August 1938 to July 1940	L4032-L4211, P1145-P1356, P4285-P4418
770	English Electric, Preston	March 1940 to March 1942	P2062-P2145, X2983-X3154, AD719-AD988 AE115-AE442, AT109-AT260
160	Canadian Associated Aircraft	Sept 1940 to March 1942	P5298-P5436, AJ988-AJ999, AN100-AN167
1430	Total number of Hampdens manufactured, plus 23 Hampdens converted from Herefords. Hampden Grand Total: 1453		
HEREFORDS			
150	Short and Harland, Belfast	May 1939 to Sept 1940	L6002-L6101, N9055-N9106

Note 1: After the "L" and "N" serial numbers, which were mainly allocated before the outbreak of war, the serial numbers do not run continuously as small groups of numbers were omitted. The object of this subterfuge was to confuse the enemy in regard to the actual number of aircraft built.

Note 2: The Canadian-built Hampdens are dealt with in Part 5K which also includes the Canadian-built Hampdens that were shipped to UK, the UK-built Hampdens that were flown out to Canada and the activities of the Hampdens of 32 OTU at Patricia Bay, near Vancouver in British Columbia.

COLOUR SCHEMES

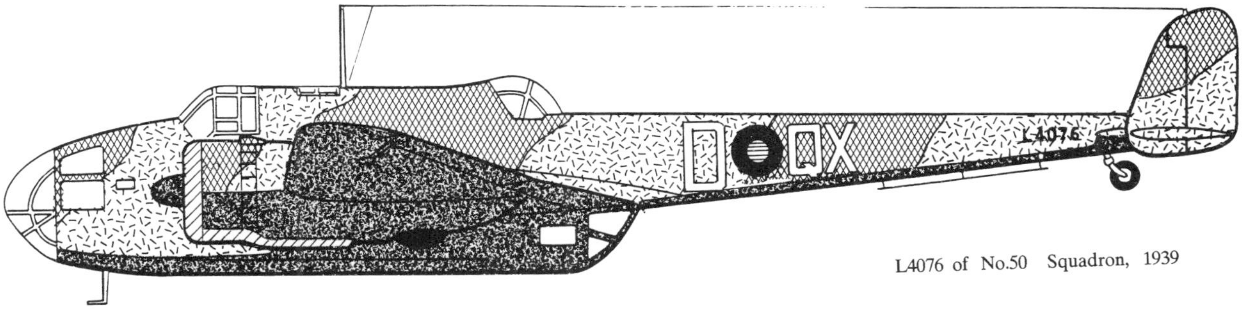

L4076 of No.50 Squadron, 1939

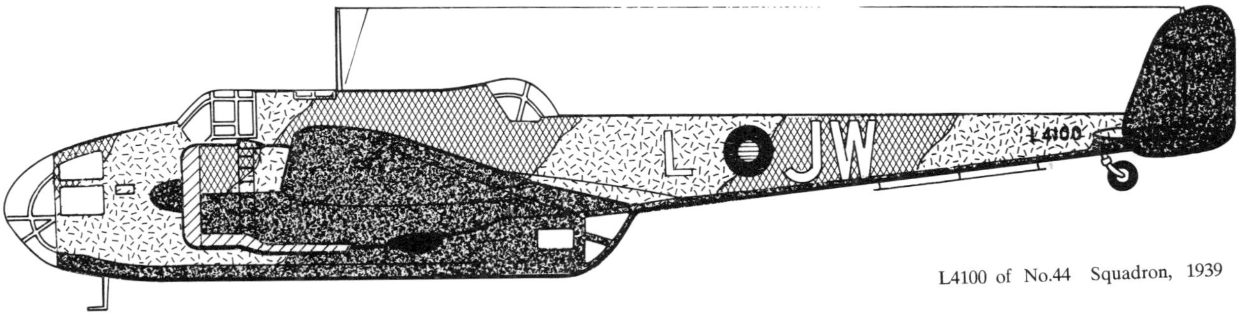

L4100 of No.44 Squadron, 1939

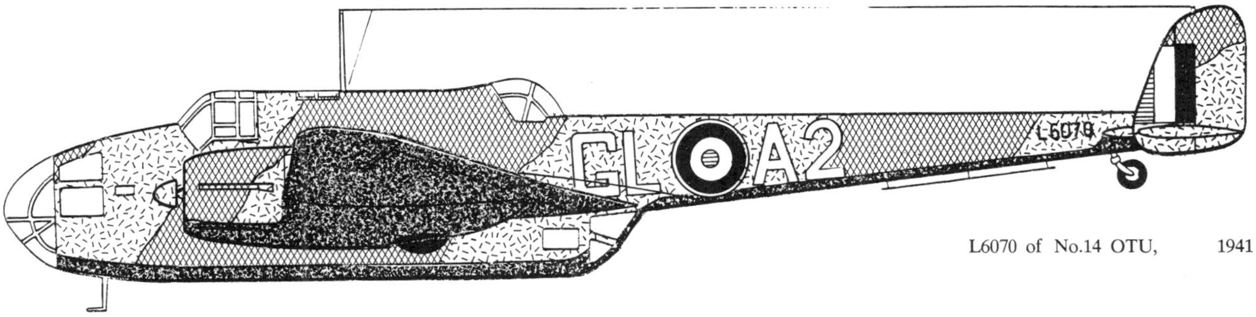

L6070 of No.14 OTU, 1941

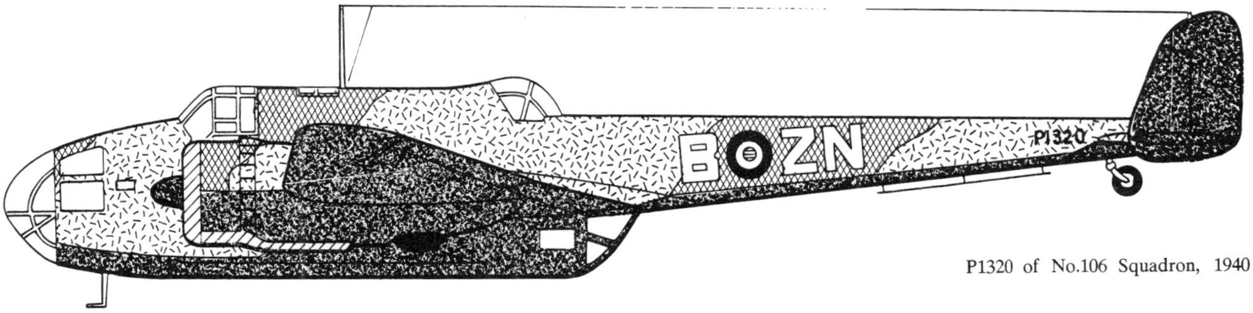

P1320 of No.106 Squadron, 1940

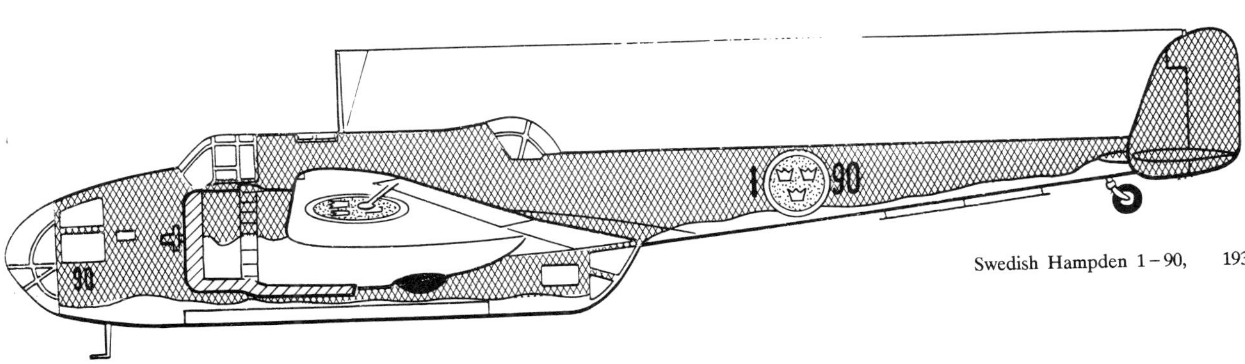

Swedish Hampden 1 – 90, 1939

Handley Page Hampden Mk I

1 De Havilland three bladed, constant speed propeller
2 Spinner
3 Bristol Pegasus XVII, 9 cylinder radial engine
4 Engine cooling air gills
5 Oil cooler intake
6 Flame suppression exhaust pipe
7 Engine bearer struts
8 Undercarriage retraction struts
9 Shock absorber undercarriage leg struts
10 Port mainwheel
11 Inboard fuel tank
12 Wing lattice ribs
13 Fabric covered flaps
14 Wing joint ribs
15 Landing and taxying lights
16 Slot guide ribs
17 Port leading edge slots
18 Port fabric covered aileron
19 Wingtip fairing
20 Port navigation light
21 Engine nacelle
22 Aerial mast
23 Oil tank
24 Dinghy stowage
25 Nacelle tail fairing
26 Wing outboard fuel tanks
27 Nose compartment glazing
28 Bomb aiming windows
29 Spare ammunition drums
30 Folding chart table

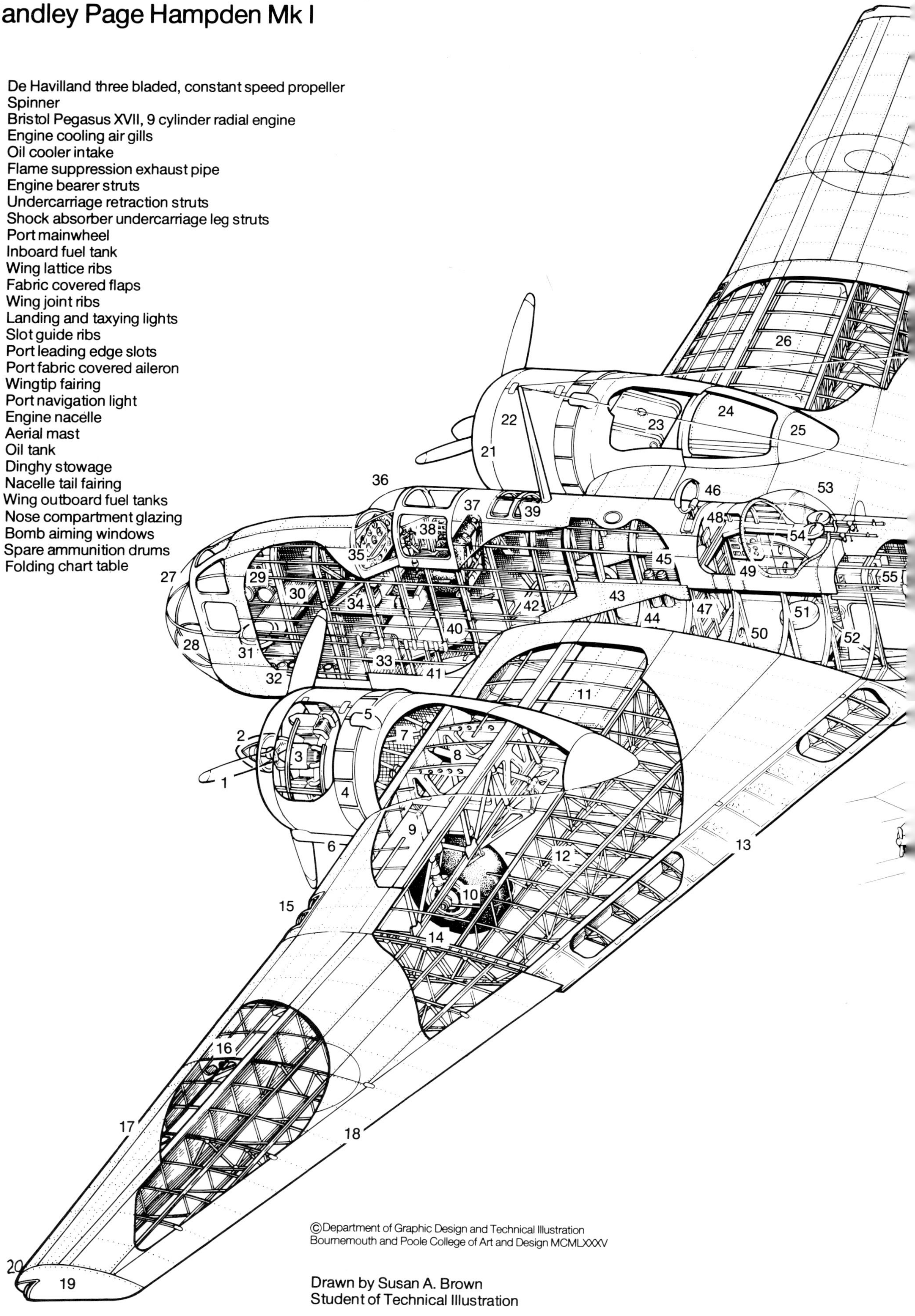

Drawn by Susan A. Brown
Student of Technical Illustration

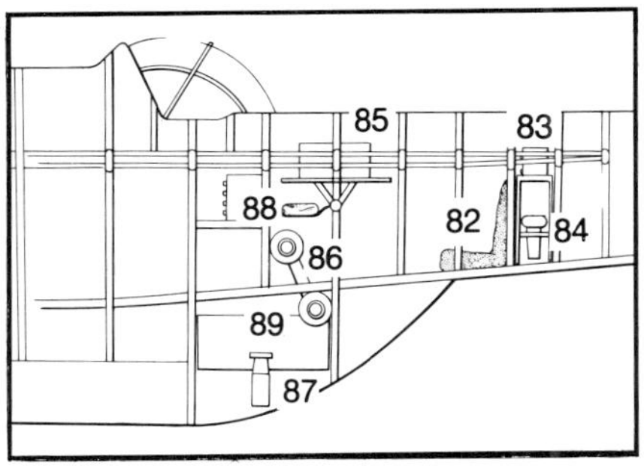

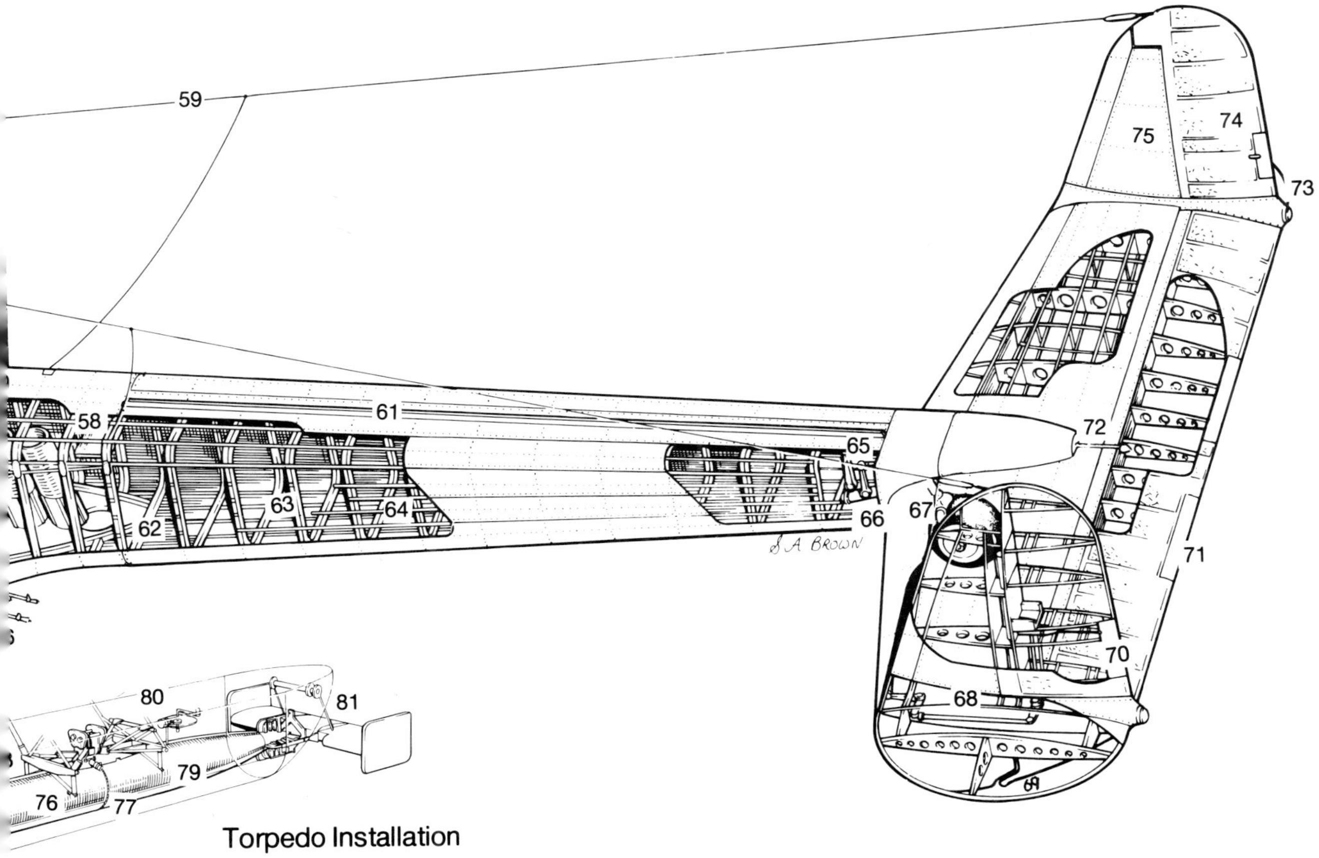

Torpedo Installation

31 Front gunners/bomb aimers seat
32 Flame floats
33 Parachute stowage
34 Engine throttle and propeller controls
35 Instrument panel
36 Windscreen panel
37 Pilots seat
38 Control column handwheel
39 Centre cabin roof hatch
40 Access to lower deck level
41 Bomb stowage
42 Wing spar carry through structure and seat
43 Wing attachment joint plate
44 Lavatory
45 Oxygen bottles
46 Retractable D/F loop aerial
47 Parachute stowage
48 Rear gunner's mirror
49 Controls
50 Wing attachment ribs
51 Rear gunner's seat
52 Trailing aerial winch
53 Upper gun hatch
54 Twin Vickers K-type machine guns
55 Radio equipment
56 Twin Vickers K-type machine guns
57 Flare chutes
58 Radio control bench
59 Aerial leading cables
60 Fuselage skin plating
61 Tail boom skin joint flange
62 Tail boom joint frame
63 Tail boom frame
64 Stringer construction
65 Tailplane control linkages
66 Retraction jack
67 Tail wheel
68 Rudder control linkage
69 Rudder mass balance
70 Rudder tab
71 Elevator tabs
72 Elevator hinge control
73 Tail lights
74 Fabric covered rudder
75 Tail fin
76 Adjusters
77 Attachment cable and slip
78 Front crutch
79 Rear crutch
80 Depth setting dial
81 Drum control attachment to air rudder
82 Operator's seat
83 Receiver
84 Camera
85 Receiver and indicator
86 Tailing aerial winch
87 Flame float stowages
88 Operator's dinghy stowage
89 Drum control gear

COLOUR SCHEMES

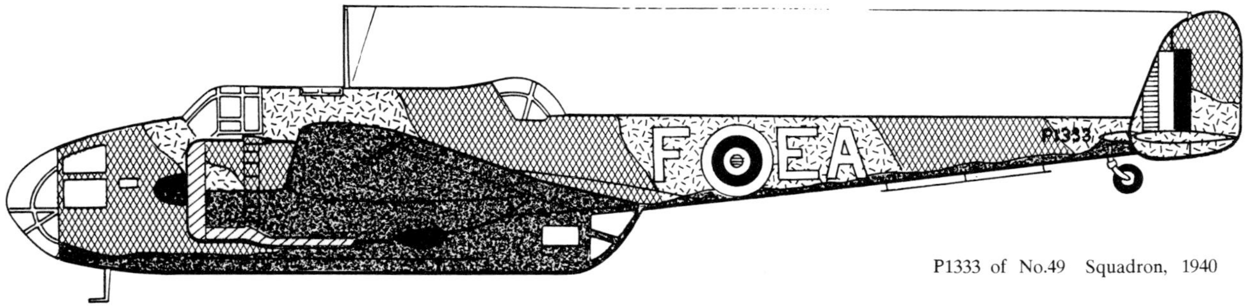

P1333 of No.49 Squadron, 1940

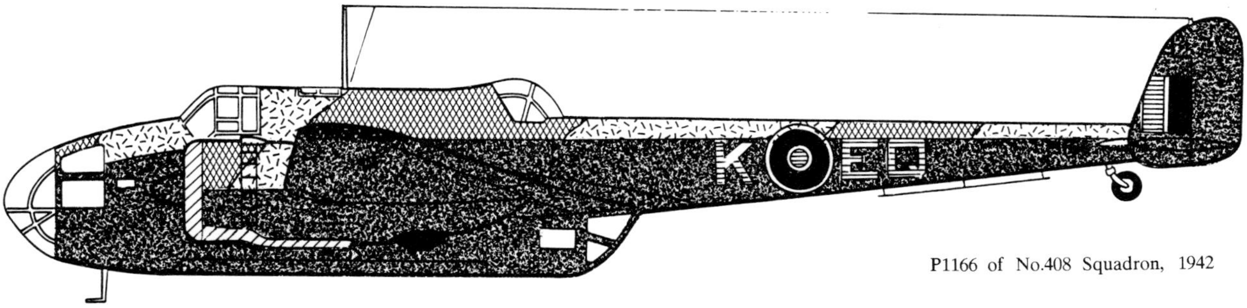

P1166 of No.408 Squadron, 1942

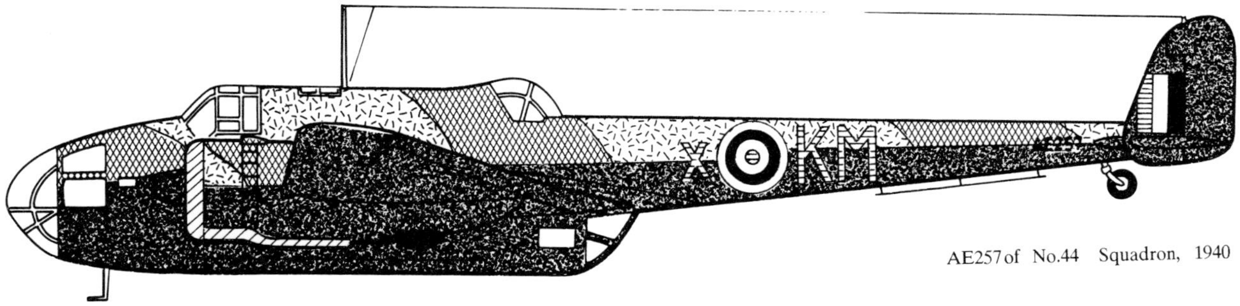

AE257 of No.44 Squadron, 1940

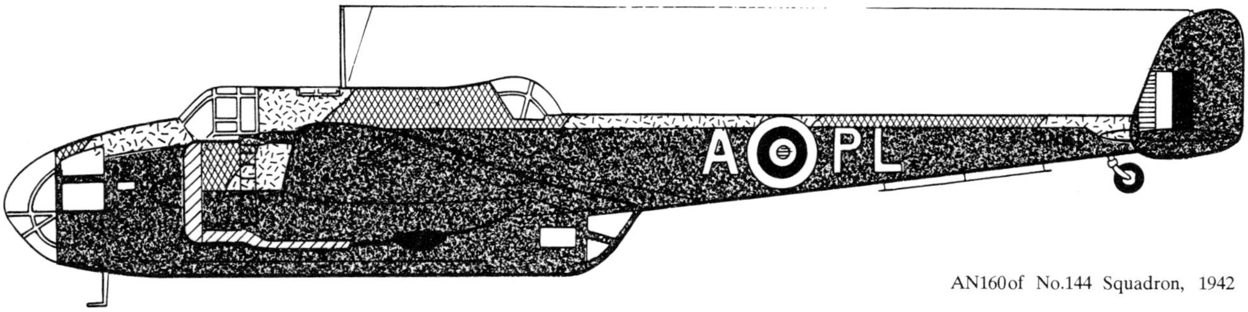

AN160 of No.144 Squadron, 1942

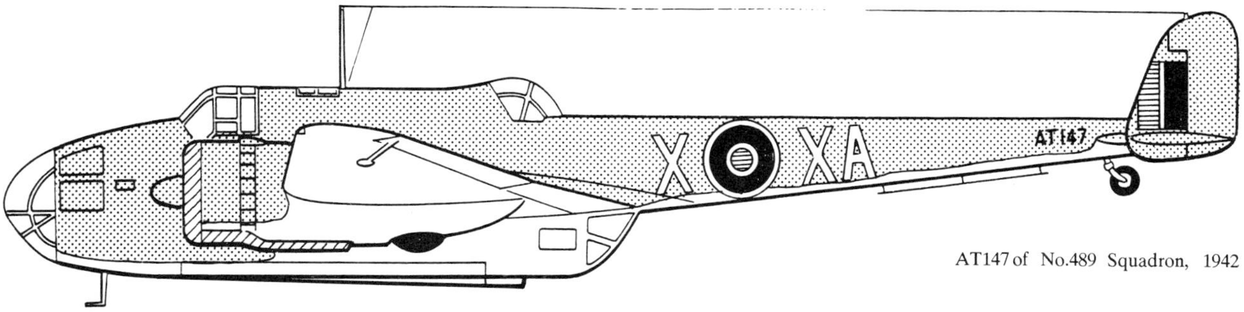

AT147 of No.489 Squadron, 1942

All are dark green/dark earth/black except the Swedish Hampden (green/blue) and No.489 (dark sea grey and Sky)

Drawings by Cliff Minney

L4098, JW – J of 44 Squadron (Photo: Harry Moyle)

AD960, EA – L of 49 Squadron, in field near Waddington after a forced landing. (Photo: Ray Leach/Eric McCabe)

APPENDIX II

HAMPDEN UNITS

No.7 Squadron (April 1939 – May 1940)
Codes: none known to be carried on Hampdens
Finningley; to Upper Heyford 23 September 1939; to 16 OTU 4 April 1940. Reformed at Finningley 30 April 1940; disbanded 20 May 1940.
Five Hampdens written off in training accidents: casualties 9 killed, 1 injured

No.44 Squadron (February 1939 – December 1941)
Codes: JW (pre – war); KM (wartime)
Waddington; converted to Lancasters December 1941.
83 Hampdens written off, 73 as operational losses.

No.49 Squadron (September 1938 – April 1942)
Codes: XU (pre – war); EA (wartime)
Scampton; converted to Manchesters April 1942
94 Hampdens written off, 82 as operational losses

No.50 Squadron (December 1938 – May 1942)
Codes: QX (pre – war); VN (wartime)
Waddington; to Lindholme 10 July 1940; to Swinderby 19 July 1941; to Skellingthorpe 26 November 1941; converted to Manchesters April 1942.
107 Hampdens written off, 91 as operational losses

No.61 Squadron (February 1939 – October 1941)
Codes: LS (pre – war); QR (wartime)
Hemswell; to Woolfox Lodge during November 1941; converted to Manchesters from July 1941.
65 Hampdens written off, 56 as operational losses

No.76 Squadron (March 1939 – April 1940)
Codes: not carried on Hampdens
Finningley; to Upper Heyford 23 September 1939; to 16 OTU 22 April 1940.
Four Hampdens written off in training.

No.83 Squadron (November 1938 – January 1942)
Codes: QQ (pre – war); OL (wartime)
Scampton; converted to Manchesters from December 1941.
85 Hampdens written off, 74 as operational losses.

No.97 Squadron (July 1941 – August 1941)
Code: OF
Coningsby; supplemented Manchesters
Three crews lost on operations, two in aircraft on 106 Squadron strength.

No.106 Squadron (May 1939 – March 1942)
Codes: XS (pre – war); ZN (wartime)
Thornaby; to Cottesmore 1 September 1939; to Finningley 6 October 1939; to Coningsby 23 February 1941; converted to Manchesters from February 1942.
84 Hampdens written off, 59 as operational losses.

No.144 Squadron (March 1939 – January 1943)
Codes: NV (pre – war); PL (wartime)
Hemswell; to North Luffenham 17 July 1941; to Leuchars 21 April 1942; to Vaenga 6 September 1942; to Leuchars 29 October 1942; re – equipped with Beaufighters
123 Hampdens written off, 109 as operational losses (9 were on torpedo – bomber sorties).

No.185 Squadron (June 1939 – April 1940)
Codes: ZM (pre – war); GL (wartime)
Thornaby; to Cottesmore 25 August 1939; to 14 OTU 5 April 1940
Six Hampdens and two Herefords written off in training.

No.207 Squadron (July 1941 – August 1941)
Code: EM
Waddington; supplemented Manchesters.
No Hampden losses.

No.408 Squadron (July 1941 – October 1942)
Code: EQ
Lindholme; to Syerston 18 July 1941; to Balderton 9 December 1941; to North Luffenham 25 January 1942; to Balderton 17 March 1942; to Leeming 20 September 1942; converted to Halifaxes.
48 Hampdens written off, 42 as operational losses.

L4076, QX – D of 50 Squadron, at Waddington pre – war. (Photo: Ray Leach)

P4324, QR – P of 61 Squadron, on Vlieland, 27 August 1940 (Photo: Dirk Bruin)

AE436, PL–J of 144 Squadron. *(Photo: RAF Museum 5932–2)*

No.415 Squadron (January 1942 – November 1943)
Code: GX
Thorney Island: to St.Eval 10 April 1942; to Thorney Island 16 May 1942; to North Coates 5 June 1942; to Wick 6 August 1942; to Leuchars 5 September 1942; to Thorney Island 9 November 1942; converted to Wellingtons.
48 Hampdens written off, 31 as operational losses.

No.420 Squadron (December 1941 – July 1942)
Code: PT
Waddington; converted to Wellingtons.
26 Hampdens written off, 31 as operational losses.

No.455 Squadron (July 1941 – December 1943)
Code: UB
Swinderby; to Wigsley 8 February 1942; to Leuchars 20 April 1942; to Vaenga 4 September 1942; to Leuchars 29 October 1942; converted to Beaufighters.
50 Hampdens written off, 35 as operational losses (15 on torpedo-bomber sorties).

No.489 Squadron (March 1942 – November 1943)
Code: XA
Thorney Island; to Skitten 5 August 1942; to Wick 24 September 1942; to Leuchars 6 October 1943; converted to Beaufighters.
20 Hampdens written off, 14 as operational losses.

No.517 Squadron (August 1943 – October 1943)
Code: none carried by Hampdens
St. Eval.

No.519 Squadron (August 1943 – October 1943)
Code: Z9
Wick; converted to Venturas

No.521 Squadron (September 1943 – December 1943)
Code: 5O
Docking; converted to Venturas

AD837, OL–E of 83 Squadron. *(Photo via Peter Green)*

P1320, ZN–B of 106 Squadron. *(Photo via Peter Green)*

L4194, ZM – B of 185 Squadron pre – war. (Air Ministry photo)

AE288, EQ – M of 408 Squadron, RCAF. (Photo via Peter Green)

No.5 Operational Training Unit (September 1942 – November 1943)
Code: none carried on Hampdens
Turnberry; to Long Kesh 4 January 1943; converted to Hudsons and Venturas.
24 Hampdens written off in training.

No.14 Operational Training Unit (April 1940 – December 1942)
Code: GL
Cottesmore; replaced by Wellingtons.
86 Hampdens written off including 11 on operations.

No.16 Operational Training Unit (April 1940 – July 1942)
Code: JS
Upper Heyford.
83 Hampdens and 7 Herefords written off in training.

No.25 Operational Training Unit (March 1941 – November 1941)
Code: none believed carried on Hampdens
Finningley; replaced by Manchesters
18 Hampdens written off in training

No.32 Operational Training Unit (May 1942 – February 1944)
Codes: DK, LB, OP, RO
Patricia Bay, BC.
48 Hampdens written off in training.

No.1 Torpedo Training Unit (January 1943 – January 1944)
Codes: TU, 9K
Turnberry.
Five Hampdens written off in training.

No.1 Air Armament School (April 1941 – November 1943)
Code: none carried
Manby.
Seven Hampdens written off in training.

No.1401 Flight (March 1943 – August 1943)
Code: none carried on Hampdens
Bircham Newton; to 521 Squadron 1 August 1943

No.1402 Flight (February 1943 – September 1943)
Code: none carried on Hampdens
Aldergrove and Gibraltar; became 520 Squadron 20 September 1943

No.1403 Flight (March 1943 – August 1943)
Code: none carried on Hampdens
Gibraltar.

No.1404 Flight (January 1943 – August 1943)
Code: none carried on Hampdens
St.Eval; to 517 Squadron 11 August 1943

No.1406 Flight (February 1943 – August 1943)
Code: none carried on Hampdens
Wick; became 519 Squadron 15 August 1943

No.1407 Flight (April 1943 – October 1943)
Code: none carried on Hampdens
Reykjavik.

AE201, GX – Q of 415 Squadron, RCAF. (Photo: J.K.Chapman)

P5330, PT – J of 420 Squadron, crashed on Funen, 25.4.42.
(Photo: Birger Hansen)

P1245, UB – J of 455 Squadron. (Photo via Dennis Thompson)

No.5 Bombing & Gunnery School (September 1940 – June 1941)
Code: none allocated
Jurby: became No.5 AOS 14 June 1941
Three Hampdens and five Herefords written off during training with 5 B&GS and 5 AOS.

No.5 Air Observers School (June 1941 – October 1942)
Code: none allocated
Jurby.

The following also flew Hampdens in small numbers:

Aeroplane & Armament Experimental Establishment, Boscombe Down;
Bombing Trials Unit, West Freugh;
Central Flying School, Upavon;
Central Gunnery School, Castle Kennedy;
Coastal Command Development Unit, Carew Cheriton and Tain;
Naval Air Fighting Development Unit, Wittering;
No.5 Group Training Flight, Coningsby;
No.6 Beam Approach Training Flight, Waddington;
No.7 Beam Approach Training Flight, Finningley;
No.10 Bombing & Gunnery School, Dumfries;
Royal Aircraft Establishment, Farnborough;
Special Duties Flight, Christchurch;
Station Flight, Farnborough;
Torpedo Development Unit, Gosport.

AT255, XA – T of 489 Squadron, RNZAF. (Photo via Roger Hayward)

14 OTU Hereford at Cottesmore. (Photo via Dennis Thompson)

P5304, JS – L of 16 OTU. (Photo via Ray Sturtivant)

P5429, HMB of 32 OTU. (Photo via Ray Sturtivant)

APPENDIX III

HAMPDEN RAIDS IN BOMBER COMMAND

Details of raids are taken from Bomber Command records and, as far as possible, details of aircraft lost on each have been allocated. Due to the haste in which records were compiled for Bomber Command, some discrepancies between unit records and the collated Station and Group records are inevitable. The letter D means "ditched", CTO is "crashed on take-off", CR is "crashed on return" and A indicates that the raid was abandoned, normally due to weather conditions when a recall signal was sent out.

DAY

September 1939			
3	Sweep	18	
26	Sweep	12	
29	Sweep	11	L4121; L4126; L4127; L4132; L4134 (all 144 Sqn)
November 1939			
20	Sweep	3	
December 1939			
1	Reconnaissance	3	
4	Reconnaissance	3	
14	Sweep	24	
21	Sweep	24	L4089/44; L4090/44 L4072/49 CR
25	Sweep	12	
January 1940			
12	Sweep	18	
March 1940			
15	Sweep	6	
April 1940			
9	Sweep	12	
	Bergen	12	
11	Warships	6	
12	Sweep	12	L4073/50; L4981/50; P1173/44; L4083/50; L4099/50; L4064/44
May 1940			
16	Met Flight	1	
March 1941			
29	Brest	6A	
30	Brest	12A	
April 1941			
1	Brest	11A	X3129/144
3	Brest	23	(22A)
5	Brest	10	(9A) AD732/61
18	Cherbourg	6	AD732/61; AD825/61 (CR)
19	Borkum	1	
	Vlissingen	1	
23	Various	6	
27	Munster	2	
	Osnabruck	2	
	Cologne	2	
28	Munster	2	
	Osnabruck	2	
	Cologne	2	
29	Emden	2	
May 1941			
1	Emmerich	3	
July 1941			
24	Brest	18	AD962/44; AE225/144
August 1941			
12	St.Omer Afld	6	
	Gosnay	6	
September 1941			
17	Marquise	6	
18	Abbeville	6	
20	Abbeville	6	
21	Lille	6	
22	Mazingarbe	6	
December 1941			
10	NW Germany	6	
	Minelaying	10	
11	NW Germany	6	
12	Brest	6	AE353/144
	N Germany	6	AE391/106
13	Brest	6	
	Minelaying	10	AD921/144; AE196/44D
21	N Germany	15A	AE151/106
24	N Germany	6	
	Brest	4	
27	Vaagso	10	AE369/50; AE428/50
January 1942			
2	N Germany	12	
4	N.Germany	12	
February 1942			
6	Minelaying	35	AE308/455
7	Minelaying	32	AE306/50; AD824/144 AE392/144
12	Warships	64	AE132/49; AE240/49; AT175/144; AT177/50; AE396/49; P5324/49; P1156/455; P4400/420; AT134/144; AE141/144
March 1942			
31	NW Germany	11	
April 1942			
4	Minelaying	21	
July 1942			
11	Bremen	7	
17	Met sortie	1	
18	Met sortie	1	
19	Met sortie	2	
	Shipping	5	AT227/408

NIGHT

January 1940			
–	Leaflets	6	
February 1940			
–	Leaflets	12	
–	Sec patrol	4	L4153/44
March 1940			
–	Leaflets	15	
–	Sec patrols	38	L4111/61 (CR); L4063/50 (CR); L4102/44 (D)
19/20	Hornum	20	
April 1940			
–	Reconnaissance and Sec Patrols	42	L4054/83 (D)
11/12	Shipping	20	
13/1	Shipping and minelaying	15	L4065/50
14/15	as above	28	L4113/61; L4152/83 (D) L4043/49 (CR)
17/18	as above	33	L4163/144 (CTO)
20/21	Minelaying	23	
	Aalborg	3	
21/22	Minelaying	36	L4088/44; P1175/49 (CR)
23/24	Borkum/Sylt	4	P1331/44 (CR)
	Minelaying	26	
25/26	Minelaying	28	P1319/49; L4092/49
	Borkum/Sylt	4	L4040/49
30/1	Fornebu	5	
	Aalborg	5	L4119/61 (CR)
May 1940			
1/2	Aalborg	6	
	Minelaying	11	
2/3	Minelaying	26	
3/4	Minelaying	20	
4/5	Minelaying	5	
6/7	Minelaying	12	
9/10	Minelaying	31	
11/12	M-Gladbach	19	P1326/144; L4068/49
13/14	Eindhoven	6	
	Minelaying	6	
14/15	Roads	12	P4286/44; L4069/83 (CR)
	Minelaying	22	
15/16	Castrop	36	
16/17	Wanne-Eickel	12	
17/18	Ivoir	6	
	Hamburg	46	
18/19	Communications	12	
19/20	Salzbergen	36	
20/21	Cambrai area	18	
21/22	Rail targets	25	L4067/144 (CR)
22/23	Roads	12	L4055/83 (CR)
	Leuna	1	34 recalled

23/24	Rail targets	50	L4171/44; L4146/61
24/25	Rail targets	18	
25/26	Rail targets	29	L4172/144; P1318/49
26/27	Meuse crossings	9	
	Rail targets	12	
27/28	Rail targets	19	
	Hamburg/Bremen	20	
	Cologne	3	
	Troop targets	7	
30/31	Hamburg	18	

June 1940

1/2	Rail targets	37	
2/3	Rail targets	6	
3/4	Frankfurt	24	
	Dusseldorf	12	
	Emmerich	12	P1178/83; P1340/44 (CR)
4/5	Frankfurt	11	
	Mannheim	12	
5/6	Hamburg/Schulau	36	P1348/83
	Minelaying	6	
6/7	Minelaying	6	
	Hamburg	18	
7/8	Hannover	24	P4349/61 (CR)
	Minelaying	6	
8/9	Minelaying	5	
	Amiens	12	L4044/49 (CR)
	Rail targets	12	
9/10	Rail targets	36	P4336/61
	Minelaying	6	
10/11	Minelaying	6	
	Rail targets	27	
11/12	Rail targets	31	P1325/44
	Minelaying	5	
12/13	Minelaying	6	
	Rail targets	19	P4345/144 (CR)
13/14	Railways	64	
14/15	Soest/Hamm	5	
16/17	Minelaying	6	
17/18	Minelaying	6	
	Rail targets	39	
18/19	Cologne	5	
	Minelaying	6	
19/20	Minelaying	6	
	Hamburg	12	
	Rail targets	12	
	Dort – Ems Can	20	
	Munster	9	
20/21	Paderborn	12	
	Rail targets	4	P4355/61
	D – Ems Canal	1	
	Minelaying	6	
21/22	Minelaying	6	
	Hannover	12	
	Sterkrade	6	
	Kassel	12	
	Schwerte	3	P4346/61
	Rail targets	9	
23/24	Wismar	20	
	Wenzendorf	18	
	Hamm	6	
	Communications	9	
	Minelaying	6	P4363/144
24/25	Communications	19	
25/26	Minelaying	6	
	Various	6	P4305/49
26/27	Minelaying	6	P4305/49
	Various	35	P1329/50 (D); L4078/50
27/28	Wismar	12	
	Nyborg	12	
	Communications	6	
	Minelaying	6	L4112/61
28/19	Minelaying	6	
	Dusseldorf	20	
	Dort – Ems Canal	10	
	Munster	4	
29/30	Hamburg	20	
	Various	10	P4341/61; P4356/61
30/1	Osnabruck	4	
	Hamburg	12	
	Dortmund	12	

July 1940

1/2	Kiel	12	P1171/83
	Homberg	12	
	Osnabruck	6	
	Minelaying	6	
2/3	Minelaying	6	
	Hamburg	10	
3/4	Dort – Ems Canal	18	
	Gremberg	6	
	Minelaying	6	P4352/44 (D)
4/5	Minelaying	6	
	Kiel	16	P4361/144
	Dort – Ems Canal	3	
5/6	Kiel	12	
	Dort – Ems Canal	3	
	Minelaying	12	
6/7	Minelaying	11	
7/8	Frankfurt	9	L4066/83 (CR);
	Soest	3	
	Duisburg	6	P4390/61
	Minelaying	12	
8/9	Minelaying	12	
9/10	Wilhelmshaven	14	
	Canals	3	
	Minelaying	12	
11/12	Stettin	1	
	Wanne – Eickel	7	P4366/144
	Mundelheim	7	
	Dort – Ems Canal	3	
13/14	Various	43	
	Minelaying	12	
14/15	Minelaying	9	
	Bremen/ Wenzendorf	12	
15/16	Paderborn	12	
	Hannover	8	
	Minelaying	12	
17/18	Minelaying	3	
18/19	Eschwege	10	
	Paderborn	10	
	Essen	15	
	Cologne	3	
19/20	Paderborn	6	
	Wilhelmshaven	8	
	Minelaying	5	L4087/44
20/21	Minelaying	12	L4077/49 (CR)
	Wilhelmshaven	15	P4343/61; P4344/61; P4358/61; P4367/144 (D)
	Eschwege	8	
21/22	Wismar	20	
	Minelaying	8	
22/23	Minelaying	4	
	Eschwege	8	
	Gelsenkirchen	8	
	Hamm	3	
23/24	Wenzendorf	14	
24/25	Wenzendorf	9	
	Wismar	4	
	Evère	6	
25/26	Wanne – Eickel	21	
	Castrop	20	P1321/50 (CR)
	Dort – Ems Canal	17	L4094/83
	Leaflets	6	
26/27	St.Nazaire	9	
	Nantes	9	
27/28	Hamburg	7	P4375/44
	Minelaying	12	
	Leaflets	6	
28/29	Minelaying	12	
29/30	Minelaying	12	
	Frankfurt	18	
	Dort – Ems Canal	4	
31/1	Various	18	L4085/44 (D); P1327/50 (CR)
	Minelaying	12	P4383/50 (D)

August 1940

2/3	Soest	2	
	Hannover	12	P4376/83 (CR)
	Minelaying	12	
3/4	Minelaying	12	
	Kiel	14	P4351/49 (D)
	Dort – Ems Canal	6	
5/6	Hamburg	17	
	Minelaying	12	
6/7	Minelaying	12	P4377/49
7/8	Kiel	17	
	Minelaying	12	
8/9	Ludwigshaven	8	L4053/83
10/11	Homberg	17	P4368/144

11/12	Dortmund	18	L4036/49
	Minelaying	12	
	Leaflets	5	
12/13	Leaflets	4	
	Salzbergen/ Dort – Ems Canal	19	P4335/61; P4379/61; P4419/83; P4340/83
	Minelaying	12	
13/14	Dessau	16	
	Bernburg	12	P2077/44
14/15	Pauillac	24	
	Cologne	6	
	Hamm	3	
	Soest	2	
	Osnabruck	2	
	Dort – Ems Canal	1	Reconnaissance
15/16	Leaflets	6	
	Gelsenkirchen	11	
	Reisholz	11	
16/17	Merseburg	63	P1333/49; P4291/144; P4365/144 (CR)
17/18	Essen	20	
	Minelaying	5	
19/20	Ambes	14	
	Minelaying	5	P2089/61 (CR)
21/22	Magdeburg	23	
	Mitteland Canal	21	P4360/144
22/23	Frankfurt	24	
	Minelaying	12	
23/24	Minelaying	40	P2117/144 (CR)
24/25	Minelaying	26	
25/26	Berlin	46	P2070/50; P2124/50 (D); P4380/83 (D); P1354/83 (D); P4416/49 (D); X2895/83 (CR)
26/27	Merseburg	18	P4324/61
	Leipzig	12	P1317/50 (CR)
	Dort – Ems Canal	2	
27/28	Ambes	9	
	Minelaying	8	
28/29	Berlin	30	X2897/83 (D)
	Kiel	2	
	Dort – Ems Canal	1	
29/30	Gelsenkirchen	20	P4372/44; P1334/83 (CR)
30/31	Magdeburg	14	
	Gelsenkirchen	9	L4079/50
	Dort – Ems Canal	2	
30/1	Berlin	15	P2123/44 (D)
	Magdeburg	15	

September 1940

1/2	Stuttgart	6	
	Ludwigshafen	2	
	Mannheim	2	
2/3	Ludwigshafen	10	P4370/144
	Lorient	12	
	Hamm	2	
	Stuttgart	3	
	Dort – Ems Canal	2	
3/4	Berlin	12	
	Magdeburg	9	
4/5	Stettin	24	P1347/49
5/6	Stettin	18	P4350/49 (D); P4290/44 (D) X2894/61 (CR)
	Hamburg	4	P1172/144 (CR); P4378/144 (D)
	Wilhelmshaven	1	
	Dort – Ems Canal	2	
6/7	Dortmund	12	
	Krefeld	3	
	Hamm	2	
	Mannheim	2	P2087/44
	Ehrang	2	
	Minelaying	4	
7/8	Ostend	29	
8/9	Hamburg	49	P4287/50
9/10	Hamburg	21	
	Minelaying	3	
	Leaflets	6	
10/11	Calais	12	P4371/44
	Ostend	9	L4097/44; P4371/44
	Minelaying	3	
11/12	Minelaying	8	
	Hamburg	21	P1338/44
	Bremerhaven	6	X2913/44
13/14	Boulogne	10	X2922/61
	Ostend	10	
14/15	Calais	10	
	Ostend	10	
	Dunkerque	10	
	Mannheim	4	
	Osnabruck	4	
15/16	Antwerp	33	
	Hamburg	2	
	Osnabruck	3	
	Minelaying	6	
17/18	Minelaying	8	
	Vlissingen	15	
	Antwerp	22	P2121/44
18/19	Le Havre	33	P1183/83
	Minelaying	8	P1259/106
19/20	Dort – Ems Canal	10	
	Ostend	15	
	Vlissingen	12	
	Minelaying	8	
20/21	Minelaying	8	
	Boulogne	8	
	Dunkerque	6	
	Antwerp	12	
	Dort – Ems Canal	6	
21/22	Ostend	8	
22/23	Le Havre	11	
	Boulogne	8	
	Ostend	4	
	Antwerp	3	
	Vlissingen	4	
23/24	Berlin	45	L4049/83
	Leaflets	6	
24/25	Calais	15	X2911/61 (CTO)
	Le Havre	9	
	Ostend	8	
25/26	Berlin	17	
	Ehrang	2	
	Osnabruck	2	
26/27	Calais	7	L4062/50 (CR)
	Kiel	12	P2090/61
	Dort – Ems Canal	6	
	Minelaying	3	X2914/106 (CR)
27/28	Lorient	24	P4392/83 (CR)
28/29	Le Havre	6	
	Fécamp	7	
	Dort – Ems Canal	6	
	Hamm	4	
	Mannheim	2	
29/30	Stuttgart	15	X2902/50
	Hamm	4	
	Gremberg	2	
	Soest	2	
	Cologne	2	
30/1	Berlin	18	P4411/50 (CR)
	Minelaying	5	

October 1940

1/2	Minelaying	6	P1324/44 (CR)
	Cologne	18	X2965/44
2/3	Hamburg	18	X2896/50 (CR)
	Minelaying	6	
5/6	Minelaying	10	P4362/144; X2920/61 (CR); X2963/144 (CR)
	Gelsenkirchen	12	X2977/83 (CR)
	Hamm	2	
	Osnabruck	2	
	Cologne	2	P4417/50
	Soest	2	
7/8	Rail yards	19	
	Minelaying	6	
	Leaflets	7	
8/9	Wilhelmshaven	19	
9/10	Essen	20	
	Minelaying	6	
	Leaflets	2	
10/11	Kiel	11	
	Minelaying	8	
	Wilhelmshaven	13	
11/12	Wilhelmshaven	7	
	Minelaying	12	
	Kiel	12	
	Hamburg	6	
12/13	Essen	11	
	Herringen	18	
	Minelaying	4	
13/14	Wilhelmshaven	35	
14/15	Berlin	20	X2910/44; X2993/50;

15/16	Magdeburg	33	X2901/83 (CR)
	Minelaying	4	
	Leaflets	2	
16/17	Merseburg	25	X2973/144; X2988/144 (CR); X2997/44 (CR); X2979/61 (CR)
	Bordeaux	12	L4129/49; L4195/44 (CR); X2900/49 (CR); P2143/49 (CR)
18/19	Hamburg	19	L4101/83 (CR)
19/20	Berlin	1	
20/21	Berlin	30	P2137/44; L4154/44 (CR); X2962/49 (CR)
23/24	Magdeburg	20	
	Hannover	10	
24/25	Leaflets	2	
	Minelaying	6	
	Hannover	22	
25/26	Kiel	18	X2971/61 (D); X2998/144 (CR)
	Minelaying	5	
26/27	Minelaying	5	X2990/83
	Berlin	17	
27/28	Hamburg	16	X3027/49 (D)
	Minelaying	5	
	Leaflets	2	
29/30	Berlin	20	X3000/50 (CR)
	Minelaying	5	L4180/106 (D)
	Leaflets	1	

November 1940

1/2	Berlin	15	X2915/144; X3047/144
	Minelaying	8	
3/4	Kiel	11	X3029/44 (CTO); L4093/83 (D)
5/6	Magdeburg	20	X2907/50
	Bremen	3	
6/7	Minelaying	8	P2082/61
7/8	Essen	30	X2994/50 (CR)
	Minelaying	3	
8/9	Leaflets	2	
	Munich	25	P4402/83 (CR); P2079/144 (CR)
	Cologne	25	
9/10	Lorient	5	X2964/83 (CTO)
10/11	Danzig	5	X2985/49 (D)
	Poznan	4	
	Mannheim	18	L4095/83; L4149/50 (D)
	Leaflets	2	
12/13	Wanne - Eickel	8	
	Dortmund	10	
13/14	Hamburg	22	P4396/61 (D); X3606/61 (CTO); X2995/44 (CTO)
14/15	Berlin	10	X2996/44; L4051/83; L4057/83
	Hamburg	17	X2967/61
	Minelaying	3	
15/16	Hamburg	25	X2908/50 (CR)
16/17	Hamburg	34	X3008/44 (CTO)
	Minelaying	3	
19/20	Lutzkendorf	13	X3024/49; X3056/144 (CR)
	Plzen	8	X2023/44 (CR)
20/21	Duisburg	18	
22/23	Bordeaux	14	
	Duisburg	7	X2987/61 (CR)
	Minelaying	3	L4194/106
23/24	Minelaying	3	
	Gelsenkirchen	5	
	Duisburg	9	
24/25	Hamburg	12	X3052/49
	Minelaying	3	
25/26	Kiel	6	X3064/61 (CR); X3125/50 (CR)
26/27	Cologne	8	P2125/83 (CTO)
	Minelaying	2	
27/28	Minelaying	4	
28/29	Mannheim/ Ludwigshafen	12	
	Le Havre	3	
29/30	Le Havre	6	X2966/44 (CR); P4407/144
	Minelaying	2	

December 1940

1/2	Wilhelmshaven	10	
4/5	Dusseldorf	20	
	Minelaying	3	
6/7	Airfields	10	X3050/49; X3028/49 (CR); P4404/49; X3065/144 (CR)
7/8	Dusseldorf	10	X3004/50
	Minelaying	4	
8/9	Dusseldorf	16	X2975/61
	Leaflets	2	
10/11	Mannheim	12	X3117/50
	Minelaying	5	X3049/44
12/13	Duisburg	6	Recalled
15/16	Berlin	30	
	Minelaying	8	
16/17	Mannheim	29	X3128/61; X3063/49 (D)
19/20	Cologne	37	
	Minelaying	8	
20/21	Minelaying	4	
	Berlin	11	
	Cologne	22	
21/22	Schornewitz	24	
	Minelaying	5	
22/23	Frankfurt	11	
	Minelaying	3	
	Leaflets	1	
23/24	Leaflets	1	
26/27	Bordeaux	6	
27/28	Bordeaux	20	P2097/83 (CR)
	Minelaying	13	P2098/106
28/19	Minelaying	7	
	Lorient	15	X3141/50

January 1941

1/2	Bremen	30	X3143/50 (CR); X3126/61 (CR)
2/3	Bremen	8	
3/4	Bremen	15	
4/5	Brest	30	
	Minelaying	6	
5/6	Minelaying	12	P4322/49
8/9	Wilhelmshaven	10	
	Minelaying	9	
9/10	Gelsenkirchen	20	
	Minelaying	4	
11/12	Wilhelmshaven	25	L4045/49 (CR)
	Minelaying	5	
12/13	Minelaying	4	
	Brest	15	
13/14	Minelaying	12	
15/16	Wilhelmshaven	17	
16/17	Wilhelmshaven	15	AD731/83
	Minelaying	3	
29/30	Wilhelmshaven	9	

February 1941

2/3	Brest	12	
	Minelaying	2	
3/4	Minelaying	11	
4/5	Dusseldorf	30	P4299/49
	Minelaying	6	AD750/106
7/8	Dunkerque	27	
8/9	Mannheim	15	P4359/144 (CR)
9/10	Wilhelmshaven	23	P4405/61 (CR)
10/11	Hannover	46	X3001/49; AD719/48 (CR)
11/12	Bremen	20	AD734/83 (CR); AD722/83 (CR)
	Hannover	11	X3007/144 (CR)
14/15	Homberg	22	
	Minelaying	11	X3048/144 (CR)
21/22	Minelaying	42	
24/25	Brest	18	
25/26	Dusseldorf	22	
26/27	Cologne	28	X3124/83 (CR)
	Minelaying	6	

March 1941

1/2	Cologne	44	AD737/144; P4394/144 (CR); AD742/50 (CR); X2984/50 (CR); AD723/61 (CR); P1253/61 (CR); X3147/61 (CR)
2/3	Brest	8	
3/4	Brest	3	
	Cologne	21	X3002/106
10/11	Cologne	19	X2918/44; AD720/144 (CTO)
12/13	Hamburg	40	
	Berlin	30	AD721/50
	Minelaying	1	
13/14	Hamburg	34	X3146/50
	Minelaying	1	
14/15	Gelsenkirchen	21	
15/16	Dusseldorf	21	
17/18	Bremen	19	
18/19	Kiel	40	
20/21	Minelaying	42	X3123/83; AD745/144 (CR)
21/22	Lorient	12	X3054/49 (CR)
	Minelaying	7	
23/24	Kiel	31	X3005/61 (CR)
	Minelaying	5	
27/28	Dusseldorf	22	
29/30	Minelaying	25	AD800/83

30/31	Minelaying	10	
	Brest	15	

April 1941

3/4	Minelaying	15	AD748/83; P4403/49 (CR)
4/5	Brest	11	AD738/106
	Minelaying	6	
6/7	Minelaying	24	X2899/83
7/8	Kiel	61	
8/9	Kiel	29	AD827/61
	Minelaying	10	AD899/44
9/10	Berlin	24	
	Minelaying	3	
10/11	Dusseldorf	29	X3153/106; X3148/106; AD789/50; AD828/50; X3066/144
12/13	Brest	12	
	Merignac Afld	5	
	Minelaying	10	
13/14	Minelaying	17	
14/15	Brest	25	
15/16	Kiel	19	
	Minelaying	5	
16/17	Bremen	21	
	Minelaying	5	AD761/144 (CTO)
17/18	Berlin	39	X2999/44; AD730/50
	Minelaying	6	
19/20	Minelaying	5	
20/21	Cologne	11	X3119/83; X2986/106
	Minelaying	9	
22/23	Minelaying	6	
23/24	Brest	10	
	Minelaying	14	
24/25	Kiel	10	
	Le Havre	6	
	Minelaying	8	
25/26	Kiel	10	
	Minelaying	7	
26/27	Hamburg	28	AD796/83; AD847/44 (CR)
28/29	Minelaying	28	AD728/50; AD834/50
29/30	Mannheim	14	
	Rotterdam	12	
30/1	Kiel	13	

May 1941

2/3	Hamburg	19	AD864/44
	Minelaying	11	
3/4	Cologne	27	
4/5	Brest	21	
5/6	Mannheim	33	
	Minelaying	4	
6/7	Hamburg	27	
	Minelaying	7	
7/8	Brest	18	
	Minelaying	2	
8/9	Hamburg	78	X3062/83 (D); AD901/144 (CR)
	Minelaying	3	
9/10	Ludwigshafen	24	
	Minelaying	2	
10/11	Minelaying	1	
	Hamburg	35	
11/12	Bremen	31	AD900/144
12/13	Mannheim	38	X2982/44 (CR)
	Minelaying	4	
15/16	Hannover	27	AD841/144
16/17	Cologne	24	
	Minelaying	2	
17/18	Cologne	23	
18/19	Kiel	18	
	Minelaying	4	
23/24	Cologne	24	
25/26	Minelaying	48	
26/27	Minelaying	38	P2083/106 (CR)
27/28	Minelaying	36	AD729/49

June 1941

2/3	Dusseldorf	43	AD797/50; P2144/61 (CR); X3120/61 (D)
10/11	Brest	39	
	Minelaying	9	
11/12	Duisburg	35	
	Minelaying	20	AD727/61
12/13	Soest	91	P4310/44; AE127/44
13/14	Brest	37	
	Minelaying	4	
14/15	Cologne	29	AE129/44 (CR)
15/16	Cologne	42	AD863/106
	Minelaying	4	
16/17	Cologne	47	
17/18	Cologne	43	AD926/144 (CR)
	Minelaying	4	
18/19	Bremen	39	
20/21	Kiel	24	
	Intruders	4	
21/22	Dusseldorf	28	
22/23	Bremen	25	AD969/83
	Dusseldorf	1	
23/24	Dusseldorf	30	
	Intruders	2	
	Minelaying	1	
24/25	Kiel	25	
25/26	Kiel	30	AD788/49
	Minelaying	1	
26/27	Dusseldorf	30	
	Minelaying	1	X3060/49 (CR)
27/28	Vegesack	28	
	Dusseldorf	1	
	Cologne	1	
	Minelaying	3	
28/29	Minelaying	34	
29/30	Bremen	30	AD895/106; X3133/50
	Minelaying	3	
30/1	Dusseldorf	14	X3134/49 (D); AD916/83

July 1941

2/3	Duisburg	39	AD862/106; AD873/106
3/4	Bremen	39	AD866/144; X3055/144 (CR)
4/5	Lorient	25	
	Cologne	2	
	Dortmund	2	L4185/106; AD914/106
	Hamburg	1	
	Dusseldorf	1	AD986/106
	Minelaying	3	
5/6	Osnabruck	39	AD739/49; AD806/61
	Minelaying	7	
6/7	Brest	88	AD871/61; AD856/49 (D) AD735/106; AD861/106 (CR)
7/8	M – Gladbach	40	AD937/61; AD735/106
8/9	Hamm	45	AD840/44; AE153/44; X3139/83
9/10	Aachen	39	AD924/144
10/11	Cologne	32	
11/12	Wilhelmshaven	36	AE119/144 (CR)
12/13	Bremen	33	AE226/50; AE230/50; AD910/49 (CR)
14/15	Hannover	44	
	Minelaying	10	
16/17	Hamburg	32	AD844/50
	Minelaying	5	
17/18	Cologne	25	
19/20	Minelaying	35	
20/21	Frankfurt	34	
	Minelaying	2	
22/23	Frankfurt	34	
	Minelaying	8	AD983/44 (CR)
23/24	Frankfurt	33	AD843/50
	Minelaying	1	
24/25	Kiel	30	AE189/61
	Minelaying	6	
25/26	Hannover	30	AD835/83; X3151/49 (CR); AE234/50 (CTO)
27/28	Minelaying	36	P4406/44
28/29	Minelaying	42	AD902/50 (CR); AE159/50
30/31	Cologne	42	AD784/144; AE252/144; P4399/61; AD755/44 (CR); AE266/61 (CR)

August 1941

2/3	Kiel	50	AE154/83 (D)
	Minelaying	5	
5/6	Mannheim	33	AE120/106
	Karlsruhe	50	AE137/50
	Minelaying	5	
6/7	Calais	21	X2917/44; AD903/144
	Karlsruhe	38	AE140/144
7/8	Essen	54	X3118/83 (D); AE303/97
	Minelaying	8	
8/9	Kiel	50	X3127/61; AE263/61; AE259/61 (CR)
	Minelaying	7	AE124/50 (CR)
11/12	Rotterdam	30	
	Krefeld	20	
12/13	Hannover	13	
	Magdeburg	36	
14/15	Braunschweig	81	AE131/83; AE262/49 (CR)
	Minelaying	5	

16/17	Dusseldorf	52	AD756/106; AE134/106; AE239/44
	Ostend	4	
17/18	Bremen	39	AE185/50; AD837/83 (D)
	Minelaying	12	
18/19	Cologne	42	AD919/106 (CR)
19/20	Kiel	41	AD907/83 (CR)
	Minelaying	3	
22/23	Mannheim	41	AE220/106
24/25	Dusseldorf	12	AD967/49; X3121/83 (collision)
	Wesel	6	AE223/83
25/26	Mannheim	38	AD918/144; AE265/144; AE320/50
26/27	Cologne	29	AE302/106
	Intruders	6	
	Minelaying	17	AE301/106
27/28	Mannheim	35	X2991/50 (CTO)
	Minelaying	17	AD917/44
28/29	Duisburg	36	AD971/49; AE126/49; AE193/106 (D)
	Ostend	1	
29/30	Frankfurt	73	AE247/61; AD839/50; AE229/50 (CR)
30/31	Minelaying	2	
31/1	Cologne	39	AD859/83; AD912/83; AD187/83; AD726/44 (D)
	Intruders	6	
	Minelaying	12	
September 1941			
1/2	Cologne	20	AD905/144
	Minelaying	4	
2/3	Frankfurt	11	AE313/44; AD913/44 (CR)
	Berlin	32	AE152/44; AE254/44; AD978/83; X2919/50 (CR); X3136/49 (CR); AE203/49 (CR); AE305/50 (CR)
	Minelaying	15	X3144/83; AE315/83
3/4	Brest	30	29 recalled
	Minelaying	6	All recalled
6/7	Huls	18	
	Minelaying	24	AE319/83 (D); AD744/49 (CR)
7/8	Berlin	43	AE299/106; AD936/144; AE304/144 (CTO)
	Kiel	18	AE318/50; AE236/49
	Minelaying	8	X2921/44 (CTO)
8/9	Kassel	27	AE188/83 (CR); AD854/50 (CR); P4285/44 (CR); AD805/49 (CR)
	Minelaying	6	
11/12	Rostock	39	AE300/106; AD981/44 (D); AE118/144 (CR); AE121/144 (CR)
	Boulogne	2	
	Minelaying	20	
12/13	Frankfurt	32	
	Minelaying	10	
13/14	Brest	38	
	Le Havre	2	
15/16	Hamburg	50	AE232/106; AE249/455; P4413/106 (CR); AD930/44 (CR); AE362/83 (CR)
	Le Havre	11	AE365/83 (CR)
	Minelaying	5	
17/18	Minelaying	14	
20/21	Berlin	26	
	Frankfurt	12	AD922/144 (CR); AD872/144 (CR); X3030/144 (CR); AD923/144 (CR)
	Ostend	4	
28/29	Frankfurt	30	AD733/49; AE376/49 (CTO)
29/30	Hamburg	48	AE143/144 (CR)
30/1	Hamburg	42	
	Cherbourg	15	
October 1941			
1/2	Karlsruhe	44	AD768/106 (CR)
	Minelaying	2	
10/11	Essen	52	AD911/83; AE144/106
	Dunkerque	23	AE382/44
11/12	Minelaying	12	
12/13	Bremen	20	
	Huls	79	AD965/144; AE267/50
	Searchlight suppression	8	
13/14	Cologne	30	AD975/44; AE251/50
	Minelaying	13	
16/17	Duisburg	31	
	Dunkerque	3	
20/21	Bremen	82	AD984/106; AE383/50; P1212/408 (CR); AD746/106 (CR);
	Minelaying	6	
21/22	Bremen	34	AE257/44
	Boulogne	2	
22/23	Mannheim	45	P1218/408; AD934/83 (CR)
23/24	Kiel	38	AE256/50
	Le Havre	10	AE290/44
24/25	Frankfurt	30	AE316/144
26/27	Hamburg	30	AD785/106 (CR); AE398/44 (CR); AE136/106 (CR)
	Cherbourg	3	
	Minelaying	5	
29/30	Schiphol	40	X3021/106
31/1	Hamburg	42	
	Minelaying	17	
November 1941			
1/2	Kiel	32	AE358/83
	Shipping	4	AE224/49
	Minelaying	5	
3/4	Shipping	6	
	Minelaying	10	
4/5	Shipping	6	
	Intruder	3	
	Minelaying	24	AE434/455 (CR)
5/6	Cherbourg	22	
	Hamburg	6	
	Shipping	5	AD846/144; AE253/144; AE424/144
	Minelaying	24	AD850/83
6/7	Shipping	1	
	Intruder	5	
	Minelaying	15	AE427/50
7/8	Cologne	69	AE238/144 (CO)
	Ostend	5	
	Intruders	6	P1201/455; AE243/455
	Minelaying	13	P1290/106; AD760/106 AD932/106 (D)
8/9	Essen	19	AE433/408
	Dunkerque	10	AE377/44
	Intruder	6	P1206/49
9/10	Hamburg	30	AE311/144 (CR)
	Ostend	4	AE438/408
	Minelaying	5	
15/16	Emden	11	
	Minelaying	5	P1152/50
23/24	Lorient	51	
26/27	Emden	20	AE317/106
27/28	Dusseldorf	34	AE437/408; AD933/44 (CR); AE440/144 (CR)
	Minelaying	48	AT115/106; P1228/106; P1272/455; AE430/455; P1202/50
	Emden	1	
	Minelaying	4	
December 1941			
7/8	Aachen	44	AD791/144; AE191/83 (D)
	Boulogne	4	
11/12	Cologne	14	
	Le Havre	15	
	Minelaying	5	AE148/408
14/15	Brest	22	AE374/83 (CO)
	Cherbourg	3	AT133/408; P5392/408 (CR)
	Minelaying	2	P5393/83 (CO)
15/16	Ostend	12	AE380/50
	Minelaying	5	
16/17	Wilhelmshaven	14	
	Dunkerque	5	
	Minelaying	18	
17/18	Brest	25	AD868/44
	Le Havre	3	
23/24	Cologne	20	
	Minelaying	17	
	Leaflets	2	
27/28	Dusseldorf	30	
	Boulogne	6	
	Minelaying	5	
28/29	Huls	81	P1295/144; P1165/408; AD840/144; AE419/49
January 1942			
2/3	Minelaying	36	P5328/455 (CR)
	Leaflets	8	
3/4	Minelaying	10	AT123/106
5/6	Brest	27	
	Minelaying	5	
	Leaflets	5	
6/7	Intruders	19	
8/9	Brest	37	AT127/83 (CR)
	Cherbourg	7	
	Minelaying	5	
	Leaflets	2	

9/10	Brest	27	
	Minelaying	5	AD909/49
10/11	Wilhelmshaven	34	AE133/83; AE286/408; AE250/50 (CR)
	Emden	7	
	Minelaying	5	
	Leaflets	1	
14/15	Hamburg	32	AE420/50; AE431/50; AE441/144 (CR)
	Emden	4	
	Minelaying	5	
15/16	Hamburg	27	P1341/106; AE393/408 (CR)
	Minelaying	3	
17/18	Bremen	20	
	Emden	3	
	Minelaying	8	
20/21	Emden	5	AT148/49
21/22	Bremen	11	AE123/106; X3051/408;
	Emden	12	AE352/455; AT130/420; AT119/455
	Minelaying	1	
	Leaflets	9	
22/23	Munster	12	AT142/50 (CR)
	Minelaying	9	AT146/106
	Leaflets	5	
25/26	Brest	35	AT129/49 (CTO); AD782/408 (CTO)
	Leaflets	3	
26/27	Hannover	17	
	Minelaying	6	
	Leaflets	2	
27/28	Brest	32	
28/29	Munster	29	P4398/106; X3058/106; AT121/106; AT122/106
	Boulogne	4	
	Leaflets	1	
31/1	St.Nazaire	5	
	Brest	41	AE359/144; AT149/144
	Leaflets	3	

February 1942

10/11	Bremen	12	
11/12	Mannheim	12	AE325/106 (CO)
	Leaflets	1	
12/13	Minelaying	11	AD783/455
14/15	Mannheim	25	AT112/49 (D)
16/17	Minelaying	37	AE397/49 (D); AT124/49
	Leaflets	9	
17/18	Leaflets	3	
18/19	Minelaying	[illegible]	AD915/420
19/20	Leaflets	1	
21/22	Intruders	20	P4414/106; AE394/50 (CR); AT181/455
22/23	Wilhelmshaven	19	
	Emden	2	
	Leaflets	2	
23/24	Minelaying	23	P4323/106
24/25	Minelaying	42	X2969/144; AT194/144
25/26	Minelaying	18	
	Leaflets	2	
26/27	Kiel	10	
	Minelaying	27	L4178/144 (CR)
	Leaflets	2	
27/28	Kiel	18	AE218/50 (CR)
	Wilhelmshaven	6	
	Minelaying	11	

March 1942

3/4	Billancourt	48	
7/8	St.Nazaire	2	
	Minelaying	17	AE400/50
8/9	Essen	37	
	Minelaying	9	AD842/408
	Leaflets	1	
9/10	Essen	15	
	Minelaying	5	
10/11	Essen	43	P2085/455; AT192/455; AT174/49; AT173/50 (CR)
	Boulogne	13	
	Leaflets	3	
12/13	Minelaying	26	
	Leaflets	1	
13/14	Cologne	22	
	Minelaying	5	
	Leaflets	7	
23/24	Minelaying	12	AT116/144 (CR)
24/25	Minelaying	24	AE429/50
25/26	Essen	9	AT151/30; AE139/408 (CR);
	Minelaying	36	AT158/50
	Leaflets	9	
26/27	Minelaying	24	AE200/144; AE298/420
	Leaflets	8	
27/28	Minelaying	18	L4140/408; AE219/408; AT176/408
28/29	Lubeck	41	AE246/420 (D)
	Minelaying	7	
	Leaflets	2	
29/30	Minelaying	18	
	Leaflets	2	

April 1942

1/2	Le Havre	22	
	Rail targets	14	AD959/144
	Leaflets	5	
2/3	Le Havre	5	
	Minelaying	23	P5325/455
5/6	Cologne	44	AT156/49; AT226/144 (CR); AT216/50 (CR)
6/7	Essen	18	P1203/455; AT126/49
8/9	Hamburg	41	
	Minelaying	16	
	Leaflets	9	
10/11	Essen	43	AE421/49; AT190/49; AE291/455; AT221/455; AT187/144 (CR); AT218/144
	Leaflets	3	
12/13	Essen	31	AT120/408; P1239/420; AT196/49 (CR)
	Minelaying	16	
13/14	Minelaying	15	
14/15	Dortmund	34	AT128/420; AT141/408; AT157/144; AT219/420 (CO); AT110/144 (CR); AD931/49
	Le Havre	5	
15/16	Dortmund	19	
	Minelaying	6	
	Leaflets	2	
16/17	Minelaying	10	
	Leaflets	5	
17/18	Minelaying	5	
19/20	Minelaying	23	AT217/49
22/23	Minelaying	22	AE237/408
23/24	Rostock	11	
24/25	Rostock	21	P5330/420
	Dunkerque	1	
25/26	Rostock	5	
	Dunkerque	2	
26/27	Rostock	19	
	Minelaying	1	
27/28	Minelaying	2	
28/29	Kiel	10	AE426/408
29/30	Gennevilliers	9	

May 1942

2/3	Minelaying	12	
	Leaflets	2	
3/4	Hamburg	5	
4/5	Stuttgart	19	P1187/420 (CR)
	Minelaying	3	
5/6	Leaflets	1	
6/7	Stuttgart	10	
	Nantes	2	
7/8	Minelaying	14	AE389/420
8/9	Warnemunde	19	AT144/420; AE288/408; AE297/408
9/10	Minelaying	1	
15/16	Minelaying	16	AD803/408; AT224/408
19/20	Mannheim	15	
	Minelaying	3	
	Leaflets	2	
29/30	Gennevilliers	3	
	Cherbourg	5	
30/31	Cologne	79	P2116/14 OTU; P5321/14 OTU; L4173/14 OTU (CR); AE399/420 (CR)

June 1942

1/2	Essen	71	AT191/408
2/3	Essen	12	AT154/408
	Minelaying	4	AE260/420
	Leaflets	4	
3/4	Bremen	9	
	Minelaying	3	
	Leaflets	1	
4/5	Dieppe	6	
	Leaflets	1	
5/6	Essen	11	
	Minelaying	3	
6/7	Emden	15	

7/8	Minelaying	3	
	Leaflets	1	
8/9	Essen	9	AT136/420
	Dieppe	3	
11/12	Minelaying	5	
16/17	Minelaying	12	
17/18	Minelaying	10	AT189/408
19/20	Emden	12	AT185/420
	Leaflets	1	
21/22	Minelaying	9	
	Leaflets	1	
22/23	Emden	8	
23/24	Minelaying	9	AD786/420
24/25	St.Nazaire	3	
26/26	Bremen	50	P5312/14 OTU
26/27	Minelaying	10	
July 1942			
2/3	Bremen	28	AE248/420; P5332/420
6/7	Minelaying	7	
8/9	Wilhelmshaven	24	
9/10	Minelaying	24	
12/13	Minelaying	12	AE390/420
14/15	Minelaying	14	
21/22	Duisburg	17	AT139/408
	Minelaying	1	
23/24	Minelaying	2	
25/26	Duisburg	14	
	Minelaying	2	
26/27	Hamburg	33	AE202/420; AE267/420
29/30	Saarbrucken	17	
31/1	Dusseldorf	54	AE244/408; AE355/420; L4117/14 OTU (CR); N9062/14 OTU (CR)
August 1942			
4/5	Minelaying	6	
5/6	Minelaying	5	
	Leaflets	2	
6/7	Duisburg	10	
9/10	Minelaying	2	
11/12	Le Havre	2	
	Minelaying	9	
12/13	Mainz	10	AE150/408; AD870/408 (CR)
15/16	Dusseldorf	7	AE432/408
17/18	Osnabruck	10	
	Minelaying	1	
18/19	Flensburg	16	
27/28	Kassel	12	P1244/408
28/29	Saarbrucken	17	AD829/408; AT143/408 (CR) AE197/408; AE227/408; AT228/408
September 1942			
1/2	Saarbrucken	7	
2/3	Karlsruhe	5	
6/7	Duisburg	4	
7/8	Minelaying	6	
8/9	Frankfurt	9	
10/11	Dusseldorf	28	L4131/14 OTU
13/14	Bremen	20	L4109/14 OTU; AD845/14 OTU (CR)
14/15	Wilhelmshaven	4	

* * * * * * * * * * * * * * * * *

Three Hampdens of No.14 OTU in formation. (CH716)

APPENDIX IV

LOSSES TO FIGHTERS

The losses suffered by RAF bombers in daylight operations from the outbreak of war until April 1940, when Norway and Denmark were invaded by the Germans, show only too well how vulnerable were the Wellingtons, Blenheims and Hampdens to fighter attack. Thirteen Hampdens were shot down during that period if the two that were shot down by Spitfires are included.

The Luftwaffe did not have things entirely its own way in these daylight attacks, however, as a total of six of about forty Bf 109s involved were shot down by return fire from Hampdens. An entirely different picture emerges from the lists of claims by night fighter pilots, very few of whom were shot down. The staggering personal scores of confirmed victories demonstrate the one – sidedness of the night air war and in fact eleven of the pilots who scored victories over Hampdens had, by the end of the war, amassed a total of 526 victories between them. Oblt Lent, as he was in the Hampden era, headed the list with 102 victories by night and eight by day; out of the grand total of 110, eight were Hampdens.

The first German night fighter unit, IV(N)/JG.2, formed late in December 1939, was equipped with Bf 109Ds from which the cockpit hoods had been removed. Fighters were directed towards their target by ground radar stations and on 25 April 1940 Ofw Forster claimed the first Hampden victim at 0110 hrs near Sylt. Three Hampdens of 49 Squadron, P1319, L4040 and L4092, failed to return from a minelaying operation off Kiel on that night and P1319 seems to be the most likely victim.

A major re – organisation of the Luftwaffe night fighter force took place on 22 June 1940 when I/NJG.1 and II/NJG.1 came into being, soon followed by III/NJG.1, I/NJG.2 and II/NJG.2. I/NJG.3 and II/NJG.3 had commenced operations by November 1940. I/NJG.2 was given the task of carrying out night intruder attacks against RAF bases in England, particularly the bomber stations. They also patrolled over the North Sea in the hope of intercepting bombers on their outward and return flights. An account of their activities against Hampdens is given in Part 5H.

"The History of the German Night Fighter Force" by Gebhard Aders is recommended reading for anyone who wishes to learn of the development, organisation, equipment and activities of the Luftwaffe's night fighters. The aircraft in use during the Hampden era were the Bf 109D, Bf 109E, Bf 110D, Bf 110E, Bf 110F, Ju 88C, Do 17Z, Do 215B and Do 217J.

A night – fighting Messerschmitt Bf 110's cumbersome radar array

DETAILS OF HAMPDEN LOSSES

Serial	Unit		Date	Details	Shot down by	Unit
L4121	144		29. 9.39	These five Hampdens comprised a complete formation	Oblt Gunther Specht	I/ZG 26
L4126	144		29. 9.39	They were shot down around 10.00 hrs between	Oblt Gunther Specht	I/ZG 26
L4127	144		29. 9.39	Heligoland and Wangerooge. Two of the Bf 109Ds	Hptm Dickore	I/ZG 26
L4132	144		29. 9.39	I/ZG26 were shot down in the sea; the pilots were	Uffz Pollack	I/ZG 26
L4134	144		29. 9.39	rescued and included Uffz Hauck. (See 2C2 to 2C5)	Uffz Pirsch	I/ZG 26
L4089	44		21.12.39	Both shot down by Spitfires of 602 Squadron		
L4090	44		21.12.39	in sea off North Berwick (see 2C8 and 2C9)		
L4064	50		12. 4.40	All six shot down in sea off southern Norway at	Fw Mengel	II/JG 77
L4073	50		12. 4.40	around 11.45 hrs. II/JG 77 lost four Bf 109Es and	Fw Mengel	II/JG 77
L4081	50		12. 4.40	their pilots. A fifth forcelanded and was later	Ofw Herfeld	II/JG 77
L4083	50		12. 4.40	wrecked (see 2C11 and 2C15). It is not possible	Uffz Opotski	II/JG 77
L4099	44		12. 4.40	to detail which pilot shot down a particular	Ofw Arnoldy	II/JG 77
P1173	44		12. 4.40	Hampden, which also applies to 29.9.39	Not known	II/JG 77
P1319	49	N	25. 4.40	Shot down at 0110/26 near Sylt (1)	Fw Forster	IV(N)/JG 2
P1178	83	N	3. 6.40	No details. Crew buried at Rheinberg	Fw Forster	IV(N)/JG 2
P4370	144	N	2. 9.40	Shot down by Bf 109s between Wijnansrade and Nuth, NL	Not known	
X2910	44	N	14.10.40	Shot down near Rowitz. German crew killed in crash	Lt Mangelsdorf	I/NJG 1
X2993	50	N	14.10.40	Shot down at 03.05 near Kalbe	Hptm Streib	I/NJG 1
?		N	20.10.40	Hampden claimed shot down near Dishforth. No possibles	Hptm Hulshoff	I/NJG 2
X3027	49	N	27.10.40	Shot down by intruder near Scampton (see 5H1)	Lt Volker	I/NJG 2
X3052	49		25.11.40	Shot down 08.46, St.Maartens Vlotberg	Lt Wubke	III/JG 54
X3064	61	N	25.11.40	Reported shot down by intruder near Hemswell at 18.45 but actually crashed on return near Hemswell at 0140	Oblt Kurt Hermann	II/NJG 1
		N	25.11.40	One of three Ju 88s shot down by Hampden in North Sea	Ofw Kurt Schlicht	II/NJG 1
X3063	49	N	16.12.40	Intruder claimed a Hereford near Scampton	Oblt Paul Bohn	I/NJG 2
?		N	4. 2.41	Do 17Z – 10 shot down by Hampden in North Sea	Oblt Otto – Heinrich Hauser	II/NJG 2
X3001	49	N	10. 2.41	Shot down 10 km N of Alkmaar at 03.50	Lt Fellerer	I/NJG 1
AD719	49	N	10. 2.41	Shot down by intruder at Scampton (see 5H2)	Oblt Hermann	I/NJG 2
	44	N	15. 2.41	X2917 and X3025 damaged near Lincoln by intruders		
X2918	44	N	10. 3.41	Shot down near Venlo at 22.18	Hptm Streib	I/NJG 1
AD721	50	N	12. 3.41	Shot down near Nordhorn	Oblt Wohlers	III/NJG 1
P2092	14 OTU	N	8. 4.41	Believed shot down by intruder. No claim corresponds		

Serial	Unit		Date	Details	Shot down by	Unit
X3066	144	N	10. 4.41	Shot down 3 km SW of Neeroeten, NL, at 23.32	Lt Frank	I/NJG 1
AD789	50	N	10. 4.41	Shot down 4 km SW of Beesel at 22.54	Lt Reese	I/NJG 1
X3148	106	N	10. 4.41	Shot down at Ittersvoort at 23.01	Hptm Streib	I/NJG 1
X3153	106	N	10. 4.41	Shot down at Kessel – Helden at 22.49	Hptm Streib	I/NJG 1
AD828(2)	50	N	10. 4.41	Shot down 6 km N of Maastricht at 22.55	Ofw Herzog	I/NJG 1
AD761	144	N	16. 4.41	Shot down near Hemswell. Two intruders claimed a Hereford (see 5H3)	Fw Hahn/Lt Volker	I/NJG 2
		N	2. 5.41	Intruder claimed a Hampden at 03.15 over Church Fenton; only AD864 lost this night and crew buried at Hamburg	Lt Feverbaum	I/NJG 2
X3062	83	N	8. 5.41	Shot down by intruder off Mablethorpe at 23.30 (5H4)	Uffz Koster	I/NJG 2
AD900	144	N	11. 5.41	Shot down 8 km W of Enkhuisen at 02.57	Oblt Woltersdorf	II/NJG 1
		N	11. 6.41	Hampden claimed NW of London by intruder at 02.07 AD727 only Hampden lost this night but one of crew buried at Kiel and others missing	Uffz Koster	I/NJG 2
P1341	16 OTU	N	11. 6.41	Hampden and two Ansons damaged in attack on U.Heyford		
		N	17. 6.41	Crashed in Channel while investigating dinghy	Uffz Alt	I/NJG 2
		N	22. 6.41	Hampden claimed at 02.13 15 km NE of Alkmaar. No Hampden fits this claim; probably a 311 Sqn Wellington	Oblt Prinz zur Lippe – Weisenfeld	II/NJG 1
?	50	N	25. 6.41	Claimed attacked at 00.15 off Yorkshire (see 5H9)	Oblt Paul Bohn	IV/NJG 2
AD895	106	N	29. 6.41	Shot down S of Uetersen at 02.55	Oblt Eckardt	II/NJG 1
X3133	50	N	29. 6.41	Shot down 3 km SE of Oldenburg at 01.53	Lt Altendorf	I/NJG 3
AD862	106	N	2. 7.41	Shot down 8 km SW of Maastricht	Lt Knacke	I/NJG 1
AD914	106	N	6. 7.41	Shot down off Cromer and claimed as Blenheim (5H12)	Ofw Beier	I/NJG 2
AD937	61	N	7. 7.41	Shot down NE of Maastricht at 01.17	Lt Knacke	I/NJG 1
AD735	106	N	7. 7.41	Shot down NE of s'Hertogenbosch at 02.10	Oblt Griese	I/NJG 1
?		N	8. 7.41	Hampden claimed 5 km NE of Feltwell at 03.01. No trace	Lt Schultz	I/NJG 2
X3139	83	N	8. 7.41	Shot down N of Groningen at 00.56	Ofw Gildner	II/NJG 1
AE226	50	N	12. 7.41	Shot down 25 km SE of Groningen at 00.55	Oblt Lent	II/NJG 1
AE230	50	N	12. 7.41	Shot down 2 km E of Weltrup at 01.05	Lt Hadeball	II/NJG 1
AE189	61	N	24. 7.41	Shot down off Den Helder at 01.05	Not known	
AD835	83	N	25. 7.41	Shot down off Schiermonnikoog at 03.57	Lt Linke	II/NJG 1
P4406	44	N	27. 7.41	Shot down off Biscay coast at 03.46	Lt Bisang	I/NJG 2
AE134	106	N	16. 8.41	Shot down 20 km W of Venlo at 01.38	Oblt Thimmig	I/NJG 1
AD756	106	N	16. 8.41	Shot down near Gembloux at 02.44	Oblt Redlich	I/NJG 1
AE239?	44	N	16. 8.41	Claimed shot down near Ophoven at 02.34 but one of crew buried at Kiel and three missing	Hptm Frank	I/NJG 1
AE185	50	N	17. 8.41	Shot down at Paterswolde 2 km S of Groningen at 01.44	Oblt Becker	II/NJG 1
AE126	49	N	28. 8.41	Shot down over North Sea at 00.43	Not known	
AD971	49	N	28. 8.41	Shot down over North Sea	Not known	
AE187	83	N	31. 8.41	Shot down 8 km NE of Maastricht at 23.47	Oblt Dimter	I/NJG 1
AD912	83	N	31. 8.41	Shot down at Meyel at 00.18	Uffz Pahler	I/NJG 1
AE232	106	N	15. 9.41	Shot down at Wilhelmsburg, Hamburg, at 22.51	Oblt Barte	I/NJG 1
AE249	455	N	15. 9.41	Shot down at Winkeldorf at 23.54	Oblt Barte	I/NJG 1
P5314	16 OTU		20. 9.41	Shot down by intruder in circuit, Croughton (see 5H5)	Oblt Semrau	I/NJG 2
AD911	83	N	10.10.41	Claimed by night fighter off Harlingen at 05.00 Crew said flak responsible	Fw Ney	II/NJG 1
AD965	144	N	12.10.41	Shot down in Ysselmeer at 00.33	Oblt Lent	II/NJG 1
AD975	44	N	13.10.41	Shot down at Waasmunster, Belgium at 23.39	Oblt Roderer	I/NJG 1
AE251	50	N	13.10.41	Shot down at 06.00; crash site not known	Lt Knacke	I/NJG 1
?		N	1.11.41	Hampden claimed shot down; possibly AE358/AE224	Oblt Schoenhert	?
P1201	455	N	7.11.41	Shot down NW of Bergen aan Zee at 22.34	Uffz Grimm	II/NJG 2
P1206	49	N	8.11.41	Shot down 20 km E of Alkmaar at 21.07	Ofw Ney	II/NJG 2
AE433	408	N	8.11.41	Shot down at Maasbree at 22.30	Oblt Dimter	III/NJG 2
AE148?	408	N	11.12.41	Place and time not known; AE148 crashed in Denmark	Lt Knacke	I/NJG 1
P1165	408	N	28.12.41	Shot down 2 km E of Winterswijk at 22.48	Oblt Woltersdorf	III/NJG 1
AT122	106	N	28.12.41	Claimed as shot down; other reports say crashed at Lievelde near Grunlo	Oblt Sawert	I/NJG 2
AT148	49	N	20. 1.42	Shot down at Roode Haan, 7 km SE of Groningen at 20.58	Ofw Gildner	II/NJG 2
AT119	455	N	21. 1.42	Shot down at Schlochteren, 15 km E of Groningen, 21.10	Ofw Ney	II/NJG 2
AT130?	420	N	21. 1.42	Claimed; crashed 6 km NNE of Harlingen at 23.30; survivors say flak	Ofw Rasper	II/NJG 2
?		N	21. 1.42	Hampden claimed NE of Lingen at 21.03. X3051 or AE352?	Uffz Frank	I/NJG 3
AE308	455		6. 2.42	Shot down 70 km W of Terschelling (see 2H1)	Hptm Lent	II/NJG 1
AD824	144		7. 2.42	Shot down W of Terschelling (see 2H1)	Ofw Luth	II/JG 1
AE392	144		7. 2.42	Shot down W of Terschelling (see 2H1)	Ofw Luth	II/JG 1
AE306	50		7. 2.42	Shot down W of Terschelling (see 2H1)	Ofw Luth	II/JG 1
P1156	455		12. 2.42	Six of these Hampdens were claimed by fighter pilots	Uffz Kirchner	II/JG 1
P4400	420		12. 2.42	covering German warships off the Dutch coast. Four	Oblt Bock (two)	II/JG 1
AE132	49		12. 2.42	pilots claimed single Hampdens and Oblt Bock claimed	Hptm Pretele	I/JG 2
AE396	49		12. 2.42	two. There were no survivors from any of the Hampdens	Oblt Schnell	III/JG 2
AT175	144		12. 2.42	and no information as to which pilot shot down which	Oblt Stolle	III/JG 2
AT177	50		12. 2.42	Hampden. P5324 and AT134 claimed by flak (see 2H2–6)		
AT194	144	N	24. 2.42	Shot down in sea N of Terschelling at 21.49	Oblt Prinz zur	
X2969	144	N	24. 2.42	Shot down in sea N of Terschelling at 22.02	Lippe – Weisenfeld	II/NJG 2
AT151	50	N	25. 3.42	Shot down in sea N of Vlieland at 00.33	Oblt Becker	II/NJG 2
AE200	144	N	26. 3.42	Shot down in sea N of Vlieland at 22.30	Ofw Gildner	II/NJG 2
AE298	420	N	26. 3.42	Shot down in sea NW of Terschelling at 22.57	Ofw Gildner	II/NJG 2

Serial	Unit		Date	Details	Shot down by	Unit
L4140	408	N	27. 3.42	Shot down in sea N of Terschelling at 22.10	Hptm Lent	II/NJG 2
AT176	408	N	27. 3.42	Shot down in sea N of Terschelling at 22.40	Hptm Lent	II/NJG 2
AE219	408	N	27. 3.42	Shot down in sea at 23.31. Position uncertain	Ofw Schoenert	II/NJG 2
		N	27. 3.42	Shot down in sea N of Petten at 00.08 Four Hampdens were claimed this night, one of which was almost certainly a Whitley of 138 Squadron. Ofw Ney was shot down and killed	Ofw Ney	II/NJG 2
AE246	420	N	28. 3.42	Shot down in sea in Heligoland Bight	Uffz Merk	II/NJG 2
AT156	49	N	5. 4.42	Shot down N of Philippeville, Belgium at 02.10	Oblt Herget	II/NJG 1
P1203	455	N	6. 4.42	Shot down S of Zwolle at 04.13	Oblt Waltersdorf	III/NJG 1
AE421	49	N	10. 4.42	Shot down 28 km NE of Den Helder at 00.23	Oblt Liedke	III/NJG 2
P1239	420	N	12. 4.42	Shot down N of Terschelling at 00.32	Hptm Lent	II/NJG 2
?		N	14. 4.42	Hampden claimed shot down; possibly AD931/AT128/AT141	Oblt Liedke	III/NJG 2
AE237	408	N	22. 4.42	Shot down in sea NW of Ameland at 00.07	Ofw Gildner	II/NJG 2
P5330	420	N	25. 4.42	Shot down NW of Ameland at 00.04	Lt Linke	II/NJG 1
P2116	14 OTU	N	30. 5.42	Shot down 3 km N of Deventer	Lt Meurer (Maurer?)	III/NJG 1
AT191	408	N	1. 6.42	Shot down N of Texel and crashed in Ysselmeer	Lt Vollkopf	II/NJG 2
AT154	408	N	2. 6.42	Shot down E of Medemblik at 01.06	Hptm Lent	II/NJG 2
AT136	420	N	8. 6.42	No details. Presumably in sea as crew all missing	Hptm Bunsch	III/NJG 2
AT185	420	N	20. 6.42	Shot down N of Ameland at 00.59	Hptm Lent	II/NJG 2
AE248	420	N	2. 7.42	Shot down S of Koudum at 00.54	Oblt Prinz zur Lippe – Weisenfeld	II/NJG 2
P5332	420	N	2. 7.42	Shot down 15 km WNW of Zwolle in Ysselmeer at 01.40	Fw Grimm	II/NJG 2
AT227	408		19. 7.42	Collided with night fighter 50 km NW of Vlieland; both crews killed	Uffz Brejczek	II/NJG 2
AE244	408	N	31. 7.42	Shot down 2 km SE of Zewerdint, Belgium, at 03.54	Fw Schellwat	II/NJG 1
N9062	14 OTU	N	31. 7.42	Shot down at Moergestel at 03.30	Oblt Prinz zur Sayn – Wittgenstein	III/NJG 2
P1185	14 OTU	N	31. 7.42	Shot down near Ostend; claimant not known		
P1244?	408	N	27. 8.42	Shot down 6 km W of Hengelo at 00.08 but one report states this was a 15 Squadron Stirling	Oblt Baver	III/NJG 1
AE197	408	N	28. 8.42	Shot down NW of Philippeville, Belgium, at 23.22	Hptm Herget	II/NJG 4
?	408	N	28. 8.42	Location not known; possibly AD829/AE227/AT228	Lt Beier	I/NJG 1
?	408	N	28. 8.42	As above	Lt Fladrich	III/NJG 4
L4109	14 OTU	N	13. 9.42	Shot down 15 km SE of Groningen at 03.00	Hptm Hulshof	Stab/NJG 2
P1258	415	N	15. 8.43	Shot down 10 km W of Egmond aan Zee at 01.15	Hptm Buchholz	IV/NJG 1

1 Possibly L4040 or L4092

2 There is still an element of doubt as to who shot down which Hampden. One collided with Lt.Reese's Bf 110, most probably AD789 (see 5C23)

3 One of the few documented defensive successes at night comes from No.408 Squadron's activities on 27/28 August 1942
On that night, David Williams of 408 Squadron was piloting P1166. The entry in his log book reads "Operations. Kassel (121 holes). Shot down Ju 88 with front gun, nearly vice – versa. Awarded DFC. Navigator F/Sgt Turnour and Wireless Op Sgt Don Leech (RAF) awarded DFMs".

David recalled how, after being attacked twice "just before Kassel came into view, Sgt Leach yelled that we had a Ju 88 dead astern and close in. I closed the throttles and damn near stopped flying. The Ju 88 broke under me, I lowered the nose, applied full throttle...found the 88 directly in front of me and opened fire with my single Browning which was loaded with alternate incendiary and semi – armour piercing bullets. I could see hits and finally there was an explosion. We flew through the debris and circled until the aircraft hit the ground then continued and bombed Kassel". P1244 of 408 Squadron failed to return from the same operation. The following night, four of No.408's Hampdens failed to return from an attack on Saarbrucken; a fifth crashed on return. Night fighters claimed three of them.

Abbreviations of Luftwaffe ranks with RAF equivalents

Gefr	Gefreiter	AC.1	Ogefr	Obergefreiter	LAC
Hgefr	Hauptgefreiter	Cpl	Uffz	Unteroffizier	Sgt
Ufw	Unterfeldwebel	Sgt	Fw	Feldwebel	F/Sgt
Ofw	Oberfeldwebel	W/O	Stfw	Stabsfeldwebel	W/O
Lt	Leutnant	P/O	Oblt	Oberleutnant	F/O
Hptm	Hauptmann	F/Lt	Maj	Major	S/Ldr

A Junkers Ju 88C night fighter

The infamous destruction of Hampden P1310 after "preservation" (Photo: Peter M.Corbell)

APPENDIX V

PRESERVATION

Two Hampden preservation projects are currently in progress, one in Canada at the Museum of Flight and Transportation near Vancouver and one in England at the Lincolnshire Aviation Heritage Centre at East Kirkby airfield where Brian Nicholls is in charge. Both groups are composed of volunteers and, as is usual with such organisations, there is always a pressing need for funds and components. Any contributions should be directed to 13527 Crescent Road, Surrey, BC, Canada V4A 2W1 and 16 Gayle Road, Castlefields, Tattershall, Lincs, LN4 4LZ, England.

The Canadians are using the remains of P5436, which they recovered from the 600-foot deep sea bed in October 1985, as the basis for their reconstruction, also using parts from the wrecks of AN132 and AN136 which crashed into local mountainsides. They also managed to persuade the RAF Museum to give them a Hampden nose section that had arrived at the Museum after spending many years in a back garden as a vegetable cloche.

The Canadians were also able to obtain a Canada Manpower grant to employ three sheet-metal workers on their project for six months.

Brian Nicholls' project at East Kirkby is based on the wreckage of AE436 which crashed into a mountainside in Sweden on 4/5 September 1942 (see Part 2J and photographs in Part 5D).

After this wreckage was discovered in August 1976 on the Tsatsa mountain, it was brought down from the wreck site and shared between the RAF Museum who put their quota into storage at Henlow and the Swedish Flygvapen Museum at Malmslatt who, in 1988, most generously agreed to give their half-ton of wreckage to East Kirkby on condition that the latter organised transport back to the UK; this was carried out by Norfreight Ltd. On 16 January 1989, the remains from Sweden were re-united with those from the RAF Museum which were already being worked upon at East Kirkby by Brian and his colleagues.

Two other Hampdens were preserved in their entirety. P1310 was one of No.455 Squadron's Hampdens on strength when the unit converted to Beaufighters. All but P1310 and AE384 were broken up soon afterwards at No.44 Maintenance Unit, Edzell. P1310 was stored by No.71 MU at Bicester until 1955 when some misguided soul decided it should be dispatched to the breaker's yard. AE384 suffered a similar fate after it had been shipped out to Australia in 1945 where it was intended to form part of the Australian War Memorial's collection in Canberra to commemorate the part played by the RAAF in Coastal and Bomber Command. Instead, it was sold for scrap in 1954.

Hampdens were not alone in being deprived of future exhibition and at least there was some attempt at preservation. No Wellington, Whitley, Stirling, Halifax or Blenheim was earmarked for museum use when the scrap merchants moved in. The sole Wellington in the RAF Museum came from private sources after being used by Vickers for experimental work. This lack of foresight was in contrast to the situation at the end of World War One when examples of most Allied and Enemy types were ordered to be preserved for the Imperial War Museum. They went on show at the Agricultural Hall at Islington, London, but instead of being incorporated into a major exhibit, only a few were placed on display in the Imperial War Museum and the rest were disposed of. Nothing seems to change!

It was most unfortunate that both the British and Australian authorities failed to appreciate the potential value to posterity of these aircraft, although at least someone did originally seem to be aware of it. Canada failed to keep any example of the Hampden, sending the survivors for breaking up at Sea Island on 21 February 1944.

The remains of a Hampden at Bovingdon in December 1955 after arrival from Bicester and presumably P1310. (Photo: Arthur Pearcy)

Initial stages of the reconstruction of the forward part of the fuselage of AE436 at East Kirkby (Photo: Brian Nicholls)

Wreckage at East Kirkby.The tail wheel assembly still functioned and there was air in the tyre. (Photo: Brian Nicholls)

Brian Nicholls and the wreckage of AE436 at East Kirkby

The first stage of fuselage rebuild on display in June 1989 (Photo: Jerry Vernon)

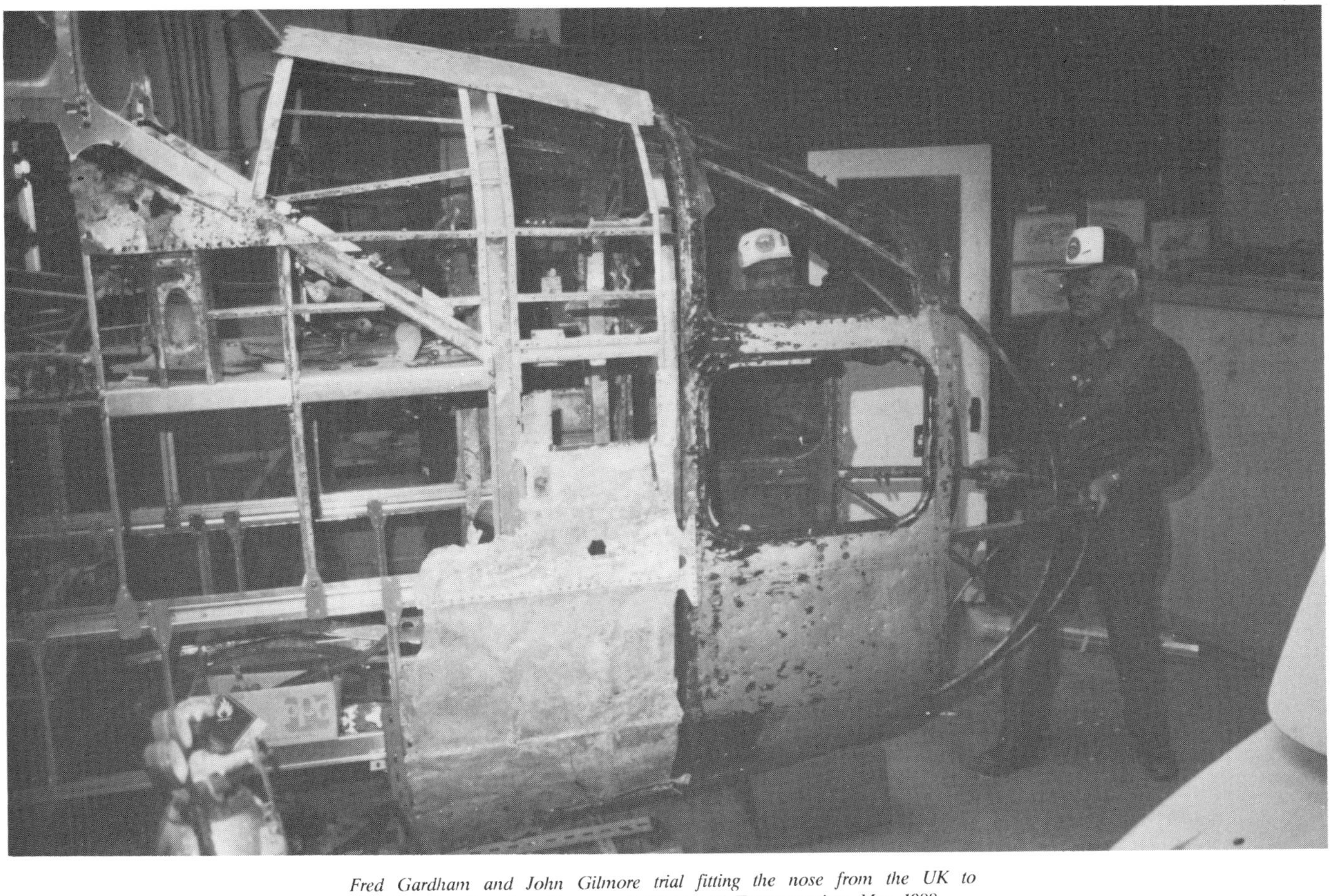

Fred Gardham and John Gilmore trial fitting the nose from the UK to P5436 at the Canadian Museum of Flight and Transportation, May 1988. (Photo: Jerry Vernon)

"The Ugly Bug's Ball". Prototype K4240 is not a pretty sight. Its coat of gloss "dirty – green" paint and the apparent covering of the inside of the glasshouse nose and rear upper gun canopy with paper, do little to enhance its appearance. At the Royal Air Force Display at Hendon it carried its "New Types Park" number "8". ("Aeroplane" photo)

APPENDIX VI

BIBLIOGRAPHY

Author	Title	Publisher	ISBN No.
Aders, Gebhard	History of the German Night Fighter Force	Janes	0 35401 247 9
Barker, Ralph	The Thousand Plan	Chatto & Windus	0 70110 527 5
Barker, Ralph	Down in the Drink	Chatto & Windus	0 70110 575 8
Barnes, C.H.	Handley Page Aircraft since 1907	Putnam	0 00030 370 7
Bowyer, Chaz	Hampden Special	Ian Allen	0 71100 683 0
Bowyer, Chaz	Bomber Group at War	Ian Allen	0 71100 087 0
Bushby, John	Gunner's Moon	Ian Allen	0 71100 476 5
Brickhill, Paul	The Great Escape	PBK Arrow Books	0 09919 020 6
Deighton, Len	Battle of Britain	Cape	0 22401 826 4
Dominey, John	The Sergeant Escapers	Ian Allen	0 71100 569 9
Finn, Sid	Lincolnshire Air War, Vols.1 and 2	Brayford Press	
Gibson, Guy	Enemy Coast Ahead	Michael Joseph	1 71811 519 8
Gomersall, Bryce	The Stirling File	Air – Britain	0 85130 152 5
Halley, James	The Squadrons of the Royal Air Force and Commonwealth, 1918 – 1988	Air – Britain	0 85130 164 9
Halley, James	The Lancaster File	Air – Britain	0 85130 107 X
Halley, James	The Role of the Fighter in Air Warfare	Profile	0 20601 214 7
Halpenny, Bruce	Action Stations Vol.2 (Lincolnshire)	Patrick Stephens	0 85059 484 7
Hastings, Max	Bomber Command	Michael Joseph	1 71811 519 8
Herington, John Air	War against Germany and Italy 1939 – 43	Australian War Memorial	
Kee, Robert	A Crowd is not Company	PBK Granada	0 58605 763 3
Mackersey, Ian	Into the Silk	Granada	0 58312 793 2
Middlebrook, Martin and Everitt, Chris	The Bomber Command War Diaries	Viking	0 80137 670 2
Philpot, Oliver	Stolen Journey	Elmfield Press	0 70570 000 3
Reid, P.R.	The Colditz Story	Coronet	0 85997 518 5
Richards, Denis	Royal Air Force 1939 – 1945 (Vols.1 and 2)	HMSO	0 0005 133 0 – 1
Roberts, R.N.	The Hampden and Hereford Crash Log	Midland Counties	0 90459 734 2
Roberts, R.N.	The Halifax File	Air – Britain	0 85130 098 7
Roskill, Stephen	The War at Sea, Vols.1 and 2	HMSO	0 00005 292 4 – 5
Schofield, B.B.	The Russian Convoys	Pan Books	0 33028 388 X
Thompson, H.L.	New Zealanders with the Royal Air Force	Dept of Internal Affairs	
Verrier, Anthony	The Bomber Offensive	Pan Books	0 33028 864.7
Verity, Hugh	We Landed by Moonlight	Hodder & Stoughton	0 71100 886 8
Wakefield, Kenneth	The First Pathfinders, the Operational History of KG 100	William Kimber	0 71830 318 0
Webster, Sir Charles and Frankland, Noble	The Strategic Air Offensive against Germany, 1939 – 1945, Vols 1 to 3	HMSO	0 00006 491 7 – 9

ACKNOWLEDGMENTS

The information contained in "The Hampden File" has been gathered from four main sources; they are: –

1. Personal Accounts from ex – aircrew who flew on Hampdens and from the "erks" who serviced them.
2. Official Records.
3. Squadron Historians, Ex – Service Organisations, Researchers in the United Kingdom.
4. Overseas Researchers.

1. Personal accounts from ex – Hampden personnel:

The assistance received from those whose accounts are used in the book is gratefully acknowledged as is that from others who made contact about the Hampdens. Their interest provided a continual source of encouragement during the compilation of "The Hampden File". The names are grouped in their different aircrew categories and squadrons, and where practicable an index number is given of the relevant incident. It is hoped that they will forgive the omission of details of rank and decorations, etc.

Pilots :

44 Sqn: Dave Penman 5H3; Tony Reid 2C9; Mike Lewis.
49 Sqn: Des Dunphy 3D6; Bob Hamer (AT112); Lewis Hodges 3D31
50 Sqn: "Abdul" Abbott (AE251); Jimmy Bennett; "Weasel" Donaldson 2C15; George Potts 5B3; Chas Stenner; Gus Walker
61 Sqn: Peter Tunstall 3D28; "Spooky" Barrett 3D13
83 Sqn: Stan Harpham 2K9; Freddie Newall 5B11; Leonard Snaith 2C; Neil Svendsen 5B2; Tony Stamer; Rod Learoyd VC.
106 Sqn: Vic Cooper (AE123); Wally Huggins; Chas Lockyer; Roy Haggar; Roger Purnell; Robert Wareing
144 Sqn: Jimmy Bennett; Bob Galloway 2C2; Max Meyer; Bob Pearman 5A16; Colin Rawlins 2D11; Gordon Walters.
408 Sqn: David Williams
415 Sqn: Bill Adams 2J36; John Enns 2J37
420 Sqn: Robert Kee 2D10
455 Sqn: Grant Lindeman
A&AEE: Tom Stanley
45 (Atlantic Transport) Group: Maurice Ratcliffe

Air Observers, Air Gunner Officers, Navigators

44 Sqn: Andrew Coveyduck 3D8; Gerrie Harris (P2077); Alan Nicoll
50 Sqn: John Day 2G; Gil Haworth 2D, 5G6; W J Johnstone (P1202); George Leonard (L4046)
83 Sqn: Julian Badcock 5A52
144 Sqn: W D Barrett 2J7; Dave Withey; "Timber" Woods 2K23; Vernon Smith
415 Sqn: Jim Chapman; E.L.Rowe
455 Sqn: Bob Anderson 2J7; Ray Lassiter (P1208)
32 OTU: W N Switzer; P Moorhead Nav W/T.

Wireless Operator/Air Gunners and Air Gunners

44 Sqn: Fred Ball; M Beale; Jim Taylor; Joe Dacey
49 Sqn: Don Foster; David Young 3D26; John Fairbrother; "Chan" Chandler 5B10
50 Sqn: Geoff Ashmore; Fred Bailey 2D8; Doug Barrie 2K26; Nairn "Robbie" Robertson 2G; Jim Mitchell; Bill Sykes
61 Sqn: "Bernie" Bernard; Neil Prendergrast 2E2; Len Walker 3D2; John Brock 3D2
83 Sqn: Eric Groves (L4057); Jack Lawrence 3D10; Hector Macdonald
106 Sqn: Tom Berry (AE123); Jim Palmer (AE123); Allen Wiseman; George Luke 3D23
144 Sqn: Bert Earl* 2J13; Bernard Sowerby * 2J12; * Both groundcrew who flew on Russia flight; Jim Rowland, groundcrew on Russia detachment
408 Sqn: Don Foster
415 Sqn: Doug Stallard
420 Sqn: David Semple
455 Sqn: George Baynes; Bill McPhee; John McKissack, groundcrew on Russia detachment
489 Sqn: Guy Gaskill 3D24; Eric Cameron

2.Official Records

Public Record Office at Kew
The Record Books of Squadrons Stations, Groups and Commands, and many files on particular aspects of Hampden operations were researched.

Air Historical Branch, London.
Permission was obtained to study Hampden "Crash" cards, "Movement" cards and "Contract" cards. The staff answered many written queries on Hampden fates.

RAF Museum, Hendon
The Museum photographic library proved to be a valuable source of photographs, and in the early stages of research Andrew Cormack offered advice and encouragement on the "Hampden File" project. The RAF Museum is continually adding to its Archive material and takes a keen interest in photographs of RAF aircraft, Log Books and general RAF memorabilia.

The Imperial War Museum, London
The staff of the photographic library assisted in the selection of suitable photographs.

The Commonwealth War Graves Commission, Maidenhead
War Cemetery Registers and the Runnymede Memorial Register provided information on crew fates. CWGC has to charge for Registers and queries on individual crew fates The service received proved well worth the financial outlay.

Public Archives of Canada, Ontario and the **Supply and Services Department, Ottawa** proved to be valuable sources of good quality photographs. The courteous assistance given by the staff of all of the aforementioned Departments is acknowledged with gratitude.

Australian War Memorial, Canberra, answered several questions regarding aircraft and crew fates.

Bournemouth and Poole College of Art and Design
Sue Brown, a student at the Department of Technical Illustration, produced the cutaway drawing and several maps.

Air – Britain
Cliff Minney, the association's graphics adviser, produced the Hampden sideviews.

3.Squadron Historians and Ex – service Associations and Researchers in the United Kingdom

Ron Low and Frank Harper of 83 Squadron; John Farley and Des Richards of 106 Squadron; Gus Dodson and Frank Priestly of 61 Squadron; Kenneth Green of 61 Squadron and Hemswell; Bob Cotterill and Bill Boorer of 144 Squadron; Alan White and Bert Dowty of 44 Squadron and Ray Leach of 50 Squadron. Peter Green provided various photographs and assisted with the reproduction of others. "Cee" Wadsworth provided photographs of 455 RAAF Squadron.

Ex – Service Organisations
Requests for information in newsletters of various ex – service organisations led to many Hampden contacts coming to light from The Aircrew Association, The Prisoner of War Association, the Royal Air Force Association, The Air Transport Auxiliary Association, The Handley Page Association, The Goldfish Club, The Air Sea Rescue/Marine Craft Club and the Lincolnshire Bomber Airfield Society. "FlyPast" and "Air – Britain Digest" were the two principal Aviation journals involved in seeking answers to particular problems.

Researchers in the United Kingdom
Brian Nicholls, who is engaged on the part – reconstruction of a Hampden at the Lincolnshire Aviation Heritage Centre at East Kirkby; Roy Nesbitt; Roger Hayward; Eric Thale; Mike Hodsman; Chaz Bowyer; John Foreman and Thomas E Willis who provided several interesting photographs, also corresponded.

4.Overseas Researchers

AUSTRALIA: Grant Lindeman, who commanded 455 RAAF Squadron during its Bomber Command days and later, in April 1942, when it became a torpedo-bomber squadron with Coastal Command and was detached to Russia for two months, provided invaluable information and photographs. Sid Finn and Don Russell also assisted with research in Australia.

CANADA: Eric Grove of Hamilton, Ontario, was the initial main source of information about the Canada Hampdens and was most helpful in initiating enquiries in Canada about RCAF personnel. Jerry Vernon of Burnaby, BC, provided a wealth of information from RCAF Hampden "crash cards", on 32 OTU and on the recovery and restoration of P5436. Two former flight commanders of 415 RCAF Squadron, Bill Adams and John Enns, assisted with details of their squadron's Hampden activities whilst R Stuart-Tate supplied information on 408 RCAF Squadron during its Hampden era. Hugh A Halliday, Douglas Harvey, Jonathan Vance, Tom Lawrence and Fred Wright also corresponded.

DENMARK: Birger Hansen, whose speciality is RAF minelaying activity, particularly in the Baltic, provided details of Hampdens that had come to grief around Denmark.

EIRE: Damian Smith from Clondalkin, Co Dublin, assisted with information on three Eire Hampden write-offs, and forwarded photographs of the military funeral of the crew of AD730.

GERMANY: Winfried Boch in West Germany provided information about the cause of many Hampden losses and details of crash sites. Friedrich A Greve answered a request in the "Wilhelmshavener Zeitung" for information and most generously forwarded many useful photographs and information that have been used in the account of the raid on 20/21 July 1940. Hans Seebrandt, Peter Soltau and Heinz Nowarra also contributed.

NETHERLANDS: Gerrie Zwanenburg, who was awarded the MBE for his services in identifying Allied aircraft that had crashed into the sea off Holland, in the area of inland water best known by the RAF at the time as the Zuider Zee, and on land in Holland. Gerrie provided a wealth of information about the fates of Hampdens and their crews. Dirk Bruin on Vlieland and Wyb Jan Groendijk on Schiermonnikoog helped with information and photographs. Ab A Jansen and several others from the Netherlands were most helpful.

NORWAY: Sten Stenersen, Knut Maesel and Per Skaugstadt each made a contribution and from their accounts of information from wartime German sources it proved possible to make an interesting comparison with contemporary RAF records of shipping sunk by RAF Hampden torpedo-bombers.

POLAND: Robert Bock from Poznan provided information about the Hampdens used by the Russians.

SWEDEN: The British Defence Attache's office in Stockholm assisted with enquiries into three Hampdens that came to grief in Sweden. Bo Widfeldt also replied to questions about these losses.

UNITED STATES OF AMERICA: Jack Fitzgerald of Houston, Texas, was one of the No.45 (Atlantic Transport) Group pilots who flew a Hampden from UK to Vancouver, whilst Herbert Thomas of the ATA, who first flew in a Bristol Fighter in 1918, detailed several ferry flights that he made in Hampdens in UK. Herbert was 83 years young when he wrote from Palm Beach and seemed a little upset because he was about to lose his pilot's licence, but was certain he would continue flying, if only as a passenger. Norm Malayney from Pittsburgh also assisted with enquiries.

The two penultimate, and most important, acknowledgements are to Dennis Thompson and Jim Halley. Dennis Thompson, Air-Britain's Handley Page specialist, had already started research on the Hampden and his generosity in forwarding the results of his labours is most gratefully acknowledged. Jim Halley, editor of Air-Britain's military aircraft publications and author of several of them, offered advice and assistance in the early stages of the Hampden File, gave encouragement during its compilation and finally attended to the editing of the completed project.

The final "Thank You" has to be given to Edna Moyle who, as Edna Melhuish, met the author when he was an 18-year-old Air Observer, in the foyer of the "Montana" dance hall in Lincoln in January 1940. Harry and Edna Moyle have travelled together along the ups and downs of life's highway ever since. Edna deserved a far better deal from the Gods of Destiny than that of being burdened with a husband who, in his more advanced years, aspired to be the author of "The Hampden File". Bless you, Mrs.M...

The Plymouth Hoe Memorial "To all airmen from all Nations", 1989
(Photo: Kath & Ray Steuart)